LIE GROUPS, LIE ALGEBRAS, AND COHOMOLOGY

LIE GROUPS, LIE ALGEBRAS, AND COHOMOLOGY

by

Anthony W. Knapp

Mathematical Notes 34

PRINCETON UNIVERSITY PRESS

PRINCETON, NEW JERSEY

1988

Printed in the United States of America by
Princeton University Press, 41 William Street,
Princeton,New Jersey 08540

ISBN 0-691-08498-x

Library of Congress Cataloguing in Publication Data
will be found on the last printed page of this book

To My Family

With Love

CONTENTS

PREFACE

This material is based on a one-semester course given at SUNY Stony Brook in Fall 1986. The audience consisted largely of graduate students knowledgeable about geometry, acquainted with tensor products of vector spaces, and having little or no background in Lie groups. The objective was to go in one semester from the beginnings of Lie theory to the frontier in algebraic constructions of group representations. The course consisted of much of the first seven chapters of the present book, done in a slightly different order. Actually the course was designed backwards from a key algebraic computation (7.79) that yields the Borel-Weil-Bott Theorem, and it ended up including whatever seemed appropriate as preliminary material. Chapter VIII was added to indicate how the computation (7.79) leads to the frontier.

The topic of interest here is the representation theory of compact Lie groups and of their natural noncompact analogs, the noncompact semisimple Lie groups. Special linear groups, symplectic groups, and isometry groups of quadratic forms give examples of noncompact semisimple Lie groups. Group representations are homomorphisms of a group into invertible linear transformations on a complex vector space, possibly infinite-dimensional and possibly with some continuity assumption. Understanding of group representations allows one to take advantage of symmetries in various problems in analysis and algebra.

For a compact group, the irreducible representations (those with no nontrivial closed subspaces invariant under the representation) are finite-dimensional. In the case of any compact connected Lie group, the Borel-Weil Theorem gives a way of realizing all such irreducible representations in spaces of holomorphic functions, and a generalization due to Bott gives alternative realizations in spaces of differential forms having only $d\bar{z}$'s present.

In 1966 Langlands conjectured that a version of the Borel-Weil-Bott Theorem should provide a realization of "discrete series" representations of noncompact semisimple Lie groups. "Discrete series" are certain irreducible infinite-dimensional representations that are known to play a fundamental role in the construction of all irreducible representations. Over a period of some years ending in the mid 1970's, Schmid proved the Langlands conjecture and several variants of it.

In 1978 Zuckerman found that he could bypass a number of difficult analytic problems of Schmid's by phrasing the Langlands conjecture in algebraic terms and using constructions in homological algebra. He found that similar algebraic constructions yielded other interesting representations for a variety of groups. His technique, now known as cohomological induction, has become a fundamental tool in representation theory.

Our goal is to start with elementary Lie theory and to arrive at the definition, elementary properties, and first applications of cohomological induction, all the while developing the computational techniques that are so important in handling Lie groups. A byproduct is that we are able to study homological algebra with a significant application in mind; we see as a consequence just what results are fundamental and what results are minor. A person who wants to study further may wish to read the general theory of roots and weights from any number of books (e.g., Knapp [1986], Chapters IV and V) and to study cohomological induction further from the book by Vogan [1981].

Prerequisites for reading the present book are a knowledge of metric spaces, an advanced course in linear algebra, and a passing acquaintance with topological groups. Also invariant integration for compact groups plays a role in Chapters III, VII, and VIII.

Chapter I is a quick introduction to Lie groups of matrices. The reader who already knows beginning Lie theory may skip this chapter except for the last section (representations). Representations of groups are the objects of interest in these notes, but the approach is by means of algebra. What we therefore study is representations of Lie algebras that are related to representations of Lie groups; the sense in which we can actually pass from the Lie algebras to the Lie groups is not addressed here very seriously. Since the group representations are the objects of interest, we constantly need examples of them in order to motivate what happens with Lie algebras.

The first half of Chapter II contains two kinds of topics alternately: a devlopment of multilinear algebra and the imposition of representations on various spaces of tensors. Most readers will be able to skip at least part of the material in §§1-4. The second half of Chapter II constructs some algebraic objects that will be used repeatedly throughout the later chapters——universal enveloping algebra, symmetrization, and tensor products over an algebra.

Chapter III treats the representation theory of compact groups, with emphasis on the unitary groups as examples. This material is included at this stage mostly for motivation and is not used until the latter part of Chapter VI.

Chapters IV and V, as well as the first half of Chapter VI, develop homological algebra as it applies to Lie algebras and their representations. The motivation that is provided comes largely from geometry and starts with the analytic Borel-Weil Theorem in §7 of Chapter III. A second line of motivation begins in §1 of Chapter IV. The two lines merge in §1 of Chapter VI and continue in §1 of Chapter VII.

Chapter VII makes a modification in the theory of the previous three chapters, so that the setting matches better what is needed for representation theory. The ingredients of cohomological induction are assembled in this chapter. Section 11 is the most important and illustrates how the theory can be applied in the ideal setting of a compact connected Lie group. This section contains the key computation (7.79) mentioned earlier.

Chapter VIII indicates the kinds of adjustments that need to be made when working with noncompact semisimple Lie groups, rather than compact groups. Specific illustrations are given for the unitary groups of indefinite quadratic forms.

At the end of the book is a chapter entitled "Notes," which gives historical comments, amplifies some notions, and points to the list of references at the end.

I am grateful to David Vogan for his advice in presenting this material and to the members of the class at SUNY Stony Brook for a number of comments and suggestions. Clifford Earle helped with some of the motivation, Chih-Han Sah helped with some of the bibliographical material, and Leticia Barchini helped with checking the mathematics. Some of the material in Chapters VII and VIII is new, and part of it is taken from the unpublished notes Knapp and Vogan [1986] and from other unpublished work joint with Vogan. This research was sponsored by the National Science Foundation under one or more of the grants DMS 85-01793, DMS 87-11593, and DMS 85-04029.

A.W.K.
December 1987

LIE GROUPS, LIE ALGEBRAS, AND COHOMOLOGY

CHAPTER I

LIE GROUPS AND LIE ALGEBRAS

1. SO(3) and its Lie algebra

We denote by $GL(n,\mathbb{R})$ the real general linear group consisting of all nonsingular n-by-n real matrices with matrix multiplication as group operation. Similarly $GL(n,\mathbb{C})$ denotes the group of nonsingular n-by-n complex matrices.

Let G be the set of matrices $SO(3)$ defined by

$$SO(3) = \{x \in GL(3,\mathbb{R}) \mid xx^{tr} = I \text{ and } \det x = 1\} .$$

This is the set of 3-by-3 rotation matrices and will be an example of a Lie group. For now we can notice that G is a group (being closed under multiplication and inversion within $GL(3,\mathbb{R})$) and is also a compact subset of $\mathbb{R}^9$ (being closed under entry-by-entry limits and having all entries ≤ 1 in absolute value). The group structure and the topological structure are related in that multiplication and inversion are continuous (being given by nice formulas).

Of more interest, however, is the fact that a kind of differentiation is available. Let us agree for now that a smooth curve $c(t)$ in $SO(3)$ is a C^∞ curve into $\mathbb{R}^9$ whose image is in $SO(3)$ for each t. We consider smooth curves $c(t)$ in $SO(3)$ with $c(0) = I$. One such is

$$c(t) = \begin{pmatrix} \cos t & \sin t & 0 \\ -\sin t & \cos t & 0 \\ 0 & 0 & 1 \end{pmatrix} .$$

For any such, we can form $c'(0)$, which will be some 3-by-3 matrix. With c as above

$$c'(t) = \begin{pmatrix} -\sin t & \cos t & 0 \\ -\cos t & -\sin t & 0 \\ 0 & 0 & 0 \end{pmatrix} , \qquad c'(0) = \begin{pmatrix} 0 & 1 & 0 \\ -1 & 0 & 0 \\ 0 & 0 & 0 \end{pmatrix} . \tag{1.1}$$

Here $c'(0)$ tells us the direction in G in which the curve extends from I. Think of it as telling how a parameter describing G is to start out from the identity. Let

$$\mathfrak{g} = \{c'(0) \mid c(t) = \text{smooth curve in } G \text{ with } c(0) = I\} .$$

Then $\mathfrak{g}$ is a subset of 3-by-3 matrices, not necessarily invertible.

Let us identify $\mathfrak{g}$. First of all, $\mathfrak{g}$ is a vector space. To see closure under addition, let $c'(0)$ and $b'(0)$ be in $\mathfrak{g}$. Forming $c(t)b(t)$, we have

$$\frac{d}{dt}[c(t)b(t)] = c(t)b'(t) + c'(t)b(t)$$

$$\frac{d}{dt}[c(t)b(t)]_{t=0} = c(0)b'(0) + c'(0)b(0) = b'(0) + c'(0) .$$

Thus $\mathfrak{g}$ is closed under addition. To see closure under scalar multiplication, let $c'(0)$ be in $\mathfrak{g}$ and k be in $\mathbb{R}$. Forming $c(kt)$, we have

$$\frac{d}{dt}[c(kt)] = c'(kt)k , \qquad \frac{d}{dt}[c(kt)]_{t=0} = kc'(0) .$$

Thus $\mathfrak{g}$ is closed under scalar multiplication.

Next let us find additional concrete matrices that are in $\mathfrak{g}$. Two such are

$$c'(0) = \begin{pmatrix} 0 & 0 & 1 \\ 0 & 0 & 0 \\ -1 & 0 & 0 \end{pmatrix} \quad \text{from} \quad c(t) = \begin{pmatrix} \cos t & 0 & \sin t \\ 0 & 1 & 0 \\ -\sin t & 0 & \cos t \end{pmatrix}$$

and

$$c'(0) = \begin{pmatrix} 0 & 0 & 0 \\ 0 & 0 & 1 \\ 0 & -1 & 0 \end{pmatrix} \quad \text{from} \quad c(t) = \begin{pmatrix} 1 & 0 & 0 \\ 0 & \cos t & \sin t \\ 0 & -\sin t & \cos t \end{pmatrix}.$$

Taking into account (1.1) and the vector space operations, we see that

$$\mathfrak{g} \supseteq \left\{ \begin{pmatrix} 0 & a & b \\ -a & 0 & c \\ -b & -c & 0 \end{pmatrix} \right\} = \{\text{skew-symmetric real matrices}\}\,. \qquad (1.2)$$

Actually equality holds in (1.2). In fact, if $c(t)^{tr} c(t) = I$, then

$$c(t)^{tr} c'(t) + c'(t)^{tr} c(t) = 0$$
$$I^{tr} c'(0) + c'(0)^{tr} I = 0$$
$$c'(0)^{tr} + c'(0) = 0\,.$$

Thus

$$\mathfrak{g} = \{\text{skew-symmetric real matrices}\}\,. \qquad (1.3)$$

The group structure of G forces $\mathfrak{g}$ to have additional structure. Indeed, G is closed under group conjugation

$x \to gxg^{-1}$. Thus if $c(t)$ is a smooth curve in G passing through I at $t = 0$, then so is $gc(t)g^{-1}$. Differentiating at $t = 0$, we see that $gc'(0)g^{-1}$ is in $\mathfrak{g}$. Thus $\mathfrak{g}$ is closed under the linear maps $\mathrm{Ad}(g)$, for $g \in G$, that carry $\mathfrak{g}$ into $\mathfrak{g}$ and are given by

$$\mathrm{Ad}(g)X = gXg^{-1}. \tag{1.4}$$

Next let X be in $\mathfrak{g}$ and let $c(t)$ be a smooth curve in G with $c(0) = I$. Then $t \to \mathrm{Ad}(c(t))X$ is a smooth function into $\mathfrak{g}$, as we see by substituting $g = c(t)$ into (1.4) and writing matters out. By definition we have

$$\frac{d}{dt} \mathrm{Ad}(c(t))X\Big|_{t=0} = \lim_{t\to 0} \frac{1}{t}[\mathrm{Ad}(c(t))X - X], \tag{1.5}$$

and we know that the left side exists. The right side involves vector space operations and a passage to the limit, all on members of $\mathfrak{g}$. Since $\mathfrak{g}$, as a 3-dimensional subspace of $\mathbb{R}^9$, is closed topologically, the limit is in $\mathfrak{g}$. Let us calculate this limit.

We need a preliminary formula, namely

$$\frac{d}{dt} c(t)^{-1} = -c(t)^{-1}c'(t)c(t)^{-1}. \tag{1.6}$$

To see this, we differentiate $c(t)c(t)^{-1} = I$ by the product rule, obtaining

$$c'(t)c(t)^{-1} + c(t)[\frac{d}{dt} c(t)^{-1}] = 0,$$

and (1.6) follows. Therefore

$$\frac{d}{dt}\,\mathrm{Ad}(c(t))X = \frac{d}{dt}[c(t)Xc(t)^{-1}] = c'(t)Xc(t)^{-1} + c(t)X[\frac{d}{dt}\,c(t)^{-1}]$$

$$= c'(t)Xc(t)^{-1} - c(t)Xc(t)^{-1}c'(t)c(t)^{-1}\,.$$

Putting $t = 0$ and taking (1.5) into account, we see that

$$c'(0)X - Xc'(0)$$

is in $\mathfrak{g}$. We conclude that $\mathfrak{g}$ is closed under the Lie bracket operation

$$[X,Y] = XY - YX\,. \tag{1.7}$$

Of course, we could have calculated directly that $\mathfrak{g}$ is closed under (1.4) and (1.7) once we had identified $\mathfrak{g}$ explicitly as in (1.3). But the point is that these closure properties did not depend upon our explicit knowledge of $\mathfrak{g}$. We shall return to this matter in §3, considering it in a wider context.

2. Exponential of a matrix

It is possible also to go backwards from $\mathfrak{g}$ to G in §1. The tool for doing so is the exponential map.

If A is an n-by-n complex matrix, then we define

$$\exp A = e^A = \sum_{N=0}^{\infty} \frac{1}{N!}\,A^N\,.$$

This definition makes sense, according to the following proposition.

Proposition 1.1. For any n-by-n complex matrix A, e^A is given by a convergent series (entry-by-entry). Moreover

(a) $e^X e^Y = e^{X+Y}$ if X and Y commute

(b) e^X is nonsingular

(c) $t \to e^{tX}$ is a smooth curve into $GL(n,\mathbb{C})$ that is I at $t = 0$

(d) $\frac{d}{dt}(e^{tX}) = Xe^{tX}$

(e) $\det e^X = e^{\mathrm{Tr}(X)}$

(f) $X \to e^X$ is a C^∞ mapping from matrix space ($\cong \mathbb{R}^{2n^2}$) into itself.

Proof. For any n-by-n matrix M, put

$$\|M\| = \sup_{\|x\| \leq 1} \|Mx\| ,$$

where $\|x\|$ and $\|Mx\|$ refer to the Euclidean norm. Then

$$\left\| \sum_{N=N_1}^{N_2} \frac{1}{N!} A^N \right\| \leq \sum_{N=N_1}^{N_2} \frac{1}{N!} \|A^N\| \leq \sum_{N=N_1}^{N_2} \frac{1}{N!} \|A\|^N ,$$

and the right side tends to 0 as N_1 and N_2 tend to infinity. Hence the series for e^A is Cauchy, entry-by-entry, and it must be convergent. This convergence is good enough to justify the manipulations in the remainder of the proof.

(a) $$e^X e^Y = \left(\sum_{r=0}^{\infty} \frac{1}{r!} X^r\right)\left(\sum_{s=0}^{\infty} \frac{1}{s!} Y^s\right) = \sum_{r,s} \frac{1}{r!s!} X^r Y^s$$

$$= \sum_{N=0}^{\infty} \sum_{k=0}^{N} \frac{X^k Y^{N-k}}{k!(N-k)!}$$

$$= \sum_{N=0}^{\infty} \frac{1}{N!} \sum_{k=0}^{N} \binom{N}{k} X^k Y^{N-k}$$

$$= \sum_{N=0}^{\infty} \frac{1}{N!} (X+Y)^N = e^{X+Y} .$$

(b) Take $Y = -X$ in (a), and use $e^0 = I$.

(c,d) $\frac{d}{dt}(e^{tX}) = \frac{d}{dt} \sum_{N=0}^{\infty} \frac{1}{N!} (tX)^N = \sum_{N=0}^{\infty} \frac{d}{dt} [\frac{1}{N!}(tX)^N]$

$$= \sum_{N=0}^{\infty} \frac{N}{N!} t^{N-1} X^N$$

$$= X \sum_{N=0}^{\infty} \frac{1}{(N-1)!}(tX)^{N-1} = Xe^{tX} .$$

(e) If X is upper triangular, then so is e^X. Moreover $\det e^X$ depends only on the diagonal entries of e^X, which depend only on the diagonal entries of X. Thus $\det e^X = e^{\operatorname{Tr} X}$ in this case. A general complex matrix is of the form gXg^{-1} with X upper triangular, and then we have

$$\det e^{gXg^{-1}} = \det(ge^X g^{-1}) = \det e^X = e^{\operatorname{Tr} X} = e^{\operatorname{Tr} gXg^{-1}} .$$

(f) This follows from standard facts about term-by-term differentiation of series of functions.

Let us return to $G = SO(3)$ and its $\mathfrak{g}$ given by (1.3). Then the exponential map actually carries $\mathfrak{g}$ into G. In fact, if X is skew-symmetric, then

$$(e^X)(e^X)^{tr} = e^X e^{X^{tr}} = e^X e^{-X} = e^0 = I ,$$

by (a) in the proposition, and

$$\det e^X = e^{\operatorname{Tr}(X)} = e^0 = 1 ,$$

by (e) in the proposition.

Thus (c) says that, for each X in $\mathfrak{g}$, e^{tX} is a smooth curve in G that is I at $t = 0$. From (d) we see that e^{tX} has X as derivative at $t = 0$. Consequently for any X in $\mathfrak{g}$, there is a "best" smooth curve in G that is I at $t=0$ and has derivative X at $t=0$, namely e^{tX}. Moreover this family of curves varies nicely as X varies. We shall see that the exponential map recaptures a number of local properties of G, even though $\mathfrak{g}$ depends only on infinitesimal data near the identity.

3. Closed linear groups

The critical property of $SO(3)$ for the developments in §1 is that $SO(3)$ is closed topologically as a subset of $GL(3,\mathbb{C})$ when $GL(3,\mathbb{C})$ is regarded as an open subset of $\mathbb{R}^{2\cdot 3^2}$. We shall refer to any closed subgroup of some $GL(n,\mathbb{C})$ as a <u>closed linear group</u>.

The reason that $SO(3)$ is closed is that it is defined by polynomial equations. The polynomials in question are the entry-by-entry relations in the 9 coordinates that amount to $xx^{tr} = I$ and $\det x = 1$. Here are some more examples of closed linear groups; they too are closed because they are defined by polynomial equations:

$$SO(n) = \{x \in GL(n,\mathbb{R}) \mid xx^{tr} = I \text{ and } \det x = 1\}$$

$$U(n) = \{x \in GL(n,\mathbb{C}) \mid x\bar{x}^{tr} = I\}$$

$$SU(n) = \{x \in U(n) \mid \det x = 1\}$$

$$SL(n,\mathbb{R}) = \{x \in GL(n,\mathbb{R}) \mid \det x = 1\}$$

$$SL(n,\mathbb{C}) = \{x \in GL(n,\mathbb{C}) \mid \det x = 1\} .$$

Further examples may be found in pp. 4-6 of Knapp [1986].

Let us generalize the calculations in §1, seeing the extent to which they can be carried through for all closed linear groups. For any closed linear group G, we can speak of smooth curves in G, and we let $\mathfrak{g}$ be the set of matrices $c'(0)$ for all smooth curves $c(t)$ in G such that $c(0) = I$.

We cannot instantly write down any nonzero members of $\mathfrak{g}$ in the way that we did in (1.1). But we can see just as in §1 that $\mathfrak{g}$ is a real vector space. The same argument as in §1 shows that $\mathfrak{g}$ is closed under the linear maps $\mathrm{Ad}(g)$, $g \in G$, given as in (1.4). And the same argument as in §1 shows that $\mathfrak{g}$ is closed under the Lie bracket operation (1.7). Thus $\mathfrak{g}$ is a Lie algebra of matrices in the sense that it is a real vector space of matrices that is closed under the bracket operation

$$[X,Y] = XY - YX .$$

Let $\mathfrak{gl}(n,\mathbb{R})$ and $\mathfrak{gl}(n,\mathbb{C})$ be the spaces of all real and complex n-by-n matrices. For the five examples above, we find that the corresponding Lie algebras are given by

$$\mathfrak{so}(n) = \{X \in \mathfrak{gl}(n,\mathbb{R}) \mid X + X^{tr} = 0\}$$

$$\mathfrak{u}(n) = \{X \in \mathfrak{gl}(n,\mathbb{C}) \mid X + \overline{X}^{tr} = 0\}$$

$$\mathfrak{su}(n) = \{X \in \mathfrak{gl}(n,\mathbb{C}) \mid X + \overline{X}^{tr} = 0 \text{ and } \mathrm{Tr}\, X = 0\}$$

$$\mathfrak{sl}(n,\mathbb{R}) = \{X \in \mathfrak{gl}(n,\mathbb{R}) \mid \mathrm{Tr}\, X = 0\}$$

$$\mathfrak{sl}(n,\mathbb{C}) = \{X \in \mathfrak{gl}(n,\mathbb{C}) \mid \operatorname{Tr} X = 0\}.$$

The Lie algebra $\mathfrak{g}$ consists of the set of matrices $X = c'(0)$ for smooth curves $c(t)$ in G that pass through I at $t = 0$. As with $SO(3)$, it will turn out that there is a "best" such curve, namely e^{tX}. Proposition 1.1 shows that e^{tX} is a smooth curve with the correct derivative at $t = 0$, but we need to see that e^{tX} lies in G. A little computation shows this to be the case for our five examples, and we establish it in general in Proposition 1.4 below.

Lemma 1.2. There exist an open cube U about 0 in $\mathfrak{gl}(n,\mathbb{C})$ and an open neighborhood V of I in $GL(n,\mathbb{C})$ such that $\exp : U \to V$ is smooth ($= C^\infty$) and one-one onto, with a smooth inverse.

Proof. Proposition 1.1f says that exp is smooth. By the Inverse Function Theorem, it is enough to prove that the derivative matrix at $X = 0$ of $X \to \exp X$ is nonsingular. Let $\{E_i\}$ be the standard $2n^2$-member basis over $\mathbb{R}$ of $\mathfrak{gl}(n,\mathbb{C})$, and let x_i be the corresponding coordinate or coordinate function. Then

$$\left.\frac{\partial(x_i(X \to e^X))}{\partial x_j}\right|_{X=0} = \frac{d}{dt} x_i(e^{tE_j})\Big|_{t=0}$$

$$= x_i(E_j e^{tE_j})\Big|_{t=0} \quad \text{since } x_i(\cdot) \text{ is linear}$$

$$= x_i(E_j) = \delta_{ij}.$$

So the derivative matrix is the identity, and the lemma follows.

Lemma 1.3. If $c(t)$ is a smooth curve in $GL(n,\mathbb{C})$ with $c(0) = I$ and $c'(0) = X$, then

$$\lim_{k\to\infty} c\left(\frac{t}{k}\right)^k = \exp tX$$

for all t in the domain of c.

Remark. The prototype of this lemma is as follows: We identify the additive reals with a closed linear group by $t \to (e^t)$. One curve in this group is $c(t) = 1+t$, and we have $c(t/k)^k = (1+t/k)^k$, which is well known to converge to e^t.

Proof. Choose U and V as in Lemma 1.2, and choose $\varepsilon > 0$ so that $c(t)$ is in V for all t with $|t| < 2\varepsilon$. Using the lemma, we can form a smooth curve

$$Z(t) = \exp^{-1} c(t) \qquad \text{for } |t| < 2\varepsilon .$$

Now $Z(0) = 0$ clearly, and the Chain Rule gives

$$Z'(0) = (\exp'(0))^{-1} c'(0) = X .$$

Thus Taylor's formula gives

$$Z(t) = tX + O(t^2) ,$$

where $O(t^2)$ is a term that is bounded for $|t| \leq \varepsilon$ and remains bounded near $t = 0$ when divided by t^2. Replacing t by t/k and regarding t as fixed gives

$$Z\left(\frac{t}{k}\right) = \frac{t}{k} X + O\left(\frac{t^2}{k^2}\right) = \frac{t}{k} X + O\left(\frac{1}{k^2}\right) .$$

Thus

$$kZ\left(\frac{t}{k}\right) = tX + O\left(\frac{1}{k}\right) \qquad \text{for any} \quad |t| < \varepsilon ,$$

and

$$c\left(\frac{t}{k}\right)^k = \left(\exp\ Z\left(\frac{t}{k}\right)\right)^k = \exp\ kZ\left(\frac{t}{k}\right) = \exp\left(tX + O\left(\frac{1}{k}\right)\right) .$$

Letting k tend to infinity and using the continuity of $\exp$, we obtain the conclusion of the lemma for $|t| < \varepsilon$.

For some other t in the domain of c, let us modify the above argument. First we choose and fix a positive integer N with $|t/N| < \varepsilon$. The above estimates apply to t/N. Instead of replacing t by t/k, we replace t/N by $t/(Nk+\ell)$ with $0 \le \ell \le N-1$. Again regarding t as fixed, we obtain

$$Z\left(\frac{t}{Nk+\ell}\right) = \frac{t}{Nk+\ell} X + O\left(\frac{t^2}{(Nk+\ell)^2}\right) = \frac{t}{Nk+\ell} X + O\left(\frac{1}{k^2}\right) .$$

Thus

$$(Nk+\ell)Z\left(\frac{t}{Nk+\ell}\right) = tX + O\left(\frac{1}{k}\right)$$

and

$$c\left(\frac{t}{Nk+\ell}\right)^{Nk+\ell} = \left(\exp\ Z\left(\frac{t}{Nk+\ell}\right)\right)^{Nk+\ell}$$

$$= \exp(Nk+\ell)Z\left(\frac{t}{Nk+\ell}\right) = \exp\left(tX + O\left(\frac{1}{k}\right)\right) .$$

Letting k tend to infinity, we obtain the conclusion of the lemma for this value of t.

Proposition 1.4. If G is a closed linear group and X is in its Lie algebra $\mathfrak{g}$, then $\exp X$ is in G. Consequently

$$\mathfrak{g} = \{X \in \mathfrak{gl}(n,\mathbb{C}) \mid \exp tX \text{ is in } G \text{ for all real } t\}.$$

Proof. Let X be in $\mathfrak{g}$, and let $c(t)$ be a smooth curve in G with $c(0) = I$ and $c'(0) = X$. Then $c(t/n)^n$ is in G for $n \geq 1$, and so is the limit on n (since G is a closed set) if the limit exists. Lemma 1.3 says that the limit does exist for all t in the domain of c and is $\exp tX$. Thus $\exp tX$ is in G for $|t|$ small, and it follows by raising to powers that $\exp tX$ is in G for all real t. The rest is clear from Proposition 1.1c.

Actually $\exp$ maps $\mathfrak{g}$ onto a neighborhood of I in G, and this fact accounts for the strong connection between $\mathfrak{g}$ and G. To get at this fact, however, requires a digression.

4. Manifolds and Lie groups

In the terminology of advanced calculus, the difficulty with handling the exponential map as well as we would like for a group like $SO(3)$ or one of the later examples is that the group is defined implicitly, and we need to define it parametrically (at least locally) in order to work with it better. Actually it is not so important to know a parametrization concretely; we just have to know that one exists. The parametric form is the one in the definition of Lie group below.

First we define the notion of "smooth manifold." Other kinds of manifolds are important for Lie groups, as well, but we shall not introduce them here. In what follows we shall use the terms "smooth" and "C^∞" interchangeably.

Let M be a separable metric space, not necessarily connected. An n-chart on M is a pair (U,φ) with U an open subset of M and φ a homeomorphism of U onto an open subset $\varphi(U)$ of $\mathbb{R}^n$. Two charts (U_1,φ_1) and (U_2,φ_2) are smoothly compatible if the mapping $\varphi_2 \circ \varphi_1^{-1}$ from the open set $\varphi_1(U_1 \cap U_2)$ in $\mathbb{R}^n$ to $\varphi_2(U_1 \cap U_2)$ is smooth and has a smooth inverse. A set of smoothly compatible n-charts (U_i,φ_i) for which the U_i cover M is called a C^∞ atlas.

Having a C^∞ atlas allows us to define smooth functions on M by referring matters back to $\mathbb{R}^n$. For example, a function $f: E \to \mathbb{R}$ on an open set E of M is smooth if for each x in E there is some chart (U,φ) in the atlas such that x is in U and $f \circ \varphi^{-1}$ is smooth on $\varphi(U \cap E)$. The compatibility of the charts makes it so that this smoothness persists for all charts about x.

We say that M as above, when endowed with a C^∞ atlas, is a smooth manifold of dimension n. (There is a small technical point here: A different atlas that leads to the same smooth functions on open sets is to yield the same smooth manifold. To handle this, one can observe that the set of all charts smoothly compatible with a C^∞ atlas is a maximal C^∞ atlas. Two C^∞ atlases lead to the same smooth functions exactly if their corresponding maximal atlases are the same. So, technically, M is a smooth manifold when it is endowed with a maximal C^∞ atlas.)

Examples. (1) Any open subset of $\mathbb{R}^n$ is a smooth manifold in a natural way. Only one chart is needed, and φ can be taken as the identity map.

(2) The unit sphere S^n in $\mathbb{R}^{n+1}$ can be made into a smooth manifold of dimension n by using two charts. One is

$$\varphi_1(x_1,\ldots,x_{n+1}) = \left(\frac{x_1}{1-x_{n+1}}, \ldots, \frac{x_n}{1-x_{n+1}}\right)$$

defined on $U_1 = S^n - \{(0,\ldots,0,1)\}$, and the other is

$$\varphi_2(x_1,\ldots,x_{n+1}) = \left(\frac{x_1}{1+x_{n+1}}, \ldots, \frac{x_n}{1+x_{n+1}}\right)$$

defined on $U_2 = S^n - \{(0,\ldots,0,-1)\}$.

(3) The closed linear group $SU(2)$, defined in §3, is a smooth manifold in a natural way since the formula

$$SU(2) = \left\{\begin{pmatrix} \alpha & \beta \\ -\bar{\beta} & \bar{\alpha}\end{pmatrix} \;\middle|\; \alpha \in \mathbb{C},\ \beta \in \mathbb{C},\ |\alpha|^2 + |\beta|^2 = 1\right\}$$

identifies it with the sphere S^3.

Every connected component of a smooth manifold is an open subset. Since the underlying metric space is assumed separable, there are at most countably many components.

If U and V are open subsets of smooth manifolds and $f : U \to V$ is a one-one smooth map with a smooth inverse, then f is called a diffeomorphism of U onto V.

We come to the definition of "Lie group." A separable topological group is a separable metric space that is a group in such a way that multiplication and inversion are continuous. A Lie group G is a separable topological group that is a smooth manifold in such a way that multiplication and inversion are smooth. (Here multiplication is a mapping from $G \times G$ into G, and we understand $G \times G$ to be a smooth manifold whose charts are products of charts in the factors.)

The term <u>analytic group</u> is used for a connected Lie group. In a Lie group G, the identity component G_0, which we know to be open, is an open closed subgroup and is an analytic group.

Both $GL(n, \mathbb{R})$ and $GL(n, \mathbb{C})$ are Lie groups. In fact, each is an open subset of Euclidean space and is a metric space for free. One chart suffices to give an atlas. $GL(n, \mathbb{R})$ turns out to have two connected components, whereas $GL(n, \mathbb{C})$ is connected. The formula for multiplication is given by polynomials and is smooth; the formula for inversion is given by Cramer's rule, using determinants, and is smooth. Thus $GL(n, \mathbb{R})$ and $GL(n, \mathbb{C})$ are Lie groups. Similarly the upper triangular subgroup in $GL(n, \mathbb{R})$ is a Lie group, and so is the subgroup of the upper triangular group with ones on the diagonal.

But for the most part, it is not so obvious how to exhibit a group as a Lie group. It turns out that all closed linear groups are Lie groups automatically. This is a hard result whose proof we give in the next section.

5. Closed linear groups as Lie groups

A relatively easy case that gives some insight into why closed linear groups are Lie groups is the case of

$$G = SL(2,\mathbb{R}) = \left\{ \begin{pmatrix} w & x \\ y & z \end{pmatrix} \mid \text{real entries and } wz - xy = 1 \right\}.$$

This G is the zero locus of the single polynomial

$$P(w,x,y,z) = wz - xy - 1,$$

and the derivative of P is the 1-by-4 matrix

$$P' = (z \quad -y \quad -x \quad w),$$

which is nowhere 0 on the locus where $P = 0$. Thus the Implicit Function Theorem allows us to solve locally about each point of G for one of the variables in terms of the other three. (We can see this explicitly here. About I, for example, we can solve for z or w: $z = w^{-1}(1+xy)$ and $w = z^{-1}(1+xy)$.)

To be concrete, let us fix attention on a neighborhood of I. The map $\begin{pmatrix} w & x \\ y & z \end{pmatrix} \to (w,x,y)$ is a local homeomorphism because its inverse is $(w,x,y) \to \begin{pmatrix} w & x \\ y & w^{-1}(1+xy) \end{pmatrix}$. Thus it defines a chart. A second choice for a chart is $\begin{pmatrix} w & x \\ y & z \end{pmatrix} \to (x,y,z)$ with inverse $(x,y,z) \to \begin{pmatrix} z^{-1}(1+xy) & x \\ y & z \end{pmatrix}$. The composition of the inverse of the first followed by the second is

$$(w,x,y) \to \begin{pmatrix} w & x \\ y & w^{-1}(1+xy) \end{pmatrix} \to (x,y,w^{-1}(1+xy))$$

and is smooth. Thus the charts are compatible. Arguing in this way for every point of G, we obtain an atlas.

Let us check that multiplication is smooth near each point. For $I \times I$, we are to check on

$$(w_1,x_1,y_1,w_2,x_2,y_2) \to \begin{pmatrix} w_1 & x_1 \\ y_1 & w_1^{-1}(1+x_1y_1) \end{pmatrix} \times \begin{pmatrix} w_2 & x_2 \\ y_2 & w_2^{-1}(1+x_2y_2) \end{pmatrix}$$

$$\to \begin{pmatrix} w_1w_2+x_1y_2 & w_1x_2+x_1w_2^{-1}(1+x_2y_2) \\ y_1w_2+w_1^{-1}(1+x_1y_1)y_2 & \text{---} \end{pmatrix}$$

$$\to (w_1w_2+x_1y_2, w_1x_2+x_1w_2^{-1}(1+x_2y_2), y_1w_2+w_1^{-1}(1+x_1y_1)y_2) .$$

This is smooth. We see that the coordinates that arise are compositions obtained from multiplication and our function produced by the Implicit Function Theorem; thus we did not need to make the explicit calculation to see the smoothness. By a similar argument, multiplication is smooth near all other points.

And in similar fashion, inversion is smooth. Thus $SL(2,\mathbb{R})$ is a Lie group.

If we try to proceed similarly with a general closed linear group G, the first problem is that the group need not be defined by polynomial equations. Even so, suitable entries of the matrices do look promising as local coordinates for the group. The difficulty is to decide what the open sets should be that yield the charts. To handle the difficulty, we shall work with the exponential mapping rather than the matrix entry functions; the advantage of the exponential mapping is that Proposition 1.4 gives us a relationship between $\mathfrak{g}$ and G.

<u>Theorem 1.5</u>. If G is a closed linear group, then G (with its relative metric) becomes a Lie group in a unique way such that

(a) the restrictions from $GL(n,\mathbb{C})$ to G of the real and imaginary parts of each entry function are smooth and

(b) whenever $\varphi : M \to GL(n,\mathbb{C})$ is a smooth function on a smooth manifold M such that $\varphi(M) \subseteq G$, then $\varphi : M \to G$ is smooth.

Moreover the dimension of the Lie algebra $\mathfrak{g}$ equals the dimension of the manifold G. And, in addition, there exist open neighborhoods U of 0 in $\mathfrak{g}$ and V of 1 in G such that $\exp : U \to V$ is a homeomorphism onto and such that $(V, \exp^{-1})$ is a compatible chart.

Remark. Part (b) explains our definition of smooth curve in §1 and §3.

Lemma 1.6. Let $\mathfrak{a}$ and $\mathfrak{b}$ be real subspaces with $\mathfrak{gl}(n,\mathbb{C}) = \mathfrak{a} \oplus \mathfrak{b}$. Then there exist open balls U_1 about 0 in $\mathfrak{a}$ and U_2 about 0 in $\mathfrak{b}$, as well as an open neighborhood V of I in $GL(n,\mathbb{C})$, such that

$$(a,b) \to \exp a \exp b$$

is a diffeomorphism from $U_1 \times U_2$ onto V.

Proof. Let $X_1, \ldots, X_r$ and $Y_1, \ldots, Y_s$ be bases of $\mathfrak{a}$ and $\mathfrak{b}$, and consider the map

$$(u_1,\ldots,u_r,v_1,\ldots,v_s) \to \exp^{-1}\{\exp(\Sigma\, u_i X_i)\exp(\Sigma\, v_j Y_j)\} \qquad (1.8)$$

defined in an open neighborhood of 0, with the result written out as a linear combination of $X_1, \ldots, X_r, Y_1, \ldots, Y_s$. We shall apply the Inverse Function Theorem to see that this map is locally invertible. Since $\exp$ is locally invertible, the lemma will follow.

Thus we are to compute the derivative matrix at 0 of (1.8). In computing the partial derivatives, we can set all variables but one equal to 0 before differentiating, and then we see that the expression to be differentiated is linear. The derivative matrix is thus seen to be the identity, the Inverse Function Theorem applies, and the proof is complete.

Proof of uniqueness in Theorem 1.5. Suppose that G and G' are two versions of G as a smooth manifold. Let $\iota : G \to G'$ be the identity function. The function $\iota : G \to GL(n,\mathbb{C})$ is smooth because smoothness of this map is detected by smoothness of each real or imaginary part of an entry function, which is given as (a). By (b), $\iota : G \to G'$ is smooth. By the same argument, $\iota^{-1} : G' \to G$ is smooth.

Proof of existence in Theorem 1.5. Choose a real vector subspace $\mathfrak{s}$ of $\mathfrak{gl}(n,\mathbb{C})$ such that $\mathfrak{gl}(n,\mathbb{C}) = \mathfrak{g} \oplus \mathfrak{s}$, and apply Lemma 1.6 to this decomposition, obtaining balls U_1 and U_2 about 0 in $\mathfrak{g}$ and $\mathfrak{s}$ and an open neighborhood V of I in $GL(n,\mathbb{C})$ such that $(X,Y) \to \exp X \exp Y$ is a diffeomorphism from $U_1 \times U_2$ onto V. Let 2ϵ be the radius of U_2. For each integer $k \geq 1$, form the set $(k+1)^{-1}U_2$. The claim is that for large k, $\exp X \exp Y$ cannot be in G if X is in U_1 and Y is in $(k+1)^{-1}U_2$ unless $Y = 0$.

Assume the contrary. Then for every $k \geq 1$ we can find X_k in U_1 and Y_k in $(k+1)^{-1}U_2$ with $Y_k \neq 0$ and with $\exp X_k \exp Y_k$ in G. By Proposition 1.4, $\exp X_k$ is in G.

Thus $\exp Y_k$ is in G for all k. Since $Y_k \neq 0$, we can choose an integer n_k such that $\epsilon/2 \leq n_k \|Y_k\| \leq \epsilon$. Passing to a subsequence if necessary, we may assume that $n_k Y_k$ converges, say to Y. Then Y is in $\mathfrak{s}$ and $Y \neq 0$. Since $\exp n_k Y_k = (\exp Y_k)^{n_k}$ is in G and G is closed, $\exp Y$ is in G.

Let us show that $\exp \frac{p}{q} Y$ is in G for all integers p and q with $q > 0$. Write $n_k p = m_k q + r_k$ with $0 \leq r_k \leq q-1$. Then $\frac{r_k}{q} Y_k \to 0$ since $Y_k \to 0$. Also $\frac{n_k p}{q} Y_k \to \frac{p}{q} Y$, so that

$$(\exp m_k Y_k)(\exp \frac{r_k}{q} Y_k) = \exp \frac{n_k p}{q} Y_k \to \exp \frac{p}{q} Y .$$

Since $\exp \frac{r_k}{q} Y_k \to I$, $\lim \exp m_k Y_k$ exists and equals $\exp \frac{p}{q} Y$. However, $\exp m_k Y_k = (\exp Y_k)^{m_k}$ is in G and G is closed, and thus $\exp \frac{p}{q} Y$ is in G.

In other words, $\exp tY$ is in G for all rational t. Since G is closed, $\exp tY$ is in G for all real t. By definition of $\mathfrak{g}$, Y is in $\mathfrak{g}$. Thus Y is a nonzero member of $\mathfrak{g} \cap \mathfrak{s}$, and this fact contradicts the directness of the sum $\mathfrak{g} \oplus \mathfrak{s}$. We conclude that for large k, $\exp X \exp Y$ cannot be in G if X is in U_1 and Y is in $(k+1)^{-1} U_2$ unless $Y = 0$.

Changing notation, we have found open balls U_1 and U_2 about 0 in $\mathfrak{g}$ and $\mathfrak{s}$ and an open neighborhood V of I in $\mathfrak{gl}(n, \mathbb{C})$ such that $(X, Y) \to \exp X \exp Y$ is a diffeomorphism of $U_1 \times U_2$ onto V and such that $\exp X \exp Y$ is in G only if $Y = 0$. In view of Proposition 1.4, $V \cap G$ is an open set in G such that $\exp$ is a homeomorphism from U_1

onto $V \cap G$.

We take $(V \cap G, \exp^{-1})$ as a chart about the identity in G. About the point g in G, we shall use also $(L_g(V \cap G), \exp^{-1} \circ L_g^{-1})$ as a chart, where L_g is left translation by g: $L_g(x) = gx$. Let us show that this system of charts is smoothly compatible. The picture is

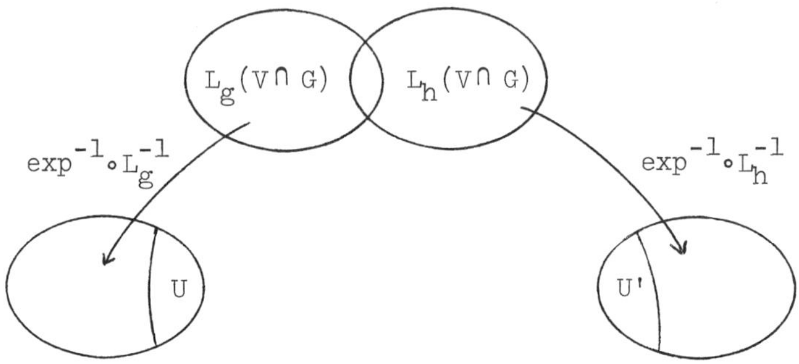

Here U and U' are open subsets of $\mathfrak{g}$. We are to check that $\exp^{-1} \circ L_{h^{-1}g} \circ \exp : U \to U'$ is smooth. This is just the restriction to a lower-dimensional Euclidean space of the smooth map $\exp^{-1} \circ L_{h^{-1}g} \circ \exp : U \times U_2 \to \mathfrak{gl}(n, \mathbb{C})$, and hence it is smooth. Thus G is a smooth manifold.

The same reasoning shows that multiplication and inversion are smooth. For example, near the identity, what needs checking in order to see that multiplication is smooth is that, for a small open neighborhood U about 0 in $\mathfrak{g}$,

$$(X, Y) \in U \times U \to \exp^{-1}(\exp X \exp Y)$$

is smooth into $\mathfrak{g}$. But this map is a restriction of the same map on $(U \times U_2) \times (U \times U_2)$, where we know it to be smooth.

In addition, this reasoning readily proves (a) and (b). Since G as a manifold has open subsets of $\mathfrak{g}$ as charts, we have $\dim G = \dim \mathfrak{g}$. Thus the theorem is completely proved.

Corollary 1.7. If G is a closed linear group and $f: G \to M$ is a function into a smooth manifold that is the restriction of a smooth map $F: U \to M$, where U is open in $GL(n,\mathbb{R})$ or $GL(n,\mathbb{C})$ and $G \subseteq U$, then $f: G \to M$ is smooth.

Proof. We can write $f = F \circ \iota$, where ι is the inclusion of G into $GL(n,\mathbb{R})$ or $GL(n,\mathbb{C})$. Then ι is smooth by Theorem 1.5a, and hence f is a composition of smooth maps.

Corollary 1.8. If G is a closed linear group and $\mathfrak{g}$ is its Lie algebra, then $\exp \mathfrak{g}$ generates the identity component G_0.

Proof. By continuity, $\exp \mathfrak{g}$ is a connected subset of G. Thus $\exp \mathfrak{g} \subseteq G_0$. Theorem 1.5 says that $\exp \mathfrak{g}$ contains a neighborhood of I in G_0. The smallest subgroup of G_0 containing a nonempty open set in G_0 is all of G_0.

Corollary 1.9. If G and G' are closed linear groups with the same Lie algebras (as sets of matrices), then the identity components of G and G' coincide.

Proof. We apply Corollary 1.8.

6. Homomorphisms

Suppose that G and H are closed linear groups. We

shall be especially interested in the case that $H = GL(n,\mathbb{C})$, but we do not make such an assumption for now. Let $\mathfrak{g}$ and $\mathfrak{h}$ be the Lie algebras of G and H, and suppose that π is a smooth homomorphism of G into H. Our objective is to associate to π a map $d\pi : \mathfrak{g} \to \mathfrak{h}$.

Before considering examples, let us comment on the case that G or H is the Lie group $\mathbb{R}$. We can always regard $\mathbb{R}$ as a closed linear group, say as the set $\{(e^t)\}$ of 1-by-1 matrices. We use this convention throughout.

Examples. (1) Regard $\mathbb{R}$ as a closed linear group, as by $t \leftrightarrow (e^t)$. If H is any closed linear group and X is in $\mathfrak{h}$, then $t \to \exp tX$ is a smooth homomorphism of $\mathbb{R}$ into H.

(2) The circle group $S^1 = \{z \in \mathbb{C} \mid |z| = 1\}$ is a closed linear group within $GL(1,\mathbb{C})$, and $t \to e^{it}$ is a smooth homomorphism of $\mathbb{R}$ into S^1.

(3) The triangular group $G = \begin{pmatrix} 1 & x & z \\ 0 & 1 & y \\ 0 & 0 & 1 \end{pmatrix}$ of real matrices is a closed linear group, and the map that sends the indicated matrix into x is a smooth homomorphism of G into $\mathbb{R}$.

We return to the general setting. If X in $\mathfrak{g}$ is given, let $c(t)$ be a smooth curve in G with $c(0) = I$ and $c'(0) = X$. (For example, we can take $c(t) = \exp tX$.) Then the composition $t \to \pi(c(t))$ is a smooth curve in H with $\pi(c(0)) = I$, and we define $d\pi(X) = (\pi \circ c)'(0)$.

Let us see that this definition is independent of the

choice of c. If $c_1(t)$ and $c_2(t)$ both have starting value I and starting derivative X, then we compute

$$(\pi\circ c_2)'(0) = \frac{d}{dt}\,\pi(c_2(t))\Big|_{t=0} = \frac{d}{dt}\,\pi(c_2(t)c_1(t)^{-1}\cdot c_1(t))\Big|_{t=0}$$

$$= \frac{d}{dt}\,\pi(c_2(t)c_1(t)^{-1})\pi(c_1(t))\Big|_{t=0} \quad \text{since } \pi \text{ is a homomorphism}$$

$$= \frac{d}{dt}\,\pi(c_2(t)c_1(t)^{-1})\Big|_{t=0}\cdot\pi(c_1(0))$$

$$+\ \pi(c_2(0)c_1(0)^{-1})\cdot\frac{d}{dt}\,\pi(c_1(t))\Big|_{t=0}$$

$$= \frac{d}{dt}\,\pi(c_2(t)c_1(t)^{-1})\Big|_{t=0} + (\pi\circ c_1)'(0)\,.$$

Thus it is enough to prove that the curve $c(t) = c_2(t)c_1(t)^{-1}$, which has $c(0) = I$ and $c'(0) = 0$, has $(\pi\circ c)'(0) = 0$. We now refer matters to local coordinates, using the exponential map. (See Theorem 1.5.) The local expression for $\pi\circ c$ is $\exp^{-1}\circ\pi\circ c$, which we write as

$$\exp^{-1}\circ\pi\circ c = (\exp^{-1}\circ\pi\circ\exp)\ (\exp^{-1}\circ c)\,;$$

here the two factors on the right are local expressions for π and c. However, the second factor is a curve in $\mathfrak{g}$, and it has $(\exp^{-1}\circ c)'(0) = 0$ by the Chain Rule, since $c'(0) = 0$. Thus $(\exp^{-1}\circ\pi\circ c)'(0) = 0$ by the Chain Rule. Applying the exponential map and using the Chain Rule once more, we see that $(\pi\circ c)'(0) = 0$. Thus our definition is independent of the choice of c.

Now we can imitate some of the development of §1 and §3. First of all, $d\pi : \mathfrak{g} \to \mathfrak{h}$ is linear. In fact, let $c_1(t)$ and

$c_2(t)$ correspond to X and Y, and let k be in $\mathbb{R}$. Then

$$d\pi(kX) = \frac{d}{dt}\,\pi(c_1(kt))\Big|_{t=0} = \frac{d}{ds}\,\pi(c_1(s))\Big|_{s=0}\cdot\frac{d}{dt}(kt)\Big|_{t=0} = k\,d\pi(X)$$

and

$$\begin{aligned} d\pi(X+Y) &= \frac{d}{dt}\,\pi(c_1(t)c_2(t))\Big|_{t=0} = \frac{d}{dt}\,\pi(c_1(t))\pi(c_2(t))\Big|_{t=0} \\ &= d\pi(X) + d\pi(Y) \end{aligned}$$

by the product rule for derivatives.

Now let g be in G and let $c(t)$ in G have $c(0) = I$ and $c'(0) = X$. Then $gc(t)g^{-1}$ has derivative $\mathrm{Ad}(g)X$ at $t = 0$. Hence

$$\begin{aligned} d\pi(\mathrm{Ad}(g)X) &= \frac{d}{dt}\,\pi(gc(t)g^{-1})\Big|_{t=0} = \frac{d}{dt}\,\pi(g)\pi(c(t))\pi(g)^{-1}\Big|_{t=0} \\ &= \pi(g)\,\frac{d}{dt}\,\pi(c(t))\Big|_{t=0}\,\pi(g)^{-1} \\ &= \pi(g)\,d\pi(X)\pi(g)^{-1}. \end{aligned} \tag{1.9}$$

If also Y is in $\mathfrak{g}$, then this formula says

$$d\pi(\mathrm{Ad}(c(t))Y) = \pi(c(t))\,d\pi(Y)\pi(c(t))^{-1}.$$

Differentiating at $t = 0$ and using the fact that $d\pi$ is linear, we obtain

$$d\pi[X,Y] = d\pi(X)d\pi(Y) - d\pi(Y)d\pi(X). \tag{1.10}$$

The right hand side of (1.10) is the definition of $[d\pi(X), d\pi(Y)]$. Thus (1.10), in the presence of the linearity of $d\pi$, says that $d\pi$ is a Lie algebra homomorphism. Thus our smooth homomorphism $\pi : G \to H$ leads to a Lie algebra homomorphism $d\pi : \mathfrak{g} \to \mathfrak{h}$.

Examples, continued. (1) If $\pi(e^t) = \exp tX$, then $d\pi(1) = X$. (To see this, we use the curve $c(t) = e^t$, which has $c'(0) = 1$, and we compute $(\pi\circ c)'(0)$ from Proposition 1.1d.

(2) If $\pi(e^t) = e^{it}$, then $d\pi(1) = (i)$ as a 1-by-1 matrix.

(3) If $\pi\begin{pmatrix} 1 & x & z \\ 0 & 1 & y \\ 0 & 0 & 1 \end{pmatrix} = (e^x)$, then $\pi\begin{pmatrix} 0 & a & c \\ 0 & 0 & b \\ 0 & 0 & 0 \end{pmatrix} = (a)$.

The fundamental relation for dealing with homomorphisms is the formula that relates π, $d\pi$, and the exponential map.

Theorem 1.10. If $\pi : G \to H$ is a smooth homomorphism between closed linear groups, then $\pi\circ\exp = \exp\circ d\pi$.

Proof. Fix X in the Lie algebra of G, and let $c_1(t)$ and $c_2(t)$ be the smooth curves of matrices (actually in H) given by

$$c_1(t) = \exp(t\, d\pi(X)) \quad \text{and} \quad c_2(t) = \pi(\exp tX).$$

Then $c_1(0) = c_2(0) = I$ and

$$\frac{d}{dt} c_1(t) = d\pi(X)\exp(t\, d\pi(X)) = d\pi(X)c_1(t)$$

by Proposition 1.1d. Also

$$\frac{d}{dt} c_2(t) = \frac{d}{dh} \pi(\exp(t+h)X)\Big|_{h=0} = \frac{d}{dh} \pi(\exp hX)\pi(\exp tX)\Big|_{h=0}$$

$$= \frac{d}{dh} \pi(\exp hX)\Big|_{h=0} \pi(\exp tX) = d\pi(X)c_2(t)$$

by definition of $d\pi(X)$. Since the jth columns of both $c_1(t)$ and $c_2(t)$ solve the initial value problem for the linear system of differential equations

$$\frac{dy}{dt} = d\pi(X)y\,, \qquad y(0) = (j^{th} \text{ column of } I)\,,$$

the uniqueness theorem for systems of ordinary differential equations says $c_1(t) = c_2(t)$. The theorem follows by taking $t = 1$.

<u>Corollary 1.11</u>. Let $\pi : G \to H$ be a smooth homomorphism between closed linear groups, and let $\mathfrak{g}$ and $\mathfrak{h}$ be the Lie algebras. If the map $x \to \pi(x)$ near $x = I$ is referred to local coordinates relative to the exponential maps of G and H, then the corresponding map is exactly $d\pi : \mathfrak{g} \to \mathfrak{h}$, which is linear. Hence $d\pi$ is also the derivative of π at the identity when π is referred to local coordinates by the exponential maps.

Proof. The map in local coordinates is $Y = \exp^{-1}(\pi(\exp(X))$ near $X = 0$, and Theorem 1.10 says this is the same as $Y = \exp^{-1}(\exp d\pi(X)) = d\pi(X)$, which is linear. A linear map is its own derivative, and the corollary follows.

<u>Corollary 1.12</u>. If π_1 and π_2 are smooth homomorphisms between two closed linear groups G and H such that $d\pi_1 = d\pi_2$, then $\pi_1 = \pi_2$ on G_0.

Proof. For X in $\mathfrak{g}$, Theorem 1.10 gives

$$\pi_1(\exp X) = \exp d\pi_1(X) = \exp d\pi_2(X) = \pi_2(\exp X) .$$

By Corollary 1.8, $\pi_1 = \pi_2$ on G_0.

Corollary 1.13. Let $\pi : G \to H$ be a smooth homomorphism between two closed linear groups, and let $d\pi : \mathfrak{g} \to \mathfrak{h}$ be the corresponding Lie algebra homomorphism. Then

(a) $d\pi$ onto implies π is onto G_0

(b) $d\pi$ one-one implies π is one-one in a neighborhood of I in G

(c) $d\pi$ one-one onto implies π is a local isomorphism on G_0.

Proof. Parts (a) and (b) are immediate from Theorems 1.10 and 1.5. Part (c) carries with it a statement about smoothness of π^{-1}, which follows from Corollary 1.11 and the Inverse Function Theorem.

7. An interesting homomorphism

This section gives an optional example of a nontrivial smooth homomorphism and illustrates a number of the points in §6. The map π will carry $SU(2)$ into $SO(3)$.

In §4 we gave particular charts to make spheres into manifolds. If we specialize to the case of S^2 and use complex variables in the notation, then the statement is that

$$z = \frac{x_1 + ix_2}{1 - x_3} \tag{1.11}$$

maps S^2 (in $\mathbb{R}^3$) one-one onto the extended complex plane $\mathbb{C} \cup \{\infty\}$. Meanwhile $SU(2)$ acts on $\mathbb{C} \cup \{\infty\}$ by

$$w = g(z) = \frac{\alpha z + \beta}{-\bar{\beta} z + \bar{\alpha}} \quad \text{if} \quad g = \begin{pmatrix} \alpha & \beta \\ -\bar{\beta} & \bar{\alpha} \end{pmatrix} \text{ is in } SU(2).$$

It is easy to check that this is a group action. We shall reinterpret this action as an action on S^2 by the identification (1.11) and shall see that each g in $SU(2)$ acts on S^2 by the restriction of a linear transformation of $\mathbb{R}^3$. It follows that each of these linear transformations is an orthogonal transformation (since S^2 is preserved); since we have a group action on S^2, we have a group action in $\mathbb{R}^3$. This means that the map

$$\pi : g \text{ in } SU(2) \longrightarrow \text{effect of } g \text{ on } \mathbb{R}^3$$

is a group homomorphism of $SU(2)$ into

$$O(3) = \{x \mid xx^{tr} = I\}.$$

Since $SU(2)$ is connected and π is continuous, the image must be contained in $SO(3)$.

Let us construct π. Afterward we shall compute $d\pi$ and examine what is happening. Thus we write

$$z = \frac{x_1 + ix_2}{1 - x_3}, \quad w = \frac{\alpha z + \beta}{-\bar{\beta} z + \bar{\alpha}}, \quad w = \frac{y_1 + iy_2}{1 - y_3},$$

and we compute (y_1, y_2, y_3) in terms of (x_1, x_2, x_3). Since $x_1^2 + x_2^2 + x_3^2 = 1$, we have

$$|z|^2 = \frac{1-x_3^2}{(1-x_3)^2} = \frac{1+x_3}{1-x_3}.$$

Then $|z|^2-1 = 2x_3/(1-x_3)$ and $|z|^2+1 = 2/(1-x_3)$. Consequently

$$\frac{|z|^2-1}{|z|^2+1} = x_3.$$

A similar computation solves for y_3 in terms of w, and we find

$$y_1 = \frac{w+\overline{w}}{1+|w|^2}, \quad y_2 = \frac{w-\overline{w}}{i(1+|w|^2)}, \quad y_3 = \frac{|w|^2-1}{|w|^2+1}.$$

To compute y_3 in terms of x_1, x_2, and x_3, we write

$$|w|^2 = \left|\frac{\alpha z+\beta}{-\overline{\beta}z+\overline{\alpha}}\right|^2 = \frac{(\alpha z+\beta)(\overline{\alpha z}+\overline{\beta})}{(\overline{\alpha}-\overline{\beta}z)(\alpha-\beta\overline{z})} = \frac{|\alpha|^2|z|^2+2\mathrm{Re}(\alpha z\overline{\beta})+|\beta|^2}{|\alpha|^2-2\mathrm{Re}(\alpha z\overline{\beta})+|\beta|^2|z|^2}.$$

Thus

$$|w|^2+1 = \frac{|z|^2+1}{|\alpha|^2-2\mathrm{Re}(\alpha z\overline{\beta})+|\beta|^2|z|^2}$$

$$|w|^2-1 = \frac{(|\alpha|^2-|\beta|^2)(|z|^2-1)+4\mathrm{Re}(\alpha z\overline{\beta})}{|\alpha|^2-2\mathrm{Re}(\alpha z\overline{\beta})+|\beta|^2|z|^2}$$

$$y_3 = \frac{|w|^2-1}{|w|^2+1} = \frac{(|\alpha|^2-|\beta|^2)(|z|^2-1)+4\mathrm{Re}(\alpha z\overline{\beta})}{|z|^2+1}$$

$$= (|\alpha|^2-|\beta|^2)\frac{|z|^2-1}{|z|^2+1} + \frac{4\mathrm{Re}(\alpha(x_1+ix_2)\overline{\beta})/(1-x_3)}{2/(1-x_3)}$$

$$= 2(\mathrm{Re}\ \alpha\overline{\beta})x_1 - 2(\mathrm{Im}\ \alpha\overline{\beta})x_2 + (|\alpha|^2-|\beta|^2)x_3. \qquad (1.12)$$

To compute y_1 and y_2, we write

$$y_1 = \frac{2\ \mathrm{Re}\ w}{1+|w|^2} = \frac{2\ \mathrm{Re}\left(\dfrac{\alpha z+\beta}{-\overline{\beta}z+\overline{\alpha}}\right)}{1+|w|^2} = \frac{2\ \mathrm{Re}[(\alpha z+\beta)(-\beta\overline{z}+\alpha)]}{|\overline{\alpha}-\overline{\beta}z|^2(1+|w|^2)}$$

$$= \frac{2\ \mathrm{Re}[(\alpha z+\beta)(-\beta\overline{z}+\alpha)]}{1+|z|^2}$$

$$= \frac{2\ \mathrm{Re}(\alpha^2 z-\beta^2\overline{z})}{1+|z|^2} + \frac{1-|z|^2}{1+|z|^2}\, 2\ \mathrm{Re}\ \alpha\beta$$

$$= \mathrm{Re}[\alpha^2(x_1+ix_2) - \beta^2(x_1-ix_2)] - 2x_3\ \mathrm{Re}\ \alpha\beta$$

$$= x_1\ \mathrm{Re}(\alpha^2-\beta^2) - x_2\ \mathrm{Im}(\alpha^2+\beta^2) - 2x_3\ \mathrm{Re}\ \alpha\beta \tag{1.13}$$

and

$$y_2 = \frac{2\ \mathrm{Im}\ w}{1+|w|^2} = \frac{2\ \mathrm{Im}\left(\dfrac{\alpha z+\beta}{-\overline{\beta}z+\overline{\alpha}}\right)}{1+|w|^2} = \frac{2\ \mathrm{Im}[(\alpha z+\beta)(-\beta\overline{z}+\alpha)]}{|\overline{\alpha}-\overline{\beta}z|^2(1+|w|^2)}$$

$$= \frac{2\ \mathrm{Im}[(\alpha z+\beta)(-\beta\overline{z}+\alpha)]}{1+|z|^2}$$

$$= \frac{2\ \mathrm{Im}(\alpha^2 z-\beta^2\overline{z})}{1+|z|^2} + \frac{1-|z|^2}{1+|z|^2}\, 2\ \mathrm{Im}\ \alpha\beta$$

$$= \mathrm{Im}[\alpha^2(x_1+ix_2) - \beta^2(x_1-ix_2)] - 2x_3\ \mathrm{Im}\ \alpha\beta$$

$$= x_1\ \mathrm{Im}(\alpha^2-\beta^2) + x_2\ \mathrm{Re}(\alpha^2+\beta^2) - 2x_3\ \mathrm{Im}\ \alpha\beta\,. \tag{1.14}$$

Combining (1.12), (1.13), and (1.14), we see that

$$\begin{pmatrix} y_1 \\ y_2 \\ y_3 \end{pmatrix} = \begin{pmatrix} \mathrm{Re}(\alpha^2-\beta^2) & -\mathrm{Im}(\alpha^2+\beta^2) & -2\ \mathrm{Re}\ \alpha\beta \\ \mathrm{Im}(\alpha^2-\beta^2) & \mathrm{Re}(\alpha^2+\beta^2) & -2\ \mathrm{Im}\ \alpha\beta \\ 2\ \mathrm{Re}\ \alpha\overline{\beta} & -2\ \mathrm{Im}\ \alpha\overline{\beta} & |\alpha|^2-|\beta|^2 \end{pmatrix} \begin{pmatrix} x_1 \\ x_2 \\ x_3 \end{pmatrix}. \tag{1.15}$$

Hence $\pi\begin{pmatrix} \alpha & \beta \\ -\overline{\beta} & \overline{\alpha} \end{pmatrix}$ is the 3-by-3 matrix in (1.15).

We compute $d\pi$ on the basis $\begin{pmatrix} i & 0 \\ 0 & -i \end{pmatrix}$, $\begin{pmatrix} 0 & 1 \\ -1 & 0 \end{pmatrix}$, $\begin{pmatrix} 0 & i \\ i & 0 \end{pmatrix}$ by using the curves

$$\begin{pmatrix} e^{it} & 0 \\ 0 & e^{-it} \end{pmatrix}, \quad \begin{pmatrix} \cos t & \sin t \\ -\sin t & \cos t \end{pmatrix}, \quad \begin{pmatrix} \cos t & i \sin t \\ i \sin t & \cos t \end{pmatrix}.$$

So, for example,

$$d\pi\begin{pmatrix} i & 0 \\ 0 & -i \end{pmatrix} = \frac{d}{dt}\, \pi\begin{pmatrix} e^{it} & 0 \\ 0 & e^{-it} \end{pmatrix}\Bigg|_{t=0} = \frac{d}{dt}\begin{pmatrix} \cos 2t & -\sin 2t & 0 \\ \sin 2t & \cos 2t & 0 \\ 0 & 0 & 1 \end{pmatrix}\Bigg|_{t=0}$$

$$= \begin{pmatrix} 0 & -2 & 0 \\ 2 & 0 & 0 \\ 0 & 0 & 0 \end{pmatrix}.$$

Similarly

$$d\pi\begin{pmatrix} 0 & 1 \\ -1 & 0 \end{pmatrix} = \begin{pmatrix} 0 & 0 & -2 \\ 0 & 0 & 0 \\ 2 & 0 & 0 \end{pmatrix}, \qquad d\pi\begin{pmatrix} 0 & i \\ i & 0 \end{pmatrix} = \begin{pmatrix} 0 & 0 & 0 \\ 0 & 0 & -2 \\ 0 & 2 & 0 \end{pmatrix}.$$

It is clear that $d\pi$ maps $\mathfrak{su}(2)$ one-one onto $\mathfrak{so}(3)$. Since $SO(3)$ is connected (because $SO(3)/SO(2) \cong S^2$ and $SO(2)$ are connected), Corollary 1.13 tells us that π maps $SU(2)$ onto $SO(3)$ and is a diffeomorphism in a neighborhood of the identity. Actually π is not globally one-one; we can check directly that $\pi\begin{pmatrix} -1 & 0 \\ 0 & -1 \end{pmatrix} = I$. The homomorphism π is everywhere two-to-one.

8. Representations

Let G be a topological group. A finite-dimensional representation of G is a homomorphism π of G into the group of invertible linear maps of a finite-dimensional complex vector space V into itself, such that the resulting map of $G \times V$ into V is continuous.

When G is given to us as a subgroup of $GL(n, \mathbb{C})$, then the identity map on G gives us what is called the standard representation of G on $\mathbb{C}^n$. It is reasonable to ask why one should study representations of a group that is already represented as matrices. The answer is that representations are often forced on us by some outside area of mathematics, and we have to deal with them.

Examples.

(1) $G = SO(n)$, V = all polynomials in n real variables with complex coefficients and all terms of degree N (i.e., homogeneous of degree N), and

$$\pi(g)P\begin{pmatrix} x_1 \\ \vdots \\ x_n \end{pmatrix} = P\left(g^{-1}\begin{pmatrix} x_1 \\ \vdots \\ x_n \end{pmatrix}\right).$$

(2) $G = U(n)$ or $SU(n)$, V = all polynomials in $z_1, \ldots, z_n, \bar{z}_1, \ldots, \bar{z}_n$ homogeneous of degree N, and

$$\pi(g)P\left(\begin{pmatrix} z_1 \\ \vdots \\ z_n \end{pmatrix}, \begin{pmatrix} \bar{z}_1 \\ \vdots \\ \bar{z}_n \end{pmatrix}\right) = P\left(g^{-1}\begin{pmatrix} z_1 \\ \vdots \\ z_n \end{pmatrix}, \bar{g}^{-1}\begin{pmatrix} \bar{z}_1 \\ \vdots \\ \bar{z}_n \end{pmatrix}\right).$$

(3) $G = U(n)$ or $SU(n)$, V = all members of the

subspace $\Lambda^k \mathbb{C}^n$ of the exterior algebra of $\mathbb{C}^n$ (see §II.4). A basis over $\mathbb{C}$ of $\Lambda^k \mathbb{C}^n$ is all alternating tensors $e_{i_1} \wedge \dots \wedge e_{i_k}$ with $i_1 < \dots < i_k$, where $\{e_j\}$ is the standard basis of $\mathbb{C}^n$ over $\mathbb{C}$. If we define

$$\pi(g)(e_{i_1} \wedge \dots \wedge e_{i_k}) = ge_{i_1} \wedge \dots \wedge ge_{i_k},$$

then $\pi(g)$ extends to a linear map of $\Lambda^k \mathbb{C}^n$ into itself, and π is a representation. (We shall see this a little more systematically in Chapter II.)

An <u>invariant subspace</u> for such a π is a (complex) vector subspace such that $\pi(g)U \subseteq U$ for all g in G; in this case we get representations of G on U and V/U in the obvious way. A finite-dimensional representation π is <u>irreducible</u> if $V \neq 0$ and if π has no invariant subspaces other than 0 and V. Two finite-dimensional representations π on V and π' on V' are <u>equivalent</u> if there is an invertible linear $E: V \to V'$ such that $E\pi(g) = \pi'(g)E$ for all g in G.

<u>Examples</u>.

(1) $G = SO(n)$, V = all polynomials in n real variables homogeneous of degree N with complex coefficients, π as before. Define the Laplacian operator by

$$\Delta = \frac{\partial^2}{\partial x_1^2} + \dots + \frac{\partial^2}{\partial x_n^2}.$$

This operator reduces degrees by two, but the operator $E(P) = |x|^2 \Delta P$ maps V into V. Moreover this operator commutes with π in the sense that $E\pi(g) = \pi(g)E$ for all g in $SO(n)$. This commutativity implies that the kernel and image of E are invariant subspaces. With some effort, one can show that the kernel is irreducible for each N if $n > 2$.

(2) $G = SU(n)$; V = homogeneous polynomials of degree N in $z_1, \ldots, z_n, \bar{z}_1, \ldots, \bar{z}_n$; π as before. The subspace of holomorphic polynomials (those with no $\bar{z}$'s) is an invariant subspace. With some effort, one can show that this subspace is irreducible.

(3) $G = SU(2)$, V = homogeneous holomorphic polynomials of degree N in z_1 and z_2, π as before, V' = holomorphic polynomials in z of degree N with

$$\pi'\begin{pmatrix} \alpha & \beta \\ -\bar{\beta} & \bar{\alpha} \end{pmatrix} Q(z) = (\bar{\beta}z + \alpha)^N \, Q\left(\frac{\bar{\alpha}z - \beta}{\bar{\beta}z + \alpha}\right) .$$

Define $E: V \to V'$ by $EP(z) = P\binom{z}{1}$. Then E exhibits π and π' as equivalent.

Our interest will be in the case that G is a Lie group, particularly when G is a closed linear group. If V is a finite-dimensional complex vector space, then we can identify V with $\mathbb{C}^n$ and the invertible linear mappings with $GL(n, \mathbb{C})$, where $n = \dim V$; all we have to do is choose a basis. If we choose a different basis, then the two identifications are related by an invertible linear mapping, which preserves all the Lie group structure of interest. Thus a result about representations on $\mathbb{C}^n$ gives us a result about representations on V. The connection between representation theory and Lie groups comes from the following theorem. We state it for closed linear groups, but it is valid for all Lie groups.

Theorem 1.14. Every finite-dimensional representation π of a closed linear group is a smooth mapping.

Lemma 1.15. If G is a closed linear group, then there exist arbitrarily small open neighborhoods U and V of I in G such that $U \subseteq V$ and each element of V has a unique square root in U.

Proof. By Theorem 1.5 we can choose a sufficiently small open ball B about 0 in $\mathfrak{g}$ and an open neighborhood V of I in G such that exp is a diffeomorphism of B onto V. Put $B' = \frac{1}{2}B$ and $U = \exp B'$. On U the map $x \to x^2$, when referred to local coordinates by the exponential map, is just $2I$ and is a diffeomorphism of B' onto B. Therefore the map $x \to x^2$ is a diffeomorphism of $U = \exp B'$ onto $V = \exp B$.

Proof of Theorem 1.14. Let $\pi : G \to GL(n,\mathbb{C})$ be given. First suppose $G = \mathbb{R} \leftrightarrow \{(e^t)\}$. By Lemmas 1.2 and 1.15 we can choose open neighborhoods U of 0 in $\mathfrak{gl}(n,\mathbb{C})$ and V of I in $GL(n,\mathbb{C})$ such that U is a ball, $\exp : U \to V$ is a diffeomorphism, and members of V have unique square roots in V. Choose $\epsilon > 0$ so that $\pi(e^t)$ is in V for $|t| \leq \epsilon$. Then we can write $\pi(e^\epsilon) = \exp \epsilon Y$ for some Y in $\epsilon^{-1}U$. Now $\pi(e^{\epsilon/2})$ and $\exp \frac{\epsilon}{2} Y$ are both square roots of $\exp \epsilon Y$ within V and hence are equal. Iterating this fact, we see that $\pi(e^{\epsilon/2^k}) = \exp (\epsilon/2^k)Y$ for all $k \geq 1$. Raising to powers shows that $\pi(e^t) = \exp tY$ for all diadic rational t. By continuity of π, $\pi(e^t) = \exp tY$ for all real t. Thus π is smooth when $G = \mathbb{R}$.

For general G, let $X_1, \ldots, X_d$ be a basis of $\mathfrak{g}$. By the previous paragraph, $t \to \pi(\exp tX_j)$ is of the form $t \to \exp tY_j$ for some Y_j in $\mathfrak{gl}(n, \mathbb{C})$. Since

$$\begin{aligned}\pi(\exp u_1X_1 \cdots \exp u_dX_d) &= \pi(\exp u_1X_1)\cdots\pi(\exp u_dX_d) \\ &= (\exp u_1Y_1)\cdots(\exp u_dY_d),\end{aligned}$$

the map

$$(u_1,\ldots,u_d) \to \pi(\exp u_1X_1 \cdots \exp u_dX_d)$$

is smooth from $\mathbb{R}^d$ into $GL(n,\mathbb{C})$. If we can show that

$$(u_1,\ldots,u_d) \to (\exp u_1X_1)\cdots(\exp u_dX_d) \tag{1.16}$$

is locally a diffeomorphism into G near the origin of $\mathbb{R}^d$, then $x \to \pi(x)$ will be smooth near the identity, and the theorem will follow.

We shall apply the Inverse Function Theorem. In local coordinates, (1.16) is given by

$$(u_1,\ldots,u_d) \to \exp^{-1}\{(\exp u_1X_1)\cdots(\exp u_dX_d)\},$$

and the derivative matrix of this at the origin is the identity (since we can set all but one u_j equal to 0 before differentiating). The Inverse Function Theorem applies, and the proof is complete.

As a consequence of this theorem, a representation

$\pi : G \to GL(n,\mathbb{C})$ automatically gives us a Lie algebra homomorphism $d\pi : \mathfrak{g} \to \mathfrak{gl}(n,\mathbb{C})$. To make the terminology match, we say that a finite-dimensional representation φ of a Lie algebra of matrices is a linear mapping of $\mathfrak{g}$ into the vector space $\mathrm{End}_{\mathbb{C}} V$ of linear maps of a finite-dimensional complex vector space V into itself such that

$$\varphi([X,Y]) = \varphi(X)\varphi(Y) - \varphi(Y)\varphi(X) .$$

(Again the right side is the bracket operation for linear maps of V into itself, and this equation says that φ respects the bracket operations.)

Anyway Theorem 1.14 implies that each finite-dimensional representation π of a closed linear group yields a corresponding finite-dimensional representation $d\pi$ of the Lie algebra. One can compute the Lie algebra representations corresponding to the group representations earlier in this section. An example of this kind of computation occurs at the end of §7. On the one hand, the resulting Lie algebra representations are simpler objects to study than the original group representations; after all, they are determined by $d\pi(X)$ for X in a basis of $\mathfrak{g}$. On the other hand, they determine the group representations completely on G_0, according to Theorem 1.10 and its corollaries. So the Lie algebra representations are worthy of study.

The terms invariant subspace, irreducible, and equivalent are applicable also to $\mathfrak{g}$, as are the representations on U and V/U that we can associate to an invariant subspace.

Let us return to a general topological group G and

consider infinite-dimensional representations. In this case it is customary to impose extra structure (such as compactness) on G or to require that the vector space be a complex linear topological space. We follow the latter course for now. For our purposes the following definition will serve nicely: A representation of G on a complex Hilbert space V is a homomorphism π of G into the group of bounded linear operators on V with bounded inverses, such that the resulting map of $G \times V$ into V is continuous.

An invariant subspace for such a π is a vector subspace U such that $\pi(g)U \subseteq U$ for all g in G. If U is a closed invariant subspace, then we obtain a subrepresentation of G on U and a quotient representation of G on V/U. A representation with $V \neq 0$ is irreducible if it has no closed invariant subspaces other than 0 and V.

Examples.

(1) $G = SL(2,\mathbb{R})$, $V = L^2(\mathbb{R}^2)$, and

$$\pi(g)f(x) = f(g^{-1}x) .$$

With the usual inner product on V, this is a representation. The operators $\pi(g)$ are in fact unitary.

(2) G = a locally compact group, $V = L^2(G, d_\ell x)$ taken with respect to a left-invariant measure $d_\ell x$, and $\pi(g)f(x) = f(g^{-1}x)$. This is called the left regular representation of G. The right regular representation of G is given by $\pi'(g)f(x) = f(xg)$ on $L^2(G, d_r x)$, where $d_r x$ is a right-invariant measure.

Theorem 1.14 does not extend to infinite-dimensional representations of closed linear groups. However, there is a substitute. If π is a representation of G on an infinite-dimensional complex Hilbert space V, then one can consider the vector subspace

$$V_0 = \{v \in V \mid g \to \pi(g)v \text{ is } C^\infty \text{ from } G \text{ into } V\} .$$

(It is not important for our purposes to understand this definition in detail; the interested reader can consult Chapter III of Knapp [1986].) If X is in $\mathfrak{g}$ and $c(t)$ is a smooth curve with $c(0) = I$ and $c'(0) = X$, we can put $d\pi(X)v = \frac{d}{dt}\pi(c(t))v\big|_{t=0}$ for v in V_0. A little checking shows that $d\pi(X)$ is well defined, mapping V_0 into itself. Then $d\pi$ is linear and satisfies

$$d\pi[X,Y] = d\pi(X)d\pi(Y) - d\pi(Y)d\pi(X) .$$

All this would not be very interesting if V_0 turned out to be 0. But in fact, it is not hard to show that V_0 is dense in V.

Accordingly we define a <u>representation</u> φ of $\mathfrak{g}$ to be a linear mapping of $\mathfrak{g}$ into the vector space of linear maps of a complex vector space into itself such that

$$\varphi([X,Y]) = \varphi(X)\varphi(Y) - \varphi(Y)\varphi(X) . \qquad (1.17)$$

<u>Invariant subspaces</u>, <u>subrepresentations</u>, <u>quotient representations</u>, <u>irreducible representations</u>, and <u>equivalent representations</u> are defined in the expected way — with no reference to topology on the vector space.

Once again one can compute the representations of $\mathfrak{g}$ that correspond to various representations of G. Again the representations of $\mathfrak{g}$ are simpler than those of G. But it turns out that Theorem 1.10 breaks down even on V_0. Thus, for example, the closure in V of a $\mathfrak{g}$-invariant subspace of V_0 is not necessarily G-invariant. However, there are representations where it is permissible to draw conclusions about G invariance from hypotheses of $\mathfrak{g}$ invariance. For this reason, one tends to study only infinite-dimensional representations of $\mathfrak{g}$ of certain types——ones where the behavior of $\mathfrak{g}$ controls the behavior of G. In any event, some Lie algebra representations are worthy of study, on this basis, and we begin a systematic treatment of Lie algebra representations in the next chapter. We return to groups in Chapter III.

CHAPTER II

REPRESENTATIONS AND TENSORS

1. Abstract Lie algebras

A <u>Lie algebra</u> $\mathfrak{g}$ is a vector space with a bilinear multiplication $[x,y]$ such that $[x,x] = 0$ for all x and such that the <u>Jacobi identity</u> holds:

$$[[x,y],z] + [[y,z],x] + [[z,x],y] = 0 .$$

We denote the underlying field by k for now. Ultimately we shall be interested only in the cases $k = \mathbb{R}$ and $k = \mathbb{C}$.

Let V be a vector space over k, and let $\mathrm{End}_k V$ be the algebra of all k-linear maps of V into itself. We readily check that $\mathfrak{g} = \mathrm{End}_k V$ becomes a Lie algebra under the definition

$$[a,b] = ab - ba . \qquad (2.1)$$

So, for example, $\mathrm{End}_{\mathbb{C}} V$ is a Lie algebra over $\mathbb{C}$. But it is also a Lie algebra over $\mathbb{R}$ if we simply forget about multiplication by nonreal scalars.

More generally, let A be any associative algebra over k. Then A becomes a Lie algebra under the definition (2.1) of the Lie bracket operation.

Within a Lie algebra $\mathfrak{g}$, if $\mathfrak{a}$ and $\mathfrak{b}$ are subsets of $\mathfrak{g}$, then $[\mathfrak{a},\mathfrak{b}]$ denotes the linear span of the brackets of a member of $\mathfrak{a}$ with a member of $\mathfrak{b}$. A Lie subalgebra $\mathfrak{h}$ of $\mathfrak{g}$ is a subspace with $[\mathfrak{h},\mathfrak{h}] \subseteq \mathfrak{h}$. An ideal $\mathfrak{h}$ is a subspace with $[\mathfrak{h},\mathfrak{g}] \subseteq \mathfrak{h}$. Some examples of Lie subalgebras of the real Lie algebra $\mathfrak{gl}(n,\mathbb{C}) \cong \text{End}_{\mathbb{C}}\, \mathbb{C}^n$ appear in §I.3.

Left bracket in $\mathfrak{g}$ by x is denoted $\text{ad}\, x$. Then $\text{ad}\, x$ is in $\text{End}_k\, \mathfrak{g}$, and the map $\text{ad} : \mathfrak{g} \to \text{End}_k\, \mathfrak{g}$ is linear. In fact, each $L = \text{ad}\, x$ is a derivation in the sense that

$$L[y,z] = [y,Lz] + [Ly,z] ;$$

in the presence of the other defining properties of a Lie algebra, this fact is equivalent with the Jacobi identity.

A homomorphism $\varphi : \mathfrak{g} \to \mathfrak{h}$ between Lie algebras is a k-linear map carrying brackets to brackets. For example, $\text{ad} : \mathfrak{g} \to \text{End}_k\, \mathfrak{g}$ is a Lie algebra homomorphism, as a consequence of the Jacobi identity; this means

$$\text{ad}[x,y] = (\text{ad}\, x)(\text{ad}\, y) - (\text{ad}\, y)(\text{ad}\, x) .$$

The kernel of a homomorphism is an ideal. If $\mathfrak{s}$ is an ideal in $\mathfrak{g}$, then the vector space $\mathfrak{g}/\mathfrak{s}$ becomes a Lie algebra under the definition

$$[x+\mathfrak{s}\, ,\, y+\mathfrak{s}] = [x,y] + \mathfrak{s} ,$$

and the quotient map of $\mathfrak{g}$ to $\mathfrak{g}/\mathfrak{s}$ is a homomorphism. An isomorphism is a one-one homomorphism onto.

Let $\mathfrak{g}$ be a Lie algebra over $\mathbb{R}$ or $\mathbb{C}$, and let V be a vector space over $\mathbb{C}$. A representation of $\mathfrak{g}$ on V is a

Lie algebra homomorphism of $\mathfrak{g}$ into $\mathrm{End}_{\mathbb{C}} V$, assumed linear over $\mathbb{R}$ if $\mathfrak{g}$ is defined over $\mathbb{R}$ and linear over $\mathbb{C}$ if $\mathfrak{g}$ is defined over $\mathbb{C}$. The Lie algebras that arise from Lie groups are normally to be regarded as real Lie algebras. In §4 we shall see how such a Lie algebra $\mathfrak{g}$ can be complexified to give a complex Lie algebra $\mathfrak{g}^{\mathbb{C}}$. A representation of $\mathfrak{g}$ on V will then yield a representation of $\mathfrak{g}^{\mathbb{C}}$ on V, and it is the latter representation that we shall work with.

We have seen examples of representations in §I.8. A representation π of $\mathfrak{g}$ on V is called <u>trivial</u> if $\pi(X) = 0$ for all X in $\mathfrak{g}$. (This definition is motivated by differentiating a group representation that is the identity for every group element.)

If π is a representation of $\mathfrak{g}$ on V, then an <u>invariant subspace</u> U is a subspace of V such that $\pi(X)U \subseteq U$ for all X in $\mathfrak{g}$. In this case we say that we have a <u>subrepresentation</u>. In addition, we get a <u>quotient representation</u> on V/U from the natural operation of each $\pi(X)$ on cosets of V/U. A representation with $V \neq 0$ is <u>irreducible</u> if the only invariant subspaces are 0 and V.

If we have representations π_α of $\mathfrak{g}$ on spaces V_α, then we can define a representation π on the direct sum $\Sigma \oplus V_\alpha$ by $\pi(X)(\Sigma v_\alpha) = \Sigma\, \pi(X)v_\alpha$, and we call π the <u>direct sum</u> of the π_α's.

If we have two representations π_1 and π_2 of $\mathfrak{g}$ on V_1 and V_2, then we say that π_1 and π_2 are <u>equivalent</u> if there is an invertible linear map $E : V_1 \to V_2$ such that $E\pi_1(X) = \pi_2(X)E$ for all X in $\mathfrak{g}$; E is an <u>equivalence</u>.

2. Tensor product of two representations

Let E and F be vector spaces over the field k. A tensor product V of E and F is a vector space over k, together with a bilinear map $\iota : E \times F \to V$, with the following universal mapping property: Whenever b is a bilinear mapping of $E \times F$ into a vector space U over k, there exists a unique linear mapping B of V into U such that the diagram

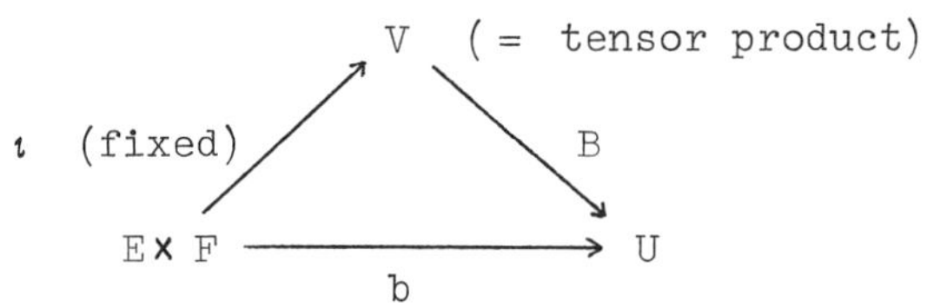

(2.2)

commutes. We call B the linear extension of b to the tensor product.

It is well known that a tensor product of E and F exists and is unique up to canonical isomorphism. The uniqueness is clear. Let us recall the proof of existence: Let V_1 be a vector space over k with a basis consisting of all pairs (e,f) with e in E and f in F. Let V_0 be the subspace generated by all elements

$$(e_1+e_2, f) - (e_1, f) - (e_2, f)$$

$$(ce, f) - c(e, f)$$

$$(e, f_1+f_2) - (e, f_1) - (e, f_2)$$

$$(e, cf) - c(e, f) .$$

Put $V = V_1/V_0$, and take $\iota(e,f)$ to be the coset $(e,f) + V_0$. To see that B always exists as in (2.2), define B_1 on V_1 by $B_1(\Sigma\, c_i(e_i,f_i)) = \Sigma\, c_i b(e_i,f_i)$. The bilinearity of b shows that B_1 maps V_0 to 0. Thus B_1 defines a linear map B from $V = V_1/V_0$ to U, and (2.2) commutes. To see that B is unique, we note that the cosets $(e,f) + V_0$ span V; since commutativity of (2.2) forces

$$B((e,f) + V_0) = B(\iota(e,f)) = b(e,f)\,,$$

B is unique.)

A tensor product of E and F is denoted $E \otimes_k F$, and the associated bilinear map ι is written $(e,f) \to e \otimes f$. The elements $e \otimes f$ generate $E \otimes_k F$, as a consequence of the proof of existence of $E \otimes_k F$.

There is a canonical isomorphism

$$E \otimes_k F \cong F \otimes_k E \tag{2.3}$$

given by taking the linear extension (the map B in (2.2)) of $(e,f) \to f \otimes e$ as the map from left to right. The linear extension of $(f,e) \to e \otimes f$ gives a two-sided inverse.

Another canonical isomorphism of interest is

$$E \otimes_k k \cong E\,. \tag{2.4}$$

Here the map from left to right is the linear extension of $(e,c) \to ce$, while the map from right to left is $e \to e \otimes 1$. In view of (2.3), we have $k \otimes_k E \cong E$ also.

Tensor product distributes over direct sums, even infinite direct sums:

$$E \otimes_k (\Sigma \oplus F_\alpha) \cong \Sigma\ (E \otimes_k F_\alpha)\ . \tag{2.5}$$

The map from left to right is the linear extension of $(e, \Sigma f_\alpha) \to \Sigma\ (e \otimes f_\alpha)$. To define the inverse, we have only to define it on a single $E \otimes_k F_\alpha$, where it is the linear extension of $(e, f_\alpha) \to e \otimes (i_\alpha(f_\alpha))$; here $i_\alpha : F_\alpha \to \Sigma \oplus F_\alpha$ is the injection corresponding to α. It follows from (2.4) and (2.5) that if $\{x_i\}$ is a basis of E and $\{y_j\}$ is a basis of F, then $\{x_i \otimes y_j\}$ is a basis of $E \otimes_k F$. Consequently

$$\dim(E \otimes_k F) = (\dim E)(\dim F)\ . \tag{2.6}$$

Let $\mathrm{Hom}_k(E,F)$ be the vector space of k-linear maps from E into F. One special case is the case $E = k$, and we have

$$\mathrm{Hom}_k(k,F) \cong F\ . \tag{2.7}$$

The map from left to right sends φ into $\varphi(1)$, while the map from right to left sends f into φ with $\varphi(c) = cf$. Another case of special interest occurs when $F = k$. Then $\mathrm{Hom}_k(E,k) = E^*$ is just the vector space <u>dual</u> of E.

We can use $\otimes_k$ and Hom_k to construct new linear mappings. Let L_1 be in $\mathrm{Hom}_k(E_1,F_1)$ and L_2 be in $\mathrm{Hom}_k(E_2,F_2)$. Then we can define

$$L_1 \otimes L_2 \quad \text{in} \quad \mathrm{Hom}_k(E_1 \otimes_k E_2, F_1 \otimes_k F_2) \tag{2.8}$$

as follows: The map $(e_1, e_2) \to L_1(e_1) \otimes L_2(e_2)$ is bilinear from $E_1 \times E_2$ into $F_1 \otimes_k F_2$, and we let $L_1 \otimes L_2$ be its linear extension to $E_1 \otimes_k E_2$. The uniqueness in the universal mapping property allows us to conclude

$$(L_1 \otimes L_2)(M_1 \otimes M_2) = L_1M_1 \otimes L_2M_2 \qquad (2.9)$$

when the domains and ranges match in the obvious way. Still with L_1 in $\mathrm{Hom}_k(E_1, F_1)$ and L_2 in $\mathrm{Hom}_k(E_2, F_2)$, we can define

$$\mathrm{Hom}(L_1, L_2) \quad \text{in} \quad \mathrm{Hom}_k(\mathrm{Hom}_k(F_1, E_2), \mathrm{Hom}_k(E_1, F_2)) \qquad (2.10a)$$

by

$$\mathrm{Hom}(L_1, L_2)(\varphi) = L_2 \circ \varphi \circ L_1 \qquad (2.10b)$$

for φ in $\mathrm{Hom}_k(F_1, E_2)$. Then it is clear that

$$\mathrm{Hom}(L_1, L_2)\mathrm{Hom}(M_1, M_2) = \mathrm{Hom}(M_1L_1, L_2M_2) \qquad (2.11)$$

when the domains and ranges match in the obvious way. The fact that L_1 and M_1 get switched in order while L_2 and M_2 do not is summarized by saying that Hom is <u>contravariant</u> in the first variable and <u>covariant</u> in the second variable.

<u>Proposition 2.1</u>. Let A, B, and C be vector spaces over k. Then there is a unique k-linear isomorphism Φ of left side to right side in

$$\mathrm{Hom}_k(A \otimes_k B, C) \cong \mathrm{Hom}_k(A, \mathrm{Hom}_k(B, C)) \qquad (2.12)$$

such that

$$\Phi(\varphi)(a)(b) = \varphi(a \otimes b) \qquad (2.13)$$

for all $\varphi \in \mathrm{Hom}_k(A \otimes_k B, C)$, $a \in A$, and $b \in B$.

Proof. The map Φ is well defined and k-linear, and it carries the left side to the right side. For ψ in the right side, let

$$\Psi(\psi)(a,b) = \psi(a)(b) .$$

Then $\Psi(\psi)$ is a bilinear map from $A \times B$ into C, and we let $\widetilde{\Psi}(\psi)$ be the linear extension from $A \otimes_k B$ into C. Then $\widetilde{\Psi}$ and Φ are two-sided inverses, and the proposition follows.

When we use this proposition, we shall often need to know that the isomorphism Φ is compatible with maps $A \to A'$, $B \to B'$, and $C \to C'$. This property is called <u>naturality</u> in the variables A, B, and C, and we make it precise for this context in the next proposition. Later "naturality" will be defined in a more general context.

<u>Proposition 2.2</u>. Let A, B, C, A', B', C' be vector spaces over k and let $L_{A'} : A' \to A$, $L_{B'} : B' \to B$, and $L_C : C \to C'$ be linear maps. Then the isomorphism Φ of Proposition 2.1 is natural in the sense that the diagram

$$\begin{array}{ccc} \mathrm{Hom}_k(A \otimes_k B, C) & \xrightarrow{\ \Phi\ } & \mathrm{Hom}_k(A, \mathrm{Hom}_k(B,C)) \\ {\scriptstyle \mathrm{Hom}(L_{A'} \otimes L_{B'}, L_C)}\Big\downarrow & & \Big\downarrow{\scriptstyle \mathrm{Hom}(L_{A'}, \mathrm{Hom}(L_{B'}, L_C))} \\ \mathrm{Hom}_k(A' \otimes_k B', C') & \xrightarrow{\ \Phi\ } & \mathrm{Hom}_k(A', \mathrm{Hom}_k(B', C')) \end{array}$$

commutes.

Remark. Observe that the first two linear maps go "backwards," while the third goes "forwards." This is a reflection of the fact that both sides of (2.12) are contravariant in the first two variables and covariant in the third variable.

Proof. For φ in $\mathrm{Hom}_k(A \otimes_k B, C)$, we have

$$
\begin{aligned}
(\mathrm{Hom}(L_{A'}, \mathrm{Hom}(L_{B'}, L_C)) \circ \Phi)(\varphi)(a')(b') \\
&= (\mathrm{Hom}(L_{B'}, L_C) \circ \Phi(\varphi) \circ L_{A'})(a')(b') \\
&= \mathrm{Hom}(L_{B'}, L_C) \circ (\Phi(\varphi) \circ L_{A'})(a')(b') \\
&= (L_C \circ \Phi(\varphi)(L_{A'}(a')) \circ L_{B'})(b') \\
&= L_C(\Phi(\varphi)(L_{A'}(a'))(L_{B'}(b'))) \\
&= L_C(\varphi(L_{A'}(a') \otimes L_{B'}(b'))) \\
&= (L_C \circ \varphi \circ (L_{A'} \otimes L_{B'}))(a' \otimes b') \\
&= (\mathrm{Hom}(L_{A'} \otimes L_{B'}, L_C)(\varphi))(a' \otimes b') \\
&= \Phi(\mathrm{Hom}(L_{A'} \otimes L_{B'}, L_C)\varphi)(a')(b') .
\end{aligned}
$$

This proves the proposition.

By way of motivation for the tensor product of two Lie algebra representations, let G be a topological group, and let σ and ρ be representations of G on finite-dimensional complex vector spaces V and W, respectively. Define a representation π of G on $V \otimes_{\mathbb{C}} W$ by means of (2.8) as

$$\pi(g) = \sigma(g) \otimes \rho(g) .$$

Relation (2.9) assures us that π is multiplicative, and we can use bases to see that π is continuous; thus π is indeed a representation. Now let us think of G as a closed linear group (as in Chapter I) and differentiate to find $d\pi$. We will be replacing g by $c(t)$ and differentiating. The product rule will then give us two terms. Accordingly we make the following definition.

Let $\mathfrak{g}$ be a Lie algebra over $\mathbb{R}$ or $\mathbb{C}$, let V and W be complex vector spaces, and let σ and ρ be representations of $\mathfrak{g}$ on V and W, respectively. Define the <u>tensor product</u> $\pi = \sigma \otimes \rho$ on $V \otimes_{\mathbb{C}} W$ by

$$(\sigma \otimes \rho)(X) = \sigma(X) \otimes I + I \otimes \rho(X) . \tag{2.14}$$

A little calculation with (2.9) shows that $\sigma \otimes \rho$ is a representation.

Again by way of motivation, let us consider group actions on $\mathrm{Hom}_{\mathbb{C}}(V,W)$. Thus let G be a topological group, and let σ and ρ be representations of G on finite-dimensional complex vector spaces V and W, respectively. Define a representation π of G on $\mathrm{Hom}_{\mathbb{C}}(V,W)$ by

$$\pi(g) = \mathrm{Hom}(\sigma(g)^{-1}, \rho(g)) .$$

Referring to (2.10a), we see that $\pi(g)$ is a linear map from $\mathrm{Hom}_{\mathbb{C}}(V,W)$ to itself. Relation (2.11) shows that $\pi(g)$ is multiplicative and explains the presence of g^{-1}. We can use a basis to see the continuity, and thus π is a representation. Concretely the formula is this: If φ is in $\mathrm{Hom}_{\mathbb{C}}(V,W)$, then

$$\pi(g)(\varphi)(v) = \rho(g)(\varphi(\sigma(g)^{-1}v)) .$$

Passing to closed linear groups and differentiating, we are led to make the following definition.

Let $\mathfrak{g}$ be a Lie algebra over $\mathbb{R}$ or $\mathbb{C}$, let V and W be complex vector spaces, and let σ and ρ be representations of $\mathfrak{g}$ on V and W, respectively. Define $\pi = \mathrm{Hom}(\sigma,\rho)$ on $\mathrm{Hom}_{\mathbb{C}}(V,W)$ by

$$(\mathrm{Hom}(\sigma,\rho)(X))(\varphi)(v) = \rho(X)(\varphi(v)) - \varphi(\sigma(X)v) . \tag{2.15}$$

A little calculation with (2.11) shows that $\mathrm{Hom}(\sigma,\rho)$ is a representation.

A special case of interest is the case that W is $\mathbb{C}$ and ρ is the trivial representation on $\mathbb{C}$. In this case we write $\sigma^* = \mathrm{Hom}(\sigma,1)$ for the resulting representation of $\mathfrak{g}$ on V^*. (The use of "1" here is a reminder that $\mathrm{Hom}(\sigma,1)$ is not necessarily zero; "0" would be a more literal symbol for the trivial representation.) Formula (2.15) specializes to

$$\sigma^*(X)(\varphi)(v) = -\varphi(\sigma(X)v) . \tag{2.16}$$

The representation σ^* is called the <u>contragredient</u> of σ.

<u>Proposition 2.3</u>. Let A, B, and C be vector spaces over $\mathbb{C}$, and suppose that representations of a Lie algebra $\mathfrak{g}$ (over $\mathbb{R}$ or $\mathbb{C}$) are given on each. Then in the vector space identity

$$\mathrm{Hom}_{\mathbb{C}}(A \otimes_{\mathbb{C}} B, C) \cong \mathrm{Hom}_{\mathbb{C}}(A, \mathrm{Hom}_{\mathbb{C}}(B,C))$$

of (2.12), the isomorphism is an equivalence of representations.

Proof. This is just a question of checking what happens in (2.13) when we plug in the representation formulas of (2.14) and (2.15). Dropping the names of all representations from the notation, we have

$$\begin{aligned}
\Phi(X\varphi)(a)(b) &= (X\varphi)(a\otimes b) = X(\varphi(a\otimes b)) - \varphi(X(a\otimes b)) \\
&= X(\varphi(a\otimes b)) - \varphi(a\otimes Xb) - \varphi(Xa\otimes b) \\
&= X(\Phi(\varphi)(a)(b)) - \Phi(\varphi)(a)(Xb) - \Phi(\varphi)(Xa)(b) \\
&= X(\Phi(\varphi)(a))(b) - \Phi(\varphi)(Xa)(b) \\
&= X(\Phi(\varphi))(a)(b) ,
\end{aligned}$$

and the result follows.

Remark. It is instructive to derive this proposition also from Proposition 2.2. We omit the details.

3. Representations on the tensor algebra

Let A, B, and C be vector spaces over k. A <u>triple tensor product</u> $V = A\otimes_k B\otimes_k C$ is a vector space over k with a trilinear map $\iota : A\times B\times C \to V$ having the following universal property: Whenever t is a trilinear mapping of $A\times B\times C$ into a vector space U over k, there exists a unique linear mapping T of V into U such that the diagram

$$\begin{array}{ccccc}
 & & V \quad (= \text{triple tensor product}) & & \\
 & \iota\ (\text{fixed}) \nearrow & & \searrow T & \\
A\times B\times C & & \xrightarrow{\quad t \quad} & & U
\end{array} \qquad (2.17)$$

commutes. It is clear that there is at most one triple tensor product up to canonical isomorphism. We shall use triple tensor products to establish an associativity formula for ordinary tensor products.

Proposition 2.4. (a) $(A \otimes_k B) \otimes_k C$ and $A \otimes_k (B \otimes_k C)$ are triple tensor products.

(b) There exists a unique isomorphism Φ from left to right in

$$(A \otimes_k B) \otimes_k C \cong A \otimes_k (B \otimes_k C) \tag{2.18}$$

such that $\Phi((a \otimes b) \otimes c) = a \otimes (b \otimes c)$ for all $a \in A$, $b \in B$, and $c \in C$.

Proof. (a) Consider $(A \otimes_k B) \otimes_k C$. Let $t : A \times B \times C \to U$ be trilinear. For c in C, define $t_c : A \times B \to U$ by $t_c(a,b) = t(a,b,c)$. Then t_c is bilinear and hence extends to a linear $T_C : A \otimes_k B \to U$. Since t is trilinear, $t_{c_1+c_2} = t_{c_1} + t_{c_2}$ and $t_{xc} = xt_c$ for scalar x; thus uniqueness of the linear extension forces $T_{c_1+c_2} = T_{c_1} + T_{c_2}$ and $T_{xc} = xT_c$. Consequently

$$t' : (A \otimes_k B) \times C \to U$$

given by $t'(d,c) = T_c(d)$ is bilinear and hence extends to a linear $T : (A \otimes_k B) \otimes_k C \to U$. This T proves existence of the linear extension of the given t. Uniqueness is trivial, since the elements $(a \otimes b) \otimes c$ generate $(A \otimes_k B) \otimes_k C$. So $(A \otimes_k B) \otimes_k C$ is a triple tensor product. In similar fashion, $A \otimes_k (B \otimes_k C)$ is a triple tensor product.

(b) In (2.17), take $V = (A \otimes_k B) \otimes_k C$, $U = A \otimes_k (B \otimes_k C)$, and $t(a,b,c) = a \otimes (b \otimes c)$. We have just seen in (a) that V is a triple tensor product with $\iota(a,b,c) = (a \otimes b) \otimes c$. Thus there exists a linear $T : V \to U$ with $T\iota(a,b,c) = t(a,b,c)$. This equation means $T((a \otimes b) \otimes c) = a \otimes (b \otimes c)$. Interchanging the roles of $(A \otimes_k B) \otimes_k C$ and $A \otimes_k (B \otimes_k C)$, we obtain a two-sided inverse for T. Thus T will serve as Φ in (b), and existence is proved. Uniqueness is trivial, since the elements $(a \otimes b) \otimes c$ generate $(A \otimes_k B) \otimes_k C$.

Just as with Proposition 2.1, Proposition 2.4 carries along with it a certain naturality that we often need to invoke in applications.

<u>Proposition 2.5</u>. Let A, B, C, A', B', C' be vector spaces over k, and let $L_A : A \to A'$, $L_B : B \to B'$, and $L_C : C \to C'$ be linear maps. Then the isomorphism Φ of Proposition 2.4 is natural in the sense that the diagram

$$\begin{array}{ccc} (A \otimes_k B) \otimes_k C & \xrightarrow{\ \Phi\ } & A \otimes_k (B \otimes_k C) \\ {\scriptstyle (L_A \otimes L_B) \otimes L_C} \downarrow & & \downarrow {\scriptstyle L_A \otimes (L_B \otimes L_C)} \\ (A' \otimes_k B') \otimes C' & \xrightarrow{\ \Phi\ } & A' \otimes_k (B' \otimes_k C') \end{array}$$

commutes.

Proof.

$$\begin{aligned} ((L_A \otimes (L_B \otimes L_C)) \circ \Phi)((a \otimes b) \otimes c) &= (L_A \otimes (L_B \otimes L_C))(a \otimes (b \otimes c)) \\ &= L_A a \otimes (L_B \otimes L_C)(b \otimes c) \\ &= L_A a \otimes (L_B b \otimes L_C c) \\ &= \Phi((L_A a \otimes L_B b) \otimes L_C c) \end{aligned}$$

$$= \Phi((L_A \otimes L_B)(a \otimes b) \otimes L_C c)$$

$$= (\Phi \circ ((L_A \otimes L_B) \otimes L_C))((a \otimes b) \otimes c) .$$

The proposition follows.

Proposition 2.3 showed that the vector space identity (2.12) is compatible with the passage to Lie algebra representations. We now prove a corresponding compatibility for the identity (2.18).

Proposition 2.6. Let A, B, and C be vector spaces over $\mathbb{C}$, and suppose that representations of a Lie algebra $\mathfrak{g}$ (over $\mathbb{R}$ or $\mathbb{C}$) are given on each. Then in the vector space identity

$$(A \otimes_{\mathbb{C}} B) \otimes_{\mathbb{C}} C \cong A \otimes_{\mathbb{C}} (B \otimes_{\mathbb{C}} C)$$

of (2.18), the isomorphism is an equivalence of representations.

Proof. Let Φ be the ismorphism implementing (2.18), and let us drop the names of all representations from the notation. Then we have

$$\begin{aligned} \Phi(X((a \otimes b) \otimes c)) &= \Phi(X(a \otimes b) \otimes c + (a \otimes b) \otimes Xc) \\ &= \Phi((Xa \otimes b) \otimes c + (a \otimes Xb) \otimes c + (a \otimes b) \otimes Xc) \\ &= \Phi((Xa \otimes b) \otimes c) + \Phi((a \otimes Xb) \otimes c) + \Phi((a \otimes b) \otimes Xc) \\ &= Xa \otimes (b \otimes c) + a \otimes (Xb \otimes c) + a \otimes (b \otimes Xc) \qquad (2.19) \\ &= Xa \otimes (b \otimes c) + a \otimes X(b \otimes c) \\ &= X(a \otimes (b \otimes c)) = X(\Phi((a \otimes b) \otimes c)) . \end{aligned}$$

Since the elements $(a \otimes b) \otimes c$ generate $(A \otimes_k B) \otimes_k C$, the equivalence follows.

Remark. The Lie algebra representation on a triple tensor product may be motivated by group representations. In a triple tensor product, a member of the group is to act in all three factors simultaneously. The differentiated action is just (2.19) above, as a consequence of the product rule.

There is no difficulty in generalizing matters to n-fold tensor products by induction. An <u>n-fold tensor product</u> is to be universal for n-multilinear maps. It is clearly unique up to canonical isomorphism. One such tensor product is the (n-1)-fold tensor product of the first $n-1$ spaces, tensored with the nth space. Proposition 2.4b allows us to regroup parentheses (inductively) in any fashion we choose, and iterated application of Proposition 2.5 shows that we get a well defined notion of the tensor product of n linear maps. Iterating Proposition 2.6 shows that representations of a Lie algebra $\mathfrak{g}$ on each factor, when the vector spaces are complex, lead to a well defined representation on the n-fold tensor product.

Fix a vector space E over k, and let $T^{(n)}(E)$ be an n-fold tensor product of E with itself. In the case $n = 0$, we let $T^{(0)}(E)$ be the field k. Define, initially as a vector space, $T(E)$ to be the direct sum

$$T(E) = \sum_{n=0}^{\infty} \oplus T^{(n)}(E) . \tag{2.20}$$

The elements that lie in one or another $T^{(n)}(E)$ are called *homogeneous*. We define a bilinear multiplication on homogeneous elements

$$T^{(m)}(E) \times T^{(n)}(E) \longrightarrow T^{(m+n)}(E)$$

to be the restriction of the above canonical isomorphism

$$T^{(m)}(E) \otimes_k T^{(n)}(E) \longrightarrow T^{(m+n)}(E) .$$

This multiplication is associative because the restriction of the isomorphism

$$T^{(\ell)}(E) \otimes_k (T^{(m)}(E) \otimes_k T^{(n)}(E)) \to (T^{(\ell)}(E) \otimes_k T^{(m)}(E)) \otimes_k T^{(n)}(E)$$

to $T^{(\ell)}(E) \times (T^{(m)}(E) \times T^{(n)}(E))$ factors through the map

$$T^{(\ell)}(E) \times (T^{(m)}(E) \times T^{(n)}(E)) \longrightarrow (T^{(\ell)}(E) \times T^{(m)}(E)) \times T^{(n)}(E)$$

given by $(r,(s,t)) \to ((r,s),t)$. Thus $T(E)$ becomes an associative algebra with identity and is known as the *tensor algebra* of E. $T(E)$ has the following universal mapping property.

Proposition 2.7. The one-one linear mapping $\iota : E \to T(E)$ with image $T^{(1)}(E)$ has the following property: Whenever $\ell : E \to A$ is a linear map of E into an associative algebra with identity, then there exists a unique associative algebra homomorphism $L : T(E) \to A$ such that $L(1) = 1$ and the diagram

$$\begin{array}{ccccc} & & T(E) & & \\ & \overset{\iota}{\nearrow} & & \searrow^{L} & \\ E & & \xrightarrow{\quad \ell \quad} & & A \end{array} \tag{2.21}$$

commutes.

Proof. Uniqueness is clear, since E generates $T(E)$ as an algebra. For existence we define $L^{(n)}$ on $T^{(n)}(E)$ to be the linear extension of the n-multilinear map

$$(v_1, v_2, \ldots, v_n) \to \ell(v_1)\ell(v_2)\cdots\ell(v_n) ,$$

and we let $L = \Sigma \oplus L^{(n)}$ in obvious notation. Let $u_1 \otimes \ldots \otimes u_m$ be in $T^{(m)}(E)$ and $v_1 \otimes \ldots \otimes v_n$ be in $T^{(n)}(E)$. Then we have

$$L^{(m)}(u_1 \otimes \ldots \otimes u_m) = \ell(u_1)\cdots\ell(u_m)$$

$$L^{(n)}(v_1 \otimes \ldots \otimes v_n) = \ell(v_1)\cdots\ell(v_n)$$

$$L^{(m+n)}(u_1 \otimes \ldots \otimes u_m \otimes v_1 \otimes \ldots \otimes v_n) = \ell(u_1)\cdots\ell(u_m)\ell(v_1)\cdots\ell(v_n) .$$

Hence

$$L^{(m)}(u_1 \otimes \ldots \otimes u_m) L^{(n)}(v_1 \otimes \ldots \otimes v_n)$$
$$= L^{(m+n)}(u_1 \otimes \ldots \otimes u_m \otimes v_1 \otimes \ldots \otimes v_n) .$$

Taking linear combinations, we see that L is a homomorphism.

Now let us consider what happens with representations. To orient ourselves, let us think about matters first in terms of groups. If π is a representation of G on V, then we can put $A = T(V)$ and $\ell = \pi(g)$ in (2.21) to obtain $\tilde{\pi}(g)$ defined on all of $T(V)$. The uniqueness of L in (2.21) allows us to see that $\tilde{\pi}$ is multiplicative in g. In fact, on $T^{(n)}(V)$, $\tilde{\pi}(g)$ is nothing more than the n-fold tensor product $\pi(g) \otimes \ldots \otimes \pi(g)$. Differentiating, we see that we should expect an action by $\mathfrak{g}$ involving n terms on $T^{(n)}(V)$.

In fact, if we start from a Lie algebra representation π of $\mathfrak{g}$ on a complex vector space V, our remarks with n-fold tensor products give us canonically a representation $\pi^{(n)}$ of $\mathfrak{g}$ on $T^{(n)}(V)$. (The representation on $T^{(0)}(V) \cong \mathbb{C}$ is to be trivial.) It is clear that the formula is

$$\pi^{(n)}(X)(v_1 \otimes \cdots \otimes v_n) = \sum_{j=1}^{n} (v_1 \otimes \cdots \otimes v_{j-1} \otimes \pi(X)v_j \otimes v_{j+1} \otimes \cdots \otimes v_n), \tag{2.22a}$$

and this agrees with what we would expect from differentiating a group action. Using direct sums, we obtain a representation $\tilde{\pi}$ of $\mathfrak{g}$ on the full tensor algebra $T(V)$.

Proposition 2.8. If π is a representation of the Lie algebra $\mathfrak{g}$ on V and if $\tilde{\pi}$ is the corresponding representation on $T(V)$, then

$$\tilde{\pi}(X)(v \otimes w) = \tilde{\pi}(X)v \otimes w + v \otimes \tilde{\pi}(X)w \tag{2.22b}$$

for all X in $\mathfrak{g}$ and for all v and w in $T(V)$.

Proof. We may assume v is in $T^{(m)}(V)$ and w is in $T^{(n)}(V)$. Then (2.22a) shows that

$$\pi^{(m)}(X) = \sum_{j=1}^{m} (1_1 \otimes \cdots \otimes 1_{j-1} \otimes \pi(X) \otimes 1_{j+1} \otimes \cdots \otimes 1_m)$$

and similarly for $\pi^{(n)}(X)$ and $\pi^{(m+n)}(X)$. Then it is clear that

$$\pi^{(m)}(X) \otimes 1 + 1 \otimes \pi^{(n)}(X) = \pi^{(m+n)}(X),$$

and this is (2.22b).

4. Representations on exterior and symmetric algebras

Let E be a vector space over k, and let $T(E)$ be its tensor algebra. We begin by defining the exterior algebra $\Lambda(E)$. The elements of $\Lambda(E)$ are to be all the alternating tensors (= skew-symmetric if k has characteristic $\neq 2$), and so we want to force $v \otimes v$ to be 0. Thus we define the exterior algebra by

$$(E) = T(E) \Big/ \left(\begin{matrix} \text{two-sided ideal } I \text{ generated by} \\ \text{all } v \otimes v \text{ with } v \text{ in } T^{(1)}(E) \end{matrix} \right) . \qquad (2.23)$$

This is an associative algebra with identity.

It is clear that the ideal I satisfies

$$I = \sum_{n=0}^{\infty} \oplus \, (I \cap T^{(n)}(E)) .$$

An ideal with this property is said to be graded. Since I is graded,

$$\Lambda(E) = \sum_{n=0}^{\infty} \oplus \, T^{(n)}(E)/(I \cap T^{(n)}(E)) .$$

We write $\Lambda^n(E)$ for the nth summand on the right side, so that

$$\Lambda(E) = \sum_{n=0}^{\infty} \oplus \Lambda^n(E) . \qquad (2.24)$$

Since $I \cap T^{(1)}(E) = 0$, the map of E into first-order elements is one-one onto. The product operation in $\Lambda(E)$ is denoted $\wedge$ rather than $\otimes$. The image in $\Lambda^n(E)$ of $v_1 \otimes \ldots \otimes v_n$ in $T^{(n)}(E)$ is thus denoted $v_1 \wedge \ldots \wedge v_n$. If a is in $\Lambda^m(E)$ and b is in $\Lambda^n(E)$, then $a \wedge b$ is in

$\Lambda^{m+n}(E)$. Moreover $\Lambda^n(E)$ is generated by elements $v_1 \wedge \ldots \wedge v_n$ with all v_j in $\Lambda^1(E) \cong E$, since $T^{(n)}(E)$ is generated by corresponding elements $v_1 \otimes \ldots \otimes v_n$. The defining relations for $\Lambda(E)$ make $v_i \wedge v_j = -v_j \wedge v_i$ for v_i and v_j in $\Lambda^1(E)$, and it follows that

$$a \wedge b = (-1)^{mn}\, b \wedge a \quad \text{if } a \in \Lambda^m(E) \text{ and } b \in \Lambda^n(E). \tag{2.25}$$

Proposition 2.9.

(a) $\Lambda^n(E)$ has the following universal property: Let ι be the map $\iota(v_1, \ldots, v_n) = v_1 \wedge \ldots \wedge v_n$ of $E \times \ldots \times E$ into $\Lambda^n(E)$. If ℓ is any alternating n-multilinear map of $E \times \ldots \times E$ into a vector space U, then there exists a unique linear map $L : \Lambda^n(E) \to U$ such that the diagram

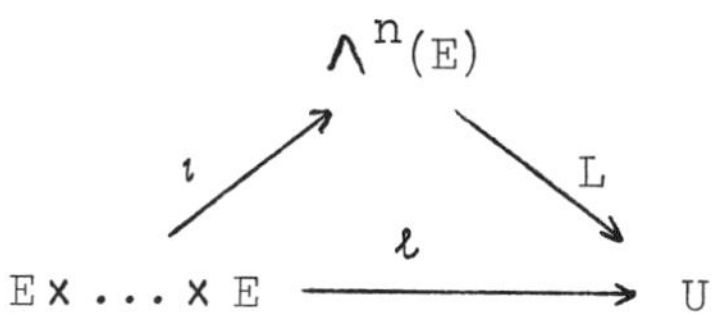

commutes.

(b) $\Lambda(E)$ has the following universal property: Let ι be the map that imbeds E as $\Lambda^1(E) \subseteq \Lambda(E)$. If ℓ is any linear map of E into an associative algebra A with identity such that $\ell(v)^2 = 0$ for all v in E, then there exists a unique algebra homomorphism $L : \Lambda(E) \to A$ with $L(1) = 1$ such that the diagram

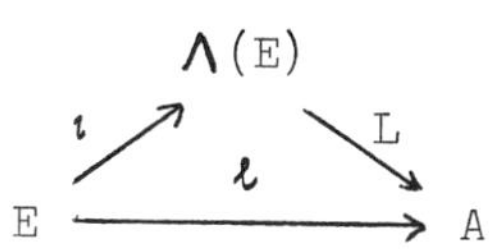

commutes.

Proof. In both cases uniqueness is trivial. For existence we use the universal mapping properties of $T^{(n)}(E)$ and $T(E)$ to produce $\widetilde{L}$ on $T^{(n)}(E)$ or $T(E)$. If we can show that $\widetilde{L}$ annihilates the appropriate subspace so as to descend to $\Lambda^n(E)$ or $\Lambda(E)$, then the resulting map can be taken as L, and we are done. For (a), we have $\widetilde{L}: T^{(n)}(E) \to U$ and we are to show $\widetilde{L}(T^{(n)}(E) \cap I) = 0$, where I is generated by all $v \otimes v$ with v in E. A member of $T^{(n)}(E) \cap I$ is thus of the form $\sum a_i \otimes (v_i \otimes v_i) \otimes b_i$ with each term in $T^{(n)}(E)$. Each term here is a sum of pure tensors

$$x_1 \otimes \ldots \otimes x_r \otimes v_i \otimes v_i \otimes y_1 \otimes \ldots \otimes y_s \tag{2.26}$$

with $r+2+s=n$. Since ℓ by assumption vanishes on

$$x_1 \times \ldots \times x_r \times v_i \times v_i \times y_1 \times \ldots \times y_s ,$$

$\widetilde{L}$ vanishes on (2.26), and it follows that $\widetilde{L}(T^{(n)}(E) \cap I) = 0$.

For (b), we are to show that $\widetilde{L}: T(E) \to A$ vanishes on I. Since $\ker \widetilde{L}$ is an ideal, it is enough to check that $\widetilde{L}$ vanishes on the generators of I. But $\widetilde{L}(v \otimes v) = \ell(v)^2 = 0$, and thus $\widetilde{L}(I) = 0$.

Corollary 2.10. If E and F are vector spaces over k, then $\mathrm{Hom}_k(\Lambda^n(E), F)$ is canonically isomorphic (via restriction to pure tensors) to the vector space of F-valued alternating n-multilinear functions on $E \times \ldots \times E$.

Proof. Restriction is linear and one-one. It is onto by Proposition 2.9a.

Proposition 2.11. Let E be a finite-dimensional vector space of dimension N. Then

(a) $\Lambda^n(E) = 0$ for $n > N$.

(b) $\dim \Lambda^n(E) = \binom{N}{n}$ for $0 \leq n \leq N$. If $\{u_i\}$ is an ordered basis of E, then $\{u_{i_1} \wedge \cdots \wedge u_{i_n} \mid i_1 < \cdots < i_n\}$ is a basis of $\Lambda^n(E)$.

(c) $\Lambda^n(E^*)$ is canonically isomorphic to $\Lambda^n(E)^*$ by

$$(f_1 \wedge \cdots \wedge f_n)(w_1, \ldots, w_n) = \det\{f_i(w_j)\} .$$

Proof. Let $u_1, \ldots, u_N$ be an ordered basis of E. The vectors $u_{i_1} \otimes \cdots \otimes u_{i_n}$ are then a basis of $T^{(n)}(E)$, and hence the corresponding $u_{i_1} \wedge \cdots \wedge u_{i_n}$ span $\Lambda^n(E)$. Using the skew symmetry in $\Lambda^n(E)$, we thus see that the elements $u_{i_1} \wedge \cdots \wedge u_{i_n}$ with $i_1 < \cdots < i_n$ span $\Lambda^n(E)$. This proves (a).

Let $u_1^*, \ldots, u_N^*$ be the dual basis of E^*, fix $r_1 < \cdots < r_n$, and define

$$\ell(w_1, \ldots, w_n) = \det\{u_{r_i}^*(w_j)\} \quad \text{for } w_1, \ldots, w_n \text{ in } E .$$

Then ℓ is alternating n-multilinear from $E \times \cdots \times E$ into k and extends by Proposition 2.9a to $L : \Lambda^n(E) \to k$. If $k_1 < \cdots < k_n$, then

$$L(u_{k_1} \wedge \cdots \wedge u_{k_n}) = \ell(u_{k_1}, \ldots, u_{k_n}) = \det\{u_{r_i}^*(u_{k_j})\} ,$$

and the right side is 0 unless $r_1 = k_1, \ldots, r_n = k_n$, in which case it is 1. This proves that the $u_{r_1} \wedge \cdots \wedge u_{r_n}$ are linearly independent in $\Lambda^n(E)$. Then (b) follows, since the number of increasing n tuples from $\{1, \ldots, N\}$ is $\binom{N}{n}$.

For (c) let $f_1, \dots, f_n$ be in E^*, and define

$$\ell_{f_1,\dots,f_n}(w_1,\dots,w_n) = \det\{f_i(w_j)\} .$$

Then $\ell_{f_1,\dots,f_n}$ is alternating n-multilinear from $E \times \dots \times E$ into k and extends by Proposition 2.9a to a linear $L_{f_1,\dots,f_n} : \Lambda^n(E) \to k$. Thus $\ell(f_1,\dots,f_n) = L_{f_1,\dots,f_n}$ defines an alternating n-multilinear map of $E^* \times \dots \times E^*$ into $\Lambda^n(E)^*$. Its linear extension L maps $\Lambda^n(E^*)$ into $\Lambda^n(E)^*$. The argument in the previous paragraph shows that L maps the basis $u^*_{r_1} \wedge \dots \wedge u^*_{r_n}$ to the dual basis of $u_{r_1} \wedge \dots \wedge u_{r_n}$. Hence L is an isomorphism.

Now let us consider representations. Let $\mathfrak{g}$ be a Lie algebra over $\mathbb{R}$ or $\mathbb{C}$, and let π be a representation of $\mathfrak{g}$ on a complex vector space V. Let $\pi^{(n)}$ and $\tilde{\pi}$ be the representations of $\mathfrak{g}$ on $T^{(n)}(V)$ and $T(V)$. Let us see that $T^{(n)}(V) \cap I$ is an invariant subspace for $\pi^{(n)}$. It is enough to apply $\tilde{\pi}(X)$ to an element (2.26), which we write in condensed form as $a \otimes v \otimes v \otimes b$ with a in $T^{(r)}(V)$, v in $T^{(1)}(V)$, and b in $T^{(s)}(V)$. By Proposition 2.8,

$$\begin{aligned}
\tilde{\pi}(X)(a \otimes v \otimes v \otimes b) &= \tilde{\pi}(X)a \otimes v \otimes v \otimes b \\
&\quad + a \otimes \tilde{\pi}(X)v \otimes v \otimes b + a \otimes v \otimes \tilde{\pi}(X)v \otimes b \\
&\quad + a \otimes v \otimes v \otimes \tilde{\pi}(X)b \\
&= \tilde{\pi}(X)a \otimes v \otimes v \otimes b + a \otimes v \otimes v \otimes \tilde{\pi}(X)b \\
&\quad + a \otimes (\tilde{\pi}(X)v + v) \otimes (\tilde{\pi}(X)v + v) \otimes b \\
&\quad - a \otimes \tilde{\pi}(X)v \otimes \tilde{\pi}(X)v \otimes b - a \otimes v \otimes v \otimes b ,
\end{aligned}$$

which is in the span of (2.26). Hence $T^{(n)}(V) \cap I$ is an invariant subspace for $\pi^{(n)}$, and I is an invariant subspace for $\tilde{\pi}$. Thus we get well defined quotient representations, also denoted $\pi^{(n)}$ and $\tilde{\pi}$ for now, on $\Lambda^n(V)$ and $\Lambda(V)$. By Proposition 2.8, we have

$$\tilde{\pi}(X)(a \wedge b) = \tilde{\pi}(X)a \wedge b + a \wedge \tilde{\pi}(X)b\,. \tag{2.27}$$

The above discussion makes precise, on the level of $\mathfrak{g}$, the third example of representations given in §I.8. On the group level, if we assume V is finite-dimensional, we obtain a representation $\pi^{(n)}$ of G on $T^{(n)}(V)$ from π on V, and this passes to the quotient as a representation $\pi^{(n)}$ on $\Lambda^n(V)$ with $\pi^{(n)}(g)(v_1 \wedge \cdots \wedge v_n) = \pi(g)v_1 \wedge \cdots \wedge \pi(g)v_n$. Its differentiated version is what we constructed above on the level of $\mathfrak{g}$.

We turn to the symmetric algebra $S(E)$, E being a vector space over k. The construction, results, and proofs are similar to those for $\Lambda(E)$. The elements of $S(E)$ are to be all symmetric tensors, and so we want to force $u \otimes v = v \otimes u$. Thus we define the <u>symmetric algebra</u> by

$$S(E) = T(E) \Big/ \left(\begin{array}{l} \text{two-sided ideal } J \text{ generated by all} \\ u \otimes v - v \otimes u \ \text{ with } u \in T^{(1)}(E),\ v \in T^{(1)}(E) \end{array} \right). \tag{2.28}$$

This is an associative algebra with identity.

It is clear that J is graded: $J = \sum_{n=0}^{\infty} \oplus\ (J \cap T^{(n)}(E))$. Thus we can write

$$S(E) = \sum_{n=0}^{\infty} \oplus\, T^{(n)}(E)/(J \cap T^{(n)}(E))\,.$$

We write $S^n(E)$ for the nth summand on the right side, so that

$$S(E) = \sum_{n=0}^{\infty} \oplus S^n(E) . \tag{2.29}$$

Since $J \cap T^{(1)}(E) = 0$, the map of E into first-order elements is one-one onto. The product operation in $S(E)$ is written without a product sign. The image in $S^n(E)$ of $v_1 \otimes \ldots \otimes v_n$ in $T^{(n)}(E)$ is thus denoted $v_1 \cdots v_n$. If a is in $S^m(E)$ and b is in $S^n(E)$, then ab is in $S^{m+n}(E)$. Moreover $S^n(E)$ is generated by elements $v_1 \cdots v_n$ with all v_j in $S^1(E) \cong E$, since $T^{(n)}(E)$ is generated by corresponding elements $v_1 \otimes \ldots \otimes v_n$. The defining relations for $S(E)$ make $v_i v_j = v_j v_i$ for v_i and v_j in $S^1(E)$, and it follows that $S(E)$ is commutative.

<u>Proposition 2.12</u>.

(a) $S^n(E)$ has the following universal property: Let ι be the map $\iota(v_1, \ldots, v_n) = v_1 \cdots v_n$ of $E \times \ldots \times E$ into $S^n(E)$. If ℓ is any symmetric n-multilinear map of $E \times \ldots \times E$ into a vector space U, then there exists a unique linear map $L : S^n(E) \to U$ such that the diagram

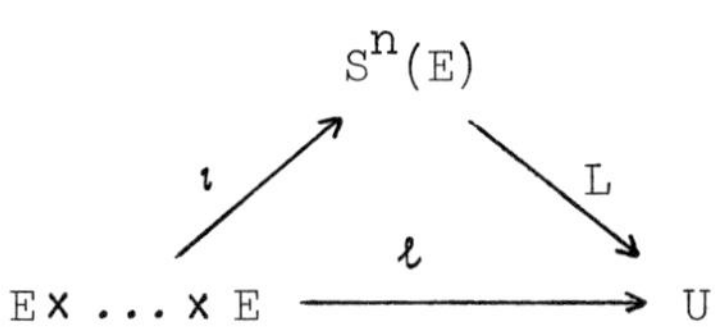

commutes.

(b) $S(E)$ has the following universal property: Let ι be the map that imbeds E as $S^1(E) \subseteq S(E)$. If ℓ is any linear map of E into a commutative associative algebra A with identity, then there exists a unique algebra homomorphism $L : S(E) \to A$ with $L(1) = 1$ such that the diagram

$$\begin{array}{ccc} & S(E) & \\ {\scriptstyle \iota}\nearrow & & \searrow{\scriptstyle L} \\ E & \xrightarrow{\ \ell\ } & A \end{array}$$

commutes.

Proof. This is completely analogous to Proposition 2.9.

<u>Corollary 2.13</u>. If E and F are vector spaces over k, then $\mathrm{Hom}_k(S^n(E), F)$ is canonically isomorphic (via restriction to pure tensors) to the vector space of F-valued symmetric n-multilinear functions on $E \times \ldots \times E$.

<u>Proposition 2.14</u>. Let E be a finite-dimensional vector space of dimension N. Then

(a) $\dim S^n(E) = \binom{n+N-1}{N-1}$ for $0 \leq n < \infty$. If $\{u_i\}$ is an ordered basis of E, then $\{u_1^{j_1} \cdots u_N^{j_N}$ with $j_1 + \ldots + j_N = n\}$ is a basis of $S^n(E)$.

(b) $S^n(E^*)$ is canonically isomorphic to $S^n(E)^*$ by

$$(f_1 \cdots f_n)(w_1, \ldots, w_n) = \sum_{\sigma \in G_n} \prod_{j=1}^{n} f_j(w_{\sigma(j)}) ,$$

where G_n is the symmetric group on n letters.

Proof. (a) Since $S(E)$ is commutative and since monomials span $T^{(n)}(E)$, the indicated set spans $S^n(E)$. To see its cardinality, we recognize that picking out $N - 1$

objects from $n+N-1$ to label as dividers is a way of assigning exponents to the u_j's ; thus the cardinality of the indicated set is $\binom{n+N-1}{N-1}$.

Let us see independence. The map $\Sigma\, c_i u_i \to \Sigma\, c_i X_i$ of E into the polynomial algebra $k[X_1,\ldots,X_N]$ is linear into a commutative algebra with identity. Its extension via Proposition 2.12b maps our spanning set for $S^n(E)$ to distinct monomials in $k[X_1,\ldots,X_N]$, which are necessarily linearly independent. Hence our spanning set is a basis.

(b) This is proved in the same way as Proposition 2.11c, and Proposition 2.12a is the tool.

Remarks. The proof of (a) suggests that $S(E)$ might be just polynomials in disguise, but this suggestion is misleading. The isomorphism with $k[X_1,\ldots,X_N]$ in (a) depended on choosing a basis. The canonical isomorphism is between $S(E^*)$ and polynomials on E ; details are left to the reader.

Finally let us consider representations. Let $\mathfrak{g}$ be a Lie algebra over $\mathbb{R}$ or $\mathbb{C}$, and let π be a representation of $\mathfrak{g}$ on a complex vector space V . Form $\pi^{(n)}$ and $\tilde{\pi}$ on $T^{(n)}(V)$ and $T(V)$, respectively. Just as with the exterior algebra, we find that $T^{(n)}(V) \cap J$ is an invariant subspace for $\pi^{(n)}$. Hence J is an invariant subspace for $\tilde{\pi}$. Thus we get well defined quotient representations, also denoted $\pi^{(n)}$ and $\tilde{\pi}$ for now, on $S^n(E)$ and $S(E)$. By Proposition 2.8 we have

$$\tilde{\pi}(X)(ab) = (\tilde{\pi}(X)a)b + a(\tilde{\pi}(X)b) . \qquad (2.30)$$

5. Extension of scalars - complexification

For the time being, extension of scalars will be a device for taking a real vector space and making from it a complex vector space of the same dimension. If the original vector space is n-dimensional real column vectors, the new space may be regarded as n-dimensional complex column vectors. In this case a real matrix gives a linear mapping between real vector spaces originally and then can be regarded as a linear mapping between the extensions to complex vector spaces. We construct the new space without using a basis.

Let K be a field with $k \subseteq K$. Then K is a vector space over k, and we can form $E \otimes_k K$ whenever E is a vector space over k. The linear map $e \to e \otimes 1$ imbeds E as a subspace of $E \otimes_k K$ (over the field k). We now make $E \otimes_k K$ into a vector space over K by defining

$$\begin{pmatrix} \text{mult by } c \\ \text{in } E \otimes_k K \end{pmatrix} = 1 \otimes \begin{pmatrix} \text{mult by } c \\ \text{in } K \end{pmatrix} \quad \text{for } c \in K .$$

We easily check that c, c_1, and c_2 in K and u, v in $E \otimes_k K$ imply

(i) $c_1(c_2 v) = (c_1 c_2) v$

(ii) $c(u + v) = cu + cv$

(iii) $(c_1 + c_2) v = c_1 v + c_2 v$

(iv) $1 v = v$

(v) $c(e \otimes 1) = ce \otimes 1$ if c is in k and e is in E.

These properties show that $E \otimes_k K$ is a vector space over K

and that its scalar multiplication, when restricted to scalars in k and the original imbedded E, agrees with scalar multiplication in E.

If we have a linear mapping $L: E \to F$ over k, then $L \otimes 1: E \otimes_k K \to F \otimes_k K$ is linear over K, as follows from the identity

$$(L \otimes 1)\begin{pmatrix} I \otimes \begin{matrix} \text{mult by} \\ c \text{ in } K \end{matrix} \end{pmatrix} = L \otimes \begin{pmatrix} \text{mult by} \\ c \text{ in } K \end{pmatrix} = \begin{pmatrix} I \otimes \begin{matrix} \text{mult by} \\ c \text{ in } K \end{matrix} \end{pmatrix}(L \otimes 1) .$$

The mapping $L \otimes 1$ is the one explained in the first paragraph of this section.

Another kind of extension of linear mappings is available when $L: E \to F$ is linear over k and F is already a vector space over K. In this case, let $\ell(e,c) = cL(e)$ for e in E and c in K. Then $\ell: E \times K \to F$ is bilinear over k and extends to a k-linear map $\tilde{L}: E \otimes_k K \to F$. The map $\tilde{L}$ reduces to L on the imbedded copy $E \otimes 1$ of E in $E \otimes_k K$. Moreover $\tilde{L}$ is linear over K; to see this linearity, we just observe that

$$\begin{aligned} \tilde{L}(c_1(e \otimes c_2)) &= \tilde{L}(e \otimes c_1 c_2) = \ell(e, c_1 c_2) = c_1 c_2 L(e) \\ &= c_1 \ell(e, c_2) = c_1 \tilde{L}(e \otimes c_2) . \end{aligned}$$

The pure tensors $e \otimes c_2$ generate $E \otimes_k K$, and the K-linearity follows.

Let us now specialize matters to the case $k = \mathbb{R}$ and $K = \mathbb{C}$. We denote $E \otimes_{\mathbb{R}} \mathbb{C}$ by $E^{\mathbb{C}}$, calling it the <u>complexification</u> of E.

Suppose that $\mathfrak{g}$ is a Lie algebra over $\mathbb{R}$. We shall make $\mathfrak{g}^{\mathbb{C}}$ into a complex Lie algebra. To obtain a complex

Lie algebra, we need to define the bracket for two members of $\mathfrak{g}^{\mathbb{C}}$. (This is very easy with bases, but it is not immediately clear that the resulting definition is independent of basis.) Thus let us form the 4-multilinear map $\mathfrak{g} \times \mathbb{C} \times \mathfrak{g} \times \mathbb{C} \to \mathfrak{g}^{\mathbb{C}}$ given by $(x, r, y, s) \to rs[x, y]$. Extend it to a linear map $\mathfrak{g} \otimes_{\mathbb{R}} \mathbb{C} \otimes_{\mathbb{R}} \mathfrak{g} \otimes_{\mathbb{R}} \mathbb{C}$ and restrict the result to an $\mathbb{R}$-bilinear map $(\mathfrak{g} \otimes_{\mathbb{R}} \mathbb{C}) \times (\mathfrak{g} \otimes_{\mathbb{R}} \mathbb{C}) \to \mathfrak{g}^{\mathbb{C}}$. Using suitable uniqueness conditions for extensions of multilinear maps, we readily check that this map is $\mathbb{C}$-bilinear from $\mathfrak{g}^{\mathbb{C}} \times \mathfrak{g}^{\mathbb{C}}$ to $\mathfrak{g}^{\mathbb{C}}$, that it is alternating, and that it satisfies the Jacobi identity. Thus $\mathfrak{g}^{\mathbb{C}}$ is a complex Lie algebra.

The $\mathbb{R}$-linear map $\operatorname{ad} X : \mathfrak{g} \to \mathfrak{g}$ gives us the $\mathbb{C}$-linear map $\operatorname{ad} X \otimes 1 : \mathfrak{g}^{\mathbb{C}} \to \mathfrak{g}^{\mathbb{C}}$, which is nothing more than $\operatorname{ad}(X \otimes 1)$. We shall write this simply as $\operatorname{ad} X$. The linear map $\operatorname{ad} : \mathfrak{g} \to \operatorname{End}_{\mathbb{C}} \mathfrak{g}^{\mathbb{C}}$ carrying X to $\operatorname{ad} X$ is a representation of $\mathfrak{g}$ on $\mathfrak{g}^{\mathbb{C}}$.

Suppose π is a representation of $\mathfrak{g}$ on a complex vector space V. Then $\pi : \mathfrak{g} \to \operatorname{End}_{\mathbb{C}} V$ is real linear into a complex vector space, and we have constructed a corresponding complex linear extension from $\mathfrak{g}^{\mathbb{C}}$ into $\operatorname{End}_{\mathbb{C}} V$. We denote this extension by π, too. It is easy to check that the extended π is a representation of $\mathfrak{g}^{\mathbb{C}}$. Thus a representation of $\mathfrak{g}$ automatically extends to a representation of $\mathfrak{g}^{\mathbb{C}}$ on the same complex vector space.

6. Universal enveloping algebra

In this section we suppose that $\mathfrak{g}$ is a complex Lie algebra of finite dimension N. (When we are studying the representation theory of a real Lie algebra $\mathfrak{g}_0$, $\mathfrak{g}$ will be $\mathfrak{g}_0^{\mathbb{C}}$.) If we have a representation π of $\mathfrak{g}$ on a complex vector space V, then the investigation of invariant subspaces in principle involves writing down all iterates $\pi(X_1)\pi(X_2)\cdots\pi(X_n)$ for members of $\mathfrak{g}$, applying them to members of V, and comparing the results. In the course of comparing results, one might be able to simplify an expression by using the identity $\pi(X)\pi(Y) = \pi(Y)\pi(X) + \pi[X,Y]$. This identity really has nothing to do with π, and our objective in this section will be to introduce an object where we can make such calculations without reference to π; to use an identity with the representation π, one simply applies π to both sides.

For a first approximation to what we want, we can use the tensor algebra $T(\mathfrak{g})$. The representation π is a linear map of $\mathfrak{g}$ into the associative algebra $\mathrm{End}_{\mathbb{C}} V$ and extends to an algebra homomorphism $\tilde{\pi}: T(\mathfrak{g}) \to \mathrm{End}_{\mathbb{C}} V$ with $\tilde{\pi}(1) = 1$. Then $\pi(X_n)\cdots\pi(X_1)$ can be replaced by $\tilde{\pi}(X_n \otimes \ldots \otimes X_1)$. The difficulty is that the tensor algebra does not take advantage of the Lie algebra structure of $\mathfrak{g}$ and does not force the identity $\pi(X)\pi(Y) = \pi(Y)\pi(X) + \pi[X,Y]$ for all X, Y in $\mathfrak{g}$ and all π. Thus instead of the tensor algebra, we use the following quotient of $T(\mathfrak{g})$:

$$U(\mathfrak{g}) = T(\mathfrak{g}) \Big/ \left(\begin{array}{l} \text{two-sided ideal generated by all} \\ X \otimes Y - Y \otimes X - [X,Y] \text{ with } X, Y \text{ in } T^{(1)}(\mathfrak{g}) \end{array} \right). \tag{2.31}$$

$U(\mathfrak{g})$ is an associative algebra with identity and is known as the <u>universal enveloping algebra</u> of $\mathfrak{g}$. Products in $U(\mathfrak{g})$ are written without multiplication signs.

The canonical map $\mathfrak{g} \to U(\mathfrak{g})$ given by imbedding $\mathfrak{g}$ into $T^{(1)}(\mathfrak{g})$ and then passing to $U(\mathfrak{g})$ is denoted ι. Because of (2.31), ι satisfies

$$\iota[X,Y] = \iota(X)\iota(Y) - \iota(Y)\iota(X) \quad \text{for } X, Y \text{ in } \mathfrak{g}. \tag{2.32}$$

$U(\mathfrak{g})$ is harder to work with than $\Lambda(\mathfrak{g})$ or $S(\mathfrak{g})$ because the ideal in (2.31) is not graded, i.e., is not generated by homogeneous elements. Thus, for example, it is not evident that the canonical map $\iota : \mathfrak{g} \to U(\mathfrak{g})$ is one-one. However, when $\mathfrak{g}$ is <u>abelian</u> (i.e., when all brackets are 0), $U(\mathfrak{g})$ reduces to $S(\mathfrak{g})$, and we have a clear notion of what to expect in this case. Even when $\mathfrak{g}$ is nonabelian, $U(\mathfrak{g})$ and $S(\mathfrak{g})$ are still related, and we shall make the relationship precise at the end of this section.

Let $U^n(\mathfrak{g})$ be the image of $\sum_{k=0}^n T^{(k)}(\mathfrak{g})$ under the passage to the quotient in (2.31). Then $U^n(\mathfrak{g})$ is a finite-dimensional subspace of $U(\mathfrak{g})$, and $U(\mathfrak{g}) = \bigcup_{n=0}^\infty U^n(\mathfrak{g})$.

<u>Proposition 2.15</u>. $U(\mathfrak{g})$ and the canonical map $\iota : \mathfrak{g} \to U(\mathfrak{g})$ have the following universal mapping property: Whenever A is a complex associative algebra with identity and $\pi : \mathfrak{g} \to A$ is a linear mapping such that

$$\pi(X)\pi(Y) - \pi(Y)\pi(X) = \pi[X,Y] \quad \text{for all } X \text{ and } Y \text{ in } \mathfrak{g}, \tag{2.33}$$

then there exists a unique algebra homomorphism $\tilde{\pi} : U(\mathfrak{g}) \to A$ such that $\tilde{\pi}(1) = 1$ and

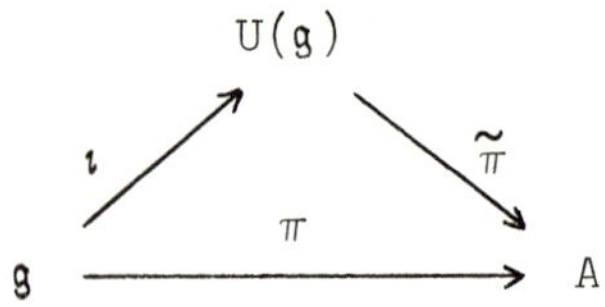

commutes.

Proof. Uniqueness follows from the fact that 1 and $\iota(\mathfrak{g})$ generate $U(\mathfrak{g})$. For existence, let $\pi_1 : T(\mathfrak{g}) \to A$ be the extension given by the universal mapping property of $T(\mathfrak{g})$. To obtain $\tilde{\pi}$, we are to show that π_1 annihilates the ideal in (2.31). It is enough to consider π_1 on a typical generator of the ideal, where we have

$$\begin{aligned}
&\pi_1(\iota X \otimes \iota Y - \iota Y \otimes iX - \iota[X,Y]) \\
&\qquad = \pi_1(\iota X)\pi_1(\iota Y) - \pi_1(\iota Y)\pi_1(\iota X) - \pi_1(\iota[X,Y]) \\
&\qquad = \pi(X)\pi(Y) - \pi(Y)\pi(X) - \pi[X,Y] = 0 .
\end{aligned}$$

<u>Corollary 2.16</u>. Representations of $\mathfrak{g}$ stand in one-one correspondence with unital left $U(\mathfrak{g})$ modules (under the correspondence $\pi \to \tilde{\pi}$ of Proposition 2.15).

Remark. <u>Unital</u> means that 1 operates as 1.

Proof. If π is a representation of $\mathfrak{g}$ on V, we apply Proposition 2.15 to $\pi : \mathfrak{g} \to \mathrm{End}_{\mathbb{C}} V$, and then V becomes a unital left $U(\mathfrak{g})$ module under $uv = \tilde{\pi}(u)v$ for u in $U(\mathfrak{g})$ and v in V. Conversely if V is a unital left $U(\mathfrak{g})$ module, then we can define $\pi(X)v = (\iota X)v$, and (2.32) implies that π is a representation of $\mathfrak{g}$. The two constructions are inverse to each other since $\tilde{\pi}\circ\iota = \pi$ in

Proposition 2.15.

Theorem 2.17 (Poincaré-Birkhoff-Witt). If $X_1, \ldots, X_N$ is an ordered basis of $\mathfrak{g}$, then the monomials

$$(\iota X_1)^{j_1} \cdots (\iota X_N)^{j_N}$$

with all $j_k \geq 0$, form a basis of $U(\mathfrak{g})$. In particular, the canonical map $\iota : \mathfrak{g} \to U(\mathfrak{g})$ is one-one.

Lemma 2.18. Let $Z_1, \ldots, Z_p$ be in $\mathfrak{g}$, and let σ be a permutation of $\{1,\ldots,p\}$. Then

$$(\iota Z_1) \cdots (\iota Z_p) - (\iota Z_{\sigma(1)}) \cdots (\iota Z_{\sigma(p)})$$

is in $U^{p-1}(\mathfrak{g})$.

Proof. Without loss of generality, let σ be the transposition of j with $j+1$. Then the lemma follows from

$$(\iota Z_j)(\iota Z_{j+1}) - (\iota Z_{j+1})(\iota Z_j) = \iota [Z_j, Z_{j+1}]$$

by multiplying through on the left by $(\iota Z_1) \cdots (\iota Z_{j-1})$ and on the right by $(\iota Z_{j+2}) \cdots (\iota Z_p)$.

For the remainder of the proof of the theorem, we shall use the following notation: Let $Y_i = \iota X_i$. For any tuple $I = (i_1, \ldots, i_p)$ of integers from 1 to N, increasing or not, we write $Y_I = Y_{i_1} \cdots Y_{i_p}$. Also $i \leq I$ means $i \leq \min\{i_1, \ldots, i_p\}$.

Lemma 2.19. The Y_I, for all increasing tuples of length $\leq p$, generate $U^p(\mathfrak{g})$.

Proof. If we use all tuples of length $\leq p$, we certainly have a set of generators, since the obvious pre-images generate $\Sigma_{k\leq p} T^{(k)}(\mathfrak{g})$. Lemma 2.18 then implies that the increasing tuples suffice.

Proof of Theorem 2.17. Let P be the polynomial algebra $\mathbb{C}[z_1,\ldots,z_N]$, and let P_p be the subspace of members of total degree $\leq p$. For a tuple $I = (i_1,\ldots,i_p)$, let $z_I = z_{i_1}\cdots z_{i_p}$ as a member of P_p. We shall construct a representation π of $\mathfrak{g}$ on P such that

$$\pi(X_i)z_I = z_i z_I \quad \text{if } i \leq I. \tag{2.34}$$

Let us see that the theorem follows from the existence of such a representation. In fact, let us use Corollary 2.16 to regard P as a unital left $U(\mathfrak{g})$ module. Then (2.34) and the identity $\pi(X)v = (\iota X)v$ imply

$$Y_i z_I = z_i z_I \quad \text{if } i \leq I.$$

If $i_1 \leq \cdots \leq i_p$, we obtain as a consequence

$$\begin{aligned}(Y_{i_1}\cdots Y_{i_p})1 &= (Y_{i_1}\cdots Y_{i_{p-1}})Y_{i_p}1 = (Y_{i_1}\cdots Y_{i_{p-1}})z_{i_p} \\ &= (Y_{i_1}\cdots Y_{i_{p-2}})Y_{i_{p-1}}z_{i_p} = (Y_{i_1}\cdots Y_{i_{p-2}})(z_{i_{p-1}}z_{i_p}) \\ &= \cdots = z_{i_1}\cdots z_{i_p}.\end{aligned}$$

Thus $\{Y_I 1 \mid I \text{ increasing}\}$ is a linearly independent set in P, and $\{Y_I \mid I$ increasing must be independent in $U(\mathfrak{g})$. The independence in Theorem 2.17 follows, and the spanning is given in Lemma 2.18.

Thus we have to construct π satisfying (2.34). We shall define linear maps $\pi(X) : P_p \to P_{p+1}$ for X in $\mathfrak{g}$, by induction on p so as to be compatible and to satisfy

(A_p) $\pi(X_i)z_I = z_i z_I$ for $i \leq I$ and z_I in P_p

(B_p) $\pi(X_i)z_I - z_i z_I$ is in P_p for all I with z_I in P_p

(C_p) $\pi(X_i)(\pi(X_j)z_J) = \pi(X_j)(\pi(X_i)z_J) + \pi[X_i, X_j]z_J$ for all J with z_J in P_{p-1}.

With $\pi(X)$ defined on P as the union of its definitions on the P_p's, π will be a representation by (C_p) and will satisfy (2.34) by (A_p). Hence we will be done.

For $p = 0$, we define $\pi(X_i)1 = z_i$. Then (A_0) holds, (B_0) is valid, and (C_0) is vacuous.

Thus now assume that $\pi(X)$ has been defined on P_{p-1} for all X in $\mathfrak{g}$ in such a way that (A_{p-1}), (B_{p-1}), and (C_{p-1}) hold. We are to define $\pi(X_i)z_I$ for an increasing sequence I of p elements in such a way that (A_p), (B_p), and (C_p) hold. If $i \leq I$, we make the definition according to (A_p). Otherwise we can write in obvious notation $I = (j, J)$ with $j < i$, $j \leq J$, $|J| = p - 1$. We are forced to define

$$\begin{aligned}
\pi(X_i)z_I &= \pi(X_i)(z_j z_J) \\
&= \pi(X_i)\pi(X_j)z_J && \text{since } \pi(X_j)z_J \text{ is already defined} \\
&= \pi(X_j)\pi(X_i)z_J + \pi[X_i, X_j]z_J && \text{by } (C_p) \\
&= \pi(X_j)(z_i z_J + w) + \pi[X_i, X_j]z_J && \text{with } w \text{ in } P_{p-1} \text{ by } (B_{p-1}) \\
&= z_j z_i z_J + \pi(X_j)w + \pi[X_i, X_j]z_J && \text{by } (A_p) \\
&= z_i z_I + \pi(X_j)w + \pi[X_i, X_j]z_J .
\end{aligned}$$

So we make this definition, and then (B_p) holds. Thus $\pi(X_i)z_I$ has now been defined in all cases on P_p, and we have to show that (C_p) holds.

Our construction above was made so that (C_p) holds if $j < i$, $j \leq J$, $|J| = p-1$. Since $[X_j, X_i] = -[X_i, X_j]$, it holds also if $i < j$, $i \leq J$, $|J| = p-1$. Also (C_p) is trivial if $i = j$. Thus it holds whenever $i \leq J$ or $j \leq J$. So we may assume that $J = (k, K)$, where $k \leq K$, $k < i$, $k < j$, $|K| = p-2$. We know that

$$\begin{aligned} \pi(X_j)z_J &= \pi(X_j)z_k z_K = \pi(X_j)\pi(X_k)z_K \\ &= \pi(X_k)\pi(X_j)z_K + \pi[X_j, X_k]z_K \quad \text{by } (C_{p-1}) \qquad (2.35) \\ &= \pi(X_k)(z_j z_K + w) + \pi[X_j, X_k]z_K \end{aligned}$$

for a certain element w in P_{p-2} given by (B_{p-2}), which is contained in (B_{p-1}). We apply $\pi(X_i)$ to both sides of this equation, calling the three terms on the right T_1, T_2, and T_3. We can use (C_p) with $\pi(X_i)$ of T_1, because $k \leq (j, K)$, and we can use (C_{p-1}) with $\pi(X_1)$ of T_2 and T_3. Reassembling T_1 and T_2 as in (2.35), we can use (C_p) with the sum $\pi(X_i)\pi(X_k)\pi(X_j)z_K$, and we can use (C_{p-1}) with $\pi(X_i)$ of T_3. Thus we have

$$\begin{aligned} \pi(X_i)\pi(X_j)z_J &= \pi(X_i)\pi(X_k)\pi(X_j)z_K + \pi(X_i)\pi[X_j, X_k]z_K \quad \text{from } (2.35) \\ &= \pi(X_k)\pi(X_i)\pi(X_j)z_K + \pi[X_i, X_k]\pi(X_j)z_K \\ &\quad + \pi[X_j, X_k]\pi(X_i)z_K + \pi[X_i, [X_j, X_k]]z_K \\ &\qquad \text{by known cases of } (C_p) \\ &= T_{1'} + T_{2'} + T_{3'} + T_{4'} . \end{aligned}$$

Interchanging i and j and subtracting, we see that the terms of type T_2, and T_3, cancel and that we get

$$\begin{aligned}
&\pi(X_i)\pi(X_j)z_J - \pi(X_j)\pi(X_i)z_J \\
&\quad = \pi(X_k)\{\pi(X_i)\pi(X_j)z_K - \pi(X_j)\pi(X_i)z_K\} \\
&\qquad + \{\pi[X_i,[X_j,X_k]] - \pi[X_j,[X_i,X_k]]\}z_K \\
&\quad = \pi(X_k)\pi[X_i,X_j]z_K + \pi[[X_i,X_j],X_k]z_K \qquad \text{by } (C_{p-1}) \text{ and Jacobi} \\
&\quad = \pi[X_i,X_j]\pi(X_k)z_K \qquad \text{by } (C_{p-1}) \\
&\quad = \pi[X_i,X_j]z_k z_K = \pi[X_i,X_j]z_J .
\end{aligned}$$

We have obtained (C_p) in the remaining case, and the proof of Theorem 2.17 is complete.

Now that ι is known to be one-one, there is no danger in dropping it from the notation. We shall freely use Corollary 2.16, identifying representations of $\mathfrak{g}$ with (unital) left $U(\mathfrak{g})$ modules. Moreover we shall feel free either to drop the name of a representation from the notation (to emphasize the module structure) or to include it even when the argument is in $U(\mathfrak{g})$ (to emphasize the representation structure).

Historically the Poincaré-Birkhoff-Witt Theorem was first proved algebraically. But there is an illuminating analytic argument, provided we are willing to ignore distinctions between $\mathbb{R}$ and $\mathbb{C}$ and to settle for a wide variety of special cases. Namely on a Lie group G, the members of the Lie algebra $\mathfrak{g}_0$ give us left-invariant vector fields: In the notation of Chapter I, if $X \leftrightarrow c(t)$, then we can define a map $\widetilde{X}: C^\infty(G) \to C^\infty(G)$ by $\widetilde{X}f(g) = \frac{d}{dt} f(gc(t))\big|_{t=0}$, and $\widetilde{X}$ is the left-invariant vector field corresponding to X. The operator $\widetilde{X}$ looks like a first-order linear partial differential operator in local coordinates, and the techniques of §I.1 and §I.3 show that $[X,Y]^\sim = \widetilde{X}\widetilde{Y} - \widetilde{Y}\widetilde{X}$. Therefore Proposition 2.15 gives us an algebra map of $U(\mathfrak{g})$ into the partial differential operators generated (under composition) by the $\widetilde{X}$'s. The use of Taylor expansions about a point shows that this map is one-one and is onto all linear left-invariant partial differential operators. Thus $U(\mathfrak{g})$ has a basis as asserted and is interpreted as the algebra of all linear left-invariant partial differential operators on G with complex coefficients.

Let us return to algebra. The Poincaré-Birkhoff-Witt Theorem appears in a number of guises. Here is one such.

Corollary 2.20. If $\mathfrak{h}$ is a Lie subalgebra of $\mathfrak{g}$, then the subalgebra of $U(\mathfrak{g})$ generated by 1 and $\mathfrak{h}$ is isomorphic to $U(\mathfrak{h})$.

Proof. If $\rho: \mathfrak{h} \to \mathfrak{g}$ denotes inclusion, then ρ yields an inclusion (also denoted ρ) of $\mathfrak{h}$ into $U(\mathfrak{g})$ such that $\rho(X)\rho(Y) - \rho(Y)\rho(X) = \rho[X,Y]$ for X and Y in $\mathfrak{h}$. By the universal mapping property of $U(\mathfrak{h})$, we obtain a corresponding algebra map $\widetilde{\rho}: U(\mathfrak{h}) \to U(\mathfrak{g})$ with $\widetilde{\rho}(1) = 1$. The image of $\widetilde{\rho}$ is certainly the subalgebra of $U(\mathfrak{g})$ generated by 1 and $\rho(\mathfrak{h})$. Theorem 2.17 says that monomials in an ordered basis of $\mathfrak{h}$ span $U(\mathfrak{h})$, and a second application of the theorem says that these monomials in $U(\mathfrak{g})$

are linearly independent. Thus $\tilde{\rho}$ is one-one and the corollary follows.

If $\mathfrak{g}$ happens to be the vector space direct sum of two Lie subalgebras $\mathfrak{a}$ and $\mathfrak{b}$, then it follows that we have a vector space isomorphism

$$U(\mathfrak{g}) \cong U(\mathfrak{a}) \otimes_{\mathbb{C}} U(\mathfrak{b}) . \tag{2.36}$$

Namely we obtain a linear map from right to left by the inclusion in Corollary 2.20. To see the map is an isomorphism, we apply Theorem 2.17 to a basis of $\mathfrak{a}$ followed by a basis of $\mathfrak{b}$. The monomials in the separate bases are identified within $U(\mathfrak{g})$ as bases for $U(\mathfrak{a})$ and $U(\mathfrak{b})$, respectively, by Corollary 2.20, while the joined-together bases give both a basis of the tensor product and a basis of $U(\mathfrak{g})$, again by Theorem 2.17. Thus our map send basis to basis and is an isomorphism.

Proposition 2.21. There exists a unique antiautomorphism $u \to u^{tr}$ of $U(\mathfrak{g})$ such that $X^{tr} = -X$ for X in $\mathfrak{g}$.

Proof. It is unique since $\mathfrak{g}$ and 1 generate $U(\mathfrak{g})$. Let us prove existence. For each $n \geq 1$, the map

$$(X_1,\ldots,X_n) \to (-1)^n\, X_n \otimes \ldots \otimes X_1$$

is n-multilinear from $\mathfrak{g} \times \ldots \times \mathfrak{g}$ into $T^{(n)}(\mathfrak{g})$ and hence extends to a linear map of $T^{(n)}(\mathfrak{g})$ into itself such that

$$X_1 \otimes \ldots \otimes X_n \to (-1)^n\, X_n \otimes \ldots \otimes X_1 .$$

Taking the direct sum of these maps as n varies, we obtain a linear map $x \to x^{tr}$ of $T(\mathfrak{g})$ into itself sending 1 into 1. It is clear that this map is an antiautomorphism and extends $X \to -X$ in $T^{(1)}(\mathfrak{g})$. Composing with passage to the quotient, we obtain an antihomomorphism of $T(\mathfrak{g})$ into $U(\mathfrak{g})$. Its kernel is an ideal. To show the map descends to $U(\mathfrak{g})$, it is enough to show that each generator

$$X \otimes Y - Y \otimes X - [X,Y]$$

maps to 0. But this element maps in $T(\mathfrak{g})$ to itself and then maps to 0 in $U(\mathfrak{g})$. Hence the transpose map descends to $U(\mathfrak{g})$. It is clearly of order two and thus is one-one onto.

The map $u \to u^{tr}$ allows us to regard left $U(\mathfrak{g})$ modules V also as right $U(\mathfrak{g})$ modules, and vice versa: To convert a left $U(\mathfrak{g})$ module into a right $U(\mathfrak{g})$ module, we just define $vu = u^{tr}v$ for u in $U(\mathfrak{g})$ and v in V. Conversion in the opposite direction is accomplished by $uv = vu^{tr}$. This conversion will allow us to use tensor products over universal enveloping algebras in investigating representations. We take up such tensor products in §8.

7. Symmetrization

Let $\mathfrak{g}$ be a finite-dimensional complex Lie algebra. We observed in (2.36) that if $\mathfrak{g} = \mathfrak{a} \oplus \mathfrak{b}$ as vector spaces and if $\mathfrak{a}$ and $\mathfrak{b}$ are Lie subalgebras, then their universal enveloping algebras satisfy $U(\mathfrak{g}) \cong U(\mathfrak{a}) \otimes_{\mathbb{C}} U(\mathfrak{b})$. In particular, the $\mathfrak{a}$ type terms and the $\mathfrak{b}$ type terms are canonical. If $\mathfrak{a}$ and $\mathfrak{b}$ are merely vector subspaces of $\mathfrak{g}$, we can still choose a basis of $\mathfrak{g}$ compatible with $\mathfrak{g} = \mathfrak{a} \oplus \mathfrak{b}$, but the linear span of the $\mathfrak{a}$ type monomials will perhaps depend on the choice of basis.

In this section we introduce a device to get around this problem. The starting point is the remark in §6 that $U(\mathfrak{g})$ reduces to the symmetric algebra $S(\mathfrak{g})$ in the case that $\mathfrak{g}$ is abelian. Lemma 2.18 indicates that $U(\mathfrak{g})$ and $S(\mathfrak{g})$ are rather similar even if $\mathfrak{g}$ is nonabelian, and we introduce a mapping of the one to the other that captures these similarities precisely.

For $p \geq 1$, define a symmetric p-multilinear map $\mathfrak{s}_p : \mathfrak{g} \times \ldots \times \mathfrak{g} \to U(\mathfrak{g})$ by

$$\mathfrak{s}_p(X_1, \ldots, X_p) = \frac{1}{p!} \sum_{\sigma \in G_p} X_{\sigma(1)} \cdots X_{\sigma(p)},$$

where G_p is the symmetric group on p letters. (The coefficients $1/p!$ are traditional but are not important.) By Proposition 2.12a we obtain a corresponding linear map, also denoted $\mathfrak{s}_p$, from $S^p(\mathfrak{g})$ into $U(\mathfrak{g})$. The direct sum of these maps for $p \geq 0$ (with $\mathfrak{s}_0(1) = 1$) is a linear map $\mathfrak{s} : S(\mathfrak{g}) \to U(\mathfrak{g})$ such that

$$\mathfrak{s}(X_1 \cdots X_p) = \frac{1}{p!} \sum_{\sigma \in G_p} X_{\sigma(1)} \cdots X_{\sigma(p)}.$$

The map $\mathfrak{s}$ is called <u>symmetrization</u>.

Proposition 2.22. Symmetrization is a vector space isomorphism of $S(\mathfrak{g})$ onto $U(\mathfrak{g})$.

Proof. The image $\mathfrak{s}_p(S^p(\mathfrak{g}))$ is in $U^p(\mathfrak{g})$, and we let $\bar{\mathfrak{s}}_p$ be the composition of $\mathfrak{s}_p$ followed by passage to the quotient $U^p(\mathfrak{g})/U^{p-1}(\mathfrak{g})$. By Lemma 2.18

$$\bar{\mathfrak{s}}_p(X_1\cdots X_p) = X_1\cdots X_p + U^{p-1}(\mathfrak{g}) .$$

By Theorem 2.17, $\bar{\mathfrak{s}}_p$ maps a basis of $S^p(\mathfrak{g})$ to a basis of $U^p(\mathfrak{g})/U^{p-1}(\mathfrak{g})$. Let us see therefore that $\mathfrak{s}$ is one-one. If $\mathfrak{s}$ carries a linear combination of monomials to 0, let p be the largest degree of such a monomial. Composing $\mathfrak{s}$ on the linear combination with the quotient

$$U^p(\mathfrak{g}) \to U^p(\mathfrak{g})/U^{p-1}(\mathfrak{g}) ,$$

we obtain $\bar{\mathfrak{s}}_p$ on the homogeneous part of degree p. Thus all the monomials of degree p in the linear combination have coefficient 0. Hence $\mathfrak{s}$ restricted to $\Sigma_{q=0}^p S^q(\mathfrak{g})$ is one-one into $U^p(\mathfrak{g})$. But these spaces have the same finite dimension, again by Theorem 2.17, and thus this restriction of $\mathfrak{s}$ is also onto $U^p(\mathfrak{g})$. Hence $\mathfrak{s}$ is one-one from $S(\mathfrak{g})$ onto $U(\mathfrak{g})$.

The canonical decomposition of $U(\mathfrak{g})$ from $\mathfrak{g} = \mathfrak{a} \oplus \mathfrak{b}$ when $\mathfrak{a}$ and $\mathfrak{b}$ are merely vector spaces is given in the following proposition.

Proposition 2.23. Suppose $\mathfrak{g} = \mathfrak{a} \oplus \mathfrak{b}$ and suppose $\mathfrak{a}$ and $\mathfrak{b}$ are subspaces of $\mathfrak{g}$. Then the mapping $(a,b) \to \mathfrak{s}(a)\mathfrak{s}(b)$

of $S(\mathfrak{a}) \otimes_{\mathbb{C}} S(\mathfrak{b})$ into $U(\mathfrak{g})$ is a vector space isomorphism onto.

Proof. We argue as in Proposition 2.22 somewhat. We restrict the map to $\sum_{m=0}^{p} S^m(\mathfrak{a}) \otimes_{\mathbb{C}} S^{p-m}(\mathfrak{b})$, with image in $U^p(\mathfrak{g})$, and follow it with passage to the quotient $U^p(\mathfrak{g})/U^{p-1}(\mathfrak{g})$. As in Proposition 2.22, this composition is one-one. We then complete the argument as in Proposition 2.22, obtaining an isomorphism onto.

<u>Corollary 2.24</u>. Suppose $\mathfrak{g} = \mathfrak{k} \oplus \mathfrak{p}$ and $\mathfrak{k}$ is a Lie subalgebra. Then the mapping $(u,p) \to u\mathcal{S}(p)$ of $U(\mathfrak{k}) \otimes_{\mathbb{C}} S(\mathfrak{p})$ into $U(\mathfrak{g})$ is a vector space isomorphism onto.

Proof. The composition

$$(k,p) \to (\mathcal{S}(k),p) \to \mathcal{S}(k)\mathcal{S}(p),$$

sending

$$S(\mathfrak{k}) \otimes S(\mathfrak{p}) \to U(\mathfrak{k}) \otimes_{\mathbb{C}} S(\mathfrak{p}) \to U(\mathfrak{g}),$$

is an isomorphism by Proposition 2.23, and the first map is an isomorphism by Proposition 2.22. Therefore the second map is an isomorphism, and the notation corresponds to the statement of the corollary when we write $u = \mathcal{S}(k)$.

We know that a representation π of $\mathfrak{g}$ on V leads to a representation of $\mathfrak{g}$ on $S(V)$ such that the product rule (2.30) holds. Applying this fact to the adjoint representation of $\mathfrak{g}$ on $\mathfrak{g}$, we obtain a representation (also called ad) of $\mathfrak{g}$ on $S(\mathfrak{g})$ such that

$$(\mathrm{ad}\ X)(ab) = a((\mathrm{ad}\ X)b) + ((\mathrm{ad}\ X)a)b \tag{2.37}$$

for all a and b in $S(\mathfrak{g})$.

The construction that gave us representations on symmetric algebras can also take the representation ad and give us a representation on $U(\mathfrak{g})$ satisfying (2.37). But there is an easier way to proceed: We simply define

$$(\mathrm{ad}\ X)u = Xu - uX \tag{2.38}$$

for X in $\mathfrak{g}$ and u in $U(\mathfrak{g})$. Clearly ad extends the usual definition on $\mathfrak{g}$, and it is easy to check that (2.37) holds. To see that it is a representation, we compute

$$(\mathrm{ad}\ X\ \mathrm{ad}\ Y)u = X((\mathrm{ad}\ Y)u) - ((\mathrm{ad}\ Y)u)X = XYu - XuY - YuX + uYX .$$

Thus

$$\begin{aligned}(\mathrm{ad}\ X\ \mathrm{ad}\ Y - \mathrm{ad}\ Y\ \mathrm{ad}\ X)u &= XYu + uYX - YXu - uXY \\ &= (XY - YX)u - u(XY - YX) \\ &= [X,Y]u - u[X,Y] \\ &= (\mathrm{ad}[X,Y])u\end{aligned}$$

as required.

<u>Proposition 2.25</u>. Symmetrization and the adjoint representations on $S(\mathfrak{g})$ and $U(\mathfrak{g})$ are related by

$$(\mathrm{ad}\ X) \circ \mathcal{S} = \mathcal{S} \circ \mathrm{ad}\ X$$

for X in $\mathfrak{g}$.

Remark. Symmetrization of course does not respect

multiplication in $S(\mathfrak{g})$ and $U(\mathfrak{g})$, and this identity compensates some for that failure.

Proof. If $X_1 \cdots X_p$ is a monomial in $S(\mathfrak{g})$, then

$$(\operatorname{ad} X)\mathfrak{s}(X_1 \cdots X_p) = (\operatorname{ad} X) \frac{1}{p!} \sum_{\sigma} X_{\sigma(1)} \cdots X_{\sigma(p)}$$

$$= \frac{1}{p!} \sum_{\sigma} \sum_{j} X_{\sigma(1)} \cdots X_{\sigma(j-1)} [X, X_{\sigma(j)}] X_{\sigma(j+1)} \cdots X_{\sigma(p)}$$

by (2.37)

$$= \frac{1}{p!} \sum_{j,k} \sum_{\substack{\sigma \\ \text{with} \\ \sigma(j)=k}} X_{\sigma(1)} \cdots X_{\sigma(j-1)} [X, X_k] X_{\sigma(j+1)} \cdots X_{\sigma(p)}. \tag{2.39}$$

Define

$$Y_i^k = \begin{cases} [X, X_k] & \text{if } i = k \\ X_i & \text{if } i \neq k. \end{cases}$$

Then

$$\mathfrak{s}(\operatorname{ad} X)(X_1 \cdots X_p) = \sum_k \mathfrak{s}(X_1 \cdots X_{k-1}[X, X_k] \cdots X_p)$$

$$= \sum_k \mathfrak{s}(Y_1^k \cdots Y_p^k)$$

$$= \frac{1}{p!} \sum_{j,k} \sum_{\substack{\sigma \\ \text{with} \\ \sigma(j)=k}} Y_{\sigma(1)}^k \cdots Y_{\sigma(p)}^k$$

$$= \frac{1}{p!} \sum_{j,k} \sum_{\substack{\sigma \\ \text{with} \\ \sigma(j)=k}} Y_{\sigma(1)}^k \cdots Y_{\sigma(j-1)}^k Y_k^k \cdots Y_{\sigma(p)}^k$$

$$= \frac{1}{p!} \sum_{j,k} \sum_{\substack{\sigma \\ \text{with} \\ \sigma(j)=k}} X_{\sigma(1)} \cdots X_{\sigma(j-1)} [X, X_k] \cdots X_{\sigma(p)}. \tag{2.40}$$

Since (2.39) and (2.40) match, the proposition follows.

8. Tensor product over an algebra

Let R be an <u>associative algebra</u> over $\mathbb{C}$. (This means only that R is a complex vector space with a bilinear multiplication that is associative.) In our applications, R at first will be a universal enveloping algebra, but later R will be a more complicated algebra that need not have an identity. When R does have an identity, we shall assume that 1 acts as 1 in all R modules; in this case every R module is a complex vector space, by restriction of the multiplication to multiples of 1. When R does not have an identity, we shall assume that an R module is also a complex vector space and that the R action commutes with scalar multiplication.

For this section let us write ${}^{R}V$ to indicate that V is a left R module and V^{R} to indicate that V is a right R module. In the situation $({}^{R}V, {}^{R}W)$, let

$$\mathrm{Hom}_R(V, W) \tag{2.41a}$$

denote the vector space of complex linear maps $\varphi : V \to W$ that are R-linear in the sense that

$$\varphi(rv) = r(\varphi(v)) \quad \text{for all } r \in R,\ v \in V. \tag{2.41b}$$

This definition works compatibly with $\mathrm{Hom}(L_1, L_2)$ defined in (2.10). Namely if L_1 is in $\mathrm{Hom}_R(V_1, W_1)$, L_2 is in $\mathrm{Hom}_R(V_2, W_2)$, and φ is in $\mathrm{Hom}_R(W_1, V_2)$, then $L_2 \circ \varphi \circ L_1$ is actually R-linear from V_1 into W_2. Thus under the definition $\mathrm{Hom}(L_1, L_2)\varphi = L_2 \circ \varphi \circ L_1$,

$\mathrm{Hom}(L_1, L_2)$ is in $\mathrm{Hom}_{\mathbb{C}}(\mathrm{Hom}_R(W_1, V_2), \mathrm{Hom}_R(V_1, W_2))$. (2.42)

Suppose that S is another associative algebra over $\mathbb{C}$ and that the left R module V is also a right S module in such a way that $(rv)s = r(vs)$ for all $r \in R$, $s \in S$, $v \in V$. (In this section we write ${}^R V^S$ to signify this situation.) Then $\mathrm{Hom}_R(V, W)$ becomes a left S module under the definition

$$(s\varphi)(v) = \varphi(vs) .$$

The only thing that requires note in this assertion is the fact that $s\varphi$ is again R-linear, and this comes down to the identity $(rv)s = r(vs)$.

A special case of interest is the situation $({}^R R^R, {}^R V)$. If R has an identity, then the map $\varphi \to \varphi(1)$ gives an isomorphism

$$\mathrm{Hom}_R(R, V) \cong V \tag{2.43}$$

of left R modules.

Similarly in the situation (V^R, W^R),

$$\mathrm{Hom}_R(V, W) \tag{2.44a}$$

denotes the subspace of complex linear maps $\varphi : V \to W$ that are R-linear in the sense that

$$\varphi(vr) = (\varphi(v))r \quad \text{for all } r \in R, \ v \in V . \tag{2.44b}$$

We still use (2.10) to define $\mathrm{Hom}(L_1, L_2)$, and (2.42) remains valid. In the situation $({}^S V^R, W^R)$, $\mathrm{Hom}_R(V, W)$ is a right S module under

$$(\varphi s)(v) = \varphi(sv).$$

In the special case $({}^R R^R, V^R)$, we obtain an isomorphism

$$\mathrm{Hom}_R(R, V) \cong V \tag{2.45}$$

of right R modules by the map $\varphi \to \varphi(1)$, provided R has an identity.

Now we shall make an analogous construction with tensor product in place of Hom. Let the situation be $(V^R, {}^R W)$. Then

$$V \otimes_R W$$

denotes the vector space quotient of $V \otimes_{\mathbb{C}} W$ by the subspace generated by all $vr \otimes w - v \otimes rw$. We still write $v \otimes w$ for the image in $V \otimes_R W$ of $v \otimes w$ in $V \otimes_{\mathbb{C}} W$. Then we have $vr \otimes w = v \otimes rw$ in $V \otimes_R W$. The definition of $V \otimes_R W$ works compatibly with $L_1 \otimes L_2$ in (2.8). Namely if L_1 is in $\mathrm{Hom}_R(V_1, W_1)$ and if L_2 is in $\mathrm{Hom}_R(V_2, W_2)$, then $L_1 \otimes L_2$ initially is a member of $\mathrm{Hom}_{\mathbb{C}}(V_1 \otimes_{\mathbb{C}} V_2, W_1 \otimes_{\mathbb{C}} W_2)$, and it passes to the quotient in the range to become a member of $\mathrm{Hom}_{\mathbb{C}}(V_1 \otimes_{\mathbb{C}} V_2, W_1 \otimes_R W_2)$. Since

$$\begin{aligned}(L_1 \otimes L_2)(v_1 r \otimes v_2 - v_1 \otimes rv_2) &= L_1(v_1 r) \otimes L_2(v_2) - L_1(v_1) \otimes L_2(rv_2) \\ &= L_1(v_1) r \otimes L_2(v_2) - L_1(v_1) \otimes rL_2(v_2) = 0,\end{aligned}$$

we see that

$$L_1 \otimes L_2 \quad \text{is in} \quad \mathrm{Hom}_{\mathbb{C}}(V_1 \otimes_R V_2, W_1 \otimes_R W_2). \tag{2.46}$$

In the situation $({}^S V^R, {}^R W)$, $V \otimes_R W$ becomes a left S

module under the definition

$$s(v \otimes w) = sv \otimes w \,. \tag{2.47a}$$

To make this precise, we define the left action by s on $V \otimes_R W$ by means of (2.46) as $L \otimes I$, where L is "left by s" in V. Formula (2.9) shows that this definition makes $V \otimes_R W$ into a left S module. Similarly in the situation $(V^R, {}^R W^S)$, $V \otimes_R W$ becomes a right S module under the definition

$$(v \otimes w)s = v \otimes ws \,. \tag{2.47b}$$

Special cases of interest are $({}^R R^R, {}^R V)$ and $(V^R, {}^R R^R)$ when R has an identity, and then we obtain the respective isomorphisms

$$R \otimes_R V \cong V \quad \text{and} \quad V \otimes_R R \cong V \tag{2.48}$$

from the maps $r \otimes v \to rv$ and $v \otimes r \to vr$.

A more general situation where two-sided modules arise is when R is a subalgebra of S. Then S is a two-sided R module, and $\mathrm{Hom}_R(S, V)$ and $S \otimes_R V$ make sense if V is an R module. Both constructions are sometimes referred to as extension of scalars, and the notion of complexification in §5 is an instance of this for real associative algebras.

The vector space $V \otimes_R W$ has a universal mapping property, given as follows.

Proposition 2.26. In the situation $(V^R, {}^R W)$, let $\iota : V \times W \to V \otimes_R W$ denote the map $\iota(v,w) = v \otimes w$. If X is any complex vector space and $b : V \times W \to X$ is R-bilinear in the sense that

$$b(vr,w) = b(v,rw) \quad \text{for all } r \in R, \quad v \in V, \quad w \in W,$$

then there exists a unique complex linear map $B : V \otimes_R W \to X$ such that

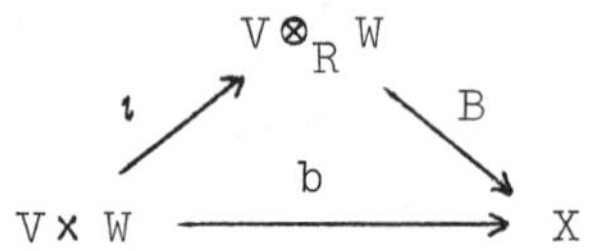

commutes.

This kind of argument is by now routine, and we omit it. Our final three results concern <u>associativity formulas</u>.

<u>Proposition 2.27</u>. In the situation $(A^R, {}^R B^S, C^S)$ for modules A, B, and C, the isomorphism

$$\mathrm{Hom}_{\mathbb{C}}(A \otimes_{\mathbb{C}} B, C) \cong \mathrm{Hom}_{\mathbb{C}}(A, \mathrm{Hom}_{\mathbb{C}}(B,C))$$

of (2.12) induces a vector space isomorphism

$$\mathrm{Hom}_S(A \otimes_R B, C) \cong \mathrm{Hom}_R(A, \mathrm{Hom}_S(B,C)) \qquad (2.49)$$

that is natural in each variable.

Remark. "Naturality" refers to the commutative diagram in Proposition 2.2, which remains commutative when we pass to the quotients and submodules here.

Proof. The map implementing (2.12) from left to right was

$$\Phi(\varphi)(a)(b) = \varphi(a \otimes b)$$

with inverse

$$\tilde{\Psi}(\psi)(a \otimes b) = \psi(a)(b) .$$

First we restrict Φ to φ's that descend to $A \otimes_R B$ and see that $\Phi(\varphi)$ is R-linear:

$$\Phi(\varphi)(ar)(b) = \varphi(ar \otimes b) = \varphi(a \otimes rb) = \Phi(\varphi)(a)(rb) = ((\Phi(\varphi)(a))r)(b).$$

So Φ induces a map of $\operatorname{Hom}_{\mathbb{C}}(A \otimes_R B, C)$ into $\operatorname{Hom}_R(A, \operatorname{Hom}_{\mathbb{C}}(B,C))$. Next we restrict the induced version of Φ to φ's that are S-linear and check that $\Phi(\varphi)$ has S-linear values:

$$\Phi(\varphi)(a)(bs) = \varphi(a \otimes bs) = \varphi((a \otimes b)s) = (\varphi(a \otimes b))s = s(\Phi(\varphi)(a)(b)) .$$

This gives us our map from left to right in (2.49). We construct a two-sided inverse in the same way from $\tilde{\Psi}$. Defining $\tilde{\Psi}(\psi)$ on elements $a \otimes b$ in $A \otimes_R B$ defines it on all of $A \otimes_R B$ as a result of Proposition 2.26. The remaining details for proving (2.49) are easy and are omitted.

Proposition 2.28. In the situation $({}^R A, {}^S B^R, {}^S C)$ for modules A, B, and C, the isomorphism

$$\operatorname{Hom}_{\mathbb{C}}(B \otimes_{\mathbb{C}} A, C) \cong \operatorname{Hom}_{\mathbb{C}}(A \otimes_{\mathbb{C}} B, C) \cong \operatorname{Hom}_{\mathbb{C}}(A, \operatorname{Hom}_{\mathbb{C}}(B,C))$$

of (2.3) and (2.12) induces a vector space isomorphism

$$\operatorname{Hom}_S(B \otimes_R A, C) \cong \operatorname{Hom}_R(A, \operatorname{Hom}_S(B,C)) \qquad (2.50)$$

that is natural in each variable.

Proposition 2.29. In the situation $(A^R, {}^RB^S, {}^SC)$ for modules A, B, and C, the isomorphism

$$(A \otimes_{\mathbb{C}} B) \otimes_{\mathbb{C}} C \cong A \otimes_{\mathbb{C}} (B \otimes_{\mathbb{C}} C)$$

of (2.18) induces a vector space isomorphism

$$(A \otimes_R B) \otimes_S C \cong A \otimes_R (B \otimes_S C) \tag{2.51}$$

that is natural in each variable.

Remarks. Naturality is by Proposition 2.5. To obtain (2.51), we use the map of (2.18) to send $(A \otimes_{\mathbb{C}} B) \otimes_{\mathbb{C}} C$ into $A \otimes_R (B \otimes_{\mathbb{C}} C)$. If $Q: B \otimes_{\mathbb{C}} C \to B \otimes_S C$ is the quotient map, then we map $A \otimes_R (B \otimes_{\mathbb{C}} C)$ into $A \otimes_R (B \otimes_S C)$ by $I \otimes Q$. The composition sends $(A \otimes_{\mathbb{C}} B) \otimes_{\mathbb{C}} C$ into $A \otimes_R (B \otimes_S C)$, and we check that it descends properly. The inverse map is obtained similarly.

CHAPTER III

REPRESENTATIONS OF COMPACT GROUPS

1. Abstract theory

We defined representations of topological groups in §I.8. In this chapter we shall assume that G is a compact topological group, and soon we shall specialize G further. Although everything we do will be valid in some form for all compact Lie groups, we shall give statements and proofs only for the unitary groups $U(n)$. In this way we shall avoid the need for an extensive digression on roots and weights.

For this section, let G be a compact topological group. Such a group has a unique Borel measure that is invariant under right and left translation and has total mass one. We refer to this measure as normalized Haar measure and write it in integrals simply as dx.

A representation Φ on a Hilbert space V is unitary if $\Phi(x)$ is a unitary operator for all x in the group. In this case, whenever U is an invariant subspace, so is the orthogonal complement $U^\perp$, since $w \in U^\perp$ and $u \in U$ imply

$$(\Phi(x)w,u) = (w,\Phi(x)^{-1}u) \in (w,U) = 0 .$$

Proposition 3.1. If Φ is a representation of G on a

finite-dimensional V, then V admits a Hermitian inner product such that Φ is unitary.

Proof. Let $(\cdot,\cdot)$ be any Hermitian inner product on V, and define

$$\langle u,v\rangle = \int_G (\Phi(x)u,v)\,dx .$$

It is straightforward to see that $\langle\cdot,\cdot\rangle$ has the required properties.

<u>Corollary 3.2</u>. If Φ is a representation of G on a finite-dimensional V, then Φ is the direct sum of irreducible representations. (That is, $V = V_1 \oplus \cdots \oplus V_k$ with each V_j an invariant subspace on which Φ acts irreducibly.)

Proof. Construct $\langle\cdot,\cdot\rangle$ as in Proposition 3.1. Find an invariant subspace $U \neq 0$ of minimal dimension and take its orthogonal complement $U^\perp$. Then $U^\perp$ is invariant. Repeating the argument with $U^\perp$ and iterating, we obtain the required decomposition.

<u>Proposition 3.3</u> (Schur's Lemma). Suppose Φ and Φ' are irreducible representations of G on finite-dimensional spaces V and V', respectively. If $L: V \to V'$ is a linear map such that $\Phi'(g)L = L\Phi(g)$ for all g in G, then L is one-one onto or $L = 0$.

Proof. We see easily that $\ker L$ and image L are invariant subspaces of V and V', respectively, and then the only possibilities are the ones listed.

Corollary 3.4. Suppose Φ is an irreducible unitary representation of G on a finite-dimensional V. If $L: V \to V$ is a linear map such that $\Phi(g)L = L\Phi(g)$ for all g in G, then L is scalar.

Proof. Let λ be an eigenvalue of L. Then $L - \lambda I$ is not one-one onto but does commute with $\Phi(g)$ for all g in G. By Proposition 3.3, $L - \lambda I = 0$.

Corollary 3.5 (Schur orthogonality).

(a) Let Φ and Φ' be inequivalent irreducible unitary representations on finite-dimensional spaces V and V', respectively. Then

$$\int_G (\Phi(x)u, v)\overline{(\Phi'(x)u', v')}\, dx = 0 \quad \text{for all } u, v \in V \text{ and } u', v' \in V'.$$

(b) Let Φ be an irreducible unitary representation on a finite-dimensional V. Then

$$\int_G (\Phi(x)u_1, v_1)\overline{(\Phi(x)u_2, v_2)}\, dx = \frac{(u_1, u_2)\overline{(v_1, v_2)}}{\dim V}$$

for all u_1, v_1, u_2, v_2 in V.

Proof. For (a), let $\ell: V' \to V$ be linear and form

$$L = \int_G \Phi(x)\ell\Phi'(x^{-1})\, dx.$$

(This integration can be regarded as occurring for matrix-valued functions and is to be handled entry-by-entry.) Then it follows that $\Phi(y)L\Phi'(y^{-1}) = L$, so that $\Phi(y)L = L\Phi'(y)$ for all y in G. By Proposition 3.3, $L = 0$. Thus

$(Lv',v) = 0$. Choose $\ell(w') = (w',u')u$, and (a) results.

For (b), we proceed in the same way, starting from $\ell : V \to V$ and obtaining $L = \lambda I$ by Corollary 3.4. Taking the trace of both sides, we find

$$\lambda \dim V = \operatorname{Tr} L = \operatorname{Tr} \ell ,$$

so that $\lambda = \operatorname{Tr} \ell/\dim V$. Thus

$$(Lv_2, v_1) = \frac{\operatorname{Tr} \ell}{\dim V} \overline{(v_1, v_2)} .$$

Choosing $\ell(w) = (w, u_2)u_1$, we obtain (b).

If Φ is a unitary representation of G, a <u>matrix coefficient</u> of Φ is any function on G of the form $(\Phi(x)u, v)$. If $\{u_i\}$ is an orthonormal basis of the space on which Φ acts, then the expressions of $\Phi(x)$ in this basis are $\Phi_{ij}(x) = (\Phi(x)u_j, u_i)$ and are matrix coefficients. The linear span of these functions is independent of the basis $\{u_i\}$.

Let Φ be unitary on V. Let $\overline{V}$ be the same vector space as V but with multiplication by i replaced by multiplication by $-i$; that is, if J_V and $J_{\overline{V}}$ are the respective multiplication-by-i maps, then $J_{\overline{V}} = -J_V$. Define an inner product in $\overline{V}$ by $(u,v)_{\overline{V}} = (v,u)_V$. It is easy to check that the linear maps $\Phi(x) : V \to V$ remain complex linear as maps $\Phi(x) : \overline{V} \to \overline{V}$, so that Φ is a representation of G on $\overline{V}$. Moreover Φ is unitary on $\overline{V}$. We call Φ on $\overline{V}$ the <u>complex conjugate</u> of Φ on V. The matrix coefficients of Φ on $\overline{V}$ are the complex conjugates of the matrix coefficients of Φ on V.

If Φ and Φ' are unitary representations on V and V', then $\Phi \otimes \Phi'$ is unitary on $V \otimes V'$ with respect to the inner product defined by

$$(u \otimes u', v \otimes v') = (u,v)(u',v') .$$

Evidently the matrix coefficients of $\Phi \otimes \Phi'$ are spanned by the products of matrix coefficients, one from Φ and one from Φ'.

We can interpret Corollary 3.5 as follows. Let $\{\Phi^{(\alpha)}\}$ be a maximal set of mutually inequivalent irreducible unitary finite-dimensional representations of G. For each $\Phi^{(\alpha)}$, choose an orthonormal basis for the underlying vector space and let $\Phi_{ij}^{(\alpha)}$ be the matrix of $\Phi^{(\alpha)}(x)$ in this basis. Then the functions $\{\Phi_{ij}^{(\alpha)}(x)\}_{i,j,\alpha}$ form an orthogonal set in $L^2(G)$. In fact, if $d^{(\alpha)}$ denotes the <u>degree</u> of $\Phi^{(\alpha)}$ (i.e., the dimension of the underlying vector space), then $\{(d^{(\alpha)})^{1/2}\Phi_{ij}^{(\alpha)}(x)\}_{i,j,\alpha}$ is an orthonormal set in $L^2(G)$. The Peter-Weyl Theorem below will say (among other things) that this orthonormal set is an orthonormal basis.

If τ is a unitary representation on a finite-dimensional space, its <u>character</u> is the function

$$\chi_\tau(x) = \operatorname{Tr} \tau(x) = \sum_i (\tau(x)u_i, u_i) , \tag{3.1}$$

where $\{u_i\}$ is an orthonormal basis. This function is independent of the basis and lies in the span of the matrix coefficients of τ.

To any representation Φ on a Hilbert space, we can associate an operator $\Phi(f)$ for any continuous $f: G \to \mathbb{C}$ by

$$(\Phi(f)u,v) = \int_G f(x)(\Phi(x)u,v)\,dx$$

or more directly by the vector-valued integral

$$\Phi(f)u = \int_G f(x)\Phi(x)u\,dx\,.$$

Then $\Phi(f)$ is linear in the f variable, and we readily check that

$$\Phi(f * h) = \Phi(f)\Phi(h)\,,$$

where $f * h$ refers to the <u>convolution</u>

$$f * h(x) = \int_G f(xy^{-1})h(y)\,dy = \int_G f(y)h(y^{-1}x)\,dy\,. \tag{3.2}$$

In addition, if Φ is unitary, then

$$\|\Phi(f)\| \leq \|f\|_1 \quad \text{and} \quad \Phi(f^*) = \Phi(f)^*\,, \tag{3.3}$$

where $f^*(x) = \overline{f(x^{-1})}$ and $\Phi(f)^*$ is the adjoint of $\Phi(f)$. We shall be especially interested in the operators $\Phi(\chi_\tau)$, where τ is an irreducible representation.

<u>Corollary 3.6</u>. Characters of finite-dimensional irreducible unitary representations τ and τ' satisfy

$$\chi_\tau(x) = \overline{\chi_\tau(x^{-1})}$$

$$\chi_\tau * \chi_{\tau'} = 0 \quad \text{if } \tau \text{ is not equivalent with } \tau'$$

$$d_\tau \chi_\tau * \chi_\tau = \chi_\tau\,.$$

Proof. The first equation follows by summing for $i = j$ the relation $(\tau(x)u_i, u_j) = \overline{(\tau(x^{-1})u_j, u_i)}$. The other two equations are routine consequences of Schur orthogonality (Corollary 3.5).

Theorem 3.7 (Peter-Weyl Theorem).

(a) The linear span of all matrix coefficients for all finite-dimensional irreducible unitary representations of G is dense in $L^2(G)$.

(b) If $\{\Phi^{(\alpha)}\}$ is a maximal set of mutually inequivalent finite-dimensional irreducible unitary representations of G and if $\{(d^{(\alpha)})^{1/2}\Phi_{ij}^{(\alpha)}(x)\}$ is a corresponding orthonormal set of matrix coefficients, then $\{(d^{(\alpha)})^{1/2}\Phi_{ij}^{(\alpha)}(x)\}$ is an orthonormal basis of $L^2(G)$.

(c) Every irreducible unitary representation of G is finite-dimensional.

(d) Let Φ be a unitary representation of G on a Hilbert space V. Then V is the orthogonal sum of finite-dimensional irreducible invariant subspaces.

(e) Let Φ be a unitary representation of G on a Hilbert space V. For each irreducible unitary representation τ of G, let E_τ be the orthogonal projection on the closure of the sum of all irreducible invariant subspaces of V that are equivalent with τ. Then E_τ is given by $d_\tau \Phi(\overline{\chi_\tau})$, where d_τ is the degree of τ and χ_τ is the character of τ. Moreover if τ and τ' are inequivalent, then $E_\tau E_{\tau'} = E_{\tau'} E_\tau = 0$. Finally every v in V satisfies

$$v = \sum_{\tau} E_{\tau} v ,$$

with the sum taken over a set of representatives τ of all equivalence classes of irreducible unitary representations of G.

Proof. (a) Although this result is valid in general, we prove it only for unitary groups $U(n)$. The linear span in question is a subspace of $C(G)$, the space of continuous complex-valued functions on G, and it is closed under multiplication (because of tensor products) and conjugation (because of complex conjugate representations). Moreover it contains the constants (because of the trivial representation) and separates points (because of the standard representation on $\mathbb{C}^n$). By the Stone-Weierstrass Theorem, it is uniformly dense in $C(G)$. Since uniform convergence implies L^2 convergence, it is dense in $L^2(G)$.

(b) The linear span of the functions in question is the linear span considered in (a). Thus (a) and general Hilbert space theory imply (b).

(c) This will follow from (d).

(d) By Zorn's Lemma, choose a maximal orthogonal set of finite-dimensional irreducible invariant subspaces. Let U be the closure of the sum. Arguing by contradiction, we suppose U is not all of V. Then $U^{\perp}$ is a nonzero closed invariant subspace. Fix $v \neq 0$ in $U^{\perp}$.

If h is a linear combination of matrix coefficients, then h lies in a finite-dimensional subspace S of $L^2(G)$

that is invariant under left translation. Let $h_1, \ldots, h_m$ be a basis of this space S. Then $g \in G$ implies

$$\Phi(g)\Phi(h)v = \Phi(g) \int_G h(x)\Phi(x)v\,dx = \int_G h(x)\Phi(gx)v\,dx$$

$$= \int_G h(g^{-1}x)\Phi(x)v\,dx = \sum_{j=1}^{m} c_j \int_G h_j(x)\Phi(x)v\,dx\,,$$

and hence the finite-dimensional subspace $\sum_j \mathbb{C}\Phi(h_j)v$ is an invariant subspace for Φ. Consequently we will obtain a contradiction if we show that $\Phi(h)v \neq 0$ for some linear combination of matrix coefficients.

To construct h, we first form $\Phi(f_N)v$, where f_N is a continuous function ≥ 0 with $f_N(1) = 1$ such that f_N vanishes off an open neighborhood N of 1 in G. Then $\Phi(f_N)v$ is in $U^\perp$ for every N. Let us see that

$$\frac{1}{\|f_N\|_1} \Phi(f_N)v \to v \tag{3.4}$$

as N shrinks to $\{1\}$. In fact,

$$\|f_N\|_1^{-1}\Phi(f_N)v - v = \int_G \|f_N\|_1^{-1} f_N(x)\Phi(x)v\,dx - v$$

$$= \int_G \|f_N\|_1^{-1} f_N(x)[\Phi(x)v - v]\,dx$$

$$= \int_N \|f_N\|_1^{-1} f_N(x)[\Phi(x)v - v]\,dx\,,$$

and so

$$\|\,\|f_N\|_1^{-1}\Phi(f_N)v - v\| \leq \sup_{x\in N} \|\Phi(x)v - v\| \int \|f_N\|_1^{-1} f_N(x)\,dx$$

$$= \sup_{x\in N} \|\Phi(x)v - v\|\,.$$

Since the right side tends to 0 as N shrinks to {1}, (3.3) follows.

It follows that $\Phi(f_N)v$ is not 0 for some N. Fix such an N. Now choose h by (a) so that

$$\|f_N - h\|_1 \leq \|f_N - h\|_2 \leq \tfrac{1}{2}\|\Phi(f_N)v\|/\|v\| . \tag{3.5}$$

Then

$$\begin{aligned}\|\Phi(f_N)v - \Phi(h)v\| &= \|\Phi(f_N - h)v\| \leq \|f_N - h\|_1\|v\| \\ &\leq \|f_N - h\|_2\|v\| \leq \tfrac{1}{2}\|\Phi(f_N)v\|\end{aligned}$$

by (3.3) and (3.5). Hence

$$\|\Phi(h)v\| \geq \|\Phi(f_N)v\| - \|\Phi(f_N)v - \Phi(h)v\| \geq \tfrac{1}{2}\|\Phi(f_N)v\| > 0 .$$

Thus h has the required property. This proves (d) and also (c).

(e) Put $E'_\tau = d_\tau \Phi(\overline{\chi}_\tau)$. By (3.3) and Corollary 3.6, the adjoint of E'_τ is given by

$$E'^{*}_\tau = d_\tau \Phi(\chi_\tau(x^{-1})) = d_\tau \Phi(\overline{\chi}_\tau) = E'_\tau ,$$

and

$$E'_\tau E'_{\tau'} = d_\tau d_{\tau'} \Phi(\overline{\chi}_\tau * \overline{\chi}_{\tau'}) = 0 \quad \text{for } \tau \text{ and } \tau' \text{ inequivalent}$$

$$E'^{2}_\tau = d^2_\tau \Phi(\overline{\chi}_\tau * \overline{\chi}_\tau) = d_\tau \Phi(\overline{\chi}_\tau) = E'_\tau .$$

Thus E'_τ is an orthogonal projection, and $E'_\tau E'_{\tau'} = E'_{\tau'} E'_\tau = 0$ for τ and τ' inequivalent.

Let U be an irreducible finite-dimensional subspace of V on which $\Phi|_U$ is equivalent with τ, and let $u_1, \ldots, u_m$ be an orthonormal basis of U. If $\Phi_{ij}(x) = (\Phi(x)u_j, u_i)$, then

$$\chi_\tau(x) = \sum_{i=1}^{m} \Phi_{ii}(x) \quad \text{and} \quad \Phi(x)u_j = \sum_{i=1}^{m} \Phi_{ij}(x)u_i .$$

Hence Schur orthogonality gives

$$E_\tau' u_j = d_\tau \int_G \overline{\chi_\tau(x)}\Phi(x)u_j\,dx = d_\tau \int_G \sum_{i,k} \overline{\Phi_{kk}(x)}\Phi_{ij}(x)u_i\,dx = u_j .$$

Thus E_τ' is the identity on every irreducible subspace of type τ.

For u in a space of type τ' with τ' and τ inequivalent, we have $E_\tau' u = E_\tau' E_{\tau'}' u = 0$. Now let us apply E_τ' to a decomposition as in (d). All terms are then annihilated except the ones of type τ, since E_τ' vanishes on spaces of type τ' with τ' not equivalent with τ. Consequently $E_\tau' = E_\tau$ and $v = \Sigma_\tau E_\tau v$ for all v in V. This completes the proof of the theorem.

The proof of (e) in the Peter-Weyl Theorem contains information even in the case that Φ is the right regular representation on $L^2(G)$. If τ is an irreducible unitary representation and $u_1, \ldots, u_{d_\tau}$ is an orthonormal basis of the space on which τ operates, then the span of a row of matrix coefficients

$$(\tau(x)u_j, u_i), \quad i \text{ fixed and } j \text{ moving}, \quad 1 \leq j \leq d_\tau ,$$

is an invariant subspace of $L^2(G)$ of type τ. By Schur orthogonality the different spaces, as i varies, are orthogonal. In the decomposition of (d) of the theorem, as made specific in (b), these d_τ spaces are the only ones of type τ, because the proof of (e) shows that E_τ annihilates the others. Thus in the case of the right regular representation on $L^2(G)$, image E_τ has dimension d_τ^2, with τ occurring d_τ linearly independent times.

The conclusion of (e) in the Peter-Weyl Theorem implies that the number of occurrences of τ in a decomposition (d) is independent of the decomposition. The number is obtained as the quotient $(\dim \text{image } E_\tau)/d_\tau$. We write $[\Phi:\tau]$ for this quantity, calling it the multiplicity of τ in Φ.

Corollary 3.8. Let Φ and τ be unitary representations of G on spaces V^Φ and V^τ, respectively, and suppose τ is irreducible. Then

$$[\Phi:\tau] = \dim \operatorname{Hom}_G(V^\Phi, V^\tau) = \dim \operatorname{Hom}_G(V^\tau, V^\Phi),$$

where the subscripts "G" refer to linear maps respecting the indicated actions by G.

Proof. By Schur's Lemma and the Peter-Weyl Theorem, any member of $\operatorname{Hom}_G(V^\Phi, V^\tau)$ annihilates $(E_\tau V^\Phi)^\perp$. Thus write $E_\tau V^\Phi$ as the orthogonal sum of irreducible subspaces V_α, by (d) of the theorem. Each V_α is equivalent with V^τ, by (e) of the theorem. Thus for each V_α the space of G maps from V_α to V^τ is at least one-dimensional. It is at most one-dimensional by Schur's Lemma. Then it follows that

$$[\Phi:\tau] = \dim \operatorname{Hom}_G(V^\Phi, V^\tau).$$

Taking adjoints, we obtain

$$\dim \operatorname{Hom}_G(V^{\Phi}, V^{\mathrm{T}}) = \dim \operatorname{Hom}_G(V^{\mathrm{T}}, V^{\Phi}) .$$

The corollary follows.

2. Irreducible representations of SU(2)

We know from §I.8 that every finite-dimensional representation of SU(2) is smooth and hence leads to a representation of $\mathfrak{su}(2)$. Since

$$\mathfrak{su}(2) \oplus i\,\mathfrak{su}(2) = \mathfrak{sl}(2,\mathbb{C}) ,$$

we obtain a complex-linear representation of $\mathfrak{sl}(2,\mathbb{C})$. The invariant subspaces for these representations correspond, and thus an irreducible representation of SU(2) leads to an irreducible complex-linear representation of $\mathfrak{sl}(2,\mathbb{C})$. We shall now classify the latter and see that they all come from representations of SU(2). As a consequence we shall see that the representations Φ_n of SU(2) on holomorphic polynomials in z_1, z_2 homogeneous of degree n (Example 2 in §I.8) exhaust the irreducible representations of SU(2), apart from equivalence.

We shall make repeated use of the basis $\{h, e, f\}$ of $\mathfrak{sl}(2,\mathbb{C})$ over $\mathbb{C}$ given by

$$h = \begin{pmatrix} 1 & 0 \\ 0 & -1 \end{pmatrix}, \qquad e = \begin{pmatrix} 0 & 1 \\ 0 & 0 \end{pmatrix}, \qquad f = \begin{pmatrix} 0 & 0 \\ 1 & 0 \end{pmatrix}.$$

These elements satisfy the bracket relations

$$[h,e] = 2e , \qquad [h,f] = -2f , \qquad [e,f] = h . \tag{3.6}$$

Theorem 3.9. For each integer $m \geq 1$, there exists up to equivalence a unique irreducible complex-linear representation π of $\mathfrak{sl}(2,\mathbb{C})$ on a space V of dimension m. In V there is a basis $\{v_0, \ldots, v_{m-1}\}$ such that (with $n = m-1$)

(1) $\pi(h)v_i = (n-2i)v_i$

(2) $\pi(e)v_0 = 0$

(3) $\pi(f)v_i = v_{i+1}$ with $v_{n+1} = 0$

(4) $\pi(e)v_i = i(n-i+1)v_{i-1}$ with $v_{-1} = 0$.

Moreover the representation π can be realized as the differentiated version of the representation Φ_n on holomorphic polynomials in z_1, z_2 homogeneous of degree n.

Remark. Property (1) gives the eigenvalues of $\pi(h)$. Notice that the smallest eigenvalue is the negative of the largest. Therefore the largest is ≥ 0.

Proof of uniqueness. Let π be a complex-linear irreducible representation of $\mathfrak{sl}(2,\mathbb{C})$ on V with $\dim V = m$. Let $v \neq 0$ be an eigenvector for $\pi(h)$, say with $\pi(h)v = \lambda v$. Then $\pi(e)v$, $\pi(e)^2 v, \ldots$ are also eigenvectors because

$$\pi(h)\pi(e)v = \pi(e)\pi(h)v + \pi([h,e])v = \pi(e)\lambda v + 2\pi(e)v = (\lambda+2)\pi(e)v.$$

Since $\lambda, \lambda+2, \lambda+4, \ldots$ are distinct, these eigenvectors are independent while nonzero. By finite-dimensionality we can find v_0 in V with (λ redefined and)

(a) $v_0 \neq 0$

(b) $\pi(h)v_0 = \lambda v_0$

(c) $\pi(e)v_0 = 0$.

Define $v_i = \pi(f)^i v_0$. Then $\pi(h)v_i = (\lambda - 2i)v_i$, by the same argument as above, and so there is a minimum integer n with $\pi(f)^{n+1}v_0 = 0$. Then $v_0, \ldots, v_n$ are independent and

(1) $\pi(h)v_i = (\lambda - 2i)v_i$

(2) $\pi(e)v_0 = 0$

(3) $\pi(f)v_i = v_{i+1}$ with $v_{n+1} = 0$.

We claim $V = \text{span}\{v_0, \ldots, v_n\}$. It is enough to show $\text{span}\{v_0, \ldots, v_n\}$ is stable under $\pi(e)$ because of the irreducibility. In fact, we show

(4) $\pi(e)v_i = i(\lambda - i + 1)v_{i-1}$ with $v_{-1} = 0$.

We proceed by induction for (4), the case $i = 0$ being (2). Assume (4) for case i. To handle $i + 1$, we write

$$\begin{aligned}\pi(e)v_{i+1} &= \pi(e)\pi(f)v_i = \pi([e,f])v_i + \pi(f)\pi(e)v_i \\ &= \pi(h)v_i + \pi(f)\pi(e)v_i \\ &= (\lambda - 2i)v_i + \pi(f)(i(\lambda - i + 1))v_{i-1} \\ &= (i + 1)(\lambda - i)v_i ,\end{aligned}$$

and the induction is complete.

To finish the proof of uniqueness, we show $\lambda = n$. We have

$$\text{Tr}\,\pi(h) = \text{Tr}(\pi(e)\pi(f) - \pi(f)\pi(e)) = 0 .$$

Thus $\Sigma_{i=0}^{n} (\lambda - 2i) = 0$, and we find $\lambda = n$.

Proof of existence. Form the differentiated version φ_n of the representation Φ_n on holomorphic polynomials in z_1, z_2 homogeneous of degree n. Here φ_n is complex linear and has dimension $m = n+1$. Also

$$\varphi_n(h)(z_2^n) = \frac{d}{dt}\Phi_n\begin{pmatrix} e^t & 0 \\ 0 & e^{-t}\end{pmatrix}(z_2^n)\Big|_{t=0} = \frac{d}{dt} e^{nt} z_2^n\Big|_{t=0} = nz_2^n ,$$

so that $\varphi_n(h)$ does have n as an eigenvalue.

Arguing by contradiction, suppose φ_n is reducible on its space V. Then we can find a chain of subspaces

$$0 = U_0 \subseteq U_1 \subseteq \cdots \subseteq U_k = V$$

such that $U_j \neq U_{j+1}$ and φ_n is irreducible on U_{j+1}/U_j for all j. Moreover $k > 1$. Since $\dim U_{j+1}/U_j \leq n$, the uniqueness argument says $\varphi_n(h)$ does not have eigenvalue n on U_{j+1}/U_j for any j. Therefore $\varphi_n(h)$ does not have eigenvalue n on V, contradiction. We conclude φ_n is irreducible.

3. Root space decomposition for U(n)

For most of the remainder of the chapter, we shall work only with unitary groups. Thus let $G = U(n)$, and let $\mathfrak{g}_0 = \mathfrak{u}(n)$ be its Lie algebra (the set of skew-Hermitian matrices). We can identify $\mathfrak{g} = \mathfrak{g}_0^{\mathbb{C}}$ with $\mathfrak{gl}(n,\mathbb{C})$. In working with $U(n)$, we shall exploit the many naturally occurring copies of $SU(2)$ that lie within $U(n)$, and the device for using the $SU(2)$'s is the root space decomposition of $\mathfrak{g}$.

Let

$$\mathfrak{h}_0 = \text{diagonal matrices in } \mathfrak{g}_0$$

$$\mathfrak{h}_{\mathbb{R}} = i\mathfrak{h}_0$$

$$\mathfrak{h} = \mathfrak{h}_0 \oplus i\mathfrak{h}_0 = \mathfrak{h}_{\mathbb{R}} \oplus i\mathfrak{h}_{\mathbb{R}} .$$

Define a matrix E_{ij} to be 1 in the $(i,j)^{\text{th}}$ place and 0 elsewhere. Define a linear functional e_i in the dual space $\mathfrak{h}^*$ by

$$e_i \begin{pmatrix} h_1 & & \\ & \ddots & \\ & & h_n \end{pmatrix} = h_i .$$

For each H in $\mathfrak{h}$, ad H is diagonalized by the basis of $\mathfrak{g}$ consisting of members of $\mathfrak{h}$ and the E_{ij} for $i \neq j$. We have

$$(\text{ad } H)E_{ij} = [H, E_{ij}] = (e_i(H) - e_j(H))E_{ij} .$$

So E_{ij} is a simultaneous eigenvector for all ad H, with eigenvalue $e_i(H) - e_j(H)$. In its dependence on H, the eigenvalue is linear. So the eigenvalue is a linear functional on $\mathfrak{h}$, namely $e_i - e_j$. The $(e_i - e_j)$'s, for $i \neq j$, are called <u>roots</u>. The set of roots is denoted Δ. We have

$$\mathfrak{g} = \mathfrak{h} \oplus \sum_{i \neq j} \mathbb{C}E_{ij} ,$$

which we can rewrite as

$$\mathfrak{g} = \mathfrak{h} \oplus \sum_{i \neq j} \mathfrak{g}_{e_i - e_j} , \tag{3.7}$$

where

$$\mathfrak{g}_{e_i - e_j} = \{X \in \mathfrak{g} \mid (\operatorname{ad} H)X = (e_i - e_j)(H)X \quad \text{for all } H \in \mathfrak{h}\} .$$

The decomposition (3.7) is called the <u>root-space decomposition</u> of $\mathfrak{g}$ with respect to $\mathfrak{h}$.

The bracket relations are easy, relative to (3.7). If α and β are roots, we can compute $[E_{ij}, E_{i'j'}]$ and see that

$$[\mathfrak{g}_\alpha, \mathfrak{g}_\beta] \begin{cases} = \mathfrak{g}_{\alpha+\beta} & \text{if } \alpha+\beta \text{ is a root} \\ = 0 & \text{if } \alpha+\beta \text{ is not a root or } 0 \\ \subseteq \mathfrak{h} & \text{if } \alpha+\beta = 0 . \end{cases} \tag{3.8}$$

In the last case, the exact formula is

$$[E_{ij}, E_{ji}] = E_{ii} - E_{jj} \in \mathfrak{h} .$$

All the roots are real on $\mathfrak{h}_{\mathbb{R}}$. We introduce an ordering on the roots and certain other linear functionals on $\mathfrak{h}_{\mathbb{R}}$. The ordering will depend on the choice of an ordered basis H_i for $\mathfrak{h}_{\mathbb{R}}$. For example, the elements $H_i = E_{ii}$, $1 \leq i \leq n$, form an ordered basis of $\mathfrak{h}_{\mathbb{R}}$. If f is in $\mathfrak{h}^*$, we say f is <u>positive</u> (relative to this ordered basis of $\mathfrak{h}_{\mathbb{R}}$) if f is real on $\mathfrak{h}_{\mathbb{R}}$ and if

$$f(H_1) > 0$$

<u>or</u> $\quad f(H_1) = 0 \quad \text{and} \quad f(H_2) > 0$

<u>or</u> $\quad f(H_1) = f(H_2) = 0 \quad \text{and} \quad f(H_3) > 0$

<u>or</u> ... <u>or</u> $\quad f(H_1) = \cdots = f(H_{n-1}) = 0 \quad \text{and} \quad f(H_n) > 0 .$

If f is not 0 but is real on each H_i, then exactly one of f and $-f$ is positive. The positive elements are closed under addition and under multiplication by positive scalars. We shall say $f > g$ or $g < f$ if $f - g$ is positive. The resulting ordering on the members of $\mathfrak{h}_{\mathbb{R}}^*$ is called the <u>lexicographic ordering</u> relative to the ordered basis $H_1, \ldots, H_n$ of $\mathfrak{h}_{\mathbb{R}}$. We denote by Δ^+ the set of positive roots.

The lexicographic ordering obtained from the choice $H_i = E_{ii}$, $1 \leq i \leq n$, will be called the "standard lexicographic ordering," and the corresponding Δ^+, which consists of all $e_i - e_j$ with $i < j$, will be called the "standard system of positive roots."

The <u>trace form</u>

$$B_0(X,Y) = \mathrm{Tr}(XY) \tag{3.9}$$

is complex bilinear on $\mathfrak{g} \times \mathfrak{g}$, and its restriction to $\mathfrak{h}_{\mathbb{R}} \times \mathfrak{h}_{\mathbb{R}}$ is real-valued and positive definite. B_0 has the invariance properties

$$B_0((\mathrm{ad}\ X)Y, Z) = -B_0(Y, (\mathrm{ad}\ X)Z) \tag{3.10a}$$

and

$$B_0(\mathrm{Ad}(g)X, \mathrm{Ad}(g)Y) = B_0(X,Y) \tag{3.10b}$$

on $\mathfrak{g}$ and G, respectively. Note also that

$$B_0(E_{ij}, E_{ji}) = 1 \quad \text{for all } i \text{ and } j. \tag{3.11}$$

4. Roots and weights for U(n)

Within $U(n)$, let T be the diagonal subgroup. This subgroup is connected and is maximal abelian in $U(n)$; it is often referred to as a maximal torus or a Cartan subgroup. Correspondingly $\mathfrak{h}_0$ is often referred to as a Cartan subalgebra of $\mathfrak{g}_0$.

Let Φ be a representation of $G = U(n)$ on a finite-dimensional complex vector space V. By Proposition 3.1 we may assume that Φ is unitary. By Theorem 1.14, Φ is smooth; let $\varphi : \mathfrak{g}_0 \to \mathrm{End}_{\mathbb{C}} V$ be its differentiated version. When we need to, we can extend φ to $U(\mathfrak{g})$, calling the extension φ. If X is in $\mathfrak{g}_0$, then differentiation at $t = 0$ of the identity

$$\Phi(\exp tX)\Phi(\exp tX)^* = I ,$$

with $(\cdot)^*$ denoting adjoint, leads to the identity

$$\varphi(X) + \varphi(X)^* = 0 .$$

Thus $\varphi(X)$ is skew-Hermitian for all X in $\mathfrak{g}_0$.

In particular, each $\varphi(H)$ for H in $\mathfrak{h}_0$ is skew-Hermitian. Hence each $\varphi(H)$ for H in $\mathfrak{h}_{\mathbb{R}}$ is Hermitian and is diagonable with real eigenvalues. Let $H_1, \ldots, H_n$ be any basis of $\mathfrak{h}_{\mathbb{R}}$, such as $H_j = E_{jj}$. These matrices commute and thus the homomorphism property

$$\varphi[X,Y] = \varphi(X)\varphi(Y) - \varphi(Y)\varphi(X)$$

says that the $\varphi(H_j)$ commute. Therefore we can find a simultaneous eigenspace decomposition of V under all the $\varphi(H_j)$. Since φ is linear on $\mathfrak{h}$, this decomposition is a simultaneous decomposition for all of $\varphi(\mathfrak{h})$, and each eigenvalue is linear. A typical eigenvalue is $\lambda(H)$, $H \in \mathfrak{h}$, and the eigenspace is V_λ:

$$V_\lambda = \{v \in V \mid \varphi(H)v = \lambda(H)v \quad \text{for all } H \in \mathfrak{h}\} .$$

These eigenvalues, which are certain linear functionals on $\mathfrak{h}$ that are real on $\mathfrak{h}_{\mathbb{R}}$, are called the <u>weights</u> of the representation Φ or φ, the spaces V_λ are called <u>weight spaces</u>, and the members of V_λ are called <u>weight vectors</u> for the weight λ. There are only finitely many weights, and we have an orthogonal direct sum <u>weight-space decomposition</u>

$$V = \sum_{\text{weights}} V_\lambda . \tag{3.12}$$

We give some examples below. Of special interest will be the <u>highest weight</u>; this is the weight that is largest in whatever lexicographic ordering we fix in §3. To be concrete, let us use the standard lexicographic ordering in the examples.

<u>Example 1</u>. For $G = SU(2)$, we can imitate the theory for $U(n)$, using $\mathfrak{h}_0 = \left\{\begin{pmatrix} i\theta & 0 \\ 0 & -i\theta \end{pmatrix}\right\}$. Put $e_1\begin{pmatrix} w & 0 \\ 0 & -w \end{pmatrix} = w$. If Φ_n is the representation on holomorphic polynomials in z_1, z_2 homogeneous of degree n, the general element of the representation space is

$$a_n z_1^n + a_{n-1} z_1^{n-1} z_2 + \cdots + a_0 z_2^n .$$

The weight spaces are $\mathbb{C}z_1^n$, $\mathbb{C}z_1^{n-1}z_2$, ..., $\mathbb{C}z_2^n$ with respective weights $-ne_1$, $-(n-2)e_1$, ..., ne_1. The highest weight is ne_1. A special feature of this example is that each weight space has dimension one.

Example 2. For $G = U(n)$, let Φ be the representation on holomorphic polynomials in $z_1, \ldots, z_n$ homogeneous of degree N. Each monomial

$$z_1^{j_1} z_2^{j_2} \cdots z_n^{j_n} \quad \text{with} \quad j_1 + \cdots + j_n = N$$

is a weight vector, and $\begin{pmatrix} h_1 & & \\ & \ddots & \\ & & h_n \end{pmatrix}$ acts by

$$\frac{d}{dt} \Phi \begin{pmatrix} e^{th_1} & & \\ & \ddots & \\ & & e^{th_n} \end{pmatrix} (z_1^{j_1} \cdots z_n^{j_n}) \Big|_{t=0}$$

$$= \frac{d}{dt} e^{-tj_1h_1 - \cdots - tj_nh_n} (z_1^{j_1} \cdots z_n^{j_n}) \Big|_{t=0}$$

$$= -(j_1h_1 + \cdots + j_nh_n)(z_1^{j_1} \cdots z_n^{j_n})$$

if $h_1, \ldots, h_n$ are imaginary. Hence the weight is $-(j_1h_1 + \cdots + j_ne_n)$. The highest weight is $-Ne_n$.

Example 3. For $G = U(n)$, let $\Phi(x) = (\det x)^k$. There is only one weight, and it is $k(e_1 + \cdots + e_n)$.

Example 4. For $G = U(n)$, let Φ be the usual representation on $\Lambda^k \mathbb{C}^n$. If $u_1, \ldots, u_n$ is the standard basis of $\mathbb{C}^n$, then the vectors

$$u_{i_1} \wedge u_{i_2} \wedge \cdots \wedge u_{i_k} \quad \text{with} \quad i_1 < \cdots < i_k$$

form a basis of weight vectors of $\Lambda^k \mathbb{C}^n$, the respective weights being $e_{i_1}+e_{i_2}+\cdots+e_{i_k}$. The highest weight is $e_1+\cdots+e_k$.

Example 5. For $G = U(n)$, let Φ be the adjoint representation Ad of G on $\mathfrak{g}$, given by conjugation. Then φ acts by ad, and the eigenvalues are as in §3, namely 0 and the roots $e_i - e_j$, $i \neq j$. The highest weight is $e_1 - e_n$. The weight-space decomposition is the root-space decomposition.

For any linear functional λ on $\mathfrak{h}$ that is real on $\mathfrak{h}_{\mathbb{R}}$, we let H_λ be the element of $\mathfrak{h}_{\mathbb{R}}$ such that $\lambda(H) = B_0(H, H_\lambda)$ for all H in $\mathfrak{h}_{\mathbb{R}}$, where B_0 is the trace form. Then we can define an inner product on $\mathfrak{h}_{\mathbb{R}}^*$ by

$$\langle \lambda, \mu \rangle = \mu(H_\lambda) = \lambda(H_\mu) = B_0(H_\lambda, H_\mu). \tag{3.13}$$

Proposition 3.10. Roots and weights have the following properties:

(a) If λ is a weight of φ with weight space V_λ and if α is a root or 0, then

$$\varphi(\mathfrak{g}_\alpha) V_\lambda \subseteq \begin{cases} V_{\lambda+\alpha} & \text{if } \lambda+\alpha \text{ is a weight} \\ 0 & \text{if not.} \end{cases}$$

(b) If α is a root and E_α and $E_{-\alpha}$ are members of $\mathfrak{g}_\alpha$ and $\mathfrak{g}_{-\alpha}$, respectively, then $[E_\alpha, E_{-\alpha}] = B_0(E_\alpha, E_{-\alpha}) H_\alpha$.

Proof. (a) Let E_α be in $\mathfrak{g}_\alpha$, H be in $\mathfrak{h}$, and X be in V_λ. Then $\varphi(E_\alpha)X$ is in $V_{\lambda+\alpha}$ because

$$\begin{aligned}\varphi(H)\varphi(E_\alpha)X &= \varphi(E_\alpha)\varphi(H)X + \varphi([H,E_\alpha])X \\ &= \lambda(H)\varphi(E_\alpha)X + \alpha(H)\varphi(E_\alpha)X .\end{aligned}$$

(b) Both $[E_\alpha, E_{-\alpha}]$ and $B_0(E_\alpha, E_{-\alpha})$ are in $\mathfrak{h}$. If H is in $\mathfrak{h}$, then (3.10a) gives

$$\begin{aligned}B_0(H,[E_\alpha,E_{-\alpha}]) &= B_0([H,E_\alpha],E_{-\alpha}) = \alpha(H)B_0(E_\alpha,E_{-\alpha}) \\ &= B_0(H,H_\alpha)B_0(E_\alpha,E_{-\alpha}) = B_0(H,B_0(E_{-\alpha},E_{-\alpha})H_\alpha) .\end{aligned}$$

Since B_0 is nondegenerate on $\mathfrak{h}\times\mathfrak{h}$, (b) follows.

Let α be a root. We consider the subspace

$$\mathbb{C}H_\alpha + \mathfrak{g}_\alpha + \mathfrak{g}_{-\alpha} \tag{3.14}$$

of $\mathfrak{g}$ in more detail. We know that $\dim \mathfrak{g}_\alpha = \dim \mathfrak{g}_{-\alpha} = 1$. Let us choose E_α and $E_{-\alpha}$ in $\mathfrak{g}_\alpha$ and $\mathfrak{g}_{-\alpha}$, respectively, with $B_0(E_\alpha, E_{-\alpha}) = 1$; for instance we can make our choices as in (3.11). Then we have

$$[H_\alpha, E_\alpha] = \alpha(H_\alpha)E_\alpha , \qquad [H_\alpha, E_{-\alpha}] = -\alpha(H_\alpha)E_{-\alpha} ,$$

and

$$[E_\alpha, E_{-\alpha}] = H_\alpha$$

by (b) of the proposition. Put

$$H'_\alpha = \frac{2}{\alpha(H_\alpha)} H_\alpha , \qquad E'_\alpha = \frac{2}{\alpha(H_\alpha)} E_\alpha , \qquad E'_{-\alpha} = E_{-\alpha} . \tag{3.15}$$

Then

$$[H'_\alpha, E'_\alpha] = 2E'_\alpha, \qquad [H'_\alpha, E'_{-\alpha}] = -2E'_{-\alpha}, \qquad [E'_\alpha, E'_{-\alpha}] = H'_\alpha .$$

Hence the map $H'_\alpha \to h$, $E'_\alpha \to e$, $E'_{-\alpha} \to f$ exhibits (3.14) as isomorphic to $\mathfrak{sl}(2,\mathbb{C})$. Moreover (3.14) is stable under adjoint, and the real subspace of skew-Hermitian members of (3.14) is isomorphic to $\mathfrak{su}(2)$. It is clear that this subspace of $\mathfrak{g}_0$ is the Lie algebra of a copy of $SU(2)$ imbedded in G.

Proposition 3.11. If λ is in $\mathfrak{h}^*$, then the following conditions on λ are equivalent:

(i) $\lambda = \sum_{i=1}^n k_i e_i$ with each k_i in $\mathbb{Z}$.

(ii) Whenever $H \in \mathfrak{h}$ satisfies $\exp H = 1$, then $\lambda(H)$ is in $2\pi i\mathbb{Z}$.

(iii) There is a one-dimensional character ξ_λ of T with $\xi_\lambda(\exp H) = e^{\lambda(H)}$ for all H in $\mathfrak{h}$.

Every weight of a finite-dimensional representation has these properties. If λ has these properties, then λ is real-valued on $\mathfrak{h}_{\mathbb{R}}$.

This is clear. A linear functional satisfying the equivalent properties in the proposition is said to be analytically integral.

Corollary 3.12. If λ in $\mathfrak{h}^*$ is analytically integral, then λ satisfies the condition

$$\frac{2\langle \lambda, \alpha\rangle}{|\alpha|^2} \text{ is in } \mathbb{Z} \text{ for each } \alpha \in \Delta .$$

Proof. Use (i) in the proposition, noting that $2/|\alpha|^2 = 1$.

5. Theorem of the Highest Weight for U(n)

We continue with $G = U(n)$ and with other notation as in §§3-4. A linear functional λ on $\mathfrak{h}$ that is real on $\mathfrak{h}_{\mathbb{R}}$ is said to be dominant if $2\langle\lambda,\alpha\rangle/|\alpha|^2 \geq 0$ for every α in Δ^+, the set of positive roots. If the lexicographic ordering is the standard one and if $\lambda = \Sigma\, k_i e_i$, this condition means $k_i \geq k_j$ whenever $i < j$.

Theorem 3.13 (Theorem of the Highest Weight). Fix a lexicographic ordering for $\mathfrak{h}_{\mathbb{R}}^*$. Apart from equivalence, the irreducible representations Φ of G stand in one-one correspondence with the dominant, analytically integral linear functionals λ on $\mathfrak{h}$, the correspondence being that λ is the highest weight (largest weight in the ordering) of Φ_λ. The highest weight λ of Φ_λ has these properties:

(a) λ depends only on Δ^+ and not on the particular lexicographic ordering that yielded Δ^+

(b) the weight space V_λ for λ is one-dimensional

(c) each E_α, for α in Δ^+, annihilates the members of V_λ, and the members of V_λ are the only vectors with this property

(d) every weight of Φ_λ is of the form $\lambda - \Sigma_{\alpha\in\Delta^+} n_\alpha \alpha$ with the n_α integers ≥ 0.

Proof of existence of the correspondence. Let Φ be given. We apply the constructions of §§3-4 and let λ be the highest weight. Then λ is analytically integral, by Proposition 3.11.

If α is in Δ^+, then $\lambda+\alpha$ exceeds λ and cannot be a weight. Thus $E_\alpha \in \mathfrak{g}_\alpha$ and $v \in V_\lambda$ imply $\varphi(E_\alpha)v = 0$, by Proposition 3.10a. This proves the first part of (c).

Since Φ is irreducible, so is φ. Thus $\varphi(U(\mathfrak{g}))v = V$ for each $v \neq 0$ in V. Let $\beta_1, \ldots, \beta_m$ be an enumeration of Δ^+, and let $H_1, \ldots, H_n$ be the basis $H_j = E_{jj}$ of $\mathfrak{h}$ over $\mathbb{C}$. By the Birkhoff-Witt Theorem, the monomials

$$E_{-\beta_1}^{i_1} \cdots E_{-\beta_m}^{i_m} H_1^{j_1} \cdots H_n^{j_n} E_{\beta_1}^{k_1} \cdots E_{\beta_m}^{k_m} \tag{3.16}$$

form a basis of $U(\mathfrak{g})$. Let us apply φ of each of these monomials to some v in the highest weight space V_λ. The E_β's give 0 (by the previous paragraph), the H's multiply by constants, and the $E_{-\beta}$'s push the weight down (by Proposition 3.10a). Consequently the only members of V_λ that can be obtained by applying φ of (3.16) to v are the vectors $\mathbb{C}v$. Thus V_λ is one-dimensional, and (b) is proved.

The effect of φ of (3.16) applied to v in V_λ is to give a weight vector with weight

$$\lambda - \sum_{j=1}^{m} i_j \beta_j , \tag{3.17}$$

and these vectors span V. Thus the weights (3.17) are the only weights of φ, and (d) follows. Also (d) implies (a).

To prove the second half of (c), let $v \notin V_\lambda$ satisfy $\varphi(E_\alpha)v = 0$ for all $\alpha \in \Delta^+$. Substracting the component in V_λ, we may assume v has 0 component in V_λ. Let λ_0 be the largest weight such that v has a nonzero component in V_{λ_0}, and let v' be the component. Then $\varphi(E_\alpha)v' = 0$ for all $\alpha \in \Delta^+$, and $\varphi(\mathfrak{h})v' \subseteq \mathbb{C}v'$. Applying φ of (3.16), we see that

$$V = \sum \mathbb{C}\varphi(E_{-\beta_1})^{i_1} \cdots \varphi(E_{-\beta_m})^{i_m} v' .$$

Every weight of vectors on the right is strictly lower than λ, and we have a contradiction to the fact that λ occurs as a weight.

Finally we prove that λ is dominant. Let α be in Δ^+, and form H'_α, E'_α, and $E'_{-\alpha}$ as in (3.15). These vectors span a subalgebra of $\mathfrak{g}$ isomorphic to $\mathfrak{sl}(2,\mathbb{C})$, and the isomorphism carries H'_α to h. For $v \neq 0$ in V_λ, the subspace of V spanned by all

$$\varphi(E'_{-\alpha})^p \varphi(H'_\alpha)^q \varphi(E'_\alpha)^r v$$

is stable under $\mathfrak{sl}(2,\mathbb{C})$, and the argument for (c) shows it is the same as the span of all $\varphi(E'_{-\alpha})^p v$. On these vectors, $\varphi(H'_\alpha)$ acts with eigenvalue

$$(\lambda - p\alpha)(H'_\alpha) = \frac{2\langle \lambda, \alpha\rangle}{|\alpha|^2} - 2p ,$$

and the largest eigenvalue of $\varphi(H'_\alpha)$ is therefore $2\langle \lambda, \alpha\rangle/|\alpha|^2$. On the other hand, this subspace is a representation space for $SU(2)$ and splits as the direct sum of irreducible representations of $SU(2)$. By the remark following the statement of Theorem 3.9, the largest eigenvalue is ≥ 0. Thus $2\langle\lambda,\alpha\rangle/|\alpha|^2$ is ≥ 0, and λ is dominant.

Proof that the correspondence is one-one. Let Φ and Φ' be irreducible on V and V', respectively, both with highest weight λ, and let φ and φ' be the corresponding representations of $U(\mathfrak{g})$. Let v_0 and v_0' be nonzero highest weight vectors. Form $\Phi\oplus\Phi'$ on $V\oplus V'$. We claim that

$$S = (\varphi\oplus\varphi')(U(\mathfrak{g}))(v_0\oplus v_0')$$

is an irreducible subspace of $V\oplus V'$.

Certainly S is invariant. Write S as the direct sum $S_1\oplus\dots\oplus S_M$ of irreducibles, and let

$$v_0\oplus v_0' = v_1+\dots+v_M \tag{3.18}$$

be the corresponding decomposition of $v_0\oplus v_0'$. We apply the formula

$$(\varphi\oplus\varphi')(H)(v_0\oplus v_0') = \varphi(H)v_0\oplus\varphi'(H)v_0' = \lambda(H)(v_0\oplus v_0') \quad \text{for } H\in\mathfrak{h}$$

to both sides of (3.18) and use the invariance of each S_j to conclude that each v_j is a weight vector in $V\oplus V'$ of weight λ. But S is given as φ of (3.16) applied to $v_0\oplus v_0'$, and S can have only a one-dimensional subspace of vectors of weight λ because

$$(\varphi\oplus\varphi')(E_\alpha)(v_0\oplus v_0') = \varphi(E_\alpha)v_0+\varphi'(E_\alpha)v_0' = 0 \quad \text{for } \alpha\in\Delta^+.$$

Thus all but one of the v_j's are 0, and S is a single

S_j. Hence $\varphi \oplus \varphi'$ is irreducible on S.

The projection of S to V commutes with the representations and is not identically 0. By Schur's Lemma, $\varphi \oplus \varphi'\big|_S$ is equivalent with φ. Similarly it is equivalent with φ'. Hence φ and φ' are equivalent.

<u>Proof that the correspondence is onto</u> when Δ^+ is the standard positive system. Suppose that a representation Φ, not necessarily irreducible, has highest weight λ. Let $v_0 \neq 0$ be in V_λ, and put $S = \varphi(U(\mathfrak{g}))v$. Then we claim that S is an irreducible subspace with highest weight λ. In fact, we argue as in the previous step, writing $S = S_1 \oplus \dots \oplus S_M$ as a sum of irreducibles and expanding $v_0 = v_1 + \dots + v_M$ accordingly. The argument in the previous step shows $v = v_j$ for some j, and thus S is irreducible. Thus if there exists a representation with highest weight λ, there exists an irreducible representation with highest weight λ.

Now let Λ be the set of all highest weights of all finite-dimensional representations of G. The examples in §4 show that all $e_1+\dots+e_k$ are in Λ, for $1 \leq k \leq n$, and so is $a(e_1+\dots+e_n)$ for any integer a. Since the weights of a tensor product $\Phi \otimes \Phi'$ are clearly all sums of a weight of Φ and a weight of Φ', Λ is closed under sums. Let $\lambda = \Sigma\, c_i e_i$ be dominant and analytically integral, and put

$$a_i = c_i - c_{i+1} \quad \text{for } 1 \leq i \leq n-1$$

$$a_n = c_n .$$

Then

$$\lambda = \sum_{i=1}^{n-1} a_i(e_1 + \dots + e_i) + a_n(e_1 + \dots + e_n)$$

exhibits λ as a sum of members of Λ, since $a_i \geq 0$ for $i \leq n-1$ by hypothesis. Hence λ is in Λ, and Λ contains all dominant, analytically integral linear functionals on $\mathfrak{h}$. The previous paragraph shows that every member of Λ occurs as the highest weight of an irreducible representation, and thus the correspondence is onto.

To complete the proof of Theorem 3.13, we need to prove that the correspondence is onto when Δ^+ is general. To do so, we introduce a finite group known as the "Weyl group."

6. Weyl group for U(n)

The Weyl group W for $G = U(n)$ is the quotient of the normalizer of T by T, namely

$$W = N_G(T)/T .$$

If $xTx^{-1} \subseteq T$, then $\mathrm{Ad}(x)\mathfrak{h} \subseteq \mathfrak{h}$ by passage to derivatives, and the converse implication is also valid (by passage to exponentials). When we factor by T, we are factoring out the elements of G that act trivially on $\mathfrak{h}$, and thus W gives the group of distinct linear transformations on $\mathfrak{h}$ that arise from $\mathrm{Ad}(G)$.

We can identify W as follows: If x is in $N_G(T)$ and $H_i = E_{ii}$, then $\mathrm{Ad}(x)H_i$ is a diagonal matrix with the same eigenvalues as H_i, namely one 1 and n-1 0's. Thus $\mathrm{Ad}(x)H_i = H_{\sigma(i)}$, where σ is a function from $\{1,\ldots,n\}$ to itself. Since $\mathrm{Ad}(x)$ is invertible, σ must be a permutation. Conversely if σ is a permutation, we can find such an x. Namely let x be 1 in the $(i,\sigma^{-1}(i))^{\text{th}}$ entry, $1 \leq i \leq n$, and 0 elsewhere. Then

$$(xH_ix^{-1})_{k\ell} = \sum_{r,s} x_{kr}(H_i)_{rs}(x^{-1})_{s\ell}$$

$$= x_{k\sigma(k)}(H_i)_{\sigma^{-1}(k)\sigma^{-1}(\ell)}(x^{-1})_{\sigma^{-1}(\ell)\ell}$$

is 0 unless $k = \ell$ and $\sigma^{-1}(k) = i$; hence $xH_ix^{-1} = H_{\sigma(i)}$. Consequently W is isomorphic to the symmetric group on n letters, which acts in the standard way on the H_i's, permuting the indices.

The Weyl group maps $\mathfrak{h}_{\mathbb{R}}$ to itself. Moreover (3.10b) says that the trace form B_0 is invariant under G, and consequently the Weyl group acts as isometries in the inner product on $\mathfrak{h}_{\mathbb{R}}$.

We can make W act on $\mathfrak{h}^*$ by $(wf)(H) = f(w^{-1}H)$. Let us check under this definition that W carries the set of roots to itself: If E_α is in $\mathfrak{g}_\alpha$ and w in W has x as a representative, then H in $\mathfrak{h}$ implies

$$[H, \mathrm{Ad}(x)E_\alpha] = \mathrm{Ad}(x)[\mathrm{Ad}(x)^{-1}H, E_\alpha] = \mathrm{Ad}(x)[w^{-1}H, E_\alpha]$$

$$= \alpha(w^{-1}H)\mathrm{Ad}(x)E_\alpha = (w\alpha)(H)\mathrm{Ad}(x)E_\alpha .$$

Hence $w\alpha$ is a root, and $\mathrm{Ad}(x)E_\alpha$ is a root vector for it.

More generally the set of weights of any

finite-dimensional representation π of G is preserved by W, along with the dimensions of the weight spaces, by the same argument: If v is a weight vector for the weight λ, then

$$\begin{aligned}\pi(H)(\pi(x)v) &= \pi(x)\{\pi(x)^{-1}\pi(H)\pi(x)v\} = \pi(x)\{\pi(Ad(x)^{-1}H)v\} \\ &= \pi(x)\pi(w^{-1}H)v = \lambda(w^{-1}H)\pi(x)v \\ &= (w\lambda)(H)(\pi(x)v) .\end{aligned}$$

Proposition 3.14. If $(\Delta^+)_1$ and $(\Delta^+)_2$ are the systems of positive roots resulting from two different lexicographic orderings, then there exists one and only one member w of W such that $w(\Delta^+)_1 = (\Delta^+)_2$.

Proof. If we arrange the e_i's in decreasing order in the lexicographic ordering that leads to $(\Delta^+)_1$, we can write them as

$$e_{\sigma(1)} > e_{\sigma(2)} > \cdots > e_{\sigma(n)}, \tag{3.19}$$

where σ is a permutation. Consequently $(\Delta^+)_1 = \sigma(\Delta^+)_0$, where $(\Delta^+)_0$ is the standard positive system. Similarly $(\Delta^+)_2 = \tau(\Delta^+)_0$ for some permutation τ. Then

$$\tau\sigma^{-1}(\Delta^+)_1 = (\Delta^+)_2,$$

and existence is proved.

If $w_1(\Delta^+)_1 = w_2(\Delta^+)_1 = (\Delta^+)_2$, then $w_1\sigma(\Delta^+)_0 = w_2\sigma(\Delta^+)_0$, and $w = \sigma^{-1}w_2^{-1}w_1\sigma$ is a permutation leaving $(\Delta^+)_0$ stable. We claim $w = 1$, so that $w_1 = w_2$ and uniqueness follows. In fact, otherwise let j be the least positive integer such that $w(j) \neq j$. Then j is also the least positive integer such that $w^{-1}(j) \neq j$, and we have $j < w(j)$ and $j < w^{-1}(j)$. Hence $e_j - e_{w(j)}$ and $e_j - e_{w^{-1}(j)}$ are in $(\Delta^+)_0$. Since w leaves $(\Delta^+)_0$ stable,

$$w(e_j - e_{w^{-1}(j)}) = e_{w(j)} - e_j = -(e_j - e_{w(j)})$$

is in $(\Delta^+)_0$ also, contradiction.

Proof of Theorem 3.13, conclusion. Let Δ^+ be a general positive system; we are left with proving that the highest-weight correspondence is onto. Let $(\Delta^+)_0$ be the standard positive system, and let λ be Δ^+ dominant and analytically integral. By Proposition 3.14 we can choose w in W so that $w\Delta^+ = (\Delta^+)_0$. Then $w\lambda$ is analytically integral, and we claim it is $(\Delta^+)_0$ dominant. In fact, if α is in $(\Delta^+)_0$, we can write $\alpha = w\beta$ with β in Δ^+, and then

$$\langle w\lambda, \alpha \rangle = \langle w\lambda, w\beta \rangle = \langle \lambda, \beta \rangle \geq 0 .$$

Since the highest-weight correspondence is onto when the positive system is $(\Delta^+)_0$, there exists an irreducible representation Φ whose highest weight is $w\lambda$ relative to $(\Delta^+)_0$. The weights of Φ are closed under the operation of W, and thus λ is a weight of Φ. We claim λ is the highest weight relative to Δ^+. In fact, let λ' be a weight. Then $w\lambda'$ is a weight, and part (d) of the theorem says

$$w\lambda' = w\lambda - \sum_{\alpha \in (\Delta^+)_0} n_\alpha \alpha = w\lambda - \sum_{w\beta \in (\Delta^+)_0} n_{w\beta} w\beta = w(\lambda - \sum_{\beta \in \Delta^+} n_{w\beta}\beta) .$$

Cancelling w, we see that $\lambda' \leq \lambda$ relative to Δ^+. Thus λ is the highest weight of Φ relative to Δ^+, and the proof is complete.

A variant of the above arguments yields the following handy lemma.

Lemma 3.15. Fix a positive system Δ^+. Then any member of $\mathfrak{h}^*$ that is real on $\mathfrak{h}_{\mathbb{R}}$ is conjugate via W to a dominant element.

Proof. Let $(\Delta^+)_0$ be the standard positive system, and let $\lambda = \Sigma\, k_i e_i$ be a member of $\mathfrak{h}^*$ that is real on $\mathfrak{h}_{\mathbb{R}}$. Then the k_i's are real. The effect of applying a member of W to λ is to permute the coefficients k_i. Since W consists of all permutations, we can arrange for the k_i's to end up nonincreasing, and the result will be $(\Delta^+)_0$ dominant.

For general Δ^+, we start with λ that may be assumed $(\Delta^+)_0$ dominant. Choose w by Proposition 3.14 so that $\Delta^+ = w(\Delta^+)_0$. We readily check that $w\lambda$ is Δ^+ dominant, and the lemma follows.

As an illustration of the use of this lemma, we prove the following proposition.

Proposition 3.16. Fix a positive system Δ^+. Let Φ_λ be an irreducible representation of G with highest weight λ, and let μ be any weight of Φ_λ. Then $|\mu|^2 \leq |\lambda|^2$.

Proof. By the lemma, there is no loss of generality in assuming that μ is dominant. By Theorem 3.13d, we have

$$\mu = \lambda - \sum_{\alpha\in\Delta^+} n_\alpha \alpha \tag{3.20}$$

with all $n_\alpha \geq 0$. Taking the inner product of both sides with μ, we obtain

$$|\mu|^2 \leq \langle\lambda,\mu\rangle - \sum n_\alpha\langle\alpha,\mu\rangle \leq \langle\lambda,\mu\rangle ,$$

the second inequality holding since μ is dominant. Taking the inner product of both sides of (3.20) with λ, we obtain

$$\langle\lambda,\mu\rangle = |\lambda|^2 - \sum n_\alpha\langle\lambda,\alpha\rangle \leq |\lambda|^2 ,$$

the second inequality holding since λ is dominant. Therefore $|\mu|^2 \leq |\lambda|^2$, as was to be proved.

The transposition $(i \quad j)$ in the symmetric group leads to a Weyl group element with a nice formula. The corresponding Weyl group element evidently sends $e_i - e_j$ to $e_j - e_i$ and fixes $\sum c_k e_k$ if $c_i = c_j$. Thus it is the linear transformation that is -1 on the root $\alpha = e_i - e_j$ and is $+1$ on the orthogonal complement in $\mathfrak{h}^*_{\mathbb{R}}$. We call such an element a <u>root reflection</u>. Its formula is

$$s_\alpha\lambda = \lambda - \frac{2\langle\lambda,\alpha\rangle}{|\alpha|^2}\alpha , \tag{3.21}$$

as we verify by checking it on α and the orthogonal complement separately. Since the transpositions generate the symmetric group, the root reflections generate the Weyl group.

If inequalities (3.19) hold relative to a positive system Δ^+, then the roots $e_{\sigma(i)} - e_{\sigma(i+1)}$ are called <u>simple roots</u> relative to Δ^+. (In the special case of the standard positive system, the simple roots are the roots $e_i - e_{i+1}$.) Evidently every positive root is uniquely a linear combination of simple roots, and the coefficients are nonnegative integers. Notice that the difference of two simple roots cannot be a root.

Relative to the standard positive system, the permutation that corresponds to the root reflection in the simple root $e_i - e_{i+1}$ is the transposition $(i \;\; i+1)$. Since the consecutive transpositions generate the symmetric group, the root reflections for the simple roots generate W. Relative to a general positive system, the root reflections in the simple roots are just the consecutive transpositions in a nonstandard enumeration of $\{1, \ldots, n\}$. Hence the same fact about permutations yields the following result.

Proposition 3.17. Fix a positive system Δ^+. Then W is generated by the root reflections for the simple roots.

If w is in W, then the proposition says that w has at least one decomposition as the product of simple reflections. Among all such decompositions we define the length of w to be the smallest number of factors necessary. We write $\ell(w)$ for the length of w.

Proposition 3.18. For w in W, $\ell(w)$ coincides with the number of elements in the set

$$\Delta^+(w) = \{\beta \in \Delta^+ \mid w^{-1}\beta < 0\}.$$

Proof. Let $\ell'(w)$ be the number of elements in $\Delta^+(w)$.

If α is a simple root $e_{\sigma(i)} - e_{\sigma(i+1)}$, then s_α sends α into $-\alpha$ and permutes the other positive roots. Let us show that such an α has

$$\ell'(ws_\alpha) = \begin{cases} \ell'(w) + 1 & \text{if } w\alpha > 0 \\ \ell'(w) - 1 & \text{if } w\alpha < 0. \end{cases}$$

In fact, we have a disjoint decomposition

$$\Delta^+(ws_\alpha) = \{\beta \in \Delta^+ \mid s_\alpha w^{-1}\beta < 0\}$$

$$= \{\beta \in \Delta^+ \mid w^{-1}\beta < 0,\ s_\alpha w^{-1}\beta < 0\} \cup \{\beta \in \Delta^+ \mid w^{-1}\beta > 0,\ s_\alpha w^{-1}\beta < 0\}.$$

In the second set, the only possible β has $w^{-1}\beta = \alpha$; thus $\beta = w\alpha$ is positive. Consequently the second set has one member $(\beta = w\alpha)$ if $w\alpha > 0$ and no members otherwsie. Meanwhile, the only way the first set can be different from $\Delta^+(w)$ is if some $\beta \in \Delta^+(w)$ has $s_\alpha w^{-1}\beta > 0$. In this case $w^{-1}\beta = -\alpha$, and $\beta = -w\alpha$, so that $w\alpha < 0$. Consequently the first set has the same members as $\Delta^+(w)$ if $w\alpha > 0$ and has one fewer otherwise. Thus our formula for $\ell'(ws_\alpha)$ follows.

Iterating our formula, we see immediately that $\ell'(w) \leq \ell(w)$ for all $w \in W$. To prove the reverse inequality, we induct on $\ell'(w)$. First assume $\ell'(w) = 0$. Then the uniqueness in Proposition 3.14 when $(\Delta^+)_1 = (\Delta^+)_2$ implies $w = 1$; hence $\ell(w) = 0$. Now assume $\ell'(w) = k$, and suppose the equality is proved for members of W with $\ell' < k$.

Since $w \neq 1$, $w\beta < 0$ for some $\beta \in \Delta^+$. Since β is a sum of simple roots, $w\alpha < 0$ for some simple root α. Then our formula says $\ell'(ws_\alpha) = \ell'(w) - 1$, and our induction says $\ell(ws_\alpha) = \ell'(w) - 1$. Writing ws_α as the product of $\ell'(w) - 1$ simple reflections and putting s_α at the right end, we see that $\ell(w) \leq (\ell'(w) - 1) + 1 = \ell'(w)$. Thus $\ell(w) = \ell'(w)$, and the proof is complete.

The sum of the members of $\Delta^+(w)$ will play an important role in Kostant's Theorem in Chapter VI, and we identify it now. Let δ be half the sum of the positive roots. In the standard positive system, δ is given by

$$\delta = \tfrac{1}{2} \sum_{\alpha\in\Delta^+} \alpha = \tfrac{1}{2}(n-1)e_1 + \tfrac{1}{2}(n-3)e_2 + \tfrac{1}{2}(n-5)e_3 + \cdots + (-\tfrac{1}{2})(n-1)e_n .$$

For general Δ^+, δ is given by a permutation of this expression.

<u>Proposition 3.19</u>. If w is in W, then

$$\sum_{\beta\in\Delta^+(w)} \beta = \delta - w\delta .$$

Proof. We write

$$\delta = \tfrac{1}{2} \Sigma \{\beta \mid \beta > 0,\ w^{-1}\beta > 0\} + \tfrac{1}{2} \Sigma \{\beta \mid \beta > 0,\ w^{-1}\beta < 0\}$$

and

$$\begin{aligned} w\delta &= \tfrac{1}{2}w \Sigma \{\alpha \mid \alpha > 0,\ w\alpha > 0\} + \tfrac{1}{2}w \Sigma \{\alpha \mid \alpha > 0,\ w\alpha < 0\} \\ &= \tfrac{1}{2} \Sigma \{w\alpha \mid \alpha > 0,\ w\alpha > 0\} + \tfrac{1}{2} \Sigma \{w\alpha \mid \alpha > 0,\ w\alpha < 0\} \\ &= \tfrac{1}{2} \Sigma \{\beta \mid w^{-1}\beta > 0,\ \beta > 0\} + \tfrac{1}{2} \Sigma \{\gamma \mid w^{-1}\gamma > 0,\ \gamma < 0\} \end{aligned}$$

under $\beta = w\alpha$, $\gamma = w\alpha$

$$= \tfrac{1}{2} \Sigma \{\beta \mid w^{-1}\beta > 0 \, , \beta > 0\} - \tfrac{1}{2} \Sigma \{\beta \mid w^{-1}\beta < 0 \, , \beta > 0\}$$

under $\beta = -\gamma$.

Subtracting, we obtain

$$\delta - w\delta = \Sigma \{\beta \mid \beta > 0 \, , w^{-1}\beta < 0\} = \sum_{\beta \in \Delta^+(w)} \beta \, .$$

7. Analytic form of Borel-Weil Theorem for U(n)

This section gives optional motivation for some of the algebraic constructions that occur later. It assumes knowledge of the definitions and elementary properties of holomorphic functions and complex manifolds. Our objective will be to give realizations of irreducible representations of $U(n)$ in terms of geometry and analysis. But for a while, we shall work with $GL(n,\mathbb{C})$ in place of $U(n)$; $GL(n,\mathbb{C})$ is a complex manifold (as an open subset of $\mathbb{C}^{n^2}$), and multiplication and inversion are holomorphic.

If we identify two nonzero members of $\mathbb{C}^n$ when one is a complex scalar multiple of the other, then the resulting quotient space $\mathbb{C}P^{n-1}$ is called <u>complex projective space</u> and is a compact complex manifold in a natural way.

The group $G = GL(n,\mathbb{C})$ acts transitively on $\mathbb{C}^n - \{0\}$ by multiplication of matrices times column vectors, and the action is holomorphic. The action respects the equivalence classes leading to $\mathbb{C}P^{n-1}$, and hence G has a natural transitive holomorphic action on $\mathbb{C}P^{n-1}$. The isotropy subgroup of G at the class of $\begin{pmatrix} 0 \\ \cdots \\ 0 \\ 1 \end{pmatrix}$ in $\mathbb{C}P^{n-1}$ consists of all g such that $g\begin{pmatrix} 0 \\ \cdots \\ 0 \\ 1 \end{pmatrix} = \begin{pmatrix} 0 \\ \cdots \\ 0 \\ \lambda \end{pmatrix}$ for some λ, hence is exactly

$$Q = \begin{pmatrix} * & 0 \\ * & * \end{pmatrix} \quad (\text{block sizes } n-1,\ 1).$$

This is a complex subgroup of G since the Lie algebra of Q is closed under multiplication by i, and G/Q becomes a complex manifold. Our group action of G on $\mathbb{C}P^{n-1}$ thus gives us a one-one holomorphic map of G/Q onto $\mathbb{C}P^{n-1}$. One can verify that the inverse map is holomorphic. Thus $G/Q \cong \mathbb{C}P^{n-1}$ as complex manifolds on which G operates.

Fix an integer $N \geq 0$. Then we have a holomorphic homomorphism $\chi : Q \to \mathbb{C}^\times$ given by

$$\chi\begin{pmatrix} * & 0 \\ * & \lambda \end{pmatrix} = \lambda^{-N} \quad (\text{block sizes } n-1,\ 1). \tag{3.22}$$

This χ gives a holomorphic action of Q on $\mathbb{C}$ by the definition $q(z) = \chi(q)z$. We form the quotient space

$$G \times_Q \mathbb{C} = (G \times \mathbb{C})/\sim ,$$

where $\sim$ is the equivalence relation $(gq,z) \sim (g,q(z))$ on $G \times \mathbb{C}$. The quotient space is a complex manifold, and G acts holomorphically by $g(x,z) = (gx,z)$ for $g \in G$, $x \in G$, $z \in \mathbb{C}$.

The space $G \times_Q \mathbb{C}$ fibers holomorphically over G/Q with projection map $e : G \times_Q \mathbb{C}$ given by $e(g,z) = gQ$. (Actually the fibering is a fiber bundle, but this fact will not explicitly concern us. The bundle $G \times_Q \mathbb{C} \to G/Q$ is called the <u>associated line bundle</u> for χ.) We consider C^∞ <u>sections</u> $\gamma : G/Q \to G \times_Q \mathbb{C}$; these are C^∞ maps such that $e \circ \gamma$ is the identity on G/Q. Let us set up a one-one correspondence between the C^∞ sections γ and the C^∞ maps $\varphi : G \to \mathbb{C}$ such that

$$\varphi(gq) = \chi(q)^{-1}\varphi(g) \quad \text{for } q \in Q, g \in G. \tag{3.23}$$

In fact, if we are given γ, then γ must be of the form $\gamma(gQ) = (g, \varphi_\gamma(g))$, and the image must be in the same equivalence class if we replace g by gq with $q \in Q$. Unraveling matters, we obtain

$$(g,\varphi_\gamma(g)) \sim (gq,\varphi_\gamma(gq)) \sim (g, q(\varphi_\gamma(gq)) = (g,\chi(q)\varphi_\gamma(gq)).$$

Then $\varphi_\gamma(g) = \chi(q)\varphi_\gamma(gq)$, and (3.23) follows. Conversely if φ is given with (3.23) holding, we define $\gamma(gQ) = (g,\varphi(g))$ and obtain a C^∞ section.

The group G acts on the C^∞ sections γ by $(g\gamma)(x) = \gamma(g^{-1}x)$, and it acts compatibly on the above functions φ by $(g\varphi)(x) = \varphi(g^{-1}x)$.

<u>Examples.</u>

(1) Let $P\begin{pmatrix} z_1 \\ \cdots \\ z_n \end{pmatrix}$ be a holomorphic polynomial homogeneous of degree N, with N as in (3.22), and define

$$\varphi(g) = P\left(g\begin{pmatrix}0\\ \cdots\\ 0\\ 1\end{pmatrix}\right), \quad g \in G.$$

If $q = \begin{pmatrix}* & 0\\ * & \lambda\end{pmatrix}$ is in Q, then

$$\varphi(gq) = P\left(g\begin{pmatrix}0\\ \cdots\\ 0\\ \lambda\end{pmatrix}\right) = P\left(\lambda\, g\begin{pmatrix}0\\ \cdots\\ 0\\ 1\end{pmatrix}\right) = \lambda^N P\left(g\begin{pmatrix}0\\ \cdots\\ 0\\ 1\end{pmatrix}\right)$$
$$= \lambda^N \varphi(g) = \chi(q)^{-1}\varphi(g).$$

Thus φ satisfies (3.23) and yields a C^∞ section. Actually this φ is holomorphic, and the corresponding section is therefore holomorphic.

(2) Let φ be as in (1), and let $f : G/Q \to \mathbb{C}$ be any C^∞ function. Define $\varphi_1(g) = f(gQ)\varphi(g)$. Then φ_1 satisfies (3.23) and yields a C^∞ section.

<u>Proposition 3.20</u>. For the line bundle $G\times_Q \mathbb{C} \to \mathbb{C}P^{n-1}$ associated to χ in (3.22), the only holomorphic sections are the ones that arise as in Example 1 from holomorphic polynomials homogeneous of degree N.

Proof. Let $\varphi : G \to \mathbb{C}$ be the holomorphic map satisfying (3.23) that arises from a given holomorphic section. We define a function $P\begin{pmatrix}z_1\\ \cdots\\ z_n\end{pmatrix}$ on $\mathbb{C}^n - \{0\}$ as follows. We find g with $g\begin{pmatrix}0\\ \cdots\\ 0\\ 1\end{pmatrix} = \begin{pmatrix}z_1\\ \cdots\\ z_n\end{pmatrix}$ and let $P\begin{pmatrix}z_1\\ \cdots\\ z_n\end{pmatrix} = \varphi(g)$. To see this is well defined, we suppose also that $g'\begin{pmatrix}0\\ \cdots\\ 0\\ 1\end{pmatrix} = \begin{pmatrix}z_1\\ \cdots\\ z_n\end{pmatrix}$. Then $g^{-1}g'$ fixes $\begin{pmatrix}0\\ \cdots\\ 0\\ 1\end{pmatrix}$, and hence is of the form $\begin{pmatrix}* & *\\ * & 1\end{pmatrix} = q$.

Writing $g' = gq$ and applying (3.23), we see that $\varphi(g') = \varphi(g)$, so that P is well defined. Moreover

$$P\left(\lambda\begin{pmatrix} z_1 \\ \cdots \\ z_n \end{pmatrix}\right) = \varphi(\lambda g) = \varphi\left(g\begin{pmatrix} \lambda & & \\ & \ddots & \\ & & \lambda \end{pmatrix}\right) = \chi\begin{pmatrix} \lambda & & \\ & \ddots & \\ & & \lambda \end{pmatrix}^{-1} \varphi(g)$$

$$= \lambda^N \varphi(g) = \lambda^N P\begin{pmatrix} z_1 \\ \cdots \\ z_n \end{pmatrix}$$

shows that P is homogeneous of degree N. Since the definition of P can be accomplished using a whole open set of g's at a time, we see that P is holomorphic on $\mathbb{C}^n - \{0\}$.

The homogeneity condition implies that P is bounded near 0. Hence P extends to be holomorphic on $\mathbb{C}^n$. The homogeneity and C^∞ behavior on the unit sphere force

$$|P(z)| \leq c|z|^N$$

and more generally

$$|\partial_z^\alpha P(z)| \leq c_\alpha |z|^{N-|\alpha|} \tag{3.24}$$

for any multi-index α and all z in $\mathbb{C}^n - \{0\}$. If $|\alpha| > N$, we see from (3.24) that the holomorphic function $\partial_z^\alpha P$ vanishes at infinity. Hence it is 0. Therefore the (convergent) Taylor expansion of P about 0 has all terms of degree $> N$ equal to 0, and P is a polynomial. This proves the result.

In terms of representation theory, Proposition 3.20 says that the natural representation of $GL(n, \mathbb{C})$ on holomorphic polynomials on $\mathbb{C}^n$ homogeneous of degree N can be realized as the space of holomorphic sections of the line bundle over $\mathbb{C}P^{n-1}$ associated to the representation χ of Q given in (3.23).

We can, of course, restrict from $GL(n,\mathbb{C})$ to $U(n)$. The representation of $U(n)$ that we are realizing concretely is irreducible and has highest weight $\Lambda = -Ne_n$ (see Example 2 in §4). The concrete realization is built from χ, which is determined by its holomorphicity and its restriction to $U(n-1) \times U(1)$; on this subgroup, χ is the irreducible representation with (highest) weight $\Lambda = -Ne_n$.

It is of interest to obtain the above geometric realization of the holomorphic polynomial representation of $U(n)$ without explicit reference to $GL(n,\mathbb{C})$. The fact that we want to exploit is that $\mathfrak{u}(n)^{\mathbb{C}} \cong \mathfrak{gl}(n,\mathbb{C})$.

Thus let $G_0 = U(n)$ and $L_0 = U(n-1) \times U(1)$. Since $L_0 = Q \cap G_0$, the C^∞ map $G_0 \to G/Q$ given by $g_0 \to g_0 Q$ descends to a one-one C^∞ map $G_0/L_0 \to G/Q$. Actually this map is onto. (The relevant fact here is that $G = G_0 Q$, which follows from

$$G = G_0 B_0, \tag{3.25}$$

where B_0 is the subgroup of the lower triangular group with positive real numbers as diagonal entries. Identity (3.25) is a unique decomposition and is a group-theoretic formulation of the Gram-Schmidt orthogonalization process for $\mathbb{C}^n$. See Knapp [1986], §V.2.)

One can show that the inverse of the map $G_0/L_0 \to G/Q$ is smooth, so that $G_0/L_0 \cong G/Q$ as smooth manifolds on which G_0 acts. We can use the isomorphism to transport the complex structure from G/Q to G_0/L_0. (Shortly we shall see how to check for holomorphicity on G_0/L_0 without referring to G/Q.)

To think of χ in real terms, we start by defining χ on $L_0 = U(n-1) \times U(1)$. We extend it holomorphically to $GL(n-1,\mathbb{C}) \times GL(1,\mathbb{C})$, and (3.22) will still hold. Finally we extend it to Q so as to remain a homomorphism, and (3.22) will still hold.

With χ defined only on L_0, we can form $G_0 \times_{L_0} \mathbb{C}$. But this is nothing new, since the relation

$$(g,z) = (g_0\ell_0,z) \sim (g_0,\ell_0(z))$$

shows that we have a diffeomorphism

$$G \times_Q \mathbb{C} \cong G_0 \times_{L_0} \mathbb{C}$$

that respects the action by G_0. The complex structure on $G \times_Q \mathbb{C}$ can be transported to $G_0 \times_{L_0} \mathbb{C}$, and we seek a way to detect holomorphicity of sections $\gamma : G_0/L_0 \to G_0 \times_{L_0} \mathbb{C}$ without reference to $GL(n,\mathbb{C})$.

Let $\mathfrak{g}_0$, $\mathfrak{g}$, $\mathfrak{l}_0$, and $\mathfrak{q}$ be the Lie algebras of G_0, G, L_0, and Q. We can regard $\mathfrak{g} = \mathfrak{gl}(n,\mathbb{C})$ as the complexification of $\mathfrak{g}_0$, but $\mathfrak{q}$ is larger than the complexification of $\mathfrak{l}_0$. As in §II.6, if X is in $\mathfrak{g}_0$, then X acts on $C^\infty(G_0)$ as a left-invariant vector field:

$$Xf(g_0) = \frac{d}{dt} f(g_0 \exp tX)\Big|_{t=0} .$$

If Z is in $\mathfrak{g}$, then Z acts, too; we write $Z = X + iY$ with X and Y in $\mathfrak{g}_0$, and then Z acts by

$$Zf = Xf + i(Yf) . \tag{3.26}$$

Left-invariant vector fields on $C^\infty(G)$ are quite another matter. If Z is in $\mathfrak{g}$, then Z acts on $C^\infty(G)$ by

$$ZF(g) = \frac{d}{dt} F(g \exp tZ)\Big|_{t=0} . \tag{3.27}$$

On $C^\infty(G)$ when $Z = X + iY$ with X and Y in $\mathfrak{g}_0$, we have $ZF = XF + (iY)F$ but we do not necessarily have $ZF = XF + i(YF)$. In fact, this kind of equality is related to the holomorphicity of F:

$$F \in C^\infty(G) \text{ is holomorphic} \iff i(ZF) = (iZ)(F) \quad \text{for all } Z \in \mathfrak{g} . \tag{3.28}$$

(To verify (3.28), we have only to realize that $\exp : \mathfrak{g} \to G$ gives us a chart about the identity in G compatible with the complex structure, and thus the right side of (3.28) says that F satisfies the Cauchy-Riemann equaltions.)

Once again we use the roots of $\mathfrak{g}$ with respect to $\mathfrak{h}$. Let Δ^+ denote the standard positive system.

Proposition 3.21. Let $\gamma : G_0/L_0 \to G_0 \times_{L_0} \mathbb{C}$ be a C^∞ section defined on an open subset U of G_0/L_0, let $\widetilde{U}$ be the inverse image of U in G_0, and let $\varphi : G_0 \to \mathbb{C}$ be the C^∞ function with domain $\widetilde{U}$ such that

$$\gamma(g_0 L_0) = (g_0, \varphi(g_0)) \tag{3.29}$$

and

$$\varphi(g_0 \ell_0) = \chi(\ell_0)^{-1}\varphi(g_0) . \tag{3.30}$$

Then γ is a holomorphic section if and only if $Z_{-\alpha}\varphi = 0$ for every root vector $Z_{-\alpha}$ of each negative root for which $Z_{-\alpha}$ is not in $\mathfrak{l}_0^{\mathbb{C}}$.

Interpretation. If we are careless about whether φ is defined on G_0 or G, we can think of φ as defined on G and transforming under Q according to (3.23). In this case φ will be right invariant under $\exp tZ_{-\alpha}$ and hence $Z_{-\alpha}\varphi$, defined by means of (3.27), will be 0. The condition of holomorphicity is precisely that we can be careless in this way.

Proof. In both halves of the proof, we can regard γ also as a C^∞ section $\gamma : G/Q \to G \times_Q \mathbb{C}$, and we can let $\tilde{\varphi} : G \to \mathbb{C}$ (defined on an open subset of G right invariant under Q) be the corresponding C^∞ function satisfying (3.23) such that

$$\gamma(gQ) = (g, \tilde{\varphi}(g)) . \tag{3.31}$$

Then $\gamma(g_0 L_0)$ means the same thing as $\gamma(g_0 Q)$. Comparing (3.31) and (3.29), we obtain

$$\tilde{\varphi}(g_0) = \varphi(g_0) . \tag{3.32}$$

Hence $\tilde{\varphi}$ is a C^∞ extension of φ to G that transforms under Q on the right.

If γ is holomorphic, then $\tilde{\varphi}$ is holomorphic. If Z

is in $\mathfrak{g}$, then the decomposition $Z = X + iY$ with X and Y in $\mathfrak{g}_0$ is given by

$$X = \frac{Z - Z^*}{2} \quad \text{and} \quad Y = \frac{Z + Z^*}{2i}, \tag{3.33}$$

where $*$ refers to adjoint. Then we have

$$\begin{aligned} Z\tilde{\varphi}(g_0) &= X\tilde{\varphi}(g_0) + (iY)\tilde{\varphi}(g_0) && \text{by real linearity} \\ &= X\tilde{\varphi}(g_0) + i(Y\tilde{\varphi})(g_0) && \text{since } \tilde{\varphi} \text{ is holomorphic} \\ &= X\varphi(g_0) + i(Y\varphi)(g_0) && \text{by (3.32)} \\ &= Z\varphi(g_0) && \text{by (3.26).} \end{aligned}$$

If $Z = Z_{-\alpha}$ with $Z_{-\alpha}$ as in the statement of the proposition, then $Z_{-\alpha}\tilde{\varphi} = 0$. Hence the above computation shows that $Z_{-\alpha}\varphi = 0$.

Conversely suppose $Z_{-\alpha}\varphi = 0$ for all $Z_{-\alpha}$ as in the statement of the proposition. We shall show $\tilde{\varphi}$ is holomorphic by using (3.28). If Z is in $\mathfrak{q}$, then

$$\begin{aligned} Z\tilde{\varphi}(g) &= \frac{d}{dt}\tilde{\varphi}(g \exp tZ)\Big|_{t=0} && \text{by (3.27)} \\ &= \frac{d}{dt}\chi(\exp tZ)^{-1}\Big|_{t=0}\tilde{\varphi}(g) && \text{by (3.23)} \\ &= Z\chi(1)\,\tilde{\varphi}(g), \end{aligned}$$

and the condition in (3.28) holds since χ is holomorphic on Q.

Next we show that

$$i(Z\tilde{\varphi})(g_0) = (iZ)\tilde{\varphi}(g_0) \tag{3.34}$$

for all $Z \in \mathfrak{g}$ and $g_0 \in G_0$. We have already handled $Z \in \mathfrak{q}$, and thus it is enough to handle $Z = Z^*_{-\alpha}$ with Z_α as in the statement of the proposition. Write $Z_{-\alpha} = X_{-\alpha} + iY_{-\alpha}$ by (3.33). Then our assumption $Z_{-\alpha}\varphi = 0$ gives

$$\begin{aligned} 0 = Z_{-\alpha}\varphi(g_0) &= X_{-\alpha}\varphi(g_0) + i(Y_{-\alpha}\varphi)(g_0) \\ &= X_{-\alpha}\tilde{\varphi}(g_0) + i(Y_{-\alpha}\tilde{\varphi})(g_0) \\ &= -\tfrac{1}{2} Z^*_{-\alpha}\tilde{\varphi}(g_0) + i((\tfrac{1}{2i} Z^*_{-\alpha})\tilde{\varphi})(g_0) \end{aligned}$$

since $Z_{-\alpha}\tilde{\varphi} = 0$. So

$$Z^*_{-\alpha}\tilde{\varphi}(g_0) = -i((iZ^*_{-\alpha})\tilde{\varphi})(g_0)$$

and

$$(iZ^*_{-\alpha})\tilde{\varphi}(g_0) = i(Z^*_{-\alpha}\tilde{\varphi})(g_0) .$$

Finally we show that

$$i(Z\tilde{\varphi})(g) = (iZ)\tilde{\varphi}(g) \tag{3.35}$$

for all $Z \in \mathfrak{g}$ and $g \in G$. We can write $g = g_0 q$ with $g_0 \in G_0$ and $q \in Q$, and hence

$$\begin{aligned} Z\tilde{\varphi}(g) &= \frac{d}{dt}\tilde{\varphi}(g_0 q(\exp tZ)q^{-1}q)\Big|_{t=0} \\ &= \chi(q)^{-1}\frac{d}{dt}\tilde{\varphi}(g_0 \exp t\,\mathrm{Ad}(q)Z)\Big|_{t=0} \\ &= \chi(q)^{-1}(\mathrm{Ad}(q)Z)\tilde{\varphi}(g_0) . \end{aligned}$$

Using (3.34) for $\mathrm{Ad}(q)Z$, we obtain (3.35) for Z. This completes the proof.

The above construction allows considerable generalization

even within $G = GL(n,\mathbb{C})$ and $G_0 = U(n)$. From any partition of n, say $(n_1,\ldots,n_k)$, we construct a block diagonal subgroup

$$L_0 = U(n_1) \times \ldots \times U(n_k) = \begin{pmatrix} * & & & 0 \\ & * & & \\ & & \ddots & \\ 0 & & & * \end{pmatrix}$$

of G_0 and a corresponding block lower triangular subgroup

$$Q = \begin{pmatrix} * & & & 0 \\ & * & & \\ & & \ddots & \\ * & & & * \end{pmatrix}$$

of G. If χ is an irreducible unitary representation of L_0 on a complex vector space V, then the proof of Theorem 3.13 shows that χ extends to a holomorphic representation of $GL(n_1) \times \ldots \times GL(n_k)$ on V. If we define $\chi(q)$ to be the identity on the elements

$$\begin{pmatrix} I & & & 0 \\ & I & & \\ & & \ddots & \\ * & & & I \end{pmatrix}$$

of Q, then χ extends to a holomorphic representation of Q on V. We shall suppose that the highest weight Λ of χ on L_0 is dominant for the standard positive system Δ^+ for G_0.

We form the quotient space $G \times_Q V = (G \times V)/\sim$, just as in the special case. The C^∞ sections $\gamma : G/Q \to G \times_Q V$ are in one-one correspondence with the C^∞ functions $\varphi : G \to V$

such that

$$\varphi(gq) = \chi(q)^{-1}\varphi(g) \quad \text{for} \quad q \in Q, \ g \in G, \tag{3.36}$$

the correspondence being $\gamma \longleftrightarrow \varphi$ when

$$\gamma(gQ) = (g, \varphi(g)) . \tag{3.37}$$

Moreover the holomorphic γ's correspond to the holomorphic φ's, and G acts on everything.

As in the special case, we can give the construction in terms of $U(n)$. We have

$$G_0/L_0 \cong G/Q \quad \text{and} \quad G_0 \times_{L_0} V \cong G \times_Q V,$$

as smooth manifolds on which G_0 operates. Thus we can transfer the complex structures to G_0/L_0 and $G_0 \times_{L_0} V$.

To be able to decide when a local C^∞ section $\gamma : G_0/L_0 \to G_0 \times_{L_0} V$ is holomorphic, we introduce Lie algebras $\mathfrak{g}$, $\mathfrak{g}_0$, $\mathfrak{l}_0$, and $\mathfrak{q}$. Then the holomorphicity condition on γ is as in Proposition 3.21: The associated function $\varphi : G_0 \to V$ (defined in analogy with (3.37)) is to satisfy $Z_{-\alpha}\varphi = 0$ for every root vector $Z_{-\alpha}$ of each negative root for which $Z_{-\alpha}$ is not in $\mathfrak{l}_0^{\mathbb{C}}$.

<u>Theorem 3.22</u>. In the above situation, the space of holomorphic sections $\gamma : G_0/L_0 \to G_0 \times_{L_0} V$, with its action by G_0, is an irreducible representation of G_0 with highest weight Λ, the same as the highest weight of χ within L_0.

The case of Theorem 3.22 in which

$$L_0 = U(1) \times U(1) \times \ldots \times U(1) = T$$

is worthy of special note. In this case every integral Λ dominant for Δ^+ gives us a concrete one-dimensional representation of T, and the theorem says that we obtain a global realization of an irreducible representation of G_0 with highest weight Λ. This result is known as the Borel-Weil Theorem.

For a proof of the Borel-Weil Theorem, see Knapp [1986], Chapter V. The Borel-Weil Theorem implies Theorem 3.22. In fact, if we are given Λ, let π be an irreducible unitary representation of G_0 with highest weight Λ, nonzero highest weight vector u_Λ, and representation space U. (We are using Theorem 3.13 for the existence of π.) Let U' be the linear span of all $\pi(\ell_0)u_\Lambda$, $\ell_0 \in L_0$, and let E be the orthogonal projection of U on U'. We can realize χ on L_0 concretely as the restriction of π to the subgroup L_0 and the subspace U'. (Thus $U' \cong V$.)

The representations π and χ extend holomorphically to the groups G and $GL(n_1) \times \ldots \times GL(n_k)$ on the spaces U and U', respectively, and χ extends to Q as before. For u in U, we obtain a holomorphic section $\gamma : G_0/L_0 \to G_0 \times_{L_0} U'$ from (3.37) and the definition

$$\varphi(g) = E(\pi(g)^{-1}u) . \tag{3.38}$$

We readily see that this φ satisfies (3.36) and that the representation of G_0 on holomorphic sections includes a copy of π.

On the other hand, let $\gamma : G_0/L_0 \to G_0 \times_{L_0} U'$ be any holomorphic section, and let φ be as in (3.36). Then

$$\varphi_T(g) = (\varphi(g), u_\Lambda)$$

is a holomorphic function $\varphi_T : G \to \mathbb{C}$ satisfying

$$\varphi_T(gb) = \xi_\Lambda(b)^{-1}\varphi_T(g) ,$$

where ξ_Λ is the character of T for Λ, extended holomorphically to the lower triangular group of G. The Borel-Weil Theorem identifies φ_T as one of the functions $(\pi(g)^{-1}u, u_\Lambda)$ for some u in U. Hence $\varphi(g)$ and $E(\pi(g)^{-1}u)$ have the same inner product with u_Λ for all g.

To see that $\varphi(g) = E(\pi(g)^{-1}u)$, we write

$$\varphi(g) = E(\pi(g)^{-1}u) + E_0(g)$$

with $E_0(g)$ in the orthogonal complement of u_Λ within U'. Replacing g by $g\ell_0$, we obtain

$$\chi(\ell_0)E_0(g\ell_0) = E_0(g) . \tag{3.39}$$

For any g with $E_0(g) \neq 0$, we see from (3.39) that $\chi(\ell_0)^{-1}E_0(g)$ is in the orthogonal complement of u_Λ. Since χ is irreducible on U', this is a contradiction. Thus $\varphi(g) = E(\pi(g)^{-1}u)$, and the only holomorphic sections are the ones in (3.38). Consequently the Borel-Weil Theorem implies the generalization in Theorem 3.22.

CHAPTER IV

COHOMOLOGY OF LIE ALGEBRAS

1. Motivation from differential forms

Historically there were two sources of motivation for cohomology of Lie algebras. One was from Lie algebra extensions and a comparison with group extensions; extensions played a prominent role because of the importance of semi-direct products in structure theory. We shall take up extensions briefly in §2.

The other source was from de Rham cohomology, the computation of classical cohomology by means of differential forms. This theory works on a manifold, hence on a Lie group. If one uses invariant differential forms in the computations, one is led to the definition of cohomology of a Lie algebra.

Let us recall some facts about de Rham cohomology. One begins with the context of an open set U in Euclidean space with coordinates $x^1, \ldots, x^N$. The tangent space at a point p consists of all first-order derivatives at p and has basis $\left(\frac{\partial}{\partial x^1}\right)_p, \ldots, \left(\frac{\partial}{\partial x^N}\right)_p$. By means of this basis we can identify the tangent space with $\mathbb{R}^N$. The dual of the tangent space is by definition the space of 1 forms at p, and the dual basis is denoted $(dx^1)_p, \ldots, (dx^N)_p$. We build alternating tensors from this vector space and are led to expressions

$$\sum_{i_1,\ldots,i_n} f_{i_1 \cdots i_n} \, dx^{i_1} \wedge \ldots \wedge dx^{i_n} \tag{4.1}$$

in which p is a variable (and f is a numerical-valued function, which we assume is smooth). Expressions (4.1) are called <u>n forms</u>; these are members of $C^\infty(U) \otimes_{\mathbb{R}} \Lambda^n((\mathbb{R}^N)^*)$ in our definition. The operator d sends n forms to $n+1$ forms and is defined by

$$d(f_{i_1 \cdots i_n} \, dx^{i_1} \wedge \ldots \wedge dx^{i_n}) = df_{i_1 \cdots i_n} \wedge dx^{i_1} \wedge \ldots \wedge dx^{i_n}, \tag{4.2}$$

where $df = \Sigma \left(\frac{\partial f}{\partial x^j}\right) dx^j$. The basic properties of d are

(i) $d^2 = 0$

(ii) $d(\xi \wedge \eta) = d\xi \wedge \eta + (-1)^{|\xi|} \xi \wedge d\eta$ if ξ is homogeneous of degree $|\xi|$

(iii) the theory respects smooth coordinate changes and so can be transplanted to a smooth manifold.

From (i) and (iii), it makes sense to speak of d globally on a manifold M and define

$$\begin{aligned} H^n(M) &= (\ker d|_{n \text{ forms}})/(\text{image } d|_{n-1 \text{ forms}}) \\ &= \{n \text{ cocycles}\}/\{n \text{ coboundaries}\} \\ &= n^{\text{th}} \text{ de Rham cohomology group}. \end{aligned}$$

Now suppose G is a connected Lie group with Lie algebra $\mathfrak{g}$ defined over $\mathbb{R}$. In the general theory $\mathfrak{g}$ is the _tangent space_ at the identity 1, consisting of all first-order derivatives at 1, regarded as acting on (germs of) smooth functions defined near 1. In the development of Chapter I, we concretized this, obtaining $\mathfrak{g}$ from curves $c(t)$ with $c(0) = 1$, with $c'(0)$ giving a member of $\mathfrak{g}$: the correspondence with the general definition is that we can make a first-order differentiation out of $X \leftrightarrow c(t)$ by using $f \to \frac{d}{dt} f(c(t))\big|_{t=0}$; it is clear from dimensional considerations that if we pass to local coordinates, we get the full tangent space this way.

A _vector field_ on G is an assignment to each p in G of a member of the tangent space at p; the result is a linear first-order differential operator on $C^\infty(G)$ annihilating constants. The vector field is said to be _smooth_ if it sends smooth functions into smooth functions; it is equivalent to say that it looks like $\Sigma\, a_i(x) \frac{\partial}{\partial x^i}$ in local coordinates with the $a_i(x)$ smooth. _Left-invariant vector fields_ are those that commute with left translations. They are automatically smooth, and they come from $\mathfrak{g}$ as follows: In the development of Chapter I, if $X \leftrightarrow c(t)$, then $\widetilde{X}f(g) = \frac{d}{dt} f(gc(t))\big|_{t=0}$ defines a left-invariant vector field. In the general theory, the formula $\widetilde{X}f(g) = Xf_g$, where f_g is the left translate of f by g, defines a left-invariant vector field. In either event the map $X \to \widetilde{X}$ gives us a vector space isomorphism of $\mathfrak{g}$ with the space of left-invariant vector fields on G.

Fix a basis $X_1, \ldots, X_N$ of $\mathfrak{g}$, regarded as left-invariant vector fields, and let $\omega^1, \ldots, \omega^N$ be the dual basis (point by point). An n form on G can then be written

$$\sum_{i_1,\ldots,i_n} f_{i_1\cdots i_n} \omega^{i_1} \wedge \cdots \wedge \omega^{i_n} . \tag{4.3}$$

The question is how to compute d. First it is pretty clear (and we could check it directly with local coordinates) that if f is in $C^\infty(G)$ then

$$df = \sum_{j=1}^{N} (X_j f)\, \omega^j . \tag{4.4}$$

To compute $d\omega^k$ (and thus to determine everything, in view of (ii)), we compute

$$0 = d^2 f = d\Big(\sum_j (X_j f)\omega^j\Big) = \sum_j d(X_j f) \wedge \omega^j + \sum_k (X_k f)\, d\omega^k \quad \text{by (ii)}$$

$$= \sum_{i,j} X_i X_j f\, \omega^i \wedge \omega^j + \sum_k (X_k f)\, d\omega^k$$

$$= \sum_{i<j} (X_i X_j - X_j X_i) f\, \omega^i \wedge \omega^j + \sum_k (X_k f)\, d\omega^k$$

$$= \sum_{i<j} [X_i, X_j] f\, \omega^i \wedge \omega^j + \sum_k (X_k f)\, d\omega^k .$$

Putting $[X_i, X_j] = \Sigma_k\, c^k_{ij} X_k$, we then have

$$0 = \sum_k X_k f \Big(\sum_{i<j} c^k_{ij}\, \omega^i \wedge \omega^j + d\omega^k \Big) .$$

Since the vector $\{X_k f\}_{k=1}^N$ can be arbitrary at a single point, $d\omega^k = -\Sigma_{i<j}\, c^k_{ij}\, \omega^i \wedge \omega^j$. Thus

$$d\omega^k = -\tfrac{1}{2} \sum_{i,j} c^k_{ij}\, \omega^i \wedge \omega^j . \tag{4.5}$$

The definition for d on n forms follows from (ii) and is

$$d(f\, \omega^{i_1} \wedge \ldots \wedge \omega^{i_n}) =$$

$$df \wedge \omega^{i_1} \wedge \ldots \wedge \omega^{i_n} + \sum_{j=1}^{n} (-1)^{j+1} f\, d\omega^{i_j} \wedge \omega^{i_1} \wedge \ldots\ \widehat{\omega}^{i_j} \wedge \ldots \wedge \omega^{i_n}, \tag{4.6}$$

where the symbol $\widehat{\omega}$ indicates a term to be omitted.

Substitution of (4.5) into (4.6) leads to the presence of the structural constants c^k_{ij} in the answer, and it is desirable to eliminate them. To do so, we use the identification $\Lambda^n(E^*) \cong \Lambda^n(E)^*$ of Proposition 2.11, with $E = \mathbb{R}^N$. This identification replaces n forms by alternating n-multilinear functions on $\mathfrak{g}$:

$$C^\infty(G) \otimes_{\mathbb{R}} \Lambda^n((\mathbb{R}^N)^*) \to C^\infty(G) \otimes_{\mathbb{R}} \Lambda^n(\mathbb{R}^N)^* \cong \operatorname{Hom}_{\mathbb{R}}(\Lambda^n(\mathbb{R}^N), C^\infty(G)) .$$

It comes from the formula

$$\omega^{i_1} \wedge \ldots \wedge \omega^{i_n}(Y_1 \wedge \ldots \wedge Y_n) = \det\{\omega^{i_s}(Y_t)\}^n_{s,t=1}. \tag{4.7}$$

We shall use (4.7) in two disguised forms. One is

$$\omega^u \wedge \omega^{i_1} \wedge \ldots \wedge \omega^{i_n}(Y_1 \wedge \ldots \wedge Y_{n+1})$$

$$= \sum_{j=1}^{n+1} (-1)^{j+1} \omega^u(Y_j)\, \omega^{i_1} \wedge \ldots \wedge \omega^{i_n}(Y_1 \wedge \ldots \wedge \widehat{Y}_j \wedge \ldots \wedge Y_{n+1}), \tag{4.8}$$

which follows by using (4.7) to write the left side of (4.8) as

$$\det \begin{pmatrix} \omega^u(Y_1) & \cdots & \omega^u(Y_{n+1}) \\ \omega^{i_1}(Y_1) & \cdots & \omega^{i_1}(Y_{n+1}) \\ & \vdots & \\ \omega^{i_n}(Y_1) & \cdots & \omega^{i_n}(Y_{n+1}) \end{pmatrix},$$

expanding in cofactors about the first row, and using (4.7) to identify each determinant. The other disguised form is

$$\omega^{i_1} \wedge \cdots \wedge \omega^{i_n}(Y \wedge Y_1 \wedge \cdots \wedge Y_{n-1})$$

$$= \sum_{k=1}^{n} (-1)^{k+1} \omega^{i_k}(Y)\, \omega^{i_1} \wedge \cdots \wedge \widehat{\omega}^{i_k} \wedge \cdots \wedge \omega^{i_n}(Y_1 \wedge \cdots \wedge Y_{n-1}), \tag{4.9}$$

which follows in the same way by a cofactor expansion about the first column.

Thus we calculate part of (4.6) as

$$d\omega^{i_j} \wedge \omega^{i_1} \wedge \cdots \wedge \widehat{\omega}^{i_j} \wedge \cdots \wedge \omega^{i_n}(X_{j_1} \wedge \cdots \wedge X_{j_{n+1}})$$

$$= -\tfrac{1}{2} \sum_{u,v} c_{uv}^{i_j}\, \omega^u \wedge \omega^v \wedge \omega^{i_1} \wedge \cdots \wedge \widehat{\omega}^{i_j} \wedge \cdots \wedge \omega^{i_n}(X_{j_1} \wedge \cdots \wedge X_{j_{n+1}})$$

$$= -\tfrac{1}{2} \sum_{u,v} c_{uv}^{i_j} \sum_{r=1}^{n+1} (-1)^{r+1} \omega^u(X_{j_r})$$

$$\times\, \omega^v \wedge \omega^{i_1} \wedge \cdots \wedge \widehat{\omega}^{i_j} \wedge \cdots \wedge \omega^{i_n}(X_{j_1} \wedge \cdots \wedge \widehat{X}_{j_r} \wedge \cdots \wedge X_{j_{n+1}})$$

by (4.8)

$$= -\tfrac{1}{2} \sum_{u,v} c_{uv}^{i_j} \sum_{r=1}^{n+1} (-1)^{r+1} \Big\{ \sum_{s<r} (-1)^{s+1} \omega^u(X_{j_r}) \omega^v(X_{j_s})$$

$$\times\, \omega^{i_1} \wedge \cdots \wedge \widehat{\omega}^{i_j} \wedge \cdots \wedge \omega^{i_n}(X_{j_1} \wedge \cdots \wedge \widehat{X}_{j_s} \wedge \cdots \wedge \widehat{X}_{j_r} \wedge \cdots \wedge X_{j_{n+1}})$$

$$+ \sum_{s>r} (-1)^s \omega^u(X_{j_r})\omega^v(X_{j_s})$$

$$\times \omega^{i_1} \wedge \ldots \wedge \hat{\omega}^{i_j} \wedge \ldots \wedge \omega^{i_n}(X_{j_1} \wedge \ldots \wedge \hat{X}_{j_r} \wedge \ldots \wedge \hat{X}_{j_s} \wedge \ldots \wedge X_{j_{n+1}})\}$$

by (4.8) again. (4.10)

Since $\omega^s(X_t) = 1$ if $s = t$ and $\omega^s(X_t) = 0$ otherwise,

$$\sum_{u,v} c^{i_j}_{uv}\omega^u(X_{j_r})\omega^v(X_{j_s}) = c^{i_j}_{j_r j_s} = \sum_k c^k_{j_r j_s}\omega^{i_j}(X_k) = \omega^{i_j}([X_{j_r}, X_{j_s}]) .$$

Thus (4.10) in an obvious shorthand is

$$= -\tfrac{1}{2} \sum_{s<r} (-1)^{r+s} \omega^{i_j}([X_{j_r}, X_{j_s}])\omega^{i_1} \wedge \ldots \wedge \hat{\omega}^{i_j} \wedge \ldots \wedge \omega^{i_n}(\hat{X}_{j_s}, \hat{X}_{j_r})$$

$$-\tfrac{1}{2} \sum_{r<s} (-1)^{r+s+1} \omega^{i_j}([X_{j_r}, X_{j_s}])\omega^{i_1} \wedge \ldots \wedge \hat{\omega}^{i_j} \wedge \ldots \wedge \omega^{i_n}(\hat{X}_{j_r}, \hat{X}_{j_s})$$

$$= \sum_{r<s} (-1)^{r+s} \omega^{i_j}([X_{j_r}, X_{j_s}])\omega^{i_1} \wedge \ldots \wedge \hat{\omega}^{i_j} \wedge \ldots \wedge \omega^{i_n}(\hat{X}_{j_r}, \hat{X}_{j_s}) .$$

Therefore the second term on the right side of (4.6) is

$$= \sum_{r<s} (-1)^{r+s} f \sum_{j=1}^{n} (-1)^{j+1} \omega^{i_j}([X_{j_r}, X_{j_s}])$$

$$\times \omega^{i_1} \wedge \ldots \wedge \hat{\omega}^{i_j} \wedge \ldots \wedge \omega^{i_n}(\hat{X}_{j_r}, \hat{X}_{j_s})$$

$$= \sum_{r<s} (-1)^{r+s} f\omega^{i_1} \wedge \ldots \wedge \omega^{i_n}([X_{j_r}, X_{j_s}]$$

$$\wedge X_{j_1} \wedge \ldots \wedge \hat{X}_{j_r} \wedge \ldots \wedge \hat{X}_{j_s} \wedge \ldots \wedge X_{j_n})$$

by (4.9). (4.11)

Meanwhile the first term on the right side of (4.6) is

$$= df \wedge \omega^{i_1} \wedge \ldots \wedge \omega^{i_n}(X_{j_1} \wedge \ldots \wedge X_{j_{n+1}})$$

$$= \sum_k (X_k f)\, \omega^k \wedge \omega^{i_1} \wedge \ldots \wedge \omega^{i_n}(X_{j_1} \wedge \ldots \wedge X_{j_{n+1}})$$

$$= \sum_{k,\ell} (X_k f)(-1)^{\ell+1} \omega^k (X_{j_\ell})\, \omega^{i_1} \wedge \ldots \wedge \omega^{i_n}(X_{j_1} \wedge \ldots \wedge \hat{X}_{j_\ell} \wedge \ldots \wedge X_{j_{n+1}})$$

by (4.8)

$$= \sum_{\ell=1}^{n+1} (-1)^{\ell+1} (X_{j_\ell} f)\omega^{i_1} \wedge \ldots \wedge \omega^{i_n}(X_{j_1} \wedge \ldots \wedge \hat{X}_{j_\ell} \wedge \ldots \wedge X_{j_{n+1}}) . \tag{4.12}$$

In this expression, $X_{j_\ell} f$ is a left-invariant derivative of f (differentiation on the right) and

$$\omega^{i_1} \wedge \ldots \wedge \omega^{i_n}(X_{j_1} \wedge \ldots \wedge \hat{X}_{j_\ell} \wedge \ldots \wedge X_{j_{n+1}})$$

is a constant function on G. Thus there is no change if we apply X_{j_ℓ} to the product

$$f\, \omega^{i_1} \wedge \ldots \wedge \omega^{i_n}(X_{j_1} \wedge \ldots \wedge \hat{X}_{j_\ell} \wedge \ldots \wedge X_{j_{n+1}})$$

instead of just to f. Making this adjustment to (4.12) and adding (4.11) to (4.12), we obtain an expression that extends by multilinearity to

$$d\omega(Y_1 \wedge \ldots \wedge Y_{n+1}) = \sum_{\ell=1}^{n+1} (-1)^{\ell+1} Y_\ell (\omega(Y_1 \wedge \ldots \wedge \hat{Y}_\ell \wedge \ldots \wedge Y_{n+1}))$$

$$+ \sum_{r<s} (-1)^{r+s} \omega([Y_r, Y_s]) \wedge Y_1 \wedge \ldots \wedge \hat{Y}_r \wedge \ldots \wedge \hat{Y}_s \wedge \ldots \wedge Y_{n+1}) . \tag{4.13}$$

Here ω is in $\mathrm{Hom}_{\mathbb{R}}(\Lambda^n\mathfrak{g}, C^\infty(G))$, and $d\omega$ is in $\mathrm{Hom}_{\mathbb{R}}(\Lambda^{n+1}\mathfrak{g}, C^\infty(G))$. In §3 we shall take this formula as a definition of a d operator in a more general context, at the same time changing the field from $\mathbb{R}$ to $\mathbb{C}$. For formal reasons we shall have $d^2 = 0$, and thus we can compute cohomology as ker d/image d, the quotient of the cocycles by the coboundaries.

2. Motivation from extensions

Let us briefly see how the cases $n = 1$ and $n = 2$ of (4.10) arise in considering Lie algebra extensions. Let $\mathfrak{g}$ and $\mathfrak{a}$ be finite-dimensional complex Lie algebras. A complex Lie algebra $\mathfrak{h}$ is an extension of $\mathfrak{g}$ by $\mathfrak{a}$ if there is an exact sequence of Lie algebras

$$0 \longrightarrow \mathfrak{a} \xrightarrow{\iota} \mathfrak{h} \xrightarrow{\rho} \mathfrak{g} \longrightarrow 0 . \qquad (4.14)$$

(Recall that this terminology means that the maps are Lie algebra homomorphisms and that for each Lie algebra the kernel of the map going out equals the image of the map coming in. In particular, the image of $\mathfrak{a}$ must be an ideal in $\mathfrak{h}$ in order to be the kernel of ρ.)

We assume that $\mathfrak{a}$ is abelian (all brackets 0). In this case, (4.14) makes $\mathfrak{a}$ into a representation space for $\mathfrak{g}$ under the definition

$$\pi(X)Y = \iota^{-1}[\rho^{-1}(X), \iota(Y)] \quad \text{for } X \in \mathfrak{g}, Y \in \mathfrak{a} ; \qquad (4.15)$$

here $\rho^{-1}(X)$ denotes any inverse image of X in $\mathfrak{h}$, and

ι^{-1} is defined on $[\rho^{-1}(X), \iota(Y)]$ since $\iota(\mathfrak{a})$ is an ideal in $\mathfrak{h}$. Fix the extension (4.14) and fix also a linear map $\tau : \mathfrak{g} \to \mathfrak{h}$ such that $\rho\tau = 1$. Then we can attach to this situation a map $\omega \in \mathrm{Hom}_{\mathbb{C}}(\Lambda^2\mathfrak{g}, \mathfrak{a})$ by the definition

$$\omega(X,Y) = \iota^{-1}([\tau(X), \tau(Y)] - \tau[X,Y]) . \tag{4.16}$$

(To see that the expression in parentheses is in $\iota\mathfrak{a}$, we apply the Lie algebra homomorphism ρ to it and check that we get 0; exactness of (4.14) then says that the expression is in $\mathfrak{a}$.)

Using (4.13), let us compute $d\omega$. First note that $\tau(X)$ for X in $\mathfrak{g}$ is a valid choice of $\rho^{-1}(X)$ in (4.15). Thus (4.16) gives

$$\begin{aligned}
&d\omega(Y_1, Y_2, Y_3) \\
&\quad = \pi(Y_1)(\omega(Y_2,Y_3)) - \pi(Y_2)(\omega(Y_1,Y_3)) + \pi(Y_3)(\omega(Y_1,Y_2)) \\
&\qquad - \omega([Y_1,Y_2],Y_3) + \omega([Y_1,Y_3],Y_2) - \omega([Y_2,Y_3],Y_1) \\
&\quad = \iota^{-1}[\tau(Y_1),[\tau(Y_2),\tau(Y_3)]] - \iota^{-1}[\tau(Y_2),[\tau(Y_1),\tau(Y_3)]] \\
&\qquad + \iota^{-1}[\tau(Y_3),[\tau(Y_1),\tau(Y_2)]] - \iota^{-1}[\tau(Y_1),\tau[Y_2,Y_3]] \\
&\qquad + \iota^{-1}[\tau(Y_2),\tau[Y_1,Y_3]] - \iota^{-1}[\tau(Y_3),\tau[Y_1,Y_2]] \\
&\qquad - \iota^{-1}[\tau[Y_1,Y_2],\tau(Y_3)] + \iota^{-1}[\tau[Y_1,Y_3],\tau(Y_2)] \\
&\qquad - \iota^{-1}[\tau[Y_2,Y_3],\tau(Y_1)] + \iota^{-1}\tau[[Y_1,Y_2],Y_3] \\
&\qquad - \iota^{-1}\tau[[Y_1,Y_3],Y_2] + \iota^{-1}\tau[[Y_2,Y_3],Y_1] .
\end{aligned}$$

On the right side, the first three terms add to 0 by the Jacobi identity in $\mathfrak{h}$, the next six terms cancel in pairs, and the last three terms add to 0 by the Jacobi identity in $\mathfrak{g}$. Therefore $d\omega = 0$, i.e., ω is a cocycle.

Suppose we change τ to τ'. Then $\alpha = \tau' - \tau$ has $\rho_\alpha = 0$, so that α maps $\mathfrak{g}$ into $\iota\mathfrak{a}$ and $\iota^{-1}\alpha$ is a member of $\mathrm{Hom}_{\mathbb{C}}(\mathfrak{g},\mathfrak{a}) = \mathrm{Hom}_{\mathbb{C}}(\Lambda^1\mathfrak{g},\mathfrak{a})$. Using (4.13), we compute $d(\iota^{-1}\alpha)$ as

$$d(\iota^{-1}\alpha)(X,Y) = \pi(X)(\iota^{-1}\alpha(Y)) - \pi(Y)(\iota^{-1}\alpha(X)) - \iota^{-1}\alpha[X,Y]. \quad (4.17)$$

On the other hand, the respective ω's for τ and τ' have

$$\begin{aligned}\omega_{\tau'}(X,Y) - \omega_\tau(X,Y) &\\ &= \iota^{-1}([\tau'(X),\tau'(Y)] - \tau'[X,Y]) - \iota^{-1}([\tau(X),\tau(Y)] - \tau[X,Y]) \\ &= \iota^{-1}([\alpha(X),\tau(Y)] + [\tau(X),\alpha(Y)] + 0 - \alpha[X,Y]) \\ &= -\pi(Y)(\iota^{-1}\alpha(X)) + \pi(X)(\iota^{-1}\alpha(Y)) - \iota^{-1}\alpha[X,Y] \qquad \text{by (4.15)} \\ &= d(\iota^{-1}\alpha)(X,Y) \qquad \text{by (4.17).}\end{aligned}$$

Thus $\omega_{\tau'}$ and ω_τ differ by a coboundary.

If we take into account the expected result $d^2 = 0$, then (4.14) leads to a well defined element of the quotient of $\ker d|_{\mathrm{Hom}(\Lambda^2\mathfrak{g},\mathfrak{a})}$ by $\mathrm{image}\ d|_{\mathrm{Hom}(\Lambda^1\mathfrak{g},\mathfrak{a})}$, and we shall soon call this quotient $H^2(\mathfrak{g},\mathfrak{a})$.

We can go backwards as well. Let us say that two extensions are <u>equivalent</u> if there exists a Lie algebra isomorphism σ such that

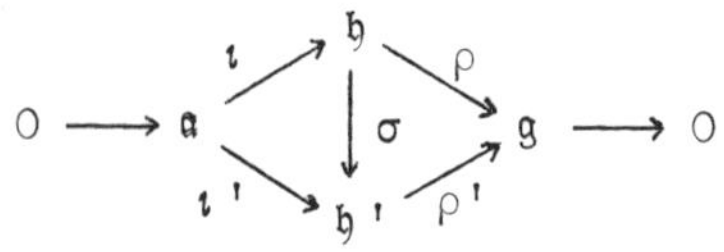

commutes. It is a simple matter to see that equivalent extensions lead to the same member of $H^2(\mathfrak{g},\mathfrak{a})$.

Equivalent extensions necessarily lead to the same π in (4.15). If we fix π, then the claim is that the map of equivalence classes of extensions for that π to $H^2(\mathfrak{g},\mathfrak{a})$ is one-one onto. In fact, if we are given a cocycle ω in $\ker d|_{\mathrm{Hom}(\Lambda^2\mathfrak{g},\mathfrak{a})}$, we introduce $\mathfrak{h}$ as a vector space. We define brackets within the subspace $\mathfrak{a}$ to be 0 and brackets between $\mathfrak{g}$ and $\mathfrak{a}$ by π. To define brackets within the subspace $\mathfrak{g}$, let τ be the injection of $\mathfrak{g}$ into $\mathfrak{h}$, and set

$$[\tau(X),\tau(Y)] = \tau[X,Y] + \iota(\omega(X,Y)),$$

where ι is the injection of $\mathfrak{a}$ into $\mathfrak{h}$. This construction allows us to form a two-sided inverse to our mapping. Hence the equivalence classes of extensions of $\mathfrak{g}$ by $\mathfrak{a}$ giving the same π are classified by $H^2(\mathfrak{g},\mathfrak{a})$.

In this context, $H^1(\mathfrak{g},\mathfrak{a})$ has an interesting interpretation, too. The starting point is the 0 element of $H^2(\mathfrak{g},\mathfrak{a})$. A representative cocycle is $\omega = 0$, and our formulas then say that the injection τ of $\mathfrak{g}$ into $\mathfrak{h}$ is a Lie algebra homomorphism. Since τ is a Lie algebra homomorphism, let us write $\mathfrak{h} = \mathfrak{g} \oplus \mathfrak{a}$ as a vector space and require that $\mathfrak{g}$ be a Lie subalgebra, $\mathfrak{a}$ be an ideal, and $\mathfrak{g}$ bracket with $\mathfrak{a}$ through π as in (4.15): $[X,Y] = \pi(X)Y$ for

$X \in \mathfrak{g}$ and $Y \in \mathfrak{a}$. We write $\mathfrak{h} = \mathfrak{g} \oplus_\pi \mathfrak{a}$ and call $\mathfrak{h}$ the <u>semidirect product</u> of $\mathfrak{g}$ and $\mathfrak{a}$ with respect to π.

Let us fix our attention on this semidirect product. Our map τ has become the inclusion $\tau : \mathfrak{g} \to \mathfrak{h}$ into the first coordinate. Let us ask about other inclusions of $\mathfrak{g}$ into $\mathfrak{h}$ that project back to the identity on $\mathfrak{g}$ and respect the bracket structure of $\mathfrak{g}$ and $\mathfrak{h}$. Such a linear map τ' must be of the form

$$\tau'(Y) = (Y, \omega(Y)), \qquad Y \in \mathfrak{g},$$

where ω is a linear function from $\mathfrak{g}$ to $\mathfrak{a}$, i.e., a member of $\mathrm{Hom}_{\mathbb{C}}(\Lambda^1 \mathfrak{g}, \mathfrak{a})$. If τ' respects brackets, then

$$\tau'[Y_1, Y_2] = ([Y_1, Y_2], \omega[Y_1, Y_2])$$

is equal to

$$\begin{aligned}[\tau'(Y_1), \tau'(Y_2)] &= [(Y_1, \omega(Y_1)), (Y_2, \omega(Y_2))] \\ &= ([Y_1, Y_2], \pi(Y_1)\omega(Y_2) - \pi(Y_2)\omega(Y_1)).\end{aligned}$$

That is, ω must satisfy

$$\omega[Y_1, Y_2] = \pi(Y_1)\omega(Y_2) - \pi(Y_2)\omega(Y_1).$$

This condition is summarized by saying that ω is a <u>derivation</u> of $\mathfrak{g}$ into $\mathfrak{a}$. Referring to (4.13), we see that

$$d\omega(Y_1, Y_2) = \pi(Y_1)(\omega(Y_2)) - \pi(Y_2)(\omega(Y_1)) - \omega[Y_1, Y_2].$$

Hence τ' is a <u>Lie algebra splitting</u> (i.e., is as above and respects brackets) if and only if the corresponding ω is a cocycle.

There is a trivial kind of derivation, namely $\omega(Y) = \pi(Y)X$ for some X in $\mathfrak{a}$. Such derivations are called <u>inner</u> and are easily seen to be the coboundaries. Thus $H^1(\mathfrak{g},\mathfrak{a})$ classifies the Lie algebra splittings modulo those that are given in terms of inner derivations.

The identifications of H^1 and H^2 above have consequences for the structure of Lie algebras. If $\mathfrak{g}$ has a property called "semisimplicity," it is not too hard to see that $H^2(\mathfrak{g},\mathfrak{a}) = 0$. With a little supplementary argument, it follows that any finite-dimensional complex Lie algebra is a semidirect product of a semisimple Lie algebra and a solvable Lie algebra (the <u>Levi decomposition</u>). The same theorem and proof apply over $\mathbb{R}$ rather than $\mathbb{C}$, and a structure theorem for Lie groups drops out. Results for H^1 then furnish a uniqueness theorem.

3. Definition and examples

Let $\mathfrak{g}$ be a finite-dimensional complex Lie algebra, and let V be a $U(\mathfrak{g})$ module, possibly infinite-dimensional. (Recall that equivalently V is the representation space for a representation of $\mathfrak{g}$.) In §1, V was $C^\infty(G)$ (except that we were working with $\mathbb{R}$ as a field), and in §2, V was the abelian Lie algebra $\mathfrak{a}$. The vector space $C^n(\mathfrak{g},V)$ of <u>n cochains</u> is defined by

$$C^n(\mathfrak{g},V) = \operatorname{Hom}_{\mathbb{C}}(\Lambda^n\mathfrak{g},V),$$

and the <u>coboundary operator</u> $d_n : C^n(\mathfrak{g},V) \to C^{n+1}(\mathfrak{g},V)$ is defined by

$$(d_n\omega)(Y_1 \wedge \cdots \wedge Y_{n+1})$$
$$= \sum_{\ell=1}^{n+1} (-1)^{\ell+1}\, Y_\ell\,(\omega(Y_1 \wedge \cdots \wedge \widehat{Y}_\ell \wedge \cdots \wedge Y_{n+1}))$$
$$+ \sum_{r<s} (-1)^{r+s}\, \omega([Y_r, Y_s] \wedge Y_1 \wedge \cdots \wedge \widehat{Y}_r \wedge \cdots \wedge \widehat{Y}_s \wedge \cdots \wedge Y_{n+1}), \tag{4.18}$$

just as in the special case (4.13).

<u>Proposition 4.1.</u> The coboundary operator satisfies $d_n d_{n-1} = 0$.

This will follow from Lemma 4.5 below. We define a space $Z^n(\mathfrak{g},V)$ of <u>cocycles</u> and a space $B^n(\mathfrak{g},V)$ of <u>coboundaries</u> by

$$Z^n(\mathfrak{g},V) = \ker d_n$$
$$B^n(\mathfrak{g},V) = \text{image } d_{n-1}.$$

Proposition 4.1 says that $B^n \subseteq Z^n$, and thus it makes sense to define a <u>cohomology space</u> $H^n(\mathfrak{g},V)$ by

$$H^n(\mathfrak{g},V) = Z^n(\mathfrak{g},V)/B^n(\mathfrak{g},V).$$

This is the nth <u>cohomology of $\mathfrak{g}$ with coefficients in V</u>. We regard $H^n(\mathfrak{g},V)$ as a vector space, not as a $U(\mathfrak{g})$ module.

<u>Examples.</u>

(1) $H^0(\mathfrak{g},V) = Z^0(\mathfrak{g},V)$ for any V, since $B^0(\mathfrak{g},V) = 0$ by convention. We can identify $C^0(\mathfrak{g},V)$ with V since $\mathrm{Hom}_{\mathbb{C}}(\mathbb{C},V) \cong V$ as vector spaces by (2.7). Under this

identification, (4.18) shows that $(d_0 v)(Y) = Yv$ for v in V and Y in $\mathfrak{g}$. Then v is in $Z^0(\mathfrak{g}, V)$ exactly when $Yv = 0$ for all Y in $\mathfrak{g}$. That is,

$$H^0(\mathfrak{g}, V) \cong V^{\mathfrak{g}},$$

with $V^{\mathfrak{g}}$ denoting the subspace of <u>invariants</u> in V.

(2) We investigate $H^1(\mathfrak{g}, V)$. The space of cochains is $\operatorname{Hom}_{\mathbb{C}}(\mathfrak{g}, V)$, the linear maps from $\mathfrak{g}$ to V. Let f be such a map. Then (4.18) gives

$$(d_1 f)(X \wedge Y) = X(f(Y)) - Y(f(X)) - f[X,Y].$$

The cocycle condition is that

$$f[X,Y] = X(f(Y)) - Y(f(X)) \quad \text{for all } X, Y \in \mathfrak{g},$$

while the coboundary condition is that

$$f(X) = Xv_0 \quad \text{for some } v_0 \in V.$$

A special case of interest is $V = \mathbb{C}$ with trivial (zero) action by $\mathfrak{g}$. In this case the coboundary condition is that $f = 0$, while the cocycle condition is that $f[X,Y] = 0$ for all X and Y in $\mathfrak{g}$; thus a cocycle f is a linear functional annihilating $[\mathfrak{g}, \mathfrak{g}]$, and we have

$$H^1(\mathfrak{g}, \mathbb{C}) \cong (\mathfrak{g}/[\mathfrak{g}, \mathfrak{g}])^*.$$

Despite the explicit nature of these examples, it is usually the case that we shall compute cohomology without identifying the cocycles and coboundaries separately. A large amount of the beginning theory will be aimed toward making such direct calculations of cohomology.

The first tool for us is the Koszul resolution for $\mathfrak{g}$, whose name anticipates properties to be listed in Theorem 4.4. Let

$$X_n = U(\mathfrak{g}) \otimes_{\mathbb{C}} \Lambda^n \mathfrak{g} , \tag{4.19a}$$

and regard X_n as a $U(\mathfrak{g})$ module by having $\mathfrak{g}$ act by left translation on $U(\mathfrak{g})$ and by 0 on $\Lambda^n \mathfrak{g}$. The <u>Koszul resolution</u> is the sequence of $U(\mathfrak{g})$ modules and maps

$$0 \longleftarrow \mathbb{C} \xleftarrow{\varepsilon} X_0 \xleftarrow{\partial_0} X_1 \xleftarrow{\partial_1} X_2 \xleftarrow{\partial_2} \cdots \tag{4.19b}$$

with ε and ∂_n given by

$$\varepsilon(u) = \text{constant term of } u , \quad u \in U(\mathfrak{g}) \otimes_{\mathbb{C}} \mathbb{C} \cong U(\mathfrak{g})$$

$$\begin{aligned} &\partial_{n-1}(u \otimes X_1 \wedge \cdots \wedge X_n) \\ &\quad = \sum_{i=1}^{n} (-1)^{i+1} (uX_i \otimes X_1 \wedge \cdots \wedge \widehat{X}_i \wedge \cdots \wedge X_n) \\ &\qquad + \sum_{k<\ell} (-1)^{k+\ell} (u \otimes [X_k, X_\ell] \wedge X_1 \wedge \cdots \wedge \widehat{X}_k \wedge \cdots \wedge \widehat{X}_\ell \wedge \cdots \wedge X_n) . \end{aligned} \tag{4.19c}$$

It is clear from (4.19c) that ε and ∂_n are indeed $U(\mathfrak{g})$ maps.

The idea is that the sequence

$$0 \longrightarrow C^0(\mathfrak{g}, V) \xrightarrow{d_0} C^1(\mathfrak{g}, V) \xrightarrow{d_1} C^2(\mathfrak{g}, V) \longrightarrow \cdots \tag{4.20}$$

(from which we compute cohomology $H^*(\mathfrak{g}, V)$) can be obtained by applying $\text{Hom}_{U(\mathfrak{g})}(\cdot, V)$ to (4.19b), modules and maps. We show this below. The significance of this fact is that the Koszul resolution is independent of V. Thus the cohomology of $\mathfrak{g}$ with coefficients in V is really a property of $\text{Hom}_{U(\mathfrak{g})}(\cdot, V)$.

Let us see that (4.20) results from (4.19b) by application of $\text{Hom}_{U(\mathfrak{g})}(\cdot, V)$. To handle the modules, we shall use Proposition 4.2 and Corollary 4.3 below.

Proposition 4.2. Let A and B be left $U(\mathfrak{g})$ modules, and regard $\text{Hom}_{\mathbb{C}}(A, B)$ as a left $U(\mathfrak{g})$ module as in §II.2. Then the subspace of $\mathfrak{g}$ invariants of $\text{Hom}_{\mathbb{C}}(A, B)$ coincides with the subspace of $U(\mathfrak{g})$ linear maps:

$$(\text{Hom}_{\mathbb{C}}(A, B))^{\mathfrak{g}} = \text{Hom}_{U(\mathfrak{g})}(A, B) .$$

Remarks. In §II.2 we made $\text{Hom}_{\mathbb{C}}(A, B)$ into a representation space for $\mathfrak{g}$ by $(X\varphi)(a) = X(\varphi(a)) - \varphi(Xa)$; then $\text{Hom}_{\mathbb{C}}(A, B)$ becomes a left $U(\mathfrak{g})$ module. Its invariants are the φ's with $X\varphi = 0$ for all X in $\mathfrak{g}$.

Proof. From the formula $(X\varphi)(a) = X(\varphi(a)) - \varphi(Xa)$, we see that $X\varphi = 0$ for all X in $\mathfrak{g}$ if and only if $X(\varphi(a)) = \varphi(Xa)$ for all X in $\mathfrak{g}$ and a in A, i.e., if and only if φ is $\mathfrak{g}$ linear. In this case, φ is $U(\mathfrak{g})$ linear.

Corollary 4.3. If A, B, and C are left $U(\mathfrak{g})$ modules, then

$$\text{Hom}_{U(\mathfrak{g})}(A \otimes_{\mathbb{C}} B, C) \cong \text{Hom}_{U(\mathfrak{g})}(A, \text{Hom}_{\mathbb{C}}(B, C))$$

under the map Φ from left to right with

$$\Phi(\varphi)(a)(b) = \varphi(a \otimes b) .$$

Proof. We apply Proposition 2.3 to obtain this identity with $\mathrm{Hom}_{\mathbb{C}}$ in place of $\mathrm{Hom}_{U(\mathfrak{g})}$, the isomorphism being as $U(\mathfrak{g})$ modules. Then the subspaces of $\mathfrak{g}$ invariants must correspond on the two sides, and the corollary results from applying Proposition 4.2.

Returning to the operation of $\mathrm{Hom}_{U(\mathfrak{g})}(\cdot, V)$ on (4.19b), we have

$$C^n(\mathfrak{g}, V) = \mathrm{Hom}_{\mathbb{C}}(\Lambda^n \mathfrak{g}, V) \cong \mathrm{Hom}_{U(\mathfrak{g})}(U(\mathfrak{g}), \mathrm{Hom}_{\mathbb{C}}(\Lambda^n \mathfrak{g}, V))$$

$$\cong \mathrm{Hom}_{U(\mathfrak{g})}(U(\mathfrak{g}) \otimes_{\mathbb{C}} \Lambda^n \mathfrak{g}, V) = \mathrm{Hom}_{U(\mathfrak{g})}(X_n, V) , \qquad (4.21)$$

by application of (2.43) and Corollary 4.3. (In applying the corollary, we should regard $\mathrm{Hom}_{\mathbb{C}}(\Lambda^n \mathfrak{g}, V)$ as a $U(\mathfrak{g})$ module with action obtained from the given representation on V and the <u>trivial</u> representation on $\Lambda^n \mathfrak{g}$.) Thus $\mathrm{Hom}_{U(\mathfrak{g})}(\cdot, V)$ has the desired effect on modules. The operation $\mathrm{Hom}_{U(\mathfrak{g})}(\cdot, V)$ is to act on maps by sending ∂ to $\mathrm{Hom}(\partial, 1)$; the next proposition says that the maps of (4.19b) go to those of (4.20).

<u>Proposition 4.4.</u> For ω in $\mathrm{Hom}_{U(\mathfrak{g})}(X_n, V)$, $d_n \omega = \omega \partial_n$. That is, $d_n = \mathrm{Hom}(\partial_n, 1)$. Here the domain of d_n is identified with $\mathrm{Hom}_{U(\mathfrak{g})}(X_n, V)$ by means of (4.21).

Proof. Formulas (4.18) and (4.19c) give

$$d_n\omega(1\otimes Y_1\wedge\ldots\wedge Y_{n+1})$$

$$= \sum_{\ell=1}^{n+1}(-1)^{\ell+1}\, Y_\ell(\omega(1\otimes Y_1\wedge\ldots\wedge\hat{Y}_\ell\wedge\ldots\wedge Y_{n+1}))$$

$$+ \sum_{r<s}(-1)^{r+s}\omega(1\otimes[Y_r,Y_s]\wedge Y_1\wedge\ldots\wedge\hat{Y}_r\wedge\ldots\wedge\hat{Y}_s\wedge\ldots\wedge Y_{n+1})$$

$$= \sum_{\ell=1}^{n+1}(-1)^{\ell+1}\,\omega(Y_\ell\otimes Y_1\wedge\ldots\wedge\hat{Y}_\ell\wedge\ldots\wedge Y_{n+1}) + \sum_{r<s}(\text{same})$$

since ω is $U(\mathfrak{g})$ linear

$$= \omega\partial_n(1\otimes Y_1\wedge\ldots\wedge Y_{n+1})\,.$$

Since $d_n\omega$, ω, and ∂_n all commute with $U(\mathfrak{g})$, the proposition follows.

<u>Lemma 4.5</u>. $\partial_{n-2}\partial_{n-1} = 0$ for all $n \geq 2$, and also $\epsilon\partial_0 = 0$.

Remarks. Proposition 4.1 follows from this, since Proposition 4.4 gives

$$d_n d_{n-1}\omega = (d_{n-1}\omega)\partial_n = \omega\partial_{n-1}\partial_n\,.$$

Proof. In (4.19c), the second term is absent for $n=1$, and thus $\epsilon\partial_0 = 0$. For $\partial_{n-2}\partial_{n-1}$, it is enough to consider

$$\partial_{n-2}\partial_{n-1}(1\otimes X_1\wedge\ldots\wedge X_n)$$

$$= \sum_{i=1}^{n}(-1)^{i+1}\,\partial_{n-2}(X_i\otimes X_1\wedge\ldots\wedge\hat{X}_i\wedge\ldots\wedge X_n)$$

$$+ \sum_{k<\ell}(-1)^{k+\ell}\partial_{n-2}(1\otimes[X_k,X_\ell]\wedge X_1\wedge\ldots\wedge\hat{X}_k\wedge\ldots\wedge\hat{X}_\ell\wedge\ldots\wedge X_n)$$

$$= \mathrm{I} + \mathrm{II}\,,\ \text{say.}$$

Here

$$I = \sum_{i=1}^{n} \sum_{j < i} (-1)^{i+1+j+1} X_i X_j \otimes X_1 \wedge (\text{omit } j\,,\, i) \wedge X_n$$

$$+ \sum_{i=1}^{n} \sum_{j > i} (-1)^{i+1+j} X_i X_j \otimes X_1 \wedge (\text{omit } i\,,\, j) \wedge X_n$$

$$+ \sum_{i} \sum_{\substack{r < s \\ \neq i}} (-1)^{r+s+\delta+i+1} X_i \otimes [X_r, X_s] \wedge X_1 \wedge (\text{omit } r\,,\, s\,,\, i) \wedge X_n$$

where $\delta = 1$ if $r < i < s$ and $\delta = 0$ otherwise

$$= \sum_{j < i} (-1)^{i+j} [X_i, X_j] \otimes X_1 \wedge (\text{omit } j\,,\, i) \wedge X_n$$

$$+ \sum_{i} \sum_{\substack{r < s \\ \neq i}} (-1)^{r+s+\delta+i+1} X_i \otimes [X_r, X_s] \wedge X_1 \wedge (\text{omit } r\,,\, s\,,\, i) \wedge X_n$$

$$= I\,A + I\,B,$$

and

$$II = \sum_{k < \ell} (-1)^{k+\ell} [X_k, X_\ell] \otimes X_1 \wedge (\text{omit } k\,,\, \ell) \wedge X_n$$

$$+ \sum_{k < \ell} \sum_{i \neq k, \ell} (-1)^{k+\ell+i+\delta} X_i \otimes [X_k, X_\ell] \wedge X_1 \wedge (\text{omit } k\,,\, \ell\,,\, i) \wedge X_n$$

where $\delta = 1$ if $r < i < s$

$$+ \sum_{k < \ell} (-1)^{k+\ell} \sum_{\substack{r < s \\ \neq k, \ell}} (-1)^{r+s+\sigma}\, 1 \otimes [X_r, X_s] \wedge [X_k, X_\ell]$$

$$\wedge X_1 \wedge (\text{omit } k\,,\, \ell\,,\, r\,,\, s) \wedge X_n$$

where $\sigma = 1$ if exactly one of

k and ℓ is between r and s

$$+ \sum_{k<\ell} (-1)^{k+\ell} \sum_{j\neq k,\ell} (-1)^{j+\tau}\ 1 \otimes [[X_k, X_\ell], X_j]$$

$$\wedge X_1 \wedge (\text{omit } j, k, \ell) \wedge X_n$$

where $\tau = 1$ if $k < j < \ell$

$$= \text{II A} + \text{II B} + \text{II C} + \text{II D}.$$

Now I A cancels with II A, and I B cancels with II B. Thus we are left with II C and II D. For II C, the (k,ℓ) and (r,s) pairs contribute together to two terms, and σ is the same for both; thus the terms of II C cancel in pairs.

In II D, fix three indices $a < b < c$. The corresponding terms come from

$k = a,\ \ell = b,\ j = c$ with term $(-1)^{a+b+c}[[X_a, X_b], X_c]$

$k = a,\ \ell = c,\ j = b$ with term $(-1)^{a+b+c+1}[[X_a, X_c], X_b]$

$k = b,\ \ell = c,\ j = a$ with term $(-1)^{a+b+c}[[X_b, X_c], X_a]$,

and these terms add to 0 by the Jacobi identity. Thus II D is 0, and the proof is complete.

Theorem 4.6 (Koszul). The sequence (4.19b) is exact, and each X_n is a free $U(\mathfrak{g})$ module.

Recall that exact means that the kernel of each map going out is equal to the image of the corresponding map coming in. Lemma 4.3 already shows that the images are contained in the kernels, and we are to prove the reverse inclusion. This step is complicated and will be carried out in §§5-6. It is related to Poincaré's Lemma for differential forms (closed implies exact in a star-shaped region), but the operators ∂ go down in dimension; Theorem 4.6 may perhaps be regarded as a pre-dual result to Poincaré's Lemma.

A free $U(\mathfrak{g})$ module is just a direct sum of a finite or infinite number of copies of $U(\mathfrak{g})$. The modules X_n are free; we simply choose a basis $\{v_j\}$ of $\Lambda^n\mathfrak{g}$, and then

$$U(\mathfrak{g}) \otimes_{\mathbb{C}} \Lambda^n\mathfrak{g} = \sum_j U(\mathfrak{g}) \otimes_{\mathbb{C}} \mathbb{C}v_j$$

exhibits X_n as free, since the action of $\mathfrak{g}$ on $\Lambda^n\mathfrak{g}$ is taken to be trivial.

4. Computation from any free resolution

The fact that Lie algebra cohomology can be handled without cocycles and coboundaries depends on having other expressions for the cohomology than (4.18). We have already seen that (4.18) can be obtained by an operation on the Koszul resolution. Theorem 4.6 will allow us to obtain the cohomology by the same operation on other sequences besides the Koszul resolution. (The historical motivation for such a change of sequence comes from a theorem of Hopf about group operations on manifolds. See the section "Notes.") We shall obtain a result in this section that allows us to change the sequence in certain ways, and we shall give a simple example of its applicability. In the next chapter, we shall abstract and generalize these matters extensively, ultimately regarding cohomology as "derived functors" of the "invariants functor"

$V \to V^{\mathfrak{g}}$. And we shall see that the same considerations apply to a number of other situations of interest.

For this section, fix an associative algebra R over $\mathbb{C}$; we should regard R as $U(\mathfrak{g})$. A complex of left R modules is a finite or infinite sequence of left R modules Y_n and R-linear maps φ_n of the form

$$\cdots \longleftarrow Y_{n-1} \xleftarrow{\varphi_{n-1}} Y_n \xleftarrow{\varphi_n} Y_{n+1} \longleftarrow \cdots$$

with $\varphi_{n-1}\varphi_n = 0$ for all n. This condition means that image $\varphi_n \subseteq \ker \varphi_{n-1}$ for all n. The complex is exact at Y_n if image $\varphi_n = \ker \varphi_{n-1}$. It is a resolution if it is eaxct at Y_n for all n. A free left R module is the finite or infinite direct sum of copies of the left R module R. A free resolution of M is a resolution

$$0 \longleftarrow M \longleftarrow Y_0 \longleftarrow Y_1 \longleftarrow Y_2 \longleftarrow \cdots$$

in which all Y_n are free. Theorem 4.6 says for $R = U(\mathfrak{g})$ that the Koszul resolution is an example of a free resolution. (Lemma 4.5 had earlier said that it was an example of a complex.)

Proposition 4.7. Suppose that

$$\begin{array}{ccccccccccc} 0 & \longleftarrow & M & \xleftarrow{\varepsilon} & X_0 & \xleftarrow{\partial_0} & X_1 & \xleftarrow{\partial_1} & X_2 & \longleftarrow & \cdots \\ & & \downarrow f & & & & & & & & \\ 0 & \longleftarrow & M' & \xleftarrow{\varepsilon'} & X'_0 & \xleftarrow{\partial'_0} & X'_1 & \xleftarrow{\partial'_1} & X'_2 & \longleftarrow & \cdots \end{array}$$

is a finite or infinite diagram of left R modules and

R-linear maps such that the top row is a complex with all X_n free over R and such that the bottom row is exact. Then there exist R-linear maps $F_n : X_n \to X_n'$ such that

$$\begin{array}{ccccccccccc} 0 & \longleftarrow & M & \xleftarrow{\varepsilon} & X_0 & \xleftarrow{\partial_0} & X_1 & \xleftarrow{\partial_1} & X_2 & \longleftarrow & \cdots \\ & & \downarrow{\scriptstyle f} & & \downarrow{\scriptstyle F_0} & & \downarrow{\scriptstyle F_1} & & \downarrow{\scriptstyle F_2} & & \\ 0 & \longleftarrow & M' & \xleftarrow{\varepsilon'} & X_0' & \xleftarrow{\partial_0'} & X_1' & \xleftarrow{\partial_1'} & X_2' & \longleftarrow & \cdots \end{array} \qquad (4.22)$$

commutes.

Lemma 4.8. In a diagram of left R modules and R-linear maps

$$\begin{array}{ccccc} & & F & & \\ & & \downarrow{\scriptstyle \tau} & \searrow & \\ A' & \xleftarrow{\varphi} & A & \xleftarrow{\psi} & A'' \end{array}$$

with $\ker \varphi = \text{image } \psi$, $\varphi\tau = 0$, and F free, there exists an R-linear map $\sigma : F \to A''$ such that $\tau = \psi\sigma$.

Proof. Let $\{x_i\}$ be free generators of F. Then $\varphi(\tau(x_i)) = 0$, $\tau(x_i)$ is in $\ker \varphi = \text{image } \psi$, and $\tau(x_i) = \psi(a_i)$. Define $\sigma(x_i) = a_i$, and extend to all of F by R linearity.

Proof of Proposition 4.7. We proceed inductively, starting with

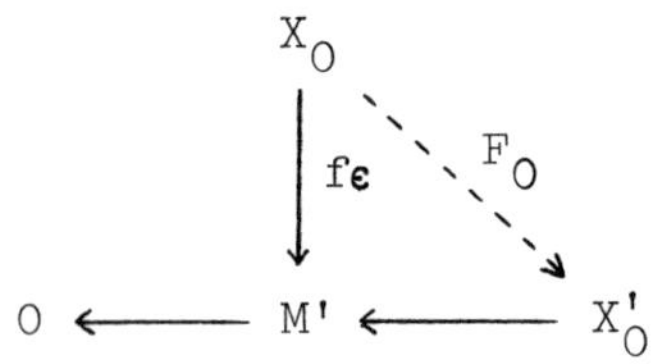

and applying the lemma to get $F_0 : X_0 \to X_0'$ with $\varepsilon' F_0 = f\varepsilon$. For the inductive step, put $X_{-1} = M$, $X_{-1}' = M'$, $\partial_{-1} = \varepsilon$, and $\partial_{-1}' = \varepsilon'$. Suppose inductively that $n \geq 0$ and that F_n has been constructed with $\partial_{n-1}' F_n = F_{n-1}\partial_{n-1}$. In the diagram

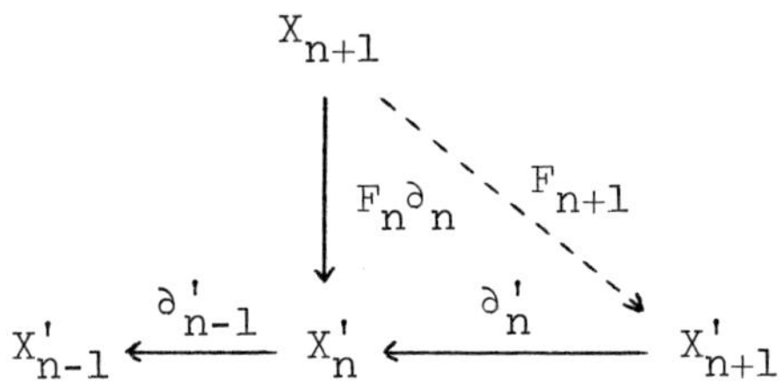

we have $\partial_{n-1}'(F_n\partial_n) = F_{n-1}\partial_{n-1}\partial_n = 0$, and the lemma gives us $F_{n+1} : X_{n+1} \to X_{n+1}'$ with $\partial_n' F_{n+1} = F_n\partial_n$. This completes the induction and the proof.

In the diagram (4.22), if $X = \{X_n\}$ and $X' = \{X_n'\}$ are complexes and if the squares commute, we say that $F = \{F_n\}$ (with $F : X \to X'$) is a chain map over f. To any such complex X and left R module V, we can associate cochains, cocycles, coboundaries, and cohomology with coefficients in V by the definitions

$$C^n(X,V) = \mathrm{Hom}_{\mathfrak{g}}(X_n, V)$$

$$d_n : C^n(X,V) \to C^{n+1}(X,V) \quad \text{by} \quad d_n(\eta) = \eta\partial_n$$

$$Z^n(X,V) = \ker d_n$$

$$B^n(X,V) = \text{image } d_{n-1}, \quad B^0(X,V) = 0$$

$$H^n(X,V) = Z^n(X,V)/B^n(X,V).$$

Here C^n, Z^n, B^n, and H^n are vector spaces (not R modules), and d_n is $\mathbb{C}$-linear. The map d_n is just $\text{Hom}(\partial_n, 1)$ in the notation of §II.2.

When we have two such complexes X and X' and a chain map $F : X \to X'$, we obtain maps $F_n^* : C^n(X',V) \to C^n(X,V)$ by the definition $F_n^* = \text{Hom}(F_n, 1)$. Since

$$\begin{aligned} d_n F_n^* &= \text{Hom}(\partial_n, 1)\text{Hom}(F_n, 1) = \text{Hom}(F_n\partial_n, 1) = \text{Hom}(\partial_n' F_{n+1}, 1) \\ &= \text{Hom}(F_{n+1}, 1)\text{Hom}(\partial_n', 1) = F_{n+1}^* d_n', \end{aligned}$$

it follows that the diagram

$$\begin{array}{ccccccccc} 0 & \longrightarrow & C^0(X,V) & \xrightarrow{d_0} & C^1(X,V) & \xrightarrow{d_1} & C^2(X,V) & \longrightarrow & \cdots \\ & & \uparrow F_0^* & & \uparrow F_1^* & & \uparrow F_2^* & & \\ 0 & \longrightarrow & C^0(X',V) & \xrightarrow{d_0'} & C^1(X',V) & \xrightarrow{d_1'} & C^2(X',V) & \longrightarrow & \cdots \end{array} \tag{4.23}$$

commutes and has complexes for rows.

From these properties of (4.23), it follows that F_n^* carries cocycles to cocycles and coboundaries to coboundaries. In fact, $d_n F_n^* = F_{n+1}^* d_n'$ implies that F_n^* of a cocycle is a cocycle and that F_{n+1}^* of a coboundary is a coboundary. Consequently F_n^* defines a map on cohomology that we denote F_n^* also: $F_n^* : H^n(X',V) \to H^n(X,V)$ is defined by

$$F_n^*(z + B^n(X',V)) = F_n^*(z) + B^n(X,V).$$

In short, a chain map $\{F_n\}$ over f as in (4.22) yields a map $\{F_n^*\}$ on cohomology with coefficients in V.

Proposition 4.7 is a device for producing chain maps. Let us address the uniqueness question for the resulting maps on cohomology with coefficients in V. In the diagram (4.22), let the rows be complexes and let the squares commute. If $G : X \to X'$ is another chain map over f, we say that F and G are homotopic if there are maps $s_n : X_n \to X'_{n+1}$ such that

$$\partial'_n s_n + s_{n-1}\partial_{n-1} = F_n - G_n$$

for $n \geq 0$ (with the second term absent for $n = 0$). The relevant diagram is

$$\begin{array}{ccccccc}
\cdots & \longleftarrow & X_n & \xleftarrow{\ \partial_n\ } & X_{n+1} & \longleftarrow & \cdots \\
 & & {\scriptstyle F_n, G_n}\big\downarrow & \overset{s_n}{\searrow} & \big\downarrow{\scriptstyle F_{n+1}, G_{n+1}} & & \\
\cdots & \longleftarrow & X'_n & \xleftarrow{\ \partial'_n\ } & X'_{n+1} & \longleftarrow & \cdots
\end{array}$$

Proposition 4.9. Under the hypotheses of Proposition 4.7 with X_n free and X' exact, if $F : X \to X'$ and $G : X \to X'$ are maps over f, then F and G are homotopic.

Proof. We construct a homotopy s_n by induction on n. To obtain s_0, we consider the diagram

$$\begin{array}{ccccc}
M & \xleftarrow{\ \varepsilon\ } & X_0 & & \\
{\scriptstyle f - f = 0}\big\downarrow & & \big\downarrow{\scriptstyle F_0 - G_0} & \overset{s_0}{\searrow} & \\
M' & \xleftarrow{\ \varepsilon'\ } & X'_0 & \xleftarrow{\ \partial'_0\ } & X'_1
\end{array}$$

We have $\epsilon'(F_0 - G_0) = f\epsilon - f\epsilon = 0$, and hence Lemma 4.6 provides s_0 with $F_0 - G_0 = \partial_0' s_0$. Suppose inductively that $s_n : X_n \to X_{n+1}'$ has been constructed with

$$\partial_n' s_n + s_{n-1}\partial_{n-1} = F_n - G_n .$$

To obtain s_{n+1}, we consider the diagram

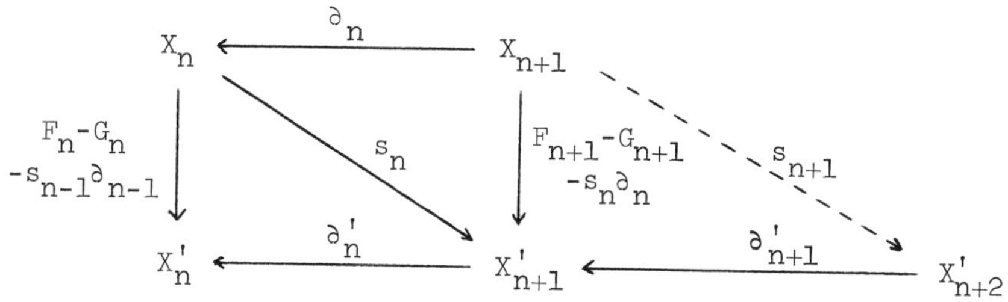

We have

$$\partial_n'(F_{n+1}-G_{n+1}-s_n\partial_n) = F_n\partial_n - G_n\partial_n - \partial_n' s_n\partial_n$$

$$= (F_n - G_n - \partial_n' s_n)\partial_n = s_{n-1}\partial_{n-1}\partial_n = 0 ,$$

and hence Lemma 4.8 provides s_{n+1} with

$$F_{n+1} - G_{n+1} - s_n\partial_n = \partial_{n+1}' s_{n+1} .$$

This completes the induction and the proof.

<u>Proposition 4.10</u>. In the diagram (4.22), let the rows be complexes and let the squares commute. Suppose that $G : X \to X'$ is another chain map over f, and suppose that F and G are homotopic. Then F_n^* and G_n^* are identical as maps on cohomology with coefficients in V.

Proof. Let $\{s_n\}$ be a homotopy. To get F_n^* and G_n^*

to coincide on $H^n(X',V)$, we want to see that ω' in $Z^n(X',V)$ implies $F_n^*(\omega') - G_n^*(\omega')$ is in $B^n(X,V)$. Thus let $d_n'\omega' = 0$. Then $\omega'\partial_n' = d_n'\omega' = 0$, and

$$F_n^*(\omega') - G_n^*(\omega') = \omega' F_n - \omega' G_n = \omega'(\partial_n' s_n + s_{n-1}\partial_{n-1})$$
$$= (\omega'\partial_n')s_n + (\omega' s_{n-1})\partial_{n-1} = 0 + d_{n-1}(\omega' s_{n-1}) .$$

The result follows.

<u>Corollary 4.11</u>. If

$$0 \longleftarrow M \xleftarrow{\ \epsilon\ } X_0 \xleftarrow{\ \partial_0\ } X_1 \xleftarrow{\ \partial_1\ } X_2 \longleftarrow \cdots$$

and

$$0 \longleftarrow M \xleftarrow{\ \epsilon'\ } X_0' \xleftarrow{\ \partial_0'\ } X_1' \xleftarrow{\ \partial_1'\ } X_2' \longleftarrow \cdots$$

are two free resolutions over the same M, then $H^n(X,V)$ and $H^n(X',V)$ are canonically isomorphic, for each n and V.

Proof. We apply Proposition 4.7 to the map $1 : M \to M$ to obtain a chain map $F : X \to X'$ over 1 and then to obtain a chain map $G : X' \to X$ over 1. Then $GF : X \to X$ is a chain map over 1 and must be homotopic to $1 : X \to X$, by Proposition 4.9. By Proposition 4.10, $F^*G^* = (GF)^* = 1^* = 1$, and similarly $G^*F^* = 1$. The maps F_n^* thus provide the required isomorphisms. They are canonical because any other choice of F is homotopic to F, by Proposition 4.9, and yields the same maps on cohomology, by Proposition 4.10.

<u>Corollary 4.12</u>. If V is a left $U(\mathfrak{g})$ module, then the cohomology of $\mathfrak{g}$ with coefficients in V may be computed as the cohomology with coefficients in V of any free resolution

$$0 \longleftarrow \mathbb{C} \longleftarrow Y_0 \longleftarrow Y_1 \longleftarrow \cdots$$

of left $U(\mathfrak{g})$ modules.

Proof. Proposition 4.4 says that the cohomology with coefficients in V may be computed from the Koszul resolution, and Theorem 4.6 says that the Koszul resolution is a free resolution. Corollary 4.11 says that any other free resolution will do as well.

Corollary 4.12 is not by itself too useful; it is the idea behind it that is useful. In the remainder of this section, we shall illustrate matters with an (admittedly artificial-looking) example. Then in the next two chapters, we shall dissect this example, generalize its parts, and obtain a wide variety of approaches to this kind of computation. For our example, we suppose that $\mathfrak{h}$ is a Lie subalgebra of $\mathfrak{g}$ and that V is a left $U(\mathfrak{h})$ module. Let

$$U = U(\mathfrak{g}) \otimes_{U(\mathfrak{h})} V .$$

Here we regard $U(\mathfrak{g})$ as a left $U(\mathfrak{g})$ module and right $U(\mathfrak{h})$ module, and consequently U is a left $U(\mathfrak{g})$ module. We shall show that the dual U^* has

$$H^n(\mathfrak{g}, U^*) = 0 \quad \text{for} \quad n > \dim \mathfrak{h} . \tag{4.24}$$

From the definitions, the vanishing is obvious only for $n > \dim \mathfrak{g}$. To obtain (4.24), we need two preliminary results.

Lemma 4.13. If F and W are left $U(\mathfrak{g})$ modules and F is free, then the left $U(\mathfrak{g})$ module $F \otimes_{\mathbb{C}} W$ is free.

Remarks. Recall that left $U(\mathfrak{g})$ modules are the same as representation spaces of $\mathfrak{g}$. In this case the representation on $F \otimes_{\mathbb{C}} W$ is the usual tensor product representation

$$X(f \otimes w) = Xf \otimes w + f \otimes Xw$$

for X in $\mathfrak{g}$, and thus $F \otimes_{\mathbb{C}} W$ becomes a left $U(\mathfrak{g})$ module.

Proof. Let $\{f_i\}$ be a basis of F over $U(\mathfrak{g})$, and let $\{w_j\}$ be a basis of W over $\mathbb{C}$. We shall prove that $\{f_i \otimes w_j\}$ is a basis of $F \otimes_{\mathbb{C}} W$ over $U(\mathfrak{g})$. First let us prove that these elements span $F \otimes_{\mathbb{C}} W$ over $U(\mathfrak{g})$. Since $\{f_i\}$ spans F over $U(\mathfrak{g})$, every element of $F \otimes_{\mathbb{C}} W$ is of the form $\Sigma\, u_k f_k \otimes w_\ell$. Thus it is enough to show that any $uf_k \otimes w$ is of the form $\Sigma_{i,j}\, u_{ij}(f_i \otimes w_j)$. We do so by induction on the degree of u.

For u of degree 0, the assertion is clear. Assume the result for degree $\leq n$, and let u have degree $n+1$. We may assume u is a monomial in $U(\mathfrak{g})$. Then $u = Xu'$ with X in $\mathfrak{g}$ and degree$(u') = n$. Hence

$$uf_k \otimes w = X(u'f_k) \otimes w = X(u'f_k \otimes w) - u'f_k \otimes Xw .$$

The induction hypothesis applies to $u'f_k \otimes w$ and $u'f_k \otimes Xw$, and we obtain the required expansion.

Now let us prove the independence of $\{f_i \otimes w_j\}$ over $U(\mathfrak{g})$. Suppose $\Sigma_{ij}\, u_{ij}(f_i \otimes v_j) = 0$ with not all $u_{ij} = 0$.

Applying the product rule for the action in the tensor product, we can write

$$u_{ij}(f_i \otimes v_j) = u_{ij}f_i \otimes v_j + \sum_r u'_{ijr}f_i \otimes u''_{ijr}v_j$$

with $\text{degree}(u'_{ijr}) < \text{degree}(u_{ij})$. If $m = \max_{i,j} \text{degree}(u_{ij})$, we find

$$0 = \sum_{i,j} u_{ij}(f_i \otimes v_j) \equiv \sum_{i,j} u_{ij}f_i \otimes v_j \mod \sum_{i,j} U^{m-1}(\mathfrak{g})f_i \otimes v_j .$$

This says u_{ij} is in $U^{m-1}(\mathfrak{g})$ for all i and j and gives a contradiction.

Lemma 4.14. Let X be a free resolution of left $U(\mathfrak{g})$ modules

$$0 \longleftarrow M \overset{\varepsilon}{\longleftarrow} X_0 \overset{\partial_0}{\longleftarrow} X_1 \longleftarrow \cdots$$

and let N and W be left $U(\mathfrak{g})$ modules. Then $X \otimes_{\mathbb{C}} N$ defined by

$$0 \longleftarrow M \otimes_{\mathbb{C}} N \overset{\varepsilon \otimes 1}{\longleftarrow} X_0 \otimes_{\mathbb{C}} N \overset{\partial_0 \otimes 1}{\longleftarrow} X_1 \otimes_{\mathbb{C}} N \longleftarrow \cdots$$

is a free resolution of left $U(\mathfrak{g})$ modules. Moreover the isomorphisms

$$\text{Hom}_{U(\mathfrak{g})}(X_n \otimes_{\mathbb{C}} N, W) \cong \text{Hom}_{U(\mathfrak{g})}(X_n, \text{Hom}_{\mathbb{C}}(N, W)) \qquad (4.25)$$

of Corollary 4.3 lead to canonical isomorphisms

$$H^n(X \otimes_{\mathbb{C}} N, W) \cong H^n(X, \text{Hom}_{\mathbb{C}}(N, W)) . \qquad (4.26)$$

Proof. Lemma 4.13 shows that $X_n \otimes_{\mathbb{C}} N$ is free. To see that $X \otimes_{\mathbb{C}} N$ is exact, it is enough to think of X as an eaxct sequence of vector spaces and to see that tensoring with N preserves exactness. This is an easy exercise, and we shall return to it in Corollary 5.11 in connection with "exact functors" in Chapter V.

Formula (4.26) comes from naturality of (4.25) in the first variable (Proposition 2.2). In fact, we apply naturality to the map $\partial_n : X_{n+1} \to X_n$ and find that the diagram

$$\begin{array}{ccc} \mathrm{Hom}_{U(\mathfrak{g})}(X_n \otimes_{\mathbb{C}} N, W) & \xrightarrow{\mathrm{Hom}(\partial_n \otimes 1, 1)} & \mathrm{Hom}_{U(\mathfrak{g})}(X_{n+1} \otimes_{\mathbb{C}} N, W) \\ \downarrow & & \downarrow \\ \mathrm{Hom}_{U(\mathfrak{g})}(X_n, \mathrm{Hom}_{\mathbb{C}}(N, W)) & \xrightarrow{\mathrm{Hom}(\partial_n, 1)} & \mathrm{Hom}_{U(\mathfrak{g})}(X_{n+1}, \mathrm{Hom}_{\mathbb{C}}(N, W)) \end{array}$$

commutes. Putting such diagrams side by side as n varies, we obtain a diagram like (4.23), from which we obtain an isomorphism on cohomology. This proves the lemma.

Let us return to our example, in which $U = U(\mathfrak{g}) \otimes_{U(\mathfrak{h})} V$. Let $X = \{X_n\}$ be the Koszul resolution. Lemma 4.14 gives us

$$H^n(X \otimes_{\mathbb{C}} U, \mathbb{C}) \cong H^n(X, \mathrm{Hom}_{\mathbb{C}}(U, \mathbb{C})) = H^n(X, U^*),$$

and we know from Proposition 4.4 that the right side is $H^n(\mathfrak{g}, U^*)$. Lemma 4.13 identifies $X \otimes_{\mathbb{C}} U$ as a free resolution of U, and Corollary 4.11 says that we can replace $X \otimes_{\mathbb{C}} U$ in $H^n(X \otimes_{\mathbb{C}} U, \mathbb{C})$ by any other free resolution and still get the same answer. Thus it is enough to find a free resolution

$$Y: \quad 0 \longleftarrow U \longleftarrow Y_0 \longleftarrow Y_1 \longleftarrow \cdots$$

with $Y_n = 0$ for $n > \dim \mathfrak{h}$.

To do so, we start with the Koszul resolution for $U(\mathfrak{h})$,

$$0 \longleftarrow \mathbb{C} \longleftarrow X_0^{\mathfrak{h}} \longleftarrow \cdots \longleftarrow X_{\dim \mathfrak{h}}^{\mathfrak{h}} \longleftarrow 0,$$

and tensor with V to get

$$0 \longleftarrow V \longleftarrow X_0^{\mathfrak{h}} \otimes_{\mathbb{C}} V \longleftarrow X_1^{\mathfrak{h}} \otimes_{\mathbb{C}} V \longleftarrow \cdots$$

This is a free resolution of V for the algebra $U(\mathfrak{h})$. Then we apply $U(\mathfrak{g}) \otimes_{U(\mathfrak{h})} (\cdot)$ to this resolution. We claim that the result is a free resolution and thus can be taken as the desired Y. To see exactness, we write $\mathfrak{g} = \mathfrak{a} \oplus \mathfrak{h}$ as vector spaces, let $U(\mathfrak{h})$ operate trivially on $S(\mathfrak{a})$, and obtain

$$\begin{aligned} U(\mathfrak{g}) \otimes_{U(\mathfrak{h})} W &\cong (S(\mathfrak{a}) \otimes_{\mathbb{C}} U(\mathfrak{h})) \otimes_{U(\mathfrak{h})} W \\ &\cong S(\mathfrak{a}) \otimes_{\mathbb{C}} (U(\mathfrak{h}) \otimes_{U(\mathfrak{h})} W) \cong S(\mathfrak{a}) \otimes_{\mathbb{C}} W . \end{aligned}$$

Hence on the vector space level, the operation amounts to tensoring with $S(\mathfrak{a})$, and exactness is preserved (cf. the first paragraph of the proof of Lemma 4.14).

To see that the resolution is free over $U(\mathfrak{g})$, we note that $X_n^{\mathfrak{h}}$ is free over $U(\mathfrak{h})$. By Lemma 4.13, $X_n^{\mathfrak{h}} \otimes_{\mathbb{C}} V$ is then free over $U(\mathfrak{h})$, say is of the form $U(\mathfrak{h}) \otimes_{\mathbb{C}} W_n$, where W_n has trivial action. Thus

$$\begin{aligned} U(\mathfrak{g}) \otimes_{U(\mathfrak{h})} (X_n^{\mathfrak{h}} \otimes_{\mathbb{C}} V) &\cong U(\mathfrak{g}) \otimes_{U(\mathfrak{h})} (U(\mathfrak{h}) \otimes_{\mathbb{C}} W_n) \\ &\cong (U(\mathfrak{g}) \otimes_{U(\mathfrak{h})} U(\mathfrak{h})) \otimes_{\mathbb{C}} W_n \cong U(\mathfrak{g}) \otimes_{\mathbb{C}} W_n \end{aligned}$$

as $U(\mathfrak{g})$ modules, and so $Y_n = U(\mathfrak{g}) \otimes_{U(\mathfrak{h})} (X_n^{\mathfrak{h}} \otimes_{\mathbb{C}} V)$ is free over $U(\mathfrak{g})$.

5. Lemmas for Koszul resolution

We turn our attention to the proof of the exactness in Theorem 4.4. The proof will involve two inductive arguments that exploit relationships among complexes. Part of the proof does not use special properties of $U(\mathfrak{g})$, and we isolate that part in this section.

For this section let R be a ring. All modules will be left R modules, and all maps will be assumed R linear.

Lemma 4.15. In a diagram of R modules and maps

$$\begin{array}{ccccc}
 & & & & 0 \\
 & & & & \downarrow \\
A_1 & \xrightarrow{\psi_1} & B_1 & \xrightarrow{\varphi_1} & C_1 \\
\psi_A \downarrow & & \psi_B \downarrow & & \psi_C \downarrow \\
A_2 & \xrightarrow{\psi_2} & B_2 & \xrightarrow{\varphi_2} & C_2 \\
\varphi_A \downarrow & & \varphi_B \downarrow & & \varphi_C \downarrow \\
A_3 & \xrightarrow{\psi_3} & B_3 & \xrightarrow{\varphi_3} & C_3 \\
\downarrow & & & & \\
0 & & & &
\end{array}$$

suppose that the squares commute, the columns are exact, the top and bottom rows are exact, and $\varphi_2\psi_2 = 0$. Then the middle row is exact.

Proof. In obvious notation, let $\varphi_2 b_2 = 0$. Then $0 = \varphi_C\varphi_2 b_2 = \varphi_3\varphi_B b_2$, $\varphi_B b_2 \in \ker \varphi_3 = \text{image } \psi_3$, $\varphi_B b_2 = \psi_3 a_3$. Since φ_A is onto, $\psi_3 a_3 = \psi_3\varphi_A a_2 = \varphi_B\psi_2 a_2$. Thus

$\varphi_B(b_2-\psi_2a_2) = \psi_3a_3-\psi_3a_3 = 0$, $b_2-\psi_2a_2 \in \ker\varphi_B = \text{image }\psi_B$, $b_2-\psi_2a_2 = \psi_Bb_1$. Now $\varphi_2b_2 = 0$ and $\varphi_2\psi_2 = 0$ make $0 = \varphi_2b_2-\varphi_2\psi_2a_2 = \varphi_2\psi_Bb_1 = \psi_C\varphi_1b_1$. Since ψ_C is one-one, $\varphi_1b_1 = 0$. Thus $b_1 \in \ker\varphi_1 = \text{image }\psi_1$, $b_1 = \psi_1a_1$, $b_2-\psi_2a_2 = \psi_B\psi_1a_1 = \psi_2\psi_Aa_1$, $b_2 = \psi_2(a_2+\psi_Aa_1)$, and $b_2 \in \text{image }\psi_2$.

Lemma 4.16. In a diagram of R modules and maps

$$\begin{array}{ccccccccccc}
 & & 0 & & 0 & & 0 & & 0 & & \\
 & & \downarrow & & \downarrow & & \downarrow & & \downarrow & & \\
0 & \longleftarrow & M' & \longleftarrow & X_0' & \longleftarrow & X_1' & \longleftarrow & X_2' & \longleftarrow & \cdots \\
 & & \downarrow & & \downarrow & & \downarrow & & \downarrow & & \\
0 & \longleftarrow & M & \longleftarrow & X_0 & \longleftarrow & X_1 & \longleftarrow & X_2 & \longleftarrow & \cdots \\
 & & \downarrow & & \downarrow & & \downarrow & & \downarrow & & \\
0 & \longleftarrow & M'' & \longleftarrow & X_0'' & \longleftarrow & X_1'' & \longleftarrow & X_2'' & \longleftarrow & \cdots \\
 & & \downarrow & & \downarrow & & \downarrow & & \downarrow & & \\
 & & 0 & & 0 & & 0 & & 0 & &
\end{array}$$

suppose that the M' and M'' rows are resolutions, the middle row is a complex, the columns are exact, and the squares commute. Then the middle row is exact.

Proof. This follows by repeated use of Lemma 4.15.

Lemma 4.17. In a diagram of R modules and maps

$$\begin{array}{ccccccccc}
0 & \longrightarrow & A_1 & \xrightarrow{\psi_A} & A_2 & \xrightarrow{\varphi_A} & A_3 & \longrightarrow & 0 \\
 & & \downarrow{\scriptstyle\varphi_1} & & \downarrow{\scriptstyle\varphi_2} & & \vdots{\scriptstyle\varphi_3} & & \\
0 & \longrightarrow & B_1 & \xrightarrow{\psi_B} & B_2 & \xrightarrow{\varphi_B} & B_3 & \longrightarrow & 0
\end{array}$$

with the rows exact and the left square commuting, there exists a unique $\varphi_3 : A_3 \to B_3$ such that the right square commutes.

All this lemma does is formalize the simple fact that the map $\varphi_2 : A_2 \to B_2$ descends to the quotients, carrying $A_2/\psi_A(A_1)$ to $B_2/\psi_B(B_1)$, if it carries the kernel $\psi_A(A_1)$ to the kernel $\psi_B(B_1)$.

6. Exactness of Koszul resolution

In this section we shall prove that the Koszul resolution is exact. We begin with the special case that $\mathfrak{g}$ is abelian. Let us take $\mathfrak{g} = \mathbb{C}^N$. Then $U(\mathbb{C}^N) = S(\mathbb{C}^N)$ is generated by a basis $x_1, \ldots, x_N$ of $\mathbb{C}^N$, and we may thus regard $U(\mathbb{C}^N)$ as the space of polynomials

$$U(\mathbb{C}^N) = \mathbb{C}[x_1, \ldots, x_N] .$$

Let us write U for the left $U(\mathbb{C}^N)$ module $\mathbb{C}[x_1, \ldots, x_N]$. Let J_k be the submodule

$$J_k = \sum_{i=1}^{k} U(\mathbb{C}^N) x_i .$$

This is the space of polynomials vanishing for $x_1 = \ldots = x_k = 0$, and thus

$$U/J_k \cong \mathbb{C}[x_{k+1}, \ldots, x_N] .$$

In particular, we have

$$U/J_N \cong \mathbb{C} .$$

The operation of ∂ on $U \otimes_{\mathbb{C}} \Lambda^n \mathbb{C}^N$ is given from (4.19c) as

$$\partial_{n-1}(u \otimes x_{i_1} \wedge \ldots \wedge x_{i_n}) = \sum_{j=1}^{n} (-1)^{j+1} u x_{i_j} \otimes x_{i_1} \wedge \ldots \wedge \hat{x}_{i_j} \wedge \ldots \wedge x_{i_n} .$$

Within $U \otimes_{\mathbb{C}} \Lambda^n \mathbb{C}^N$, the elements whose $\Lambda^n \mathbb{C}^N$ part involves only $x_1, \ldots, x_k$ form a $U(\mathbb{C}^N)$ submodule, and ∂_{n-1} carries this into the corresponding submodule of $U \otimes_{\mathbb{C}} \Lambda^{n-1} \mathbb{C}^N$. Thus we obtain a sequence

$$0 \longleftarrow U/J_k \xleftarrow{\epsilon_k} U \otimes_{\mathbb{C}} \mathbb{C} \xleftarrow{\partial_0} U \otimes_{\mathbb{C}} \Lambda^1(x_1, \ldots, x_k)$$

$$\xleftarrow{\partial_1} \cdots \longleftarrow U \otimes_{\mathbb{C}} \Lambda^k(x_1, \ldots, x_k) \longleftarrow 0 , \qquad (4.27)$$

which is a complex by Lemma 4.3. Here the map ϵ_k carries $u \otimes 1$ to $u + J_k$. To prove the theorem for $\mathfrak{g}$ abelian, we shall prove by induction on k that the complex (4.27) is exact, the result for $k = N$ being the result we seek.

If $k = 0$, the complex in question is

$$0 \longleftarrow U/J_0 \xleftarrow{\epsilon_0} U \otimes_{\mathbb{C}} \mathbb{C} \longleftarrow 0$$

with $J_0 = \mathbb{C}$. Since ϵ_0 is an isomorphism, this is exact.

Suppose inductively that (4.27) is exact with $k-1$ in place of k. We are to prove it is exact as it stands. Let

$$X_n^{(k)} = U \otimes_{\mathbb{C}} \Lambda^n(x_1, \ldots, x_k) .$$

In (4.27), ϵ_k is defined so as to be onto. For exactness at $U \otimes_{\mathbb{C}} \mathbb{C}$, we note that

$$\ker \epsilon_k = J_k = \{\sum_{i=1}^{k} u_i x_i\}$$

and

$$\text{image } \partial_0 = \{\partial_0(\sum_{i=1}^{k} u_i \otimes x_i)\} = \{\sum_{i=1}^{k} u_i x_i\} .$$

Thus the exactness of (4.27) is equivalent with the exactness of

$$0 \longleftarrow J_k \quad (\subseteq X_0^{(k)}) \xleftarrow{\partial_0} X_1^{(k)} \xleftarrow{\partial_1} \cdots \xleftarrow{\partial_{k-1}} X_k^{(k)} \longleftarrow 0 . \tag{4.28}$$

We set up the diagram

$$\begin{array}{ccccccccccc}
 & & 0 & & 0 & & & 0 & & 0 & \\
 & & \downarrow & & \downarrow & & & \downarrow & & \downarrow & \\
0 & \longleftarrow & J_{k-1} & \longleftarrow & X_1^{(k-1)} & \longleftarrow & \cdots & X_{k-1}^{(k-1)} & \longleftarrow & 0 & \longleftarrow 0 \\
 & & \hookdownarrow & & \hookdownarrow & & & \hookdownarrow & & \hookdownarrow & \\
0 & \longleftarrow & J_k & \longleftarrow & X_1^{(k)} & \longleftarrow & \cdots & X_{k-1}^{(k)} & \longleftarrow & X_k^{(k)} & \longleftarrow 0 \\
 & & \downarrow & & \downarrow & & & \downarrow & & \downarrow & \\
0 & \longleftarrow & J_k/J_{k-1} & \xleftarrow{\bar{\partial}_0} & X_1^{(k)}/X_1^{(k-1)} & \xleftarrow{\bar{\partial}_1} & \cdots & X_{k-1}^{(k)}/X_{k-1}^{(k-1)} & \xleftarrow{\bar{\partial}_{k-1}} & X_k^{(k)} & \longleftarrow 0 \\
 & & \downarrow & & \downarrow & & & \downarrow & & \downarrow & \\
 & & 0 & & 0 & & & 0 & & 0 &
\end{array} \tag{4.29}$$

The first row of horizontal maps is (4.28) for $k-1$ and is exact. The second row of horizontal maps is (4.28) for k and is a complex. In each column the vertical maps are the inclusion and quotient mappings. Thus the columns are exact and the top row of squares commutes. The third row of horizontal maps exists by Lemma 4.17 (applied to the diagram tilted on end) so that the second row of squares commutes, and it is clear from the construction that this row is a complex.

We shall prove that this row is exact, and then the exactness of (4.28) for k will follow by Lemma 4.16.

Thus consider the diagram

$$\begin{array}{ccccccccccc}
0 & \longleftarrow & U/J_{k-1} & \xleftarrow{\epsilon_{k-1}} & X_0^{(k-1)} & \xleftarrow{\partial_0} \cdots & X_{k-2}^{(k-1)} & \xleftarrow{\partial_{k-2}} & X_{k-1}^{(k-1)} & \longleftarrow & 0 \\
 & & \downarrow{\scriptstyle R_k} & & \downarrow{\scriptstyle R_k} & & \downarrow{\scriptstyle R_k} & & \downarrow{\scriptstyle R_k} & & \\
0 & \longleftarrow & J_k/J_{k-1} & \xleftarrow{\bar{\partial}_0} & X_1^{(k)}/X_1^{(k-1)} & \xleftarrow{\bar{\partial}_1} \cdots & X_{k-1}^{(k)}/X_{k-1}^{(k-1)} & \xleftarrow{\bar{\partial}_{k-1}} & X_k^{(k)} & \longleftarrow & 0
\end{array} \tag{4.30}$$

in which each vertical mapping R_k is a version of right multiplication by x_k:

$$R_k(u + J_{k-1}) = ux_k + J_{k-1} \qquad \text{on } U/J_{k-1}$$

$$R_k(u \otimes x_{i_1} \wedge \cdots \wedge x_{i_n}) = u \otimes x_{i_1} \wedge \cdots \wedge x_{i_n} \wedge x_k + X_{n+1}^{(k-1)} \qquad \text{on } X_n^{(k-1)} .$$

The top row of (4.30) is (4.27) for $k-1$, and the bottom row is the third horizontal of (4.29), whose exactness is to be proved. Anyway we know this row is a complex.

We shall prove that the vertical maps of (4.30) are isomorphisms onto and that the squares of (4.30) commute. Then it will follow that the second row of (4.30) is exact, and our induction will go through.

First we consider the vertical maps. The elements $ux_k + \Sigma_{i=1}^{k-1} u_i x_i$ exhaust J_k and thus $R_k(u) = ux_k$ carries U/J_{k-1} onto J_k/J_{k-1}. This map is one-one because if ux_k is in J_{k-1}, then ux_k vanishes for $x_1 = \cdots = x_{k-1} = 0$; dividing by x_k, we see that u vanishes for

$x_1 = \ldots = x_{k-1} = 0$, so that u is in J_{k-1}. The other vertical maps involve

$$\begin{array}{c} x_{i_1} \wedge \ldots \wedge x_{i_n}, \quad 1 \leq i_1 < \ldots < i_n \leq k-1 \\ \downarrow \\ x_{i_1} \wedge \ldots \wedge x_{i_n} \wedge x_k . \end{array}$$

The images here, together with all alternating tensors from $\Lambda^{n+1}(x_1, \ldots, x_{k-1})$, give all of $X_{n+1}^{(k)}$, so that R_k carries $X_n^{(k-1)}$ onto $X_{n+1}^{(k)}/X_{n+1}^{(k-1)}$. Clearly it is one-one. Thus R_k is an isomorphism in all cases.

Now we examine the squares in (4.30). For the first square, we have

$$R_k \circ \varepsilon_{k-1}(u \otimes 1) = R_k(u + J_{k-1}) = ux_k + J_{k-1}$$

and

$$\delta_0 \circ R_k(u \otimes 1) = \delta_0(u \otimes x_k + X_1^{(k-1)}) = ux_k + J_{k-1} .$$

Thus the first square commutes. For the other squares, we have

$$\begin{aligned} &R_k \circ \partial_{n-1}(u \otimes x_{i_1} \wedge \ldots \wedge x_{i_n}) \\ &\quad = R_k\Big(\sum_{j=1}^{n} (-1)^{j+1} ux_{i_j} \otimes x_{i_1} \wedge \ldots \wedge \hat{x}_{i_j} \wedge \ldots \wedge x_{i_n}\Big) \\ &\quad = \sum_{j=1}^{n} (-1)^{j+1} ux_{i_j} \otimes x_{i_1} \wedge \ldots \wedge \hat{x}_{i_j} \wedge \ldots \wedge x_{i_n} \wedge x_k + X_n^{(k-1)} \end{aligned}$$

and

$$\bar{\partial}_{n-1} \circ R_k(u \otimes x_{i_1} \wedge \cdots \wedge x_{i_n})$$

$$= \bar{\partial}_{n-1}(u \otimes x_{i_1} \wedge \cdots \wedge x_{i_n} \wedge x_k + X_{n+1}^{(k-1)})$$

$$= \sum_{j=1}^{n} (-1)^{j+1}\, ux_{i_j} \otimes x_{i_1} \wedge \cdots \wedge \hat{x}_{i_j} \wedge \cdots \wedge x_{i_n} \wedge x_k$$

$$+ (-1)^n\, ux_k \otimes x_{i_1} \wedge \cdots \wedge x_{i_n} + X_n^{(k-1)}$$

$$= \sum_{j=1}^{n} (-1)^{j+1}\, ux_{i_j} \otimes x_{i_1} \wedge \cdots \wedge \hat{x}_{i_j} \wedge \cdots \wedge x_{i_n} \wedge x_k + X_n^{(k-1)} .$$

Thus the other squares commute, our induction goes through, and the proof is complete when $\mathfrak{g}$ is abelian.

We turn to the case that $\mathfrak{g}$ is not necessarily abelian. For $p \geq 0$, let

$$X_n^{(p)} = U^{p-n}(\mathfrak{g}) \otimes_{\mathbb{C}} \Lambda^n \mathfrak{g} , \qquad 0 \leq n \leq p .$$

Then ∂_{n-1} maps $X_n^{(p)}$ into $X_{n-1}^{(p)}$, and we obtain the following complex of vector spaces by restricting each ∂_{n-1} accordingly:

$$0 \longleftarrow \mathbb{C} \xleftarrow{\varepsilon} X_0^{(p)} \xleftarrow{\partial_0} X_1^{(p)} \xleftarrow{\partial_1} \cdots \longleftarrow X_p^{(p)} \longleftarrow 0 . \tag{4.31}$$

We shall prove by induction on p that this complex is exact, and then the exactness of the Koszul resolution will follow. (In fact, the Koszul resolution is known to be a complex by Lemma 4.3. If x is in $\ker \partial_{n-1}$, then x is in $X_n^{(p)} \cap \ker \partial_{n-1}$ for some p, and the exactness of (4.31) will exhibit x as ∂_n of an element in $X_{n+1}^{(p)}$.)

Fix a basis $x_1, \ldots, x_N$ of $\mathfrak{g}$. Then $\ker \varepsilon|_{X_0^{(p)}}$ is generated by monomials $x_1^{j_1} \cdots x_N^{j_N}$ with $0 < \Sigma\, j_i \leq p$. If j_k is the last nonzero exponent, then

$$\partial_0(x_1^{j_1} \cdots x_{k-1}^{j_{k-1}} x_k^{j_k - 1} \otimes x_k) = x_1^{j_1} \cdots x_k^{j_k},$$

so that image $\partial_0|_{X_1^{(p)}} = \ker \varepsilon|_{X_0^{(p)}}$. Thus the exactness of (4.31) is equivalent with the exactness of

$$0 \longleftarrow X_0^{(p)}/\mathbb{C} \xleftarrow{\partial_0} X_1^{(p)} \xleftarrow{\partial_1} \cdots \xleftarrow{\partial_{p-1}} X_p^{(p)} \longleftarrow 0. \tag{4.32}$$

Here the factor $\mathbb{C}$ in $X_0^{(p)}/\mathbb{C}$ refers to the subspace of constants. For $p=0$, this complex is all 0 and is exact. For $p=1$ it is

$$0 \longleftarrow \mathfrak{g} \xleftarrow{\partial_0} \mathbb{C} \otimes_{\mathbb{C}} \mathfrak{g} \longleftarrow 0$$

and is exact.

Suppose inductively that (4.32) is exact with $p-1$ in place of p. We are to prove it is exact as it stands. We set up the diagram that follows, in which the columns are exact by construction, the top row of horizontal maps is exact by inductive hypothesis, the middle row of horizontal maps is a complex by Lemma 4.3, the top row of squares commutes by construction, and the maps $\overline{\partial}_{n-1}$ are obtained from Lemma 4.17 so that the bottom row of squares commutes. Certainly the bottom row of maps is a complex. We shall prove that this row is exact, and then the exactness of the middle row will follow by Lemma 4.16.

$$\begin{array}{ccccccccccc}
 & & 0 & & 0 & & & & 0 & & 0 \\
 & & \downarrow & & \downarrow & & & & \downarrow & & \downarrow \\
0 & \leftarrow & X_0^{(p-1)}/\mathbb{C} & \xleftarrow{\partial_0} & X_1^{(p-1)} & \leftarrow & \cdots & & X_{p-1}^{(p-1)} & \leftarrow & 0 & \leftarrow 0 \\
 & & \int & & \int & & & & \int & & \int \\
0 & \leftarrow & X_0^{(p)}/\mathbb{C} & \xleftarrow{\partial_0} & X_1^{(p)} & \leftarrow & \cdots & & X_{p-1}^{(p)} & \leftarrow & X_p^{(p)} & \leftarrow 0 \\
 & & \downarrow & & \downarrow & & & & \downarrow & & \downarrow \\
0 & \leftarrow & X_0^{(p)}/X_0^{(p-1)} & \xleftarrow{\bar{\partial}_0} & X_1^{(p)}/X_1^{(p-1)} & \leftarrow & \cdots & & X_{p-1}^{(p)}/X_{p-1}^{(p-1)} & \leftarrow & X_p^{(p)} & \leftarrow 0 \\
 & & \downarrow & & \downarrow & & & & \downarrow & & \downarrow \\
 & & 0 & & 0 & & & & 0 & & 0
\end{array}$$

(4.33)

Recall the space $S^q(\mathfrak{g})$ of symmetric tensors homogeneous of degree q. The Poincaré-Birkhoff-Witt Theorem gives

$$U^p(\mathfrak{g})/U^{p-1}(\mathfrak{g}) \cong S^p(\mathfrak{g}) .$$

Replacing p by p-n and tensoring with $\Lambda^n\mathfrak{g}$ gives

$$X_n^{(p)}/X_n^{(p-1)} \cong S^{p-n}(\mathfrak{g}) \otimes_{\mathbb{C}} \Lambda^n\mathfrak{g} , \qquad 0 \leq n \leq p-1$$

$$X_p^{(p)} \cong \mathbb{C} \otimes_{\mathbb{C}} \Lambda^p\mathfrak{g} .$$

If $j_1+\ldots+j_N = p-n$, then

$$\bar{\partial}_{n-1}(x_1^{j_1}\cdots x_N^{j_N} \otimes x_{i_1} \wedge \ldots \wedge x_{i_n})$$

$$= \sum_{\ell=1}^{n} (-1)^{\ell+1}(x_1^{j_1}\cdots x_N^{j_N} x_{i_\ell} \otimes x_{i_1} \wedge \ldots \wedge \widehat{x}_{i_\ell} \wedge \ldots \wedge x_{i_n})$$

$$+ \sum_{r<s} (-1)^{r+s}(x_1^{j_1}\cdots x_N^{j_N} \otimes [x_{i_r}, x_{i_s}]\, x_{i_1} \wedge \ldots \wedge \widehat{x}_{i_r} \wedge \ldots \wedge \widehat{x}_{i_s} \wedge \ldots \wedge x_{i_n})$$

$$+ X_{n-1}^{(p-1)}$$

$$= \sum_{\ell=1}^{n} (-1)^{\ell+1} (x_1^{j_1} \cdots x_N^{j_N} x_{i_\ell} \otimes x_{i_1} \wedge \cdots \wedge \hat{x}_{i_\ell} \wedge \cdots \wedge x_{i_n}) + X_{n-1}^{(p-1)} .$$

Thus the bottom row of (4.33) is isomorphic to that part of the standard complex for $\mathfrak{g}$ abelian in which the $U(\mathfrak{g}) = S(\mathfrak{g})$ coefficients are homogeneous of degree $p-n$. This sequence is exact by our special case (since the whole Koszul resolution respects degrees when $\mathfrak{g}$ is abelian), and the induction is complete.

CHAPTER V

HOMOLOGICAL ALGEBRA

1. Projectives and injectives

In Chapter IV we saw that the Koszul sequence is a free resolution and that $H^n(\mathfrak{g},V)$ can be computed from it by applying $\mathrm{Hom}_{U(\mathfrak{g})}(\cdot,V)$ to the sequence and forming successive quotients $\ker d_n/\mathrm{image}\ d_{n-1}$. Moreover there was evidence that $H^n(\mathfrak{g},V)$ could be computed in other ways, and we saw an example of such an alternate computation. In this chapter and the next, we shall examine this example in detail, extracting and generalizing a number of its parts. The present chapter will concentrate on results that do not use that our ring is $U(\mathfrak{g})$; special features that come from universal enveloping algebras will be taken up in the next chapter.

Thus let R be a ring, with or without identity. A good category $\mathcal{C}$ of R modules consists of

(i) some class of left R modules closed under passage to submodules, quotients, and finite direct sums (the modules of the category)

(ii) the sets $\mathrm{Hom}_R(A,B)$ of all R-linear maps from A to B, for each A and B in (i) (the maps of the category).

In Chapter IV, the modules in the category were all of the

left R modules, but this will not necessarily be the case later on: If $R = \mathbb{C}$, $\mathcal{C}$ could consist of all finite-dimensional complex vector spaces and $\mathbb{C}$-linear maps.

The definition of "good category" specifies <u>left</u> R modules. One could make a definition also for right R modules. But this distinction will not be important for us, since right $U(\mathfrak{g})$ modules can be made into left $U(\mathfrak{g})$ modules by means of the transpose mapping.

Fix the ring R and a good category $\mathcal{C}$. A <u>complex</u> in $\mathcal{C}$ is a finite or infinite sequence of modules and maps in $\mathcal{C}$ of the form

$$X: \quad \cdots \longleftarrow X_{n-1} \xleftarrow{\varphi_{n-1}} X_n \xleftarrow{\varphi_n} X_{n+1} \longleftarrow \cdots \tag{5.1}$$

such that $\varphi_{n-1}\varphi_n = 0$ for all n. This condition means that image $\varphi_n \subseteq \ker \varphi_{n-1}$. The <u>homology</u> of the complex (5.1) is is given by

$$H_n(X) = \ker \varphi_{n-1}/\text{image } \varphi_n .$$

We shall allow also complexes with indices increasing:

$$Y: \quad \cdots \longrightarrow Y_{n-1} \xrightarrow{d_{n-1}} Y_n \xrightarrow{d_n} Y_{n+1} \longrightarrow \cdots \tag{5.2}$$

Here $d_n d_{n-1} = 0$ for all n. When the indices are increasing, we shall speak of cohomology. The <u>cohomology</u> of the complex (5.2) is given by

$$H^n(Y) = \ker d_n/\text{image } d_{n-1} .$$

The two notions come to the same thing since we can always

replace n by $-n$ in (5.1) and obtain a complex of the form (5.2). Thus we often need to formulate a result about just one of (5.1) and (5.2).

The complex (5.1) is <u>exact</u> if image $\varphi_n = \ker \varphi_{n-1}$ for all n, i.e., if its homology is 0 in every dimension. An exact complex also is called a <u>resolution</u>.

In a diagram in $\mathcal{C}$ of the form

$$\begin{array}{lccccccc}
X: & \cdots \longleftarrow & X_{n-1} & \xleftarrow{\varphi_{n-1}} & X_n & \xleftarrow{\varphi_n} & X_{n+1} & \longleftarrow \cdots \\
 & & \downarrow{\scriptstyle F_{n-1}} & & \downarrow{\scriptstyle F_n} & & \downarrow{\scriptstyle F_{n+1}} & \\
X': & \cdots \longleftarrow & X'_{n-1} & \xleftarrow{\varphi'_{n-1}} & X'_n & \xleftarrow{\varphi'_n} & X'_{n+1} & \longleftarrow \cdots
\end{array}$$

in which X and X' are complexes and the squares commute, we say that $F = \{F_n\}$ is a <u>chain map</u> (or <u>cochain map</u> if the indices are increasing). A chain map $F = \{F_n\}$ yields maps on homology that we denote also by F:

$$F_n : H_n(X) \to H_n(X') .$$

In fact, $\varphi'_{n-1}F_n = F_{n-1}\varphi_{n-1}$ shows $F_n(\ker \varphi_{n-1}) \subseteq \ker \varphi'_{n-1}$, and $\varphi'_n F_{n+1} = F_n\varphi_n$ shows $F_n(\text{image } \varphi_n) \subseteq \text{image } \varphi'_n$. In the diagram of a chain map

$$\begin{array}{ccccccc}
0 & \longleftarrow & M & \longleftarrow & X_0 & \longleftarrow & \cdots \\
 & & \downarrow{\scriptstyle f} & & \downarrow{\scriptstyle F_0} & & \\
0 & \longleftarrow & M' & \longleftarrow & X'_0 & \longleftarrow & \cdots
\end{array}$$

or a cochain map

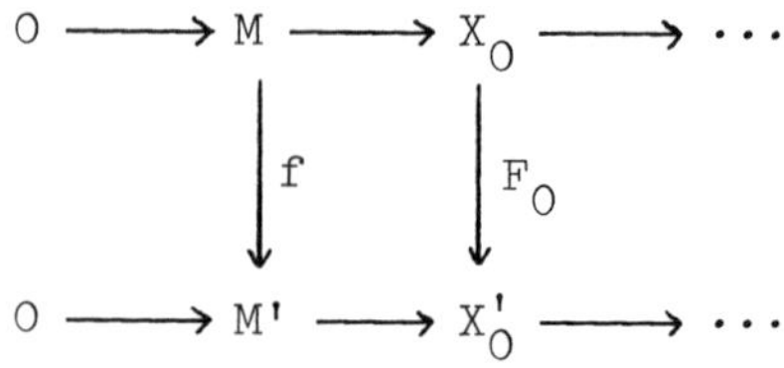

we may refer to a chain map over f or a cochain map over f .

Let $F : X \to X'$ and $G : X \to X'$ be chain maps between complexes as in (5.1). A system $s = \{s_n\}$ of maps in C with $s_n : X_n \to X'_{n+1}$ is a homotopy if

$$\varphi'_n s_n + s_{n-1}\varphi_{n-1} = F_n - G_n$$

for all n . This definition is consistent with the one in § IV.4, even though we did not insist on an s_{-1}; the reason is that $F_{-1} = G_{-1} = f$ in that chapter, so that we could always take $s_{-1} = 0$. The relevant diagram is

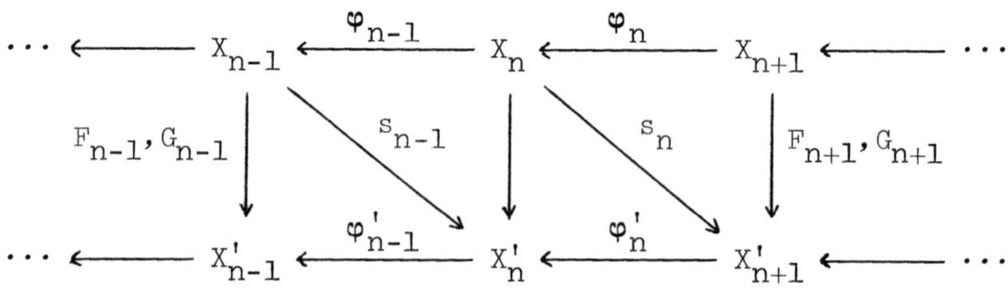

In the case of cochain maps between complexes as in (5.2), a homotopy $s = \{s_n\}$ is to have $s_n : X_n \to X'_{n-1}$ with

$$d_{n-1}s_n + d_n s_{n+1} = F_n - G_n$$

for all n . The relevant diagram is

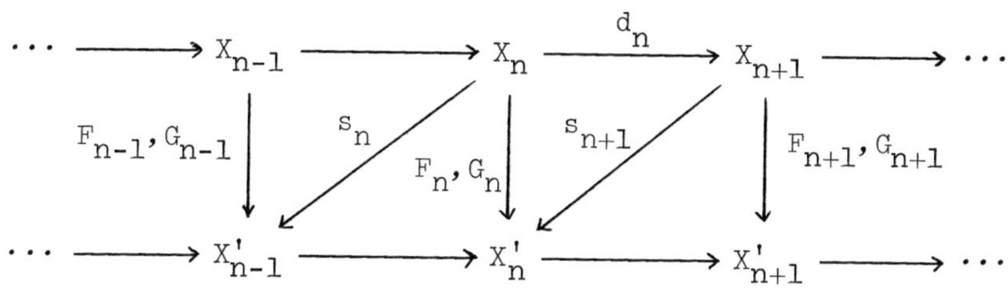

<u>Proposition 5.1</u>. If $F : X \to X'$ and $G : X \to X'$ are homotopic chain maps or cochain maps, then F and G induce the same maps on homology or cohomology.

Proof. For the case of chain maps, let x be in $\ker \varphi_{n-1}$. Then

$$F_n(x) - G_n(x) = \varphi'_n s_n(x) + s_{n-1}\varphi_{n-1}(x) = \varphi'_n(s_n(x)) \in \text{image } \varphi'_n .$$

Hence F_n and G_n agree on $H_n(X)$. The argument for cochain maps is similar.

In §IV.4, we exploited the fact that certain resolutions were free to obtain chain maps and homotopies. The hypothesis "free" entered the arguments only through Lemma 4.8. Accordingly we make the property in Lemma 4.8 into a definition. Actually the full property in Lemma 4.8 follows from a special case that is a little easier to check in practice. We say that a module P in a good category $\mathcal{C}$ is <u>projective</u> for $\mathcal{C}$ if whenever a diagram

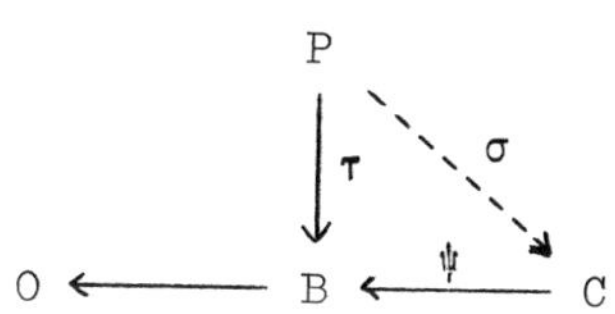

(5.3)

in $\mathcal{C}$ has ψ mapping onto B, then there exists $\sigma : P \to C$ such that the diagram commutes. We shall give examples shortly. Let us first prove an analog of Lemma 4.8.

Lemma 5.2. If P is projective in a good category $\mathcal{C}$ and if the diagram

$$\begin{array}{ccccc} & & P & & \\ & & \downarrow{\scriptstyle\tau} & \searrow{\scriptstyle\sigma} & \\ A' & \xleftarrow{\ \varphi\ } & A & \xleftarrow{\ \psi\ } & A'' \end{array} \qquad (5.4)$$

in $\mathcal{C}$ has $\ker\varphi = \text{image}\,\psi$ and $\varphi\tau = 0$, then there exists $\sigma : P \to A''$ such that the diagram commutes.

Proof. The hypotheses force $\text{image}\,\tau \subseteq \ker\varphi = \text{image}\,\psi$. Thus if we put $B = \text{image}\,\psi$ and $C = A''$, the diagram (5.4) leads to the diagram (5.3). Then the hypothesis "projective" gives us σ in (5.3) with $\tau = \psi\sigma$, and the same σ is the required map for (5.4).

Examples.

(1) If $R = \mathbb{C}$ and $\mathcal{C}$ is the category of all complex vector spaces, then every module is free, hence projective, since every vector space has a basis.

(2) For general R, if $\mathcal{C}$ is the category of all left R modules, then the projective modules are the direct summands of free modules. This fact is easily verified from (5.3): In one direction if $F = P \oplus P'$ is free and (5.3) is given, extend τ to F as 0 on P', find σ, and restrict to P. In the other direction, if P is

projective, find F free mapping onto P by ψ, and use $B = P$, $C = F$, and $\tau = 1$; then $P \cong$ image σ and $F =$ image $\sigma \oplus \ker \psi$.

(3) For $R = \mathbb{Z}$, the category $\mathcal{C}$ of all R modules is the category of all abelian groups. Then the projective modules are the free abelian groups, since a subgroup of a free abelian group is free abelian.

(4) For $R = U(\mathfrak{g})$, let $\mathcal{C}$ be the category of all left $U(\mathfrak{g})$ modules. Suppose that $\mathfrak{h}$ is a Lie subalgebra of $\mathfrak{g}$ and that V is projective for the category of all left $U(\mathfrak{h})$ modules. Then we shall see in Chapter VI that $U(\mathfrak{g}) \otimes_{U(\mathfrak{h})} V$ is projective for $\mathcal{C}$. In the special case that $\mathfrak{h} = 0$, $U(\mathfrak{h})$ is just $\mathbb{C}$, and $U(\mathfrak{g}) \otimes_{\mathbb{C}} V$ is actually free.

Within a good category $\mathcal{C}$, the resolution

$$0 \longleftarrow M \longleftarrow X_0 \longleftarrow X_1 \longleftarrow X_2 \longleftarrow \cdots$$

is called a <u>projective resolution</u> of M if $X_0, X_1, X_2, \ldots$ are all projective. This notion generalizes "free resolution" usefully from Chapter IV. With it we have corresponding generalizations of Propositions 4.7 and 4.9, as follows.

<u>Proposition 5.3</u>. Suppose that

$$\begin{array}{llllllllll} X: & 0 \longleftarrow & M & \longleftarrow & X_0 & \longleftarrow & X_1 & \longleftarrow & X_2 & \longleftarrow \cdots \\ & & \Big\downarrow f & & \Big\downarrow F_0 & & \Big\downarrow F_1 & & \Big\downarrow F_2 & \\ X': & 0 \longleftarrow & M' & \longleftarrow & X_0' & \longleftarrow & X_1' & \longleftarrow & X_2' & \longleftarrow \cdots \end{array}$$

is a diagram in $\mathcal{C}$ in which X and X' are complexes, X'

is exact, and all X_n are projective for $n \geq 0$. Then there exists a chain map $F: X \to X'$ over f.

Proof. The argument is the same as for Proposition 4.7.

<u>Proposition 5.4</u>. Suppose that

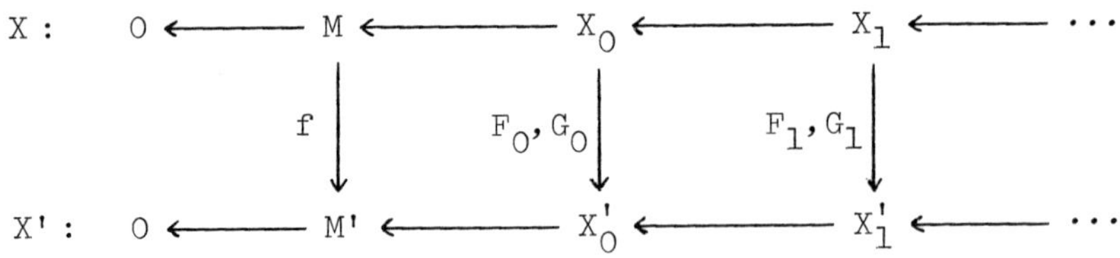

is a diagram in $\mathcal{C}$ in which X and X' are complexes, X' is exact, all X_n are projective for $n \geq 0$, and $F: X \to X'$ and $G: X \to X'$ are chain maps over f. Then F and G are homotopic.

Proof. The argument is the same as for Proposition 4.9.

There is a kind of dual notion to "projective" that we shall find extremely useful. We say that a module I in a good category $\mathcal{C}$ is <u>injective</u> for $\mathcal{C}$ if whenever a diagram

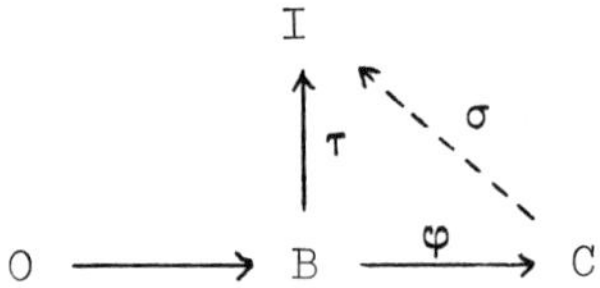

(5.5)

has φ mapping one-one into C, then there exists $\sigma: C \to I$ such that the diagram commutes. (We can think of the condition as saying that we can always extend such a τ from B to C, the extension being σ.) We give some examples after proving an analog of Lemma 5.2.

Lemma 5.5. If I is injective in a good category $\mathcal{C}$ and if the diagram

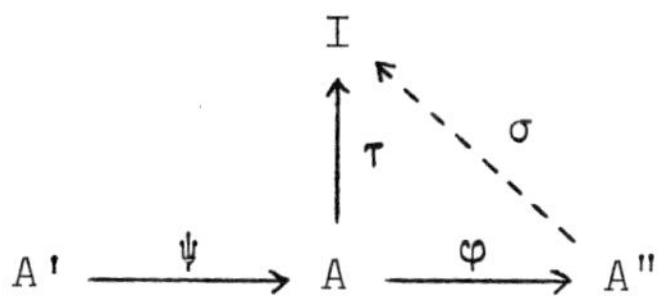

(5.6)

in $\mathcal{C}$ has $\ker \varphi = \text{image } \psi$ and $\tau\psi = 0$, then there exists $\sigma : A'' \to I$ such that the diagram commutes.

Proof. The hypotheses force $\ker \tau \supseteq \text{image } \psi = \ker \varphi$. Thus $\tau : A \to I$ and $\varphi : A \to A''$ both descend to maps $\bar{\tau} : A/\ker \varphi \to I$ and $\bar{\varphi} : A/\ker \varphi \to A''$. If we put $B = A/\ker \varphi$ and $C = A''$, the diagram (5.6) leads to the diagram (5.5) with $\bar{\tau}$ and $\bar{\varphi}$ in place of τ and φ. Then the hypothesis "injective" gives us σ in (5.5) with $\bar{\tau} = \sigma\bar{\varphi}$, and the same σ is the required map for (5.6).

Examples.

(1) If $R = \mathbb{C}$ and $\mathcal{C}$ is the category of all complex vector spaces, then every module is injective. In fact, in (5.5) we write $C = \text{image } \varphi \oplus B'$ and let $\eta : \text{image } \varphi \to B$ be the inverse of $\varphi : B \to \text{image } \varphi$. Then we can define σ to be 0 on B' and to be $\tau\eta$ on image φ.

(2) For $R = U(\mathfrak{g})$, let $\mathcal{C}$ be the category of all left $U(\mathfrak{g})$ modules. Suppose that $\mathfrak{h}$ is a Lie subalgebra of $\mathfrak{g}$ and that V is injective for the category of all left $U(\mathfrak{h})$ modules. Then we shall see in Chapter VI that

$\mathrm{Hom}_{U(\mathfrak{h})}(U(\mathfrak{g}),V)$ is injective for $\mathcal{C}$. Taking $\mathfrak{h} = 0$ and $V = \mathbb{C}$, we see from (1) that the contragredient $U(\mathfrak{g})^*$ is an example of an injective in $\mathcal{C}$.

(3) For $R = \mathbb{Z}$, let $\mathcal{C}$ be the category of all abelian groups. An abelian group G is <u>divisible</u> if for each integer $n \neq 0$ and each x in G, there exists y in G with $ny = x$. The additive rationals and the additive rationals modulo one are examples of divisible groups. In $\mathcal{C}$, injective is equivalent with divisible. To see that injective implies divisible, we set up (5.5) with $B = C = \mathbb{Z}$, $\tau(1) = x$, and $\varphi(1) = n$; then $\sigma(1) = y$. The proof that divisible implies injective involves Zorn's Lemma and is omitted.

Within a good category, the resolution

$$0 \longrightarrow M \longrightarrow X_0 \longrightarrow X_1 \longrightarrow X_2 \longrightarrow \cdots$$

is called an <u>injective resolution</u> of M if $X_0, X_1, X_2, \ldots$ are all injective. With it we have analogs of Propositions 5.3 and 5.4 as follows.

<u>Proposition 5.6</u>. Suppose that

$$\begin{array}{lccccccccccc}
X: & 0 & \longrightarrow & M & \xrightarrow{\ \varepsilon\ } & X_0 & \xrightarrow{\ d_0\ } & X_1 & \xrightarrow{\ d_1\ } & X_2 & \longrightarrow & \cdots \\
 & & & \downarrow f & & \downarrow F_0 & & \downarrow F_1 & & \downarrow F_2 & & \\
X': & 0 & \longrightarrow & M' & \xrightarrow{\ \varepsilon'\ } & X_0' & \xrightarrow{\ d_0'\ } & X_1' & \xrightarrow{\ d_1'\ } & X_2' & \longrightarrow & \cdots
\end{array}$$

is a diagram in $\mathcal{C}$ in which X and X' are complexes, X

is exact, and all X'_n are injective for $n \geq 0$. Then there exists a cochain map $F : X \to X'$ over f.

Proof. We proceed inductively, starting with

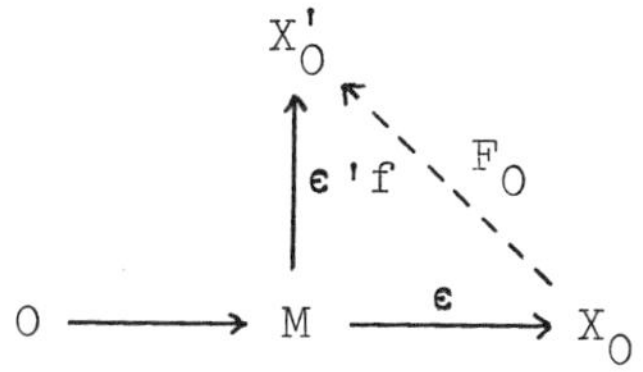

and applying the definition of "injective" to get $F_0 : X_0 \to X'_0$ with $F_0\varepsilon = \varepsilon' f$. For the inductive step, put $X_{-1} = M$, $X'_{-1} = M'$, $d_{-1} = \varepsilon$, and $d'_{-1} = \varepsilon'$. Suppose inductively that $n \geq 0$ and that F_n has been constructed with $F_n d_{n-1} = d'_{n-1}F_{n-1}$. In the diagram

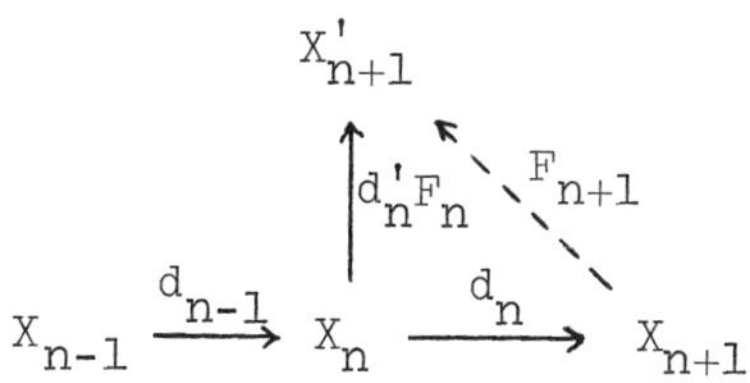

we have $d'_n F_n d_{n-1} = d'_n d'_{n-1} F_{n-1} = 0$, and Lemma 5.5 gives us $F_{n+1} : X_{n+1} \to X'_{n+1}$ with $F_{n+1}d_n = d'_n F_n$. This completes the induction and the proof.

Proposition 5.7. Suppose that

$$\begin{array}{lccccccccc}
X: & 0 & \longrightarrow & M & \longrightarrow & X_0 & \longrightarrow & X_1 & \longrightarrow & \cdots \\
 & & & \downarrow f & & \downarrow F_0, G_0 & & \downarrow F_1, G_1 & & \\
X': & 0 & \longrightarrow & M' & \longrightarrow & X'_0 & \longrightarrow & X'_1 & \longrightarrow & \cdots
\end{array}$$

is a diagram in $\mathbb{C}$ in which X and X' are complexes, X is exact, all X_n are injective for $n \geq 0$, and $F: X \to X'$ and $G: X \to X'$ are cochain maps over f. Then F and G are homotopic.

Proof. Combine the techniques of Propositions 5.4 and 5.6.

2. Functors

Let $\mathbb{C}$ be a good category of R modules, and let $\mathbb{C}'$ be a good category of S modules. A covariant functor $F: \mathbb{C} \to \mathbb{C}'$ is an assignment of a module $F(A)$ of $\mathbb{C}'$ to each module A of $\mathbb{C}$ and of a map $F(\varphi)$ in $\mathrm{Hom}_S(F(A), F(B))$ to each map φ in $\mathrm{Hom}_R(A,B)$ in such a way that

(i) the identity map of $\mathrm{Hom}_R(A,A)$ is carried to the identity map of $\mathrm{Hom}_S(F(A),F(A))$, and

(ii) $F(\varphi_2\varphi_1) = F(\varphi_2)F(\varphi_1)$ whenever φ_1 is in $\mathrm{Hom}_R(A,B)$ and φ_2 is in $\mathrm{Hom}_R(B,C)$.

A contravariant functor $F: \mathbb{C} \to \mathbb{C}'$ is an assignment of a module $F(A)$ of $\mathbb{C}'$ to each module A of $\mathbb{C}$ and of a map $F(\varphi)$ in $\mathrm{Hom}_S(F(B),F(A))$ to each map φ in $\mathrm{Hom}_R(A,B)$ in such a way that (i) holds above and also

(ii') $F(\varphi_2\varphi_1) = F(\varphi_1)F(\varphi_2)$ whenever φ_1 is in $\mathrm{Hom}_R(A,B)$ and φ_2 is in $\mathrm{Hom}_R(B,C)$.

Examples with R and S algebras over $\mathbb{C}$.

(1) Let V be a left R module. Then $\mathrm{Hom}_R(\cdot, V)$ is a

contravariant functor from the category of left R modules to the category of complex vector spaces. The module assigned to U is $\mathrm{Hom}_R(U,V)$, and the map assigned to φ is $\mathrm{Hom}(\varphi,1)$. Condition (ii') is the property

$$\mathrm{Hom}(\varphi\varphi',1) = \mathrm{Hom}(\varphi',1)\mathrm{Hom}(\varphi,1).$$

(2) Let the situation be ${}^R U^S$, with notation as in §II.8. Then $\mathrm{Hom}_R(U,\cdot)$ is a covariant functor from the category of left R modules to the category of left S modules. The module assigned to V is $\mathrm{Hom}_R(U,V)$, and the map assigned to φ is $\mathrm{Hom}(1,\varphi)$.

(3) Let the situation be ${}^S U^R$. Then $U \otimes_R (\cdot)$ is a covariant functor from the category of left R modules to the category of left S modules. The module assigned to V is $U \otimes_R V$, and the map assigned to φ is $1 \otimes \varphi$.

(4) The invariants functor $V \to V^{\mathfrak{g}}$ (see §IV.3) is a covariant functor from the category of all $U(\mathfrak{g})$ modules to the category of all vector spaces. The map φ in $\mathrm{Hom}_{U(\mathfrak{g})}(V,W)$ gets assigned to $\varphi\big|_{V^{\mathfrak{g}}}$ in $\mathrm{Hom}_{\mathbb{C}}(V^{\mathfrak{g}},W^{\mathfrak{g}})$. The invariants functor has the same effect as $\mathrm{Hom}_{U(\mathfrak{g})}(\mathbb{C},\cdot)$, where $\mathbb{C}$ is the trivial module.

(5) Let $\mathcal{C}$ be the category of all $U(\mathfrak{g})$ modules, and fix V in $\mathcal{C}$. Then $\mathrm{Hom}_{\mathbb{C}}(\cdot,V)$ is a contravariant functor from $\mathcal{C}$ to $\mathcal{C}$, $\mathrm{Hom}_{\mathbb{C}}(V,\cdot)$ is a covariant functor from $\mathcal{C}$ to $\mathcal{C}$, and $V \otimes_{\mathbb{C}} (\cdot)$ is a covariant functor from $\mathcal{C}$ to $\mathcal{C}$.

(6) Let $\mathcal{C}$ be the category of all complex vector spaces. Assign to a vector space V the tensor product $V \otimes_{\mathbb{C}} V$, and assign to a map φ in $\mathrm{Hom}_{\mathbb{C}}(V,W)$ the map $\varphi \otimes \varphi$

in $\mathrm{Hom}_{\mathbb{C}}(V \otimes_{\mathbb{C}} V, W \otimes_{\mathbb{C}} W)$. The result is a covariant functor from $\mathcal{C}$ to $\mathcal{C}$.

Any functor carries one-one onto maps to one-one onto maps. For example, let $F : \mathcal{C} \to \mathcal{C}'$ be covariant, and let $\varphi \in \mathrm{Hom}_R(A, B)$ be one-one onto; then

$$1_{F(A)} = F(1_A) = F(\varphi^{-1}\varphi) = F(\varphi^{-1})F(\varphi)$$

and

$$1_{F(B)} = F(1_B) = F(\varphi\varphi^{-1}) = F(\varphi)F(\varphi^{-1})$$

show $F(\varphi^{-1})$ is a two-sided inverse of $F(\varphi)$. Examples shortly will show that one-one maps do not necessarily go to one-one maps, nor do onto maps necessarily go to onto maps.

A functor $F : \mathcal{C} \to \mathcal{C}'$ is <u>additive</u> if

$$F(\varphi_1 + \varphi_2) = F(\varphi_1) + F(\varphi_2)$$

whenever φ_1 and φ_2 are in the same $\mathrm{Hom}_R(A, B)$. Examples (1) to (5) above are additive, but example (6) is not. We shall be concerned only with additive functors and henceforth shall use the terms "functor" and "additive functor" interchangeably.

An additive functor sends any 0 map to the corresponding 0 map. Consequently it always sends complexes to complexes. Moreover the 0 map from the 0 module to itself coincides with the identity map, and property (i) of functors thus implies that the 0 module gets sent to the 0 module.

It is true also that additive functors carry finite direct sums to finite direct sums. (Recall that good categories are closed under finite direct sums.) This fact needs proper formulation, and we need first to express direct sums in terms of modules and maps. If $C \cong A \oplus B$, then there are projections $p_A : C \to A$ and $p_B : C \to B$ and injections $\iota_A : A \to C$ and $\iota_B : B \to C$ such that

$$p_A \iota_A = 1_A \quad \text{and} \quad p_B \iota_B = 1_B \tag{5.7a}$$

$$p_B \iota_A = 0 \quad \text{and} \quad p_A \iota_B = 0 \tag{5.7b}$$

$$\iota_A p_A + \iota_B p_B = 1_C . \tag{5.7c}$$

Conversely if we have maps p_A, p_B, ι_A, and ι_B with these properties, then C is the (internal) direct sum $C = \iota_A A \oplus \iota_B B$, and ι_A and ι_B are one-one. In fact, (5.7c) shows that $\iota_A A + \iota_B B = C$. To see that $\iota_A A$ and $\iota_B B$ have 0 intersection, let x be in the intersection. Then (5.7b) gives $p_B x \in p_B \iota_A A = 0$ and $p_A x \in p_A \iota_B B = 0$; hence (5.7c) gives $0 = \iota_A p_A x + \iota_B p_B x = x$. Equations (5.7a) imply that ι_A and ι_B are one-one.

Returning to the effect of additive functors on direct sums, let $C \cong A \oplus B$ and let p_A, p_B, ι_A, ι_B be as above. Suppose that F is covariant. Applying F to the identities (5.7), we see that $F(p_A)$, $F(p_B)$, $F(\iota_A)$, $F(\iota_B)$ have the properties that allow us to recognize a direct sum. Hence

$$F(C) = F(\iota_A)F(A) \oplus F(\iota_B)F(B)$$

with $F(\iota_A)$ and $F(\iota_B)$ one-one. Thus

$$F(C) \cong F(A) \oplus F(B) .$$

If instead F is contravariant, the roles of the projections and injections get interchanged, but we still obtain $F(C) \cong F(A) \oplus F(B)$.

A functor $F: \mathbb{C} \to \mathbb{C}'$ is <u>exact</u> if it transforms exact sequences into exact sequences. Proposition 5.12 will show that exact covariant functors preserve kernels, images, submodules, quotients, and more.

<u>Examples</u>. In the category $\mathbb{C}$ of all left R modules (with R an algebra), let

$$0 \longrightarrow A_1 \xrightarrow{\psi} A_2 \xrightarrow{\varphi} A_3 \longrightarrow 0 \tag{5.8}$$

be exact.

(1) If we apply $\mathrm{Hom}_R(\cdot, V)$ to (5.8), we shall see below that

$$0 \longrightarrow \mathrm{Hom}_R(A_3, V) \xrightarrow{\mathrm{Hom}(\varphi,1)} \mathrm{Hom}_R(A_2, V) \xrightarrow{\mathrm{Hom}(\psi,1)} \mathrm{Hom}_R(A_1, V) \tag{5.9}$$

is exact. But $\mathrm{Hom}(\psi,1)$ need not be onto. In fact, let $R = U(\mathfrak{g})$ for $\mathfrak{g} = \mathbb{C}$ one-dimensional abelian. Our modules are representation spaces for the one-dimensional Lie alegbra. Let $V = \mathbb{C}$ with 0 action, and let $A_2 = \mathbb{C}^2 = \left\{\begin{pmatrix} x \\ y \end{pmatrix}\right\}$ with action

$$\pi(z)\begin{pmatrix} x \\ y \end{pmatrix} = \begin{pmatrix} 0 & z \\ 0 & 0 \end{pmatrix}\begin{pmatrix} x \\ y \end{pmatrix} = \begin{pmatrix} zy \\ 0 \end{pmatrix} \quad \text{for } z \in \mathbb{C} \ (= \mathfrak{g}) .$$

Let A_1 be the invariant subspace $\left\{\begin{pmatrix} x \\ 0 \end{pmatrix}\right\}$, and let A_3 be the quotient A_2/A_1. Then ψ is inclusion, and $\mathrm{Hom}(\psi,1)$ is restriction. Let us see that restriction does not map

$\mathrm{Hom}_{U(\mathfrak{g})}(A_2,\mathbb{C})$ onto $\mathrm{Hom}_{U(\mathfrak{g})}(A_1,\mathbb{C})$. If σ is in $\mathrm{Hom}_{U(\mathfrak{g})}(A_2,\mathbb{C})$, write $\sigma\begin{pmatrix}x\\y\end{pmatrix} = ax+by$ for complex a and b. Then

$$0 = z(\sigma\begin{pmatrix}x\\y\end{pmatrix}) = \sigma(\pi(z)\begin{pmatrix}x\\y\end{pmatrix}) = \sigma\begin{pmatrix}zy\\0\end{pmatrix} = azy$$

implies $a = 0$. So $\sigma\begin{pmatrix}x\\y\end{pmatrix} = by$. The restriction of σ to A_1 is 0. Thus $\mathrm{Hom}(\psi,1)$ has image 0. But $\mathrm{Hom}_{U(\mathfrak{g})}(A_1,\mathbb{C})$ is one-dimensional, since $A_1 \cong \mathbb{C}$ with 0 action. Thus $\mathrm{Hom}(\psi,1)$ is not onto, and $\mathrm{Hom}_R(\cdot,V)$ need not be exact.

(2) If we apply $\mathrm{Hom}_R(U,\cdot)$ to (5.8), we shall see below that

$$0 \longrightarrow \mathrm{Hom}_R(U,A_1) \xrightarrow{\mathrm{Hom}(1,\psi)} \mathrm{Hom}_R(U,A_2) \xrightarrow{\mathrm{Hom}(1,\varphi)} \mathrm{Hom}_R(U,A_3) \tag{5.10}$$

is exact. But $\mathrm{Hom}(1,\varphi)$ need not be onto. In fact, again let $R = U(\mathfrak{g})$ for $\mathfrak{g} = \mathbb{C}$ one-dimensional abelian, and take $U = \mathbb{C}$ with 0 action. Then we can identify $\mathrm{Hom}_{U(\mathfrak{g})}(\mathbb{C},V)$ with $V^{\mathfrak{g}}$ and $\mathrm{Hom}(1,\varphi)$ with restriction of φ to the subspace of invariants. With the same A_1, A_2, and A_3 as in Example (1), the invariants of A_2 are $\left\{\begin{pmatrix}x\\0\end{pmatrix}\right\}$, which map to 0 in A_3. But the invariants of A_3 comprise all of A_3 since A_3 is one-dimensional with 0 action. Thus $\mathrm{Hom}(1,\varphi)$ is not onto, and $\mathrm{Hom}_R(U,\cdot)$ need not be exact.

(3) If we apply $U\otimes_R(\cdot)$ to (5.8), we shall see below that

$$U\otimes_R A_1 \xrightarrow{1\otimes\psi} U\otimes_R A_2 \xrightarrow{1\otimes\varphi} U\otimes_R A_3 \longrightarrow 0 \tag{5.11}$$

is exact. But $1 \otimes \psi$ need not be one-one. In fact, let $\mathfrak{g}$, A_1, A_2, A_3, and U be as in Example (2). For any Lie algebra $\mathfrak{g}$ and $U(\mathfrak{g})$ module V, we have

$$\mathbb{C} \otimes_{U(\mathfrak{g})} V \cong V/\mathfrak{g}V$$

because, in $\mathbb{C} \otimes_{U(\mathfrak{g})} V$, $1 \otimes Xv = -X(1) \otimes v = 0 \otimes v = 0$. In our example, $A_1/\mathfrak{g}A_1 = A_1/\{0\} = A_1$ and $\mathfrak{g}A_2 = \left\{\begin{pmatrix} x \\ 0 \end{pmatrix}\right\} = \psi(A_1)$. Since $\mathfrak{g}A_2$ is the 0 coset in $A_2/\mathfrak{g}A_2$, $1 \otimes \psi$ maps A_1 to 0 and is not one-one, and $U \otimes_R (\cdot)$ need not be exact.

(4) If $R = U(\mathfrak{g})$, the functors $\mathrm{Hom}_{\mathbb{C}}(\cdot, V)$, $\mathrm{Hom}_{\mathbb{C}}(V, \cdot)$, and $V \otimes_{\mathbb{C}} (\cdot)$ are all exact. In fact, the Lie algebra structure is irrelevant for the question of exactness, and we shall observe below that all these functors are exact in the category of complex vector spaces.

Proposition 5.8. A functor $F : \mathcal{C} \to \mathcal{C}'$ is exact if and only if it carries all exact sequences of the form (5.8) into exact sequences.

Remark. This result is convenient for testing a functor for exactness.

Proof. Necessity is obvious. For sufficiency, let

$$B_1 \xrightarrow{\ \psi\ } B_2 \xrightarrow{\ \varphi\ } B_3$$

be exact, and let F be covariant. Put $C_1 = \ker \psi$, $C_2 = \ker \varphi$, and $C_3 = \text{image } \varphi$. Since $\varphi\psi = 0$, we can factor ψ as $\psi = \psi_2\psi_1$, where $\psi_1 : B_1 \to C_2$ is ψ with the range regarded as C_2 and where $\psi_2 : C_2 \to B_2$ is the inclusion. Similarly we can factor φ as $\varphi = \varphi_2\varphi_1$, where $\varphi_1 : B_2 \to C_3$ is φ with the range regarded as C_3 and where $\varphi_2 : C_3 \to B_3$ is the inclusion. Of the sequences

$$0 \longrightarrow C_1 \longrightarrow B_1 \xrightarrow{\psi_1} C_2 \longrightarrow 0$$

$$0 \longrightarrow C_2 \xrightarrow{\psi_2} B_2 \xrightarrow{\varphi_1} C_3 \longrightarrow 0$$

$$0 \longrightarrow C_3 \xrightarrow{\varphi_2} B_3 \longrightarrow B_3/C_3 \longrightarrow 0$$

the first and the third are trivially exact, and the second is exact since $\ker \varphi_1 = \ker \varphi = \text{image } \psi = \text{image } \psi_2$. Our hypothesis thus implies that the three sequences

$$F(B_1) \xrightarrow{F(\psi_1)} F(C_2) \longrightarrow 0$$

$$F(C_2) \xrightarrow{F(\psi_2)} F(B_2) \xrightarrow{F(\varphi_1)} F(C_3)$$

$$0 \longrightarrow F(C_3) \xrightarrow{F(\varphi_2)} F(B_3)$$

are exact. From these, $\ker F(\varphi_1) = \text{image } F(\psi_2)$. Also $F(\varphi_2)$ is one-one, so that

$$\ker F(\varphi_1) = \ker F(\varphi_2)F(\varphi_1) = \ker F(\varphi) .$$

And $F(\psi_1)$ is onto, so that

$$\text{image } F(\psi_2) = \text{image } F(\psi_2)F(\psi_1) = \text{image } F(\psi) .$$

Hence $\ker F(\varphi) = \text{image } F(\psi)$, and

$$F(B_1) \xrightarrow{F(\psi)} F(B_2) \xrightarrow{F(\varphi)} F(B_3)$$

is exact. This proves the result in the covariant case, and

the contravariant case is similar.

Proposition 5.9. Let R be an algebra over $\mathbb{C}$, and let $\mathcal{C}$ be a good category of R modules. Suppose (5.8) is an exact sequence in $\mathcal{C}$.

(a) If V is in $\mathcal{C}$, then application of $\mathrm{Hom}_R(\cdot, V)$ yields the exact sequence (5.9).

(b) If U is in $\mathcal{C}$, then application of $\mathrm{Hom}_R(U, \cdot)$ yields the exact sequence (5.10).

(c) If U is a right R module, then application of $U \otimes_R (\cdot)$ yields the exact sequence (5.11).

Proof. Since the functors in question are additive, the sequences in question are complexes.

(a) Let $\mathrm{Hom}(\varphi, 1)\sigma = 0$ with $\sigma \in \mathrm{Hom}_R(A_3, V)$. Then $\sigma(\varphi(a_2)) = 0$ for all a_2. Since φ is onto A_3, $\sigma = 0$.

Let $\mathrm{Hom}(\psi, 1)\sigma = 0$ with $\sigma \in \mathrm{Hom}_R(A_2, V)$. Then we have $\sigma(\psi(a_1)) = 0$ for all a_1. So σ vanishes on image ψ $= \ker \varphi$ and descends to $\bar{\sigma}$ on A_3. That is, $\sigma = \bar{\sigma}\varphi$ $= \mathrm{Hom}(\varphi, 1)\bar{\sigma}$. Hence σ is in the image of $\mathrm{Hom}(\varphi, 1)$.

(b) Let $\mathrm{Hom}(1, \psi)\sigma = 0$ with $\sigma \in \mathrm{Hom}_R(U, A_1)$. Then $\psi(\sigma(u)) = 0$ for all u, and $\sigma(u) = 0$ since ψ is one-one. Thus $\sigma = 0$.

Let $\mathrm{Hom}(1, \varphi)\sigma = 0$ with $\sigma \in \mathrm{Hom}_R(U, A_2)$. Then $\varphi(\sigma(u)) = 0$ for all u, and $\sigma(u) = \psi(u')$ with u' in A_1 depending on u. (Since ψ is one-one, u' is unique.) Define $\sigma' \in \mathrm{Hom}_R(U, A_1)$ by $\sigma'(u) = (\text{this } u')$. Then $\sigma(u) = \psi(u') = \psi(\sigma'(u)) = \mathrm{Hom}(1, \psi)\sigma'(u)$. Hence σ is in the image of $\mathrm{Hom}(1, \psi)$.

(c) Let $u \otimes a_3$ be given. Since $\varphi : A_2 \to A_3$ is onto, choose a_2 with $\varphi(a_2) = a_3$. Then $(1 \otimes \varphi)(u \otimes a_2) = u \otimes a_3$, and $1 \otimes \varphi$ is onto.

To show that $\ker(1 \otimes \varphi) = \text{image}(1 \otimes \psi)$, we prove that φ onto implies $\ker(1 \otimes \varphi) = U \otimes_R \ker \varphi$. Put $W = U \otimes_R \ker \varphi \subseteq U \otimes_R A_2$, and let $p : U \otimes_R A_2 \to (U \otimes_R A_2)/W$ be the quotient map. Define an "R-bilinear" map $q : U \times A_3 \to (U \otimes_R A_2)/W$ by

$$q(u, a_3) = u \otimes a_2 \quad \text{where } a_2 \text{ is chosen so that } \varphi(a_2) = a_3 .$$

This is well defined since another choice a_2' will differ by a member of $\ker \varphi$ and will affect our definition only by a member of W. The universal mapping property in Proposition 2.26 allows us to extend q to a linear map $q : U \otimes_R A_3 \to (U \otimes_R A_2)/W$, and it is clear that $p = q(1 \otimes \varphi)$. Therefore $\ker(1 \otimes \varphi) \subseteq \ker p = W$. Since the reverse inclusion is obvious, we have $\ker(1 \otimes \varphi) = W$, and the proof is complete.

Proposition 5.10. Let R be an algebra over $\mathbb{C}$, and let $\mathcal{C}$ be a good category of R modules. Then

(a) a module P in $\mathcal{C}$ is projective if and only if $\text{Hom}_R(P, \cdot)$ is an exact functor, and

(b) a module I in $\mathcal{C}$ is injective if and only if $\text{Hom}_R(\cdot, I)$ is an exact functor.

Proof. We shall use the characterization of exactness in Proposition 5.8. Thus suppose (5.8) is exact.

(a) Applying $\text{Hom}_R(P, \cdot)$ and using Proposition 5.9b, we see that the question is whether $\text{Hom}(1, \varphi)$ is onto. Suppose

τ is in $\mathrm{Hom}_R(P, A_3)$. Then we set up the diagram

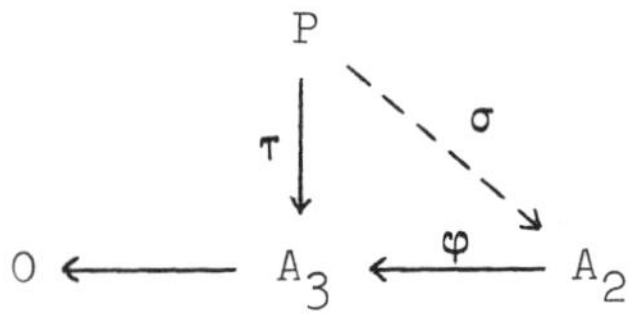

which says that $\mathrm{Hom}(1,\varphi)(\sigma) = \tau$. Thus $\mathrm{Hom}(1,\varphi)$ is onto in every such situation if and only if σ exists in every such situation, and the latter condition is the definition of "projective."

(b) Applying $\mathrm{Hom}_R(\cdot, I)$ and using Proposition 5.9a, we see that the question is whether $\mathrm{Hom}(\psi, 1)$ is onto. Suppose τ is in $\mathrm{Hom}_R(A_1, I)$. Then we set up the diagram

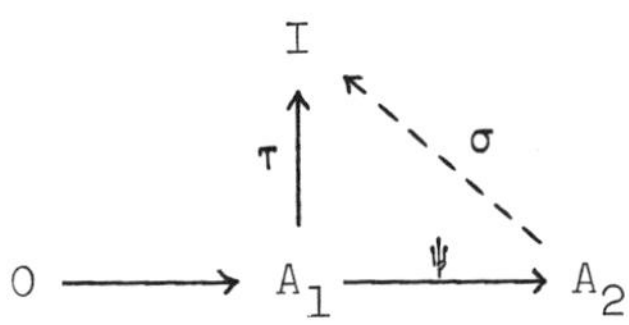

which says that $\mathrm{Hom}(\psi, 1)\sigma = \tau$. Thus $\mathrm{Hom}(\psi, 1)$ is onto in every such situation if and only if σ exists in every such situation, and the latter condition is the definition of "injective."

Corollary 5.11. In the category of all complex vector spaces, $\mathrm{Hom}_{\mathbb{C}}(\cdot, V)$, $\mathrm{Hom}_{\mathbb{C}}(V, \cdot)$, and $V \otimes_{\mathbb{C}} (\cdot)$ are all exact functors.

Proof. The exactness of the first two follows from Proposition 5.10, since every module is projective and injective. Let us apply the third functor to the exact sequence in (5.8). Then Proposition 5.9c shows that the question is whether $1 \otimes \psi : V \otimes_{\mathbb{C}} A_1 \to V \otimes_{\mathbb{C}} A_2$ is one-one, and we readily see that it is by choosing bases.

Proposition 5.12. Let $F : \mathcal{C} \to \mathcal{C}'$ be a functor between good categories, let X be a complex in $\mathcal{C}$, and let $F(X)$ be the corresponding complex in $\mathcal{C}'$. If F is exact, then F carries the homology or cohomology of X to the homology or cohomology of $F(X)$.

Remarks. (1) Our convention is to use homology when the indexing goes down and cohomology when the indexing goes up. If F is covariant, it preserves this indexing, while if F is contravariant, it reverses it.

(2) It follows from the proposition that an exact covariant functor maps kernel to kernel, image to image, submodule to submodule, quotient to quotient, and so on.

Proof in covariant case. Let

$$B_1 \xrightarrow{\psi} B_2 \xrightarrow{\varphi} B_3$$

be given with $\varphi\psi = 0$, and form

$$F(B_1) \xrightarrow{F(\psi)} F(B_2) \xrightarrow{F(\varphi)} F(B_3) .$$

We are to exhibit an isomorphism

$$F(\ker \varphi/\text{image } \psi) \cong \ker F(\varphi)/\text{image } F(\psi) . \qquad (5.12)$$

Let $i : \text{image } \psi \to \ker \varphi$ and $j : \ker \varphi \to B_2$ be the inclusions, and let $q : \ker \varphi \to \ker \varphi/\text{image } \psi$ be the quotient map. Applying F to the exact sequence

$$0 \longrightarrow \text{image } \psi \xrightarrow{\ i\ } \ker \varphi \xrightarrow{\ q\ } \ker \varphi/\text{image } \psi \longrightarrow 0$$

and using exactness, we obtain an isomorphism via $F(q)$:

$$F(\ker \varphi/\text{image } \psi) \cong F(\ker \varphi)/F(i)F(\text{image } \psi) . \qquad (5.13)$$

Since j is one-one and F is exact, $F(j)$ is one-one. Thus (5.13) is

$$\cong F(j)F(\ker \varphi)/F(ji)F(\text{image } \psi) \qquad (5.14)$$

via $F(j)$. Let $\overline{\psi}$ be ψ with range regarded as image ψ, so that $\psi = ji\overline{\psi}$. Applying F to the exact sequences

$$B_1 \xrightarrow{\ \overline{\psi}\ } \text{image } \psi \longrightarrow 0$$

$$\ker \varphi \xrightarrow{\ j\ } B_2 \xrightarrow{\ \varphi\ } B_3$$

gives us $F(\overline{\psi})F(B_1) = F(\text{image } \psi)$ and $\ker F(\varphi) = F(j)F(\ker \varphi)$. Applying $F(ji)$ to the first of these and substituting both into (5.14), we obtain the right side of (5.12).

Proof in contravariant case. Let

$$B_1 \xrightarrow{\ \psi\ } B_2 \xrightarrow{\ \varphi\ } B_3$$

be given with $\varphi\psi = 0$, and form

$$F(B_1) \xleftarrow{\ F(\psi)\ } F(B_2) \xleftarrow{\ F(\varphi)\ } F(B_3) .$$

We are to exhibit an isomorphism

$$F(\ker \varphi/\text{image } \psi) \cong \ker F(\psi)/\text{image } F(\varphi) . \qquad (5.15)$$

Let $j : \ker \varphi \to B_2$ be the inclusion, let

$$\bar{j} : \ker \varphi/\text{image } \psi \to B_2/\text{image } \psi$$

be the induced map between quotients, and let q, q', q'' be the quotient maps

$$q : B_2 \to B_2/\ker \varphi$$

$$q' : B_2 \to B_2/\text{image } \psi$$

$$q'' : B_2/\text{image } \psi \to B_2/\ker \varphi ,$$

so that $q = q''q'$. Applying F to the exact sequence

$$0 \longrightarrow \ker \varphi/\text{image } \psi \xrightarrow{\bar{j}} B_2/\text{image } \psi \xrightarrow{q''} B_2/\ker \varphi \longrightarrow 0$$

and using exactness, we obtain an isomorphism via $F(\bar{j})$:

$$F(\ker \varphi/\text{image } \psi) \cong F(B_2/\text{image } \psi)/F(q'')F(B_2/\ker \varphi) . \qquad (5.16)$$

Since q' is onto and F is exact, $F(q')$ is one-one. Thus (5.16) is

$$\cong F(q')F(B_2/\text{image } \psi)/F(q)F(B_2/\ker \varphi) \qquad (5.17)$$

via $F(q')$. Applying F to the exact sequences

$$\ker \varphi \xrightarrow{j} B_2 \xrightarrow{q} B_2/\ker \varphi$$

$$\ker \varphi \xrightarrow{j} B_2 \xrightarrow{\varphi} B_3$$

$$B_1 \xrightarrow{\psi} B_2 \xrightarrow{q'} B_2/\text{image } \psi$$

gives us $F(q)F(B_2/\ker \varphi) = \ker F(j) = \text{image } F(\varphi)$ and $F(q')F(B_2/\text{image } \psi) = \ker F(\psi)$. Substituting into (5.17), we obtain the right side of (5.15).

Proposition 5.9 and the examples before Proposition 5.8 indicate that Hom_R and $\otimes_R$ need not yield exact functors. In fact, exact functors are not very common. The more common behavior is that exhibited by Hom_R and $\otimes_R$ in Proposition 5.9, and we shall give it a name. We introduce definitions of left and right exactness so that Hom_R is left exact in each variable and $\otimes_R$ is right exact. An additive functor F is <u>left exact</u> if the exactness of (5.8) implies the exactness of

$$0 \longrightarrow F(A_1) \xrightarrow{F(\psi)} F(A_2) \xrightarrow{F(\varphi)} F(A_3) \qquad (F \text{ covariant})$$

$$0 \longrightarrow F(A_3) \xrightarrow{F(\varphi)} F(A_2) \xrightarrow{F(\psi)} F(A_1) \qquad (F \text{ contravariant}).$$

We say F is <u>right exact</u> if the exactness of (5.8) implies the exactness of

$$F(A_1) \xrightarrow{F(\psi)} F(A_2) \xrightarrow{F(\varphi)} F(A_3) \longrightarrow 0 \qquad (F \text{ covariant})$$

$$F(A_3) \xrightarrow{F(\varphi)} F(A_2) \xrightarrow{F(\psi)} F(A_1) \longrightarrow 0 \qquad (F \text{ contravariant}).$$

The words "left" and "right" refer to the part of the <u>target</u> sequence that is exact. Note that these definitions imply things about one-one or onto maps; for example, a left exact covariant F carries one-one maps to one-one maps.

<u>Proposition 5.13</u>. If F is a covariant left exact functor, then F carries an exact sequence

$$0 \longrightarrow A_1 \xrightarrow{\psi} A_2 \xrightarrow{\varphi} A_3 \tag{5.18}$$

into an exact sequence

$$0 \longrightarrow F(A_1) \xrightarrow{F(\psi)} F(A_2) \xrightarrow{F(\varphi)} F(A_3)\,. \tag{5.19}$$

Remark. The obvious analogs of this result are valid also if F is contravariant or if F is right exact.

Proof. Let (5.18) be exact, let $i : \text{image}\, \varphi \to A_3$ be the inclusion, and let $\bar{\varphi} : A_2 \to \text{image}\, \varphi$ be φ with its range space changed. Then $\varphi = i\bar{\varphi}$, and the sequences

$$0 \longrightarrow A_1 \xrightarrow{\psi} A_2 \xrightarrow{\bar{\varphi}} \text{image}\, \varphi \longrightarrow 0$$

$$0 \longrightarrow \text{image}\, \varphi \xrightarrow{i} A_3 \longrightarrow A_3/\text{image}\, \varphi \longrightarrow 0$$

are exact. Applying F and using its left exactness, we see that

$$0 \longrightarrow F(A_1) \xrightarrow{F(\psi)} F(A_2) \xrightarrow{F(\bar{\varphi})} F(\text{image}\, \varphi) \tag{5.20}$$

$$0 \longrightarrow F(\text{image}\, \varphi) \xrightarrow{F(i)} F(A_3)$$

are exact. Thus $F(i)$ is one-one, and $F(\varphi) = F(i\bar{\varphi}) = F(i)F(\bar{\varphi})$ has the same kernel as $F(\bar{\varphi})$. Since (5.20) is exact, we conclude that (5.19) is exact.

The last topic of this section has to do with isomorphic functors. Just as with functions in calculus, many nice functors are given by simple formulas. The functors $F(\cdot) = \text{Hom}_{\mathbb{C}}(\cdot \otimes_{\mathbb{C}} B, C)$ and $G(\cdot) = \text{Hom}_{\mathbb{C}}(\cdot, \text{Hom}_{\mathbb{C}}(B,C))$ on the category of vector spaces are examples. For these two expressions, we have an isomorphism

$$\text{Hom}_{\mathbb{C}}(A \otimes_{\mathbb{C}} B, C) \cong \text{Hom}_{\mathbb{C}}(A, \text{Hom}_{\mathbb{C}}(B,C)) \tag{5.21}$$

for each A. But this formula is more than a simple isomorphism; it is really an identity, and the isomorphisms for different A's are obtained in the same way. We called this property "naturality" in Chapter II: The isomorphisms are compatible with maps between different A's. In this circumstance we shall want to say that our functors F and G are isomorphic.

Let $\mathcal{C}$ and $\mathcal{C}'$ be good categories, and let $F : \mathcal{C} \to \mathcal{C}'$ and $G : \mathcal{C} \to \mathcal{C}'$ be functors, either both covariant or both contravariant. A <u>natural map</u> $T = \{T_A\}$ from F to G is a system of maps in $\mathcal{C}'$, namely $T_A : F(A) \to G(A)$ for each module A in $\mathcal{C}$, such that whenever $\varphi : A \to A'$ is a map in $\mathcal{C}$, then the appropriate one of the following diagrams commutes:

$$\begin{array}{ccc} F(A) & \xrightarrow{T_A} & G(A) \\ {\scriptstyle F(\varphi)}\downarrow & & \downarrow{\scriptstyle G(\varphi)} \\ F(A') & \xrightarrow{T_{A'}} & G(A') \end{array} \qquad \text{(F covariant)}$$

$$\begin{array}{ccc} F(A) & \xrightarrow{T_A} & G(A) \\ {\scriptstyle F(\varphi)}\uparrow & & \uparrow{\scriptstyle G(\varphi)} \\ F(A') & \xrightarrow{T_{A'}} & G(A') \end{array} \qquad \text{(F contravariant)}$$

The natural map is a <u>natural isomorphism</u> if each T_A is an isomorphism. Thus, for example, the isomorphism (5.21) is natural in this sense in each of the three variables, by Proposition 2.2.

Natural isomorphisms provide an additional tool for establishing isomorphisms. For an example, suppose that we know the two associativity formulas for Hom given in (2.49) and (2.50). From these we can derive the associativity formula for tensor products

$$(A \otimes_R B) \otimes_S C \cong A \otimes_R (B \otimes_S C) \tag{5.22}$$

given in (2.51). We write down the following string of isomorphisms, all natural in the variable V:

$$\begin{aligned} \mathrm{Hom}_{\mathbf{C}}((A \otimes_R B) \otimes_S C, V) &\cong \mathrm{Hom}_S(C, \mathrm{Hom}_{\mathbf{C}}(A \otimes_R B, V)) \\ &\cong \mathrm{Hom}_S(C, \mathrm{Hom}_R(B, \mathrm{Hom}_{\mathbf{C}}(A, V))) \\ &\cong \mathrm{Hom}_R(B \otimes_S C, \mathrm{Hom}_{\mathbf{C}}(A, V)) \\ &\cong \mathrm{Hom}_{\mathbf{C}}(A \otimes_R (B \otimes_S C), V) . \end{aligned} \tag{5.23}$$

Then we shall see that the member of the left side that corresponds to the identity on the right when $V = A \otimes_R (B \otimes_S C)$ will be the isomorphism from left to right in (5.22); moreover this isomorphism will be natural in any of the variables A, B, C for which we establish that (5.23) is natural. This conclusion follows from the proposition below. Note that without the naturality in the V variable, we get no useful result.

<u>Proposition 5.14</u>. Let $\mathbf{C}'$ be a good category of S modules, and suppose U and U' are modules in $\mathbf{C}'$ such that

$$\mathrm{Hom}_S(U, V) \cong \mathrm{Hom}_S(U', V) \tag{5.24}$$

naturally in V. Then $U \cong U'$, the isomorphism from left to right being given by the member of the left side of (5.24) that corresponds to 1 on the right side when $V = U'$. Moreover if $F : \mathcal{C} \to \mathcal{C}'$ and $G : \mathcal{C} \to \mathcal{C}'$ are functors and there are isomorphisms

$$\mathrm{Hom}_S(F(A),V) \cong \mathrm{Hom}_S(G(A),V) \tag{5.25}$$

natural for A in $\mathcal{C}$ and V in $\mathcal{C}'$, then $F(A) \cong G(A)$ naturally in A.

Remark. A corresponding result is valid if the roles of the two variables in Hom are interchanged.

Proof. Let T_V be the natural isomorphism from left to right in (5.24). Put $\varphi = T_{U'}^{-1}(1_{U'})$ and $\psi = T_U(1_U)$. Then φ is in $\mathrm{Hom}_S(U,U')$ and ψ is in $\mathrm{Hom}_S(U',U)$. We shall prove that $U \cong U'$ by showing that $\varphi\psi = 1_{U'}$ and $\psi\varphi = 1_U$.

If σ maps V to V', then naturality of (5.24) gives us the commutative diagram

$$\begin{array}{ccc} \mathrm{Hom}_S(U,V) & \xrightarrow{T_V} & \mathrm{Hom}_S(U',V) \\ {\scriptstyle \mathrm{Hom}(1,\sigma)}\downarrow & & \downarrow{\scriptstyle \mathrm{Hom}(1,\sigma)} \\ \mathrm{Hom}_S(U,V') & \xrightarrow{T_{V'}} & \mathrm{Hom}_S(U',V') \end{array}$$

Taking $V = U$, $V' = U'$, and $\sigma = \varphi$, we obtain

$$\mathrm{Hom}(1,\varphi)T_U = T_{U'}\mathrm{Hom}(1,\varphi) .$$

Applying both sides to 1_U gives

$$\varphi\psi = \mathrm{Hom}(1,\varphi)\psi = \mathrm{Hom}(1,\varphi)T_U(1_U) = T_{U'}\mathrm{Hom}(1,\varphi)(1_U) = T_{U'}\varphi = 1_{U'} .$$

Similarly the choice $V = U'$, $V' = U$, and $\sigma = \psi$ gives

$$\mathrm{Hom}(1,\psi)T_{U'} = T_U \mathrm{Hom}(1,\psi) .$$

Applying both sides to φ gives

$$\psi = \mathrm{Hom}(1,\psi)1_{U'} = \mathrm{Hom}(1,\psi)T_{U'}(\varphi) = T_U\mathrm{Hom}(1,\psi)(\varphi) = T_U(\psi\varphi) .$$

If we apply T_U^{-1} to both sides, we obtain $\psi\varphi = 1_U$. This proves the first conclusion of the proposition.

If (5.25) holds, let $T_{V,A}$ be the natural isomorphism from left to right. Then the above argument shows that

$$T^{-1}_{G(A),A}(1_{G(A)}) : F(A) \to G(A)$$

is an isomorphism. We have to prove naturality in A. Let $\sigma : A \to A'$ be given, and suppose F and G are covariant. The naturality of $T_{V,A}$ in the V variable says that

$$\begin{array}{ccc} \mathrm{Hom}_S(F(A),G(A)) & \xrightarrow{T_{G(A),A}} & \mathrm{Hom}_S(G(A),G(A)) \\ {\scriptstyle \mathrm{Hom}(1,G(\sigma))}\downarrow & & \downarrow{\scriptstyle \mathrm{Hom}(1,G(\sigma))} \\ \mathrm{Hom}_S(F(A),G(A')) & \xrightarrow{T_{G(A'),A}} & \mathrm{Hom}_S(G(A),G(A')) \end{array}$$

commutes. Tracing around this commutativity for $T^{-1}_{G(A),A}(1_{G(A)})$ in the upper left, we obtain

$$G(\sigma)T^{-1}_{G(A),A}(1_{G(A)}) = T^{-1}_{G(A'),A}(G(\sigma)) . \qquad (5.26)$$

The naturality of $T_{V,A}$ in the A variable says that

$$\begin{array}{ccc} \mathrm{Hom}_S(F(A'),G(A')) & \xrightarrow{T_{G(A'),A'}} & \mathrm{Hom}_S(G(A'),G(A')) \\ {\scriptstyle \mathrm{Hom}(F(\sigma),1)}\downarrow & & \downarrow{\scriptstyle \mathrm{Hom}(G(\sigma),1)} \\ \mathrm{Hom}_S(F(A),G(A')) & \xrightarrow{T_{G(A'),A}} & \mathrm{Hom}_S(G(A),G(A')) \end{array}$$

commutes. Tracing around this commutativity for $T^{-1}_{G(A'),A'}(1_{G(A')})$ in the upper left, we obtain

$$T^{-1}_{G(A'),A'}(1_{G(A')})F(\sigma) = T^{-1}_{G(A'),A}(G(\sigma)) . \qquad (5.27)$$

Comparing (5.26) and (5.27), we see that

$$G(\sigma)T^{-1}_{G(A),A}(1_{G(A)}) = T^{-1}_{G(A'),A'}(1_{G(A')})F(\sigma) .$$

In other words, if $\widetilde{T}_A : F(A) \to G(A)$ is given by $\widetilde{T}_A = T^{-1}_{G(A),A}(1_{G(A)})$, then

$$\begin{array}{ccc} F(A) & \xrightarrow{\widetilde{T}_A} & G(A) \\ {\scriptstyle F(\sigma)}\downarrow & & \downarrow{\scriptstyle G(\sigma)} \\ F(A') & \xrightarrow{\widetilde{T}_{A'}} & G(A') \end{array}$$

commutes. Thus $F(A) \cong G(A)$ naturally in A. This handles the covariant case, and the contravariant case is similar.

3. Derived functors

Recall the example in §IV.4 of the utility of different free resolutions. We saw that if X is the Koszul resolution and U is any $U(\mathfrak{g})$ module, then

$$H^n(X\otimes_{\mathbb{C}} U, \mathbb{C}) \cong H^n(\mathfrak{g}, U^*) .$$

Since $X\otimes_{\mathbb{C}} U$ is a free (hence projective) resolution of U, it may be replaced by any other projective resolution, essentially because of Propositions 5.3 and 5.4. Thus $H^n(\mathfrak{g}, U^*)$ may be computed by taking any projective resolution of U, applying the functor $\mathrm{Hom}_{\mathfrak{g}}(\cdot, \mathbb{C}) = ((\cdot)^*)^{\mathfrak{g}}$, and taking the cohomology of the resulting complex. In the terminology of this section, we shall say that the functors $H^n(\mathfrak{g}, (\cdot)^*)$ are the "derived functors" of the invariants-of-the-contragredient functor. Lie algebra cohomology $H^n(\mathfrak{g}, \cdot)$ itself will be the derived functors of the invariants functor.

In order to proceed further, we need to ensure that every module in our good category $\mathcal{C}$ of R modules has a projective resolution and an injective resolution. We say that $\mathcal{C}$ has <u>enough projectives</u> if every module in $\mathcal{C}$ is a quotient of a projective in $\mathcal{C}$. In this case, let M be in $\mathcal{C}$ and let X_0 be a projective that maps onto M, say by ε. Then $\ker \varepsilon$ is in $\mathcal{C}$ (since $\mathcal{C}$ is good), and we let X_1 be a projective that maps onto $\ker \varepsilon$, say by ∂_0. We let X_2 be a projective that maps onto $\ker \partial_0$ in X_1, and so on, and the result is a projective resolution

$$0 \longleftarrow M \xleftarrow{\ \varepsilon\ } X_0 \xleftarrow{\ \partial_0\ } X_1 \longleftarrow X_2 \longleftarrow \cdots$$

Hence "enough projectives" implies that every M in $\mathcal{C}$ has a projective resolution in $\mathcal{C}$. Similarly we say that $\mathcal{C}$ has <u>enough injectives</u> if every module in $\mathcal{C}$ is a submodule of an injective in $\mathcal{C}$. In this case, let M be in $\mathcal{C}$ and let X_0

be an injective into which M imbeds, say by ε. Then $X_0/\text{image } \varepsilon$ is in $\mathbf{C}$ (since $\mathbf{C}$ is good), and we let X_1 be an injective into which $X_0/\text{image } \varepsilon$ imbeds, say by $\bar{d}_0$. Let d_0 be the composition of the quotient map to $X_0/\text{image } \varepsilon$ followed by $\bar{d}_0$, so that d_0 maps X_0 to X_1 with $\ker d_0 = \text{image } \varepsilon$. We let X_2 be an injective into which $X_1/\text{image } d_0$ imbeds, and so on, and the result is an injective resolution

$$0 \longrightarrow M \xrightarrow{\varepsilon} X_0 \xrightarrow{d_0} X_1 \longrightarrow X_2 \longrightarrow \cdots$$

Hence "enough injectives" implies that every M in $\mathbf{C}$ has an injective resolution in $\mathbf{C}$.

We assume throughout this section that our good category $\mathbf{C}$ of R modules has enough projectives and enough injectives. We shall consider functors $F : \mathbf{C} \to \mathbf{C}'$, where $\mathbf{C}'$ is a good category of S modules.

Derived functors of F will be useful when F is left exact or right exact, and they will be denoted F_n or F^n. To compute them on M, we start with a projective or injective resolution of M. (Which one is indicated in Table 5.1 below.) We apply F to the resolution, drop the $F(M)$ term, and compute the homology or cohomology of the resulting complex. If the resulting complex has decreasing indices, we write $F_n(M)$ for its nth homology; otherwise we write $F^n(M)$ for its nth cohomology. The choice of projective or injective resolution at the start is made so that F_0 or F^0 is isomorphic with F. There are several things that need elaboration in this definition, and we take them up right away.

Exactness	Covariant Contravariant	Resolution	Homology Cohomology	Example
right	covariant	projective	homology	$U \otimes_R (\cdot)$
right	contravariant	injective	homology	$\mathrm{Hom}_{U(\mathfrak{g})}(U, (\cdot)^*)$
left	covariant	injective	cohomology	$\mathrm{Hom}_R(U, \cdot)$
left	contravariant	projective	cohomology	$\mathrm{Hom}_R(\cdot, U)$

Table 5.1. Formation of derived functors

First there is the fact that $F_n(M)$ or $F^n(M)$ is well defined. Suppose we start with two resolutions X and X' of M (projective or injective as in Table 5.1). Proposition 5.3 or 5.6 gives us chain or cochain maps $G : X \to X'$ and $H : X' \to X$ over $1 : M \to M$. Then $H \circ G : X \to X$ is homotopic to $1 : X \to X$, by Proposition 5.4 or 5.7, and $G \circ H : X' \to X'$ is homotopic to $1 : X' \to X'$. For definiteness, let us suppose that F is covariant and right exact, so that the derived functors of F are to be denoted F_n. Applying F to our chain maps, we obtain chain maps $F(G) : F(X) \to F(X')$, $F(H) : F(X') \to F(X)$, $F(H \circ G) : F(X) \to F(X)$, and $F(G \circ H) : F(X') \to F(X')$. The last two of these are homotopic to $1 : F(X) \to F(X)$ and $1 : F(X') \to F(X')$, respectively, by F of the respective homotopies. By Proposition 5.1, $F(H \circ G)$ and $F(G \circ H)$ induce the identity on homology, and it follows that $F(G)$ and $F(H)$ are two-sided inverses of each other on homology. Thus $F_n(M)$ computed from X is isomorphic to $F_n(M)$ computed from X'. Moreover this isomorphism is canonical: If $G' : X \to X'$ is another chain map, then the same calculation shows that $F(G')$ and $F(H)$ are two-sided of each other on homology, so that $F(G) = F(G')$ on homology. Thus $F_n(M)$ is well defined up to canonical isomorphism. (The other choices from Table 5.1 give analogous conclusions.)

Next we make F_n or F^n into a functor. To do so, let $\varphi : M \to M'$ be given. For definiteness, again let us suppose that F is covariant and right exact. Let X and X' be projective resolutions of M and M', respectively, and let $\Phi : X \to X'$ be a chain map over φ. Then $F(\Phi) : F(X) \to F(X')$ is a chain map and induces maps on homology that we denote $F_n(\varphi)$. Here $F_n(\varphi)$ maps $F_n(M)$ into $F_n(M')$.

Let us see that $F_n(\varphi)$ is well defined. If X is replaced by $\overline{X}$, let $G : X \to \overline{X}$ and $H : \overline{X} \to X$ be chain maps over $1 : M \to M$, and let $\overline{\Phi} : \overline{X} \to X'$ be a chain map over φ. Then $\overline{\Phi} \circ G$ is homotopic to Φ, and $\Phi \circ H$ is homotopic to $\overline{\Phi}$, by Proposition 5.4. By Proposition 5.1, $F(\overline{\Phi} \circ G) = F(\Phi)$ on homology, and $F(\Phi \circ H) = F(\overline{\Phi})$ on homology. Thus on homology, $F(\overline{\Phi})$ corresponds to $F(\Phi)$ under the canonical isomorphism $F(G)$ (with inverse $F(H)$). In short, $F_n(\varphi)$ is well defined up to the previously obtained canonical isomorphisms. (The other choices from Table 5.1 give analogous conclusions.)

Tracing through the definition of how derived functors affect maps, we see that the map 1 goes to the map 1 and compositions go to compositions, in the same order as for F. Thus the derived functors of F are indeed functors. The derived functors of a covariant functor are covariant, and the derived functors of a contravariant functor are contravariant.

Proposition 5.15. In the situations of Table 5.1, the 0th derived functor of F is naturally isomorphic with F.

Proof if F is covariant and right exact. Let

$$0 \longleftarrow M \xleftarrow{\varepsilon} X_0 \xleftarrow{\partial_0} X_1$$

be the first four terms of a projective resolution of M. By Proposition 5.13 and the accompanying remark, the right exactness and covariance of F imply that

$$0 \longleftarrow F(M) \xleftarrow{F(\varepsilon)} F(X_0) \xleftarrow{F(\partial_0)} F(X_1)$$

is exact. Now $F_0(M)$ is computed as the 0th homology of

$$0 \longleftarrow F(X_0) \xleftarrow{F(\partial_0)} F(X_1) .$$

Thus

$$F_0(M) = F(X_0)/\text{image } F(\partial_0) = F(X_0)/\ker F(\varepsilon) .$$

Since $F(\varepsilon)$ is onto $F(M)$, the right side here is $\cong F(M)$ via $F(\varepsilon)$.

This establishes the isomorphism. Let us prove it is natural. If $\varphi : M \to M'$ is given, we are to prove that

$$\begin{array}{ccc} F_0(M) & \xrightarrow{\text{via } F(\varepsilon)} & F(M) \\ {\scriptstyle F_0(\varphi)}\downarrow & & \downarrow{\scriptstyle F(\varphi)} \\ F_0(M') & \xrightarrow{\text{via } F(\varepsilon')} & F(M') \end{array} \qquad (5.28)$$

commutes. Thus we form a chain map as indicated:

$$\begin{array}{ccccccc}
0 & \longleftarrow & M & \xleftarrow{\ \varepsilon\ } & X_0 & \xleftarrow{\ \partial_0\ } & X_1 \\
 & & \Big\downarrow \varphi & & \Big\downarrow G_0 & & \Big\downarrow G_1 \\
0 & \longleftarrow & M' & \xleftarrow{\ \varepsilon'\ } & X_0' & \xleftarrow{\ \partial_0'\ } & X_1'
\end{array}$$

Application of F gives a commutative diagram

$$\begin{array}{ccc}
F(X_0) & \xrightarrow{\ F(\varepsilon)\ } & F(M) \\
F(G_0)\Big\downarrow & & \Big\downarrow F(\varphi) \\
F(X_0') & \xrightarrow{\ F(\varepsilon')\ } & F(M')
\end{array}$$

and this becomes (5.28) upon passage to the quotients $F(X_0)/\ker F(\varepsilon)$ and $F(X_0')/\ker F(\varepsilon')$.

It makes sense to speak of the <u>composition of two functors</u>. Derived functors of such compositions arise frequently in applications. Two cases where we can compute these derived functors without the aid of any fancy machinery are given in the following two propositions.

<u>Proposition 5.16</u>. Let $G : \mathbf{C} \to \mathbf{C}'$ be a one-sided exact functor with derived functors G_n or G^n, and let $F : \mathbf{C}' \to \mathbf{C}''$ be an exact functor.

(a) If F is covariant, then $F \circ G$ is one-sided exact, and its derived functors satisfy $(F \circ G)_n = F \circ G_n$ or $(F \circ G)^n = F \circ G^n$.

(b) If F is contravariant, then $F \circ G$ is one-sided exact, and its derived functors satisfy $(F \circ G)^n = F \circ G_n$ or $(F \circ G)_n = F \circ G^n$.

Proof. Given a module M in $\mathcal{C}$, we form the appropriate kind of resolution X of M needed to compute the derived functors of G, and the same X will be appropriate for computing the derived functors of $F\circ G$. The derived functors of G come from the homology or cohomology of $G(X)$ with $G(M)$ removed, and the derived functors of $F\circ G$ come similarly from $F(G(X))$. Thus the result follows from Proposition 5.12.

<u>Proposition 5.17</u>. Let $\widetilde{\mathcal{C}}$ and $\mathcal{C}$ be good categories with enough projectives and enough injectives, let $F:\widetilde{\mathcal{C}}\to\mathcal{C}$ be an exact functor, and let $G:\mathcal{C}\to\mathcal{C}'$ be a one-sided exact functor with derived functors G_n or G^n.

(a) Suppose that F is covariant.

(a1) Suppose G_n or G^n is defined from projective resolutions and F carries projectives to projectives. Then $G\circ F$ is one-sided exact, and its derived functors satisfy $(G\circ F)_n = G_n\circ F$ or $(G\circ F)^n = G^n\circ F$.

(a2) Suppose G_n or G^n is defined from injective resolutions and F carries injectives to injectives. Then $G\circ F$ is one-sided exact, and its derived fucntors satisfy $(G\circ F)_n = G_n\circ F$ or $(G\circ F)^n = G^n\circ F$.

(b) Suppose that F is contravariant.

(b1) Suppose G_n or G^n is defined from projective resolutions and F carries injectives to projectives. Then $G\circ F$ is one-sided exact, and its derived functors satisfy $(G\circ F)^n = G^n\circ F$ or $(G\circ F)_n = G_n\circ F$.

(b2) Suppose G_n or G^n is defined from injective

resolutions and F carries projectives to injectives. Then $G \circ F$ is one-sided exact, and its derived functors satisfy $(G \circ F)^n = G^n \circ F$ or $(G \circ F)_n = G_n \circ F$.

Proof. Let M be a module in $\widetilde{C}$, and let X be the appropriate kind of resolution for computing the derived functors of $G \circ F$ on M. Then $F(X)$ is the appropriate kind of resolution for computing the derived functors of G on $F(M)$, and the result follows.

4. Connecting homomorphisms and long exact sequences

When we apply, for example, a right exact covariant functor F to an exact sequence

$$0 \longleftarrow A \longleftarrow B \longleftarrow C \longleftarrow 0,$$

we obtain an exact sequence

$$0 \longleftarrow F(A) \longleftarrow F(B) \longleftarrow F(C),$$

and it is reasonable to ask how this exact sequence can be continued beyond $F(C)$. The answer will be that it continues forever as an exact sequence

$$0 \longleftarrow F(A) \longleftarrow F(B) \longleftarrow F(C) \longleftarrow F_1(A) \longleftarrow F_1(B)$$
$$\longleftarrow F_1(C) \longleftarrow F_2(A) \longleftarrow \cdots$$

We shall state and prove this result in §5. In the present section, we shall show how a long exact sequence of this type arises whenever we have a certain kind of diagram. Of particular interest are the mappings that above carry $F_n(A)$ to $F_{n-1}(C)$ for $n \geq 1$; these are the "connecting homomorphisms." For simplicity of notation we shall state the main result of this section in terms of complexes with decreasing indices; since the result allows negative indices, this convention involves no loss of generality.

Lemma 5.18. In a good category $\mathbf{C}$, a diagram

$$\begin{array}{ccccccccc} 0 & \longrightarrow & A & \xrightarrow{\psi} & B & \xrightarrow{\varphi} & C & \longrightarrow & 0 \\ & & \downarrow{\scriptstyle \varphi_A} & & \downarrow{\scriptstyle \varphi_B} & & \downarrow{\scriptstyle \varphi_C} & & \\ 0 & \longrightarrow & A' & \xrightarrow{\psi'} & B' & \xrightarrow{\varphi'} & C' & \longrightarrow & 0 \end{array}$$

with exact rows and commuting squares induces a map

$$\rho : \ker \varphi_C \to A'/\text{image } \varphi_A$$

with $\ker \rho = \varphi(\ker \varphi_B)$ and with

$$\text{image } \rho = \psi'^{-1}(\text{image } \varphi_B)/\text{image } \varphi_A .$$

Remarks. The map ρ is called a connecting homomorphism. Its definition occurs early in the proof and is needed on other occasions (e.g., to prove naturality later).

Proof. In obvious notation, let c be in $\ker \varphi_C$. Write $c = \varphi(b)$. Then $0 = \varphi_C(c) = \varphi_C\varphi(b) = \varphi'(\varphi_B b)$, so that $\varphi_B(b) \in \ker \varphi' = \text{image } \psi'$. Thus $\varphi_B(b) = \psi'(a')$ and a' is unique (since ψ' is one-one). We set $\rho(c) = a'$. Then ρ is well defined: If $\varphi(b) = 0$, then $b = \psi(a)$ and the unique a' is $\varphi_A(a)$ because $\psi'\varphi_A(a) = \varphi_B\psi(a) = \varphi_B(b)$.

We compute $\ker \rho$. Suppose $\rho(c) = 0$. Then $c = \varphi(b)$ with $\varphi_B(b) = \psi'(a')$ and with $a' = \varphi_A(a)$. Then $\varphi_B(b) = \psi'\varphi_A(a) = \varphi_B\psi(a)$, and $b-\psi(a)$ is in $\ker \varphi_B$. Consequently $c = \varphi(b) = \varphi(b)-\varphi\psi(a)$ is in $\varphi(\ker \varphi_B)$, and $\ker \rho \subseteq \varphi(\ker \varphi_B)$. For the reverse inclusion, if c is in $\varphi(\ker \varphi_B)$, choose $b \in \ker \varphi_B$ with $\varphi(b) = c$. Then $\varphi_C(c) = \varphi_C\varphi(b) = \varphi'\varphi_B(b) = 0$ says $\rho(c)$ is defined. Then $0 = \varphi_B(b) = \psi'(a')$ says $a' = 0$, since ψ' is one-one. Hence $\rho(c) = a' = 0$, and $\varphi(\ker \varphi_B) \subseteq \ker \rho$.

Finally we compute image ρ. Our definition of ρ makes image $\rho \subseteq \psi'^{-1}(\text{image } \varphi_B)/\text{image } \varphi_A$. For the reverse inclusion, suppose $a' \in A'$ has $\psi'(a') = \varphi_B(b)$. Then the element $c = \varphi(b)$ of C has $\varphi_C(c) = \varphi_C\varphi(b) = \varphi'\varphi_B(b)$ $= \varphi'\psi'(a') = 0$, so that $\rho(c)$ is defined. Our definition of ρ makes $\rho(c) = a'$, and thus image ρ is equal to $\psi'^{-1}(\text{image } \varphi_B)/\text{image } \varphi_A$.

<u>Lemma 5.19</u>. In a good category $\mathcal{C}$, a diagram

$$\begin{array}{ccc}
B_{n+1} & \xrightarrow{\varphi_{n+1}} & C_{n+1} \\
\downarrow{\scriptstyle \psi_B} & & \downarrow{\scriptstyle \psi_C} \\
B_n & \xrightarrow{\varphi_n} & C_n \\
\downarrow{\scriptstyle \varphi_B} & & \downarrow{\scriptstyle \varphi_C} \\
B_{n-1} & \xrightarrow{\varphi_{n-1}} & C_{n-1}
\end{array}$$

in which the columns B and C are complexes and the squares commute has $\varphi_n|_{\ker \varphi_B}$ mapping to $\ker \varphi_C/\text{image } \psi_C$ with image given by $\varphi_n(\ker \varphi_B)/\text{image } \psi_C$ and with kernel given by $\ker \varphi_B \cap \varphi_n^{-1}(\text{image } \psi_C)$. The kernel contains image ψ_B.

Remark. This is a sharpening of the statement that a chain map induces a map on homology.

Proof. If b is in $\ker \varphi_B$, then $0 = \varphi_{n-1}\varphi_B(b)$ $= \varphi_C\varphi_n(b)$, so that $\varphi_n(b)$ is in $\ker \varphi_C$. Thus $\varphi_n|_{\ker \varphi_B}$ maps into $\ker \varphi_C/\text{image } \psi_C$. The image and kernel are clear, and image ψ_B is in the kernel because

$$\varphi_n\psi_B(B_{n+1}) = \psi_C\varphi_{n+1}(B_{n+1}) \subseteq \text{image } \psi_C .$$

<u>Theorem 5.20</u>. In a good category $\mathcal{C}$, an infinite diagram

$$\begin{array}{ccccccccc}
 & & \vdots & & \vdots & & \vdots & & \\
 & & \downarrow & & \downarrow & & \downarrow & & \\
0 & \longrightarrow & A_{n+1} & \xrightarrow{\psi_{n+1}} & B_{n+1} & \xrightarrow{\varphi_{n+1}} & C_{n+1} & \longrightarrow & 0 \\
 & & \downarrow{\scriptstyle \alpha_n} & & \downarrow{\scriptstyle \beta_n} & & \downarrow{\scriptstyle \gamma_n} & & \\
0 & \longrightarrow & A_n & \xrightarrow{\psi_n} & B_n & \xrightarrow{\varphi_n} & C_n & \longrightarrow & 0 \\
 & & \downarrow{\scriptstyle \alpha_{n-1}} & & \downarrow{\scriptstyle \beta_{n-1}} & & \downarrow{\scriptstyle \gamma_{n-1}} & & \\
0 & \longrightarrow & A_{n-1} & \xrightarrow{\psi_{n-1}} & B_{n-1} & \xrightarrow{\varphi_{n-1}} & C_{n-1} & \longrightarrow & 0 \\
 & & \downarrow & & \downarrow & & \downarrow & & \\
 & & \vdots & & \vdots & & \vdots & &
\end{array} \tag{5.29}$$

(in which the columns A, B, and C are complexes, the rows are exact, and the squares commute) induces an exact sequence on homology

$$\cdots \longrightarrow H_{n+1}(C) \xrightarrow{\rho_n} H_n(A) \xrightarrow{\psi_n} H_n(B) \xrightarrow{\varphi_n} H_n(C) \xrightarrow{\rho_{n-1}} H_{n-1}(A) \longrightarrow \cdots \tag{5.30}$$

Here ψ_n and φ_n refer to the maps on homology induced by the chain maps ψ and φ, and ρ_n is induced by the connecting homomorphism for the top two indicated rows.

Proof. First let us check that ρ_n descends from $\ker \gamma_n$ to the quotient $H_{n+1}(C)$. The question is whether image $\gamma_{n+1} \subseteq \ker \rho_n$. Lemma 5.18 says $\ker \rho_n = \varphi_{n+1}(\ker \beta_n)$. We have

$$\begin{aligned} \text{image } \psi_{n+1} &= \text{image } \psi_{n+1}\varphi_{n+2} = \text{image } \varphi_{n+1}\beta_{n+1} \\ &= \varphi_{n+1}(\text{image } \beta_{n+1}) \subseteq \varphi_{n+1}(\ker \beta_n), \end{aligned}$$

and thus ρ_n does descend.

Next let us check that (5.30) is a complex. To see that $\psi_n\rho_n = 0$, we need image $\rho_n \subseteq \ker \psi_n$. In view of Lemmas 5.18 and 5.19, we need

$$\psi_n^{-1}(\text{image } \beta_n) \subseteq \ker \alpha_{n-1} \cap \psi_n^{-1}(\text{image } \beta_n). \tag{5.31}$$

Let a_n be in the left side, and write $\psi_n(a_n) = \beta_n(b_{n+1})$. Then the facts that $\psi_{n-1}\alpha_{n-1}(a_n) = \beta_{n-1}\psi_n(a_n) = \beta_{n-1}\beta_n(b_{n+1}) = 0$ and ψ_{n-1} is one-one give $\alpha_{n-1}(a_n) = 0$. Thus (5.31) holds, and $\psi_n\rho_n = 0$.

Certainly $\varphi_n\psi_n = 0$ in (5.30), because $\varphi_n\psi_n = 0$ in (5.29). Let us see that $\rho_{n-1}\varphi_n = 0$. In view of Lemmas 5.18 and 5.19, we need

$$\varphi_n(\ker \beta_{n-1}) \subseteq \ker \psi_{n-1} .$$

But b_{n-1} in $\ker \beta_{n-1}$ implies $\psi_{n-1}\varphi_n(b_{n-1}) = \varphi_{n-1}\beta_{n-1}(b_{n-1}) = 0$, and so $\rho_{n-1}\varphi_n = 0$. Thus (5.30) is a complex.

Finally we check exactness of (5.30), starting with $H_n(A)$. The kernel of ψ_n is

$$[\ker \alpha_{n-1} \cap \psi_n^{-1}(\text{image } \beta_n)]/\text{image } \alpha_n ,$$

by Lemma 5.19, and the image of ρ_n is

$$\psi_n^{-1}(\text{image } \beta_n)/\text{image } \alpha_n ,$$

by Lemma 5.18. So $\ker \psi_n \subseteq \text{image } \rho_n$ trivially.

Consider exactness at $H_n(B)$. If b_n is in $\ker \varphi_n \cap \ker \beta_{n-1}$, then $\ker \varphi_n = \text{image } \psi_n$ says $b_n = \psi_n(a_n)$. Now the facts that

$$\psi_{n-1}\alpha_{n-1}(a_n) = \beta_{n-1}\psi_n(a_n) = \beta_{n-1}(b_n) = 0$$

and ψ_{n-1} is one-one say a_n is in $\ker \alpha_{n-1}$. Thus $b_n = \psi_n(a_n)$ is ψ_n of a member of $\ker \alpha_{n-1}$, and $\ker \varphi_n \subseteq \text{image } \psi_n$.

Consider exactness at $H_n(C)$. The kernel of ρ_n is $\varphi_n(\ker \beta_{n-1})/\text{image } \gamma_n$, and the image of φ_n is $\varphi_n(\ker \beta_{n-1})/\text{image } \gamma_n$, so that $\ker \rho_n = \text{image } \varphi_n$ trivially. This completes the proof of exactness.

Proposition 5.21. In a good category $\mathbf{C}$, the long exact sequence (5.30) obtained from a diagram (5.29) is natural in the following sense: If there are two diagrams (5.29), one with twiddles (~) over all modules and maps and the other as is, and if there are vertical maps $f_{A_{n+1}}$, etc., from the twiddle version of (5.29) to the original version such that all vertical squares commute, then the squares of the diagram

$$\begin{array}{ccccccccccc}
\cdots \longrightarrow & H_{n+1}(\widetilde{C}) & \xrightarrow{\widetilde{\rho}_n} & H_n(\widetilde{A}) & \xrightarrow{\widetilde{\psi}_n} & H_n(\widetilde{B}) & \xrightarrow{\widetilde{\varphi}_n} & H_n(\widetilde{C}) & \xrightarrow{\widetilde{\rho}_{n-1}} & H_{n-1}(\widetilde{A}) & \longrightarrow \cdots \\
& \Big\downarrow f_{C_{n+1}} & & \Big\downarrow f_{A_n} & & \Big\downarrow f_{B_n} & & \Big\downarrow f_{C_n} & & \Big\downarrow f_{A_{n-1}} & \\
\cdots \longrightarrow & H_{n+1}(C) & \xrightarrow{\rho_n} & H_n(A) & \xrightarrow{\psi_n} & H_n(B) & \xrightarrow{\varphi_n} & H_n(C) & \xrightarrow{\rho_{n-1}} & H_{n-1}(A) & \longrightarrow \cdots
\end{array}$$

all commute.

Proof. The commutativity of the squares involving ψ_n and φ_n follows immediately from the commutativity of the vertical squares involving them and built over (5.29). Thus we consider the square

$$\begin{array}{ccc}
\ker \widetilde{\gamma}_n & \xrightarrow{\widetilde{\rho}_n} & \ker \widetilde{\alpha}_{n-1}/\text{image } \widetilde{\alpha}_n \\
\Big\downarrow f_{C_{n+1}} & & \Big\downarrow f_{A_n} \\
\ker \gamma_n & \xrightarrow{\rho_n} & \ker \alpha_{n-1}/\text{image } \alpha_n
\end{array} \tag{5.32}$$

Let $\widetilde{c}$ be in $\ker \widetilde{\gamma}_n$, choose $\widetilde{b}$ in $\widetilde{B}_{n+1}$ with $\widetilde{\varphi}_{n+1}(\widetilde{b}) = \widetilde{c}$, and write $\widetilde{\beta}_n(\widetilde{b}) = \widetilde{\psi}_n(\widetilde{a})$ as in the proof of Lemma 5.18. Then $\widetilde{\rho}_n(\widetilde{c})$ is defined by $\widetilde{\rho}_n(\widetilde{c}) = \widetilde{a}$. Now we know $f_{C_{n+1}}(\widetilde{c})$ is in $\ker \gamma_n$, and vertical commutativity gives us

$$\varphi_{n+1} f_{B_{n+1}}(\widetilde{b}) = f_{C_{n+1}} \widetilde{\varphi}_{n+1}(\widetilde{b}) = f_{C_{n+1}}(\widetilde{c})$$

$$\psi_n f_{A_n}(\widetilde{a}) = f_{B_n} \widetilde{\psi}_n(\widetilde{a}) = f_{B_n} \widetilde{\beta}_n(\widetilde{b}) = \beta_n f_{B_{n+1}}(\widetilde{b}) .$$

Thus $\rho_n(f_{C_{n+1}}(\tilde{c}))$ can be taken as $f_{A_n}(\tilde{a}) = f_{A_n}(\tilde{\rho}_n(\tilde{a}))$. This fact says that (5.32) commutes and completes the proof.

5. Long exact sequence of derived functors

Now let $\mathbf{C}$ and $\mathbf{C}'$ be good categories, let $F : \mathbf{C} \to \mathbf{C}'$ be a one-sided exact functor, and suppose that $\mathbf{C}$ has enough projectives and injectives. Our goal in this section is to construct from an exact sequence of modules

$$0 \longrightarrow R \xrightarrow{\;\psi\;} S \xrightarrow{\;\varphi\;} T \longrightarrow 0 \qquad (5.33)$$

in $\mathbf{C}$ compatible resolutions of R, S, and T. If the resolutions are to be projective, then the diagram is to be

$$\begin{array}{ccccccccc}
 & & 0 & & 0 & & 0 & & \\
 & & \uparrow & & \uparrow & & \uparrow & & \\
0 & \longrightarrow & R & \longrightarrow & S & \longrightarrow & T & \longrightarrow & 0 \\
 & & \uparrow & & \uparrow & & \uparrow & & \\
0 & \longrightarrow & X_0 & \longrightarrow & Y_0 & \longrightarrow & Z_0 & \longrightarrow & 0 \\
 & & \uparrow & & \uparrow & & \uparrow & & \\
0 & \longrightarrow & X_1 & \longrightarrow & Y_1 & \longrightarrow & Z_1 & \longrightarrow & 0 \\
 & & \uparrow & & \uparrow & & \uparrow & & \\
 & & \vdots & & \vdots & & \vdots & &
\end{array} \qquad (5.34)$$

If the resolutions are to be injective, then the vertical arrows will be reversed in (5.34). Anyway we can apply the functor F to the relevant two-dimensional diagram, and the result will be a diagram as in Theorem 5.20. Invoking that theorem, we shall obtain a long exact sequence relating the derived functors of F on R, S, and T.

The first lemma is included only for motivation. It says that if we are trying to construct (5.34), we may as well insist that $Y_n = X_n \oplus Z_n$ for all n. We say that an exact sequence (5.33) splits if there exists a map $\bar{\varphi} : T \to S$ in $\mathcal{C}$ with $\varphi\bar{\varphi} = 1_T$. In this case, there exists a map $\bar{\psi} : S \to R$ with $\bar{\psi}\psi = 1_R$, and S is a direct sum:

$$S = \ker \varphi \oplus \text{image } \bar{\varphi} \cong R \oplus T .$$

In fact, $\varphi(s - \bar{\varphi}\varphi s) = \varphi(s) - (\varphi\bar{\varphi})\varphi(s) = 0$ says $s - \bar{\varphi}\varphi(s) = \psi(r)$ for a unique r in R, and we define $\bar{\psi}(s) = r$. To compute $\bar{\psi}\psi$, let $s = \psi(r_0)$; then $s - \bar{\varphi}\varphi(s) = \psi(r_0) - \bar{\varphi}\varphi\psi(r_0) = \psi(r_0)$, so that $\bar{\psi}\psi = 1_R$.

Conversely if there exists $\bar{\psi} : S \to R$ with $\bar{\psi}\psi = 1_R$, then we can construct $\bar{\varphi} : T \to S$ with $\varphi\bar{\varphi} = 1_T$ and therefore exhibit the sequence as split: $\bar{\varphi}(t) = \varphi(s) - \psi\bar{\psi}(s)$, where s is any member of $\varphi^{-1}(t)$.

Lemma 5.22. Suppose that (5.33) is exact. If R is injective or T is projective, then the sequence splits.

Proof. If R is injective, we set up the diagram

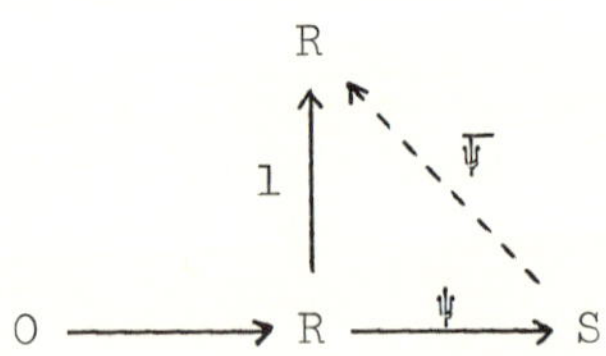

and form $\bar{\psi}$. Then $\bar{\psi}\psi = 1_R$ as required. If T is

projective, we set up the diagram

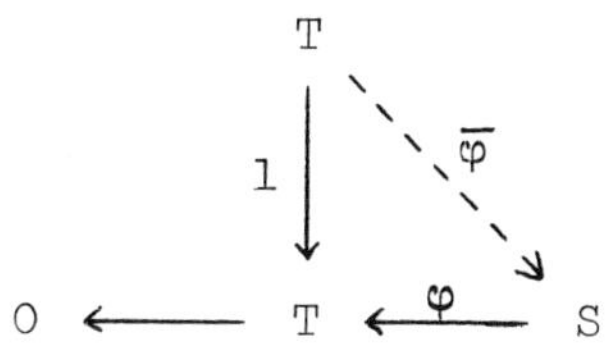

and form $\bar{\varphi}$. Then $\varphi\bar{\varphi} = 1_T$ as required.

Lemma 5.23. In the good category $\mathbf{C}$, suppose that the diagram

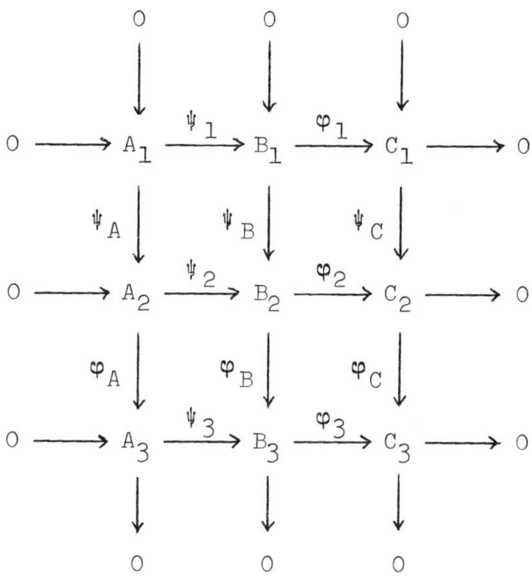

is given, and suppose that the columns are exact, the squares commute, the middle row is exact, and one other row is exact. Then the third row is exact.

Proof of exactness of first row. Since ψ_C is one-one and $\psi_C(\varphi_1\psi_1 a_1) = \varphi_2\psi_B\psi_1(a_1) = \varphi_2\psi_2\psi_A(a_1) = 0$, we see that

$\varphi_1\psi_1 = 0$ and the row is a complex. If $\psi_1(a_1) = 0$, then $0 = \psi_B\psi_1 a_1 = \psi_2\psi_A a_1$; since ψ_2 and ψ_A are one-one, $a_1 = 0$ and we conclude that ψ_1 is one-one.

Let us prove φ_1 is onto. If c_1 in C_1 is given, write $\psi_C(c_1) = \varphi_2(b_2)$. Then $0 = \varphi_C\psi_C c_1 = \varphi_C\varphi_2 b_2 = \varphi_3\varphi_B b_2$, $\varphi_B(b_2) \in \ker\varphi_3 = \text{image}\,\psi_3$, and $\varphi_B(b_2) = \psi_3(a_3) = \psi_3\varphi_A a_2$ $= \varphi_B(\psi_2 a_2)$. Hence $b_2 - \psi_2 a_2 \in \ker\varphi_B = \text{image}\,\psi_B$, $b_2 - \psi_2 a_2 = \psi_B b_1$, and $\psi_C\varphi_1 b_1 = \varphi_2\psi_B b_1 = \varphi_2(b_2 - \psi_2 a_2) = \varphi_2 b_2$ $= \psi_C(c_1)$. Since ψ_C is one-one, $\varphi_1 b_1 = c_1$. Thus φ_1 is onto.

Finally let us prove $\ker\varphi_1 \subseteq \text{image}\,\psi_1$. Let $\varphi_1(b_1) = 0$. Then $0 = \psi_C\varphi_1 b_1 = \varphi_2\psi_B b_1$, $\psi_B b_1 \in \ker\varphi_2$ $= \text{image}\,\psi_2$, and $\psi_B b_1 = \psi_2 a_2$. Hence $\psi_3\varphi_A a_2 = \varphi_B\psi_2 a_2$ $= \varphi_B\psi_B b_1 = 0$, $\varphi_A a_2 = 0$, $a_2 = \psi_A a_1$, and $\psi_B\psi_1 a_1 = \psi_2\psi_A a_1$ $= \psi_2 a_2 = \psi_B b_1$. Since ψ_B is one-one, $\psi_1 a_1 = b_1$. Thus $\ker\varphi_1 \subseteq \text{image}\,\psi_1$.

Proof of exactness of third row. Since $\varphi_3\psi_3 a_3 = \varphi_3\psi_3\varphi_A a_2$ $= \varphi_3\varphi_B\psi_2 a_2 = \varphi_C\varphi_2\psi_2 a_2 = \varphi_C 0 = 0$, the row is a complex. If c_3 in C_3 is given, then $c_3 = \varphi_C c_2 = \varphi_C\varphi_2 b_2 = \varphi_3(\varphi_B b_2)$, and we conclude that φ_3 is onto.

Let us prove that ψ_3 is one -one. Let $\psi_3 a_3 = 0$. Then $0 = \psi_3 a_3 = \psi_3\varphi_A a_2 = \varphi_B\psi_2 a_2$, $\psi_2 a_2 \in \ker\varphi_B = \text{image}\,\psi_B$, and $\psi_2 a_2 = \psi_B b_1$. Then $0 = \varphi_2\psi_2 a_2 = \varphi_2\psi_B b_1 = \psi_C\varphi_1 b_1$. Since ψ_C is one-one, $\varphi_1 b_1 = 0$; thus $b_1 \in \ker\varphi_1$ $= \text{image}\,\psi_1$, and $b_1 = \psi_1 a_1$. Hence $\psi_2 a_2 = \psi_B b_1 = \psi_B\psi_1 a_1$ $= \psi_2\psi_A a_1$. Since ψ_2 is one-one, $a_2 = \psi_A a_1$. Thus $a_3 = \varphi_A a_2 = \varphi_A\psi_A a_1 = 0$, and we conclude that ψ_3 is one-one.

Finally let us prove $\ker \varphi_3 \subseteq \text{image } \psi_3$. Let $\varphi_3(b_3) = 0$. Then $b_3 = \varphi_B b_2$, $0 = \varphi_3 \varphi_B b_2 = \varphi_C \varphi_2 b_2$, $\varphi_2 b_2 \in \ker \varphi_C = \text{image } \psi_C$, and $\varphi_2 b_2 = \psi_C c_1 = \psi_C \varphi_1 b_1 = \varphi_2 \psi_B b_1$. Hence $b_2 - \psi_B b_1$ is in $\ker \varphi_2 = \text{image } \psi_2$, and $b_2 - \psi_B b_1 = \psi_2 a_2$. Thus $b_3 = \varphi_B b_2 = \varphi_B \psi_B b_1 + \varphi_B \psi_2 a_2 = 0 + \varphi_B \psi_2 a_2 = \psi_3(\varphi_A a_2)$. Thus $\ker \varphi_3 \subseteq \text{image } \psi_3$.

Proposition 5.24. In the good category $\mathbf{C}$, suppose that P_A and P_C are projective and that the diagram

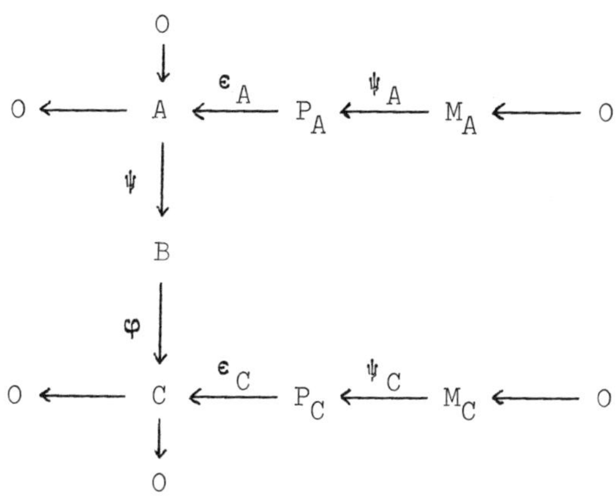

has the column and two rows exact. Then there exist a module M_B and maps in $\mathbf{C}$ so that the diagram that follows has exact rows and columns and has commuting squares. In the diagram $P_A \oplus P_C$ is a projective in $\mathbf{C}$, i_A is the usual inclusion, and p_C is the usual projection.

$$
\begin{array}{ccccccccc}
 & & 0 & & 0 & & 0 & & \\
 & & \downarrow & & \downarrow & & \downarrow & & \\
0 & \longleftarrow & A & \xleftarrow{\epsilon_A} & P_A & \xleftarrow{\psi_A} & M_A & \longleftarrow & 0 \\
 & & \psi\downarrow & & \downarrow i_A & & \downarrow \psi_1 & & \\
0 & \longleftarrow & B & \xleftarrow{\epsilon_B} & P_A \oplus P_C & \xleftarrow{\psi_B} & M_B & \longleftarrow & 0 \\
 & & \varphi\downarrow & & \downarrow p_C & & \downarrow \varphi_1 & & \\
0 & \longleftarrow & C & \xleftarrow{\epsilon_C} & P_C & \xleftarrow{\psi_C} & M_C & \longleftarrow & 0 \\
 & & \downarrow & & \downarrow & & \downarrow & & \\
 & & 0 & & 0 & & 0 & &
\end{array}
$$

Proof. $P_A \oplus P_C$ is in $\mathbf{C}$ since $\mathbf{C}$ is good, and it is easy to see $P_A \oplus P_C$ is projective. We define ϵ_B. Since P_C is projective, choose $h : P_C \to B$ so that $\varphi h = \epsilon_C$, and put $\epsilon_B(x_A, x_C) = \psi\epsilon_A x_A + hx_C$. Then

$$\psi\epsilon_A x_A = \epsilon_B(x_A, 0) = \epsilon_B i_A x_A$$

says the upper left square commutes, and

$$\varphi\epsilon_B(x_A, x_C) = \varphi\psi\epsilon_A x_A + \varphi h x_C = 0 + \epsilon_C x_C$$

says the lower left square commutes.

To see that ϵ_B is onto B, let $b \in B$ be given. Since p_C and ϵ_C are onto, so is $\epsilon_C p_C = \varphi\epsilon_B$. Thus we can choose (x_A, x_C) in $P_A \oplus P_C$ with $\varphi(b) = \varphi\epsilon_B(x_A, x_C)$. Hence $b - \epsilon_B(x_A, x_C)$ is in $\ker \varphi = \text{image } \psi$, and we can write

$$b - \epsilon_B(x_A, x_C) = \psi(a) = \psi\epsilon_A(x_A') = \epsilon_B i_A(x_A') = \epsilon_B(x_A', 0) .$$

Then $b = \epsilon_B(x_A + x_A', x_C)$, and so ϵ_B is onto.

Let $M_B = \ker \varepsilon_B$, and let $\psi_B : M_B \to P_A \oplus P_C$ be the inclusion. For $m_A \in M_A$, let $\psi_1(m_A) = (\psi_A m_A, 0)$. Then $\psi_1(m_A)$ is in M_B since

$$\varepsilon_B(\psi_A m_A, 0) = \psi\varepsilon_A \psi_A m_A + h0 = \psi 0 + h0 = 0 .$$

Moreover this definition of ψ_1 makes the upper right square commute.

To define φ_1, let (x_A, x_C) be in M_B, so that $\varepsilon_B(x_A, x_C) = 0$. Then $0 = \varphi\varepsilon_B(x_A, x_C) = \varepsilon_C p_C(x_A, x_C) = \varepsilon_C(x_C)$, $x_C \in \ker \varepsilon_C = \text{image } \psi_C$, and $x_C = \psi_C(m_C)$ for a unique m_C in M_C. So we put $\varphi_1(x_A, x_C) = m_C$. Then

$$\psi_C\varphi_1(x_A, x_C) = \psi_C(m_C) = x_C = p_C(x_A, x_C) = p_C\psi_B(x_A, x_C)$$

shows the lower right square commutes.

Now all the squares commute, and all the rows and columns are exact except possibly the third column. Lemma 5.23 allows us to conclude that the third column is exact, and the proof is complete.

<u>Corollary 5.25</u>. In the good category $\mathbf{C}$, let (5.33) be an exact sequence, and let X and Z be projective resolutions of R and T, respectively. Then there exists a projective resolution Y of S such that (5.34) has commuting squares and exact rows. Moreover Y_n can be taken as $X_n \oplus Z_n$, and the horizontal maps can be taken as the inclusion of X_n into $X_n \oplus Z_n$ and the projection of $X_n \oplus Z_n$ on Z_n.

Proof. Proposition 5.24 allows us to construct

$$\begin{array}{ccccccccc}
 & & 0 & & 0 & & 0 & & \\
 & & \downarrow & & \downarrow & & \downarrow & & \\
0 & \longleftarrow & R & \xleftarrow{\epsilon_R} & X_0 & \xleftarrow{\text{inc}} & \ker \epsilon_R & \longleftarrow & 0 \\
 & & \psi\downarrow & & \downarrow & & \downarrow\psi_1 & & \\
0 & \longleftarrow & S & \xleftarrow{\epsilon_S} & X_0 \oplus Z_0 & \xleftarrow{\text{inc}} & \ker \epsilon_S & \longleftarrow & 0 \\
 & & \varphi\downarrow & & \downarrow & & \downarrow\varphi_1 & & \\
0 & \longleftarrow & T & \longleftarrow & Z_0 & \xleftarrow{\text{inc}} & \ker \epsilon_T & \longleftarrow & 0 \\
 & & \downarrow & & \downarrow & & \downarrow & & \\
 & & 0 & & 0 & & 0 & &
\end{array} \qquad (5.35)$$

with commuting squares and with exact rows and columns. Applying the proposition to

$$\begin{array}{ccccccccc}
 & & 0 & & & & & & \\
 & & \downarrow & & & & & & \\
0 & \longleftarrow & \ker \epsilon_R & \xleftarrow{\partial_X} & X_1 & \xleftarrow{\text{inc}} & \ker \partial_X & \longleftarrow & 0 \\
 & & \downarrow & & & & & & \\
 & & \ker \epsilon_S & & & & & & \\
 & & \downarrow & & & & & & \\
0 & \longleftarrow & \ker \epsilon_T & \xleftarrow{\partial_Z} & Z_1 & \xleftarrow{\text{inc}} & \ker \partial_Z & \longleftarrow & 0 \\
 & & \downarrow & & & & & & \\
 & & 0 & & & & & &
\end{array}$$

we construct

$$\begin{array}{ccccccccc}
 & & 0 & & 0 & & 0 & & \\
 & & \downarrow & & \downarrow & & \downarrow & & \\
0 & \longleftarrow & \ker \varepsilon_R & \xleftarrow{\partial_X} & X_1 & \xleftarrow{\text{inc}} & \ker \partial_X & \longleftarrow & 0 \\
 & & \downarrow{\scriptstyle \psi_1} & & \downarrow & & \downarrow{\scriptstyle \psi_2} & & \\
0 & \longleftarrow & \ker \varepsilon_S & \xleftarrow{\partial_Y} & X_1 \oplus Z_1 & \xleftarrow{\text{inc}} & \ker \partial_Y & \longleftarrow & 0 \\
 & & \downarrow{\scriptstyle \varphi_1} & & \downarrow & & \downarrow{\scriptstyle \varphi_2} & & \\
0 & \longleftarrow & \ker \varepsilon_T & \xleftarrow{\partial_Z} & Z_1 & \xleftarrow{\text{inc}} & \ker \partial_Z & \longleftarrow & 0 \\
 & & \downarrow & & \downarrow & & \downarrow & & \\
 & & 0 & & 0 & & 0 & &
\end{array} \qquad (5.36)$$

with commuting squares and with exact rows and columns. We can put (5.35) and (5.36) together as

$$\begin{array}{ccccccccccc}
 & & 0 & & 0 & & 0 & & 0 & & \\
 & & \downarrow & & \downarrow & & \downarrow & & \downarrow & & \\
0 & \longleftarrow & R & \xleftarrow{\varepsilon_R} & X_0 & \xleftarrow{\partial_X} & X_1 & \xleftarrow{\text{inc}} & \ker \partial_X & \longleftarrow & 0 \\
 & & \downarrow{\scriptstyle \psi} & & \downarrow & & \downarrow & & \downarrow{\scriptstyle \psi_2} & & \\
0 & \longleftarrow & S & \xleftarrow{\varepsilon_S} & X_0 \oplus Z_0 & \xleftarrow{\partial_Y} & X_1 \oplus Z_1 & \xleftarrow{\text{inc}} & \ker \partial_Y & \longleftarrow & 0 \\
 & & \downarrow{\scriptstyle \varphi} & & \downarrow & & \downarrow & & \downarrow{\scriptstyle \varphi_2} & & \\
0 & \longleftarrow & T & \xleftarrow{\varepsilon_T} & Z_0 & \xleftarrow{\partial_Z} & Z_1 & \xleftarrow{\text{inc}} & \ker \partial_Z & \longleftarrow & 0 \\
 & & \downarrow & & \downarrow & & \downarrow & & \downarrow & & \\
 & & 0 & & 0 & & 0 & & 0 & &
\end{array} \qquad (5.37)$$

We can repeat the use of Proposition 5.24, starting from the last column of (5.37) and adjoining the resulting diagram in similar fashion. Continuing in this way, we arrive at (5.34) as required.

<u>Proposition 5.26</u>. In the good category $\mathbf{C}$, suppose that I_A and I_C are injective and that the diagram

$$\begin{array}{ccccccccc}
 & & 0 & & & & & & \\
 & & \downarrow & & & & & & \\
0 & \longrightarrow & A & \xrightarrow{\epsilon_A} & I_A & \xrightarrow{\varphi_A} & Q_A & \longrightarrow & 0 \\
 & & \psi\downarrow & & & & & & \\
 & & B & & & & & & \\
 & & \varphi\downarrow & & & & & & \\
0 & \longrightarrow & C & \xrightarrow{\epsilon_C} & I_C & \xrightarrow{\varphi_C} & Q_C & \longrightarrow & 0 \\
 & & \downarrow & & & & & & \\
 & & 0 & & & & & &
\end{array}$$

has the column and two rows exact. Then there exist a module Q_B and maps in $\mathbf{C}$ so that the diagram

$$\begin{array}{ccccccccc}
 & & 0 & & 0 & & 0 & & \\
 & & \downarrow & & \downarrow & & \downarrow & & \\
0 & \longrightarrow & A & \xrightarrow{\epsilon_A} & I_A & \xrightarrow{\varphi_A} & Q_A & \longrightarrow & 0 \\
 & & \psi\downarrow & & \downarrow i_A & & \downarrow\psi_1 & & \\
0 & \longrightarrow & B & \xrightarrow{\epsilon_B} & I_A \oplus I_C & \xrightarrow{\varphi_B} & Q_B & \longrightarrow & 0 \\
 & & \varphi\downarrow & & \downarrow p_C & & \downarrow\varphi_1 & & \\
0 & \longrightarrow & C & \xrightarrow{\epsilon_C} & I_C & \xrightarrow{\varphi_C} & Q_C & \longrightarrow & 0 \\
 & & \downarrow & & \downarrow & & \downarrow & & \\
 & & 0 & & 0 & & 0 & &
\end{array}$$

has exact rows and columns and has commuting squares. Here

$I_A \oplus I_C$ is an injective in $\mathcal{C}$, i_A is the usual inclusion, and p_C is the usual projection.

Proof. $I_A \oplus I_C$ is in $\mathcal{C}$ since $\mathcal{C}$ is good, and it is easy to see that $I_A \oplus I_C$ is injective. We define ϵ_B. Since I_A is injective, choose $h : B \to I_A$ so that $h\psi = \epsilon_A$, and put $\epsilon_B(b) = (h(b), \epsilon_C\varphi(b))$. Then

$$\epsilon_B\psi(a) = (h\psi a, \epsilon_C\varphi\psi a) = (\epsilon_A(a), 0) = i_A\epsilon_A(a)$$

says the upper left square commutes, and

$$\epsilon_C\varphi(b) = p_C(h(b), \epsilon_C\varphi(b)) = p_C\epsilon_B(b)$$

says the lower left square commutes.

To see that ϵ_B is one-one, let $\epsilon_B(b) = 0$. Then $0 = p_C\epsilon_B(b) = \epsilon_C\varphi(b)$. Since ϵ_C is one-one, $\varphi(b) = 0$, b is in $\ker\varphi = \text{image}\ \psi$, and $b = \psi(a)$. Then $0 = \epsilon_B(b) = \epsilon_B\psi(a) = i_A\epsilon_A(a)$, and $a = 0$ since i_A and ϵ_A are one-one. Hence $b = \psi(a) = 0$, and ϵ_B is one-one.

Let $Q_B = (I_A \oplus I_C)/B$, and let $\varphi_B : I_A \oplus I_C \to Q_B$ be the quotient map. Lemma 4.17 allows us to fill in ψ_1 and φ_1 so that the upper right and lower right squares commute.

Now all the squares commute, and all the rows and columns are exact except possibly the third column. Lemma 5.23 allows us to conclude that the third column is exact, and the proof is complete.

Corollary 5.27. In the good category $\mathcal{C}$, let (5.33) be an exact sequence, and let X and Z be injective resolutions of R and T, respectively. Then there exists an injective resolution Y of S such that (5.38) below has commuting squares and exact rows. Moreover Y_n can be taken as $X_n \oplus Z_n$, and the horizontal maps can be taken as the inclusion of X_n into $X_n \oplus Z_n$ and the projection of $X_n \oplus Z_n$ on Z_n.

$$
\begin{array}{ccccccccc}
 & & 0 & & 0 & & 0 & & \\
 & & \downarrow & & \downarrow & & \downarrow & & \\
0 & \longrightarrow & R & \longrightarrow & S & \longrightarrow & T & \longrightarrow & 0 \\
 & & \downarrow & & \downarrow & & \downarrow & & \\
0 & \longrightarrow & X_0 & \longrightarrow & Y_0 & \longrightarrow & Z_0 & \longrightarrow & 0 \\
 & & \downarrow & & \downarrow & & \downarrow & & \\
0 & \longrightarrow & X_1 & \longrightarrow & Y_1 & \longrightarrow & Z_1 & \longrightarrow & 0 \\
 & & \downarrow & & \downarrow & & \downarrow & & \\
 & & \vdots & & \vdots & & \vdots & &
\end{array} \qquad (5.38)
$$

Proof. This follows from Proposition 5.26 in the same way that Corollary 5.25 followed from Proposition 5.24.

Now we are ready to obtain a long exact sequence for our derived functors. Suppose $\mathbf{C}$ is a good category with enough projectives and enough injectives. Let $\mathbf{C}'$ be a good category, let $F : \mathbf{C} \to \mathbf{C}'$ be a one-sided exact functor, and let the exact sequence (5.33) in $\mathbf{C}$ be given. We form projective or injective resolutions of R and T, whichever is appropriate for the computation of the derived functors of F (see Table 5.1 in §3). Using Corollary 5.25 or Corollary 5.27, we fill out the diagram to (5.34) or (5.38). Then we apply F to the two-dimensional diagram and drop the row containing $F(R)$, $F(S)$, and $F(T)$.

The resulting diagram has complexes for columns and has commuting squares. But also its rows are exact. The reason is that the exact sequence

$$0 \longrightarrow X_n \xrightarrow{\psi_n} Y_n \xrightarrow{\varphi_n} Z_n \longrightarrow 0$$

is split. If, for example, F is covariant and left exact, choose $\bar{\varphi}_n : Z_n \to Y_n$ with $\varphi_n \bar{\varphi}_n = 1_{Z_n}$. Then

$$0 \longrightarrow F(X_n) \xrightarrow{F(\psi_n)} F(Y_n) \xrightarrow{F(\varphi_n)} F(Z_n)$$

is exact since F is left exact, and $F(\varphi_n)F(\bar{\varphi}_n) = F(1_{Z_n}) = 1_{F(Z_n)}$ says $F(\varphi_n)$ is onto. Thus the rows in F of (5.34) or (5.38) are exact, and Theorem 5.20 is applicable. We therefore obtain the following result.

<u>Theorem 5.28</u>. Let $\mathcal{C}$ and $\mathcal{C}'$ be good categories, let

$$0 \longrightarrow R \xrightarrow{\psi} S \xrightarrow{\varphi} T \longrightarrow 0$$

be an exact sequence in $\mathcal{C}$, let $F : \mathcal{C} \to \mathcal{C}'$ be a one-sided exact functor, and suppose $\mathcal{C}$ has enough projectives and enough injectives. Then the derived functors of F on R, S, and T form a long exact sequence as follows:

(a) If F is covariant and right exact, the long exact sequence is

$$0 \longleftarrow F(T) \longleftarrow F(S) \longleftarrow F(R) \longleftarrow F_1(T) \longleftarrow F_1(S) \longleftarrow F_1(R)$$

$$\longleftarrow F_2(T) \longleftarrow F_2(S) \longleftarrow F_2(R) \longleftarrow F_3(T) \longleftarrow \cdots .$$

(b) If F is contravariant and right exact, the long exact sequence is

$$0 \longleftarrow F(R) \longleftarrow F(S) \longleftarrow F(T) \longleftarrow F_1(R) \longleftarrow F_1(S) \longleftarrow F_1(T)$$
$$\longleftarrow F_2(R) \longleftarrow F_2(S) \longleftarrow F_2(T) \longleftarrow F_3(R) \longleftarrow \cdots .$$

(c) If F is covariant and left exact, the long exact sequence is

$$0 \longrightarrow F(R) \longrightarrow F(S) \longrightarrow F(T) \longrightarrow F^1(R) \longrightarrow F^1(S) \longrightarrow F^1(T)$$
$$\longrightarrow F^2(R) \longrightarrow F^2(S) \longrightarrow F^2(T) \longrightarrow F^3(R) \longrightarrow \cdots .$$

(d) If F is contravariant and left exact, the long exact sequence is

$$0 \longrightarrow F(T) \longrightarrow F(S) \longrightarrow F(R) \longrightarrow F^1(T) \longrightarrow F^1(S) \longrightarrow F^1(R)$$
$$\longrightarrow F^2(T) \longrightarrow F^2(S) \longrightarrow F^2(R) \longrightarrow F^3(T) \longrightarrow \cdots .$$

6. Naturality of long exact sequence of derived functors

Theorem 5.28 has an accompanying naturality result as follows. We include this result for completeness but will have no occasion to use it.

Theorem 5.29. Let $\mathcal{C}$ and $\mathcal{C}'$ be good categories, let

$$\begin{array}{ccccccccc}
0 & \longrightarrow & \widetilde{R} & \xrightarrow{\widetilde{\psi}} & \widetilde{S} & \xrightarrow{\widetilde{\varphi}} & \widetilde{T} & \longrightarrow & 0 \\
 & & \downarrow{\scriptstyle f_R} & & \downarrow{\scriptstyle f_S} & & \downarrow{\scriptstyle f_T} & & \\
0 & \longrightarrow & R & \xrightarrow{\psi} & S & \xrightarrow{\varphi} & T & \longrightarrow & 0
\end{array}$$

be a diagram in $\mathcal{C}$ with exact rows and commuting squares, let $F : \mathcal{C} \to \mathcal{C}'$ be a one-sided exact functor, and suppose $\mathcal{C}$ has enough projectives and enough injectives. Then the long exact sequences of derived functors of F on $\widetilde{R}, \widetilde{S}, \widetilde{T}$ and R, S, T make commutative squares with the maps induced by the derived functors on f_R, f_S, and f_T.

<u>Lemma 5.30</u>. Let $\mathcal{C}$ be a good category, and let

$$\begin{array}{ccccccccc}
0 & \longrightarrow & A' & \xrightarrow{\psi'} & B' & \xrightarrow{\varphi'} & C' & \longrightarrow & 0 \\
 & & \uparrow{\scriptstyle \varepsilon_A} & & \uparrow{\scriptstyle \varepsilon_B} & & \uparrow{\scriptstyle \varepsilon_C} & & \\
0 & \longrightarrow & A & \xrightarrow{\psi} & B & \xrightarrow{\varphi} & C & \longrightarrow & 0 \\
 & & \uparrow{\scriptstyle \partial_A} & & \uparrow{\scriptstyle \partial_B} & & \uparrow{\scriptstyle \partial_C} & & \\
0 & \longrightarrow & X & \xrightarrow{i_X} & X \oplus Z & \xrightarrow{p_Z} & Z & \longrightarrow & 0
\end{array} \qquad (5.39)$$

be a diagram in $\mathcal{C}$ with exact rows and columns, with commuting squares, and with X and Z projective. Suppose also that a second copy of (5.39), with twiddles ($\sim$) over everything has the same properties and lies over (5.39). In addition, suppose that there are eight vertical maps in $\mathcal{C}$

$$f_{A'}, f_{B'}, f_{C'}, f_A, f_B, f_C, f_X, f_Z$$

from the twiddle diagram to the original diagram such that all the vertical squares commute. Then there exists a vertical map $f_{X\oplus Z} : \widetilde{X} \oplus \widetilde{Z}$ in $\mathcal{C}$ such that all three vertical squares commute that involve $f_{X\oplus Z}$.

Proof. For $\tilde{z} \in \widetilde{Z}$, we apply φ to $\partial_B(0, f_Z\tilde{z}) - f_B\partial_{\widetilde{B}}(0, \tilde{z})$ and get

$$\begin{aligned}\varphi\partial_B(0,f_Z\tilde{z}) - \varphi f_B\partial_{\tilde{B}}(0,\tilde{z}) &= \partial_C p_Z(0,f_Z\tilde{z}) - f_C\tilde{\varphi}\partial_{\tilde{B}}(0,\tilde{z}) \\ &= \partial_C f_Z\tilde{z} - f_C\partial_{\tilde{C}}p_{\tilde{Z}}(0,\tilde{z}) \\ &= f_C\partial_{\tilde{C}}\tilde{z} - f_C\partial_{\tilde{C}}\tilde{z} = 0 .\end{aligned}$$

Thus $\partial_B(0,f_Z\tilde{z})-f_B\partial_{\tilde{B}}(0,\tilde{z}) = \psi(a)$ for a unique $a \in A$, and we define $\tau : \tilde{Z} \to A$ by saying that $\tau(\tilde{z})$ should be this a. This makes

$$\psi\tau(\tilde{z}) = \partial_B(0,f_Z\tilde{z}) - f_B\partial_{\tilde{B}}(0,\tilde{z}) .$$

Setting up

$$\begin{array}{ccccc} & & \tilde{Z} & & \\ & & \downarrow{\scriptstyle\tau} & \overset{\sigma}{\searrow} & \\ A' & \xleftarrow{\epsilon_A} & A & \xleftarrow{\partial_A} & X \end{array}$$

we prepare to invoke Lemma 5.2. We have

$$\begin{aligned}\psi'\epsilon_A\tau(\tilde{z}) = \epsilon_B\psi\tau(\tilde{z}) &= \epsilon_B\partial_B(0,f_Z\tilde{z}) - \epsilon_B f_B\partial_{\tilde{B}}(0,\tilde{z}) \\ &= 0 - f_{B'}\epsilon_{\tilde{B}}\partial_{\tilde{B}}(0,\tilde{z}) = 0 .\end{aligned}$$

Since ψ' is one-one, $\epsilon_A\tau = 0$ and Lemma 5.2 applies. Thus we obtain $\sigma : \tilde{Z} \to X$ with $\partial_A\sigma = \tau$, and σ satisfies

$$\psi\partial_A\sigma(\tilde{z}) = \partial_B(0,f_Z\tilde{z}) - f_B\partial_{\tilde{B}}(0,\tilde{z}) . \tag{5.40}$$

Define

$$f_{X\oplus Z}(\tilde{x},\tilde{z}) = (f_X(\tilde{x})-\sigma(\tilde{z}), f_Z(\tilde{z})) . \tag{5.41}$$

The three commutativity relations that are to be proved are

$$i_X f_X(\tilde{x}) = f_{X\oplus Z} i_{\tilde{X}}(\tilde{x}) \tag{5.42}$$

$$p_Z f_{X\oplus Z}(\tilde{x},\tilde{z}) = f_Z p_{\tilde{Z}}(\tilde{x},\tilde{z}) \tag{5.43}$$

$$f_B \partial_{\tilde{B}}(\tilde{x},\tilde{z}) = \partial_B f_{X\oplus Z}(\tilde{x},\tilde{z}) . \tag{5.44}$$

Relation (5.42) follows from (5.41) by taking $\tilde{z} = 0$, and (5.43) is just the second coordinate of (5.41). For (5.44) we observe from (5.42) that

$$\begin{aligned}\partial_B f_{X\oplus Z}(\tilde{x},0) &= \partial_B f_{X\oplus Z} i_{\tilde{X}}(\tilde{x}) = \partial_B i_X f_X(\tilde{x}) \\ &= \psi\partial_A f_X(\tilde{x}) = \psi f_A \partial_{\tilde{A}}(\tilde{x}) \\ &= f_B \tilde{\psi}\partial_{\tilde{A}}(\tilde{x}) = f_B \partial_{\tilde{B}} i_{\tilde{X}}(\tilde{x}) \\ &= f_B \partial_{\tilde{B}}(\tilde{x},0)\end{aligned} \tag{5.45}$$

and that

$$\begin{aligned}\partial_B f_{X\oplus Z}(0,\tilde{z}) &= -\partial_B i_X \sigma(\tilde{z}) + \partial_B(0, f_Z(\tilde{z})) && \text{by (5.41)} \\ &= -\psi\partial_A \sigma(\tilde{z}) + \partial_B(0, f_Z(\tilde{z})) && \text{by commutativity} \\ &= f_B \partial_{\tilde{B}}(0,\tilde{z}) && \text{by (5.40).}\end{aligned} \tag{5.46}$$

Adding (5.45) and (5.46), we obtain (5.44).

<u>Lemma 5.31</u>. Let $\mathbf{C}$ be a good category, and let

$$\begin{array}{ccccccccc} 0 & \longrightarrow & A' & \xrightarrow{\psi'} & B' & \xrightarrow{\varphi'} & C' & \longrightarrow & 0 \\ & & \downarrow{\scriptstyle \varepsilon_A} & & \downarrow{\scriptstyle \varepsilon_B} & & \downarrow{\scriptstyle \varepsilon_C} & & \\ 0 & \longrightarrow & A & \xrightarrow{\psi} & B & \xrightarrow{\varphi} & C & \longrightarrow & 0 \\ & & \downarrow{\scriptstyle d_A} & & \downarrow{\scriptstyle d_B} & & \downarrow{\scriptstyle d_C} & & \\ 0 & \longrightarrow & X & \xrightarrow{i_X} & X\oplus Z & \xrightarrow{p_Z} & Z & \longrightarrow & 0 \end{array} \tag{5.47}$$

be a diagram in $\mathcal{C}$ with exact rows and columns, with commuting squares, and with X and Z injective. Suppose also that a second copy of (5.47), with twiddles ($\sim$) over everything has the same properties and lies over (5.47). In addition, suppose that there are eight vertical maps in $\mathcal{C}$

$$f_{A'},\ f_{B'},\ f_{C'},\ f_A,\ f_B,\ f_C,\ f_X,\ f_Z$$

from the twiddle diagram to the original diagram such that all the vertical squares commute. Then there exists a vertical map $f_{X\oplus Z}: \tilde{X} \oplus \tilde{Z} \to X \oplus Z$ in $\mathcal{C}$ such that all three vertical squares commute that involve $f_{X\oplus Z}$.

Proof. For $\tilde{b}$ in $\tilde{B}$, we claim that the element

$$d_B f_B(\tilde{b}) - (f_X p_{\tilde{X}} d_{\tilde{B}}(\tilde{b}), f_Z p_{\tilde{Z}} d_{\tilde{B}}(\tilde{b})) \quad \text{in} \quad X \oplus Z \tag{5.48}$$

depends only on $\tilde{\varphi}(\tilde{b})$ in $\tilde{C}$. In fact, if $\tilde{\varphi}(\tilde{b}) = 0$, then $\tilde{b} = \tilde{\psi}(\tilde{a})$, and (5.48) is

$$= d_B f_B \tilde{\psi}(\tilde{a}) - (f_X p_{\tilde{X}} d_{\tilde{B}} \tilde{\psi}(\tilde{a}), f_Z p_{\tilde{Z}} d_{\tilde{B}} \tilde{\psi}(\tilde{a}))$$

$$= d_B \psi f_A(\tilde{a}) - (f_X p_{\tilde{X}} i_{\tilde{X}} d_{\tilde{A}}(\tilde{a}), f_Z p_{\tilde{Z}} i_{\tilde{X}} d_{\tilde{A}}(\tilde{a}))$$

$$= i_X d_A f_A(\tilde{a}) - (f_X d_{\tilde{A}}(\tilde{a}), 0) = 0 .$$

Thus (5.48) depends only on $\tilde{\varphi}(\tilde{b})$. If we apply p_Z to (5.48), we get

$$p_Z d_B f_B(\tilde{b}) - f_Z p_{\tilde{Z}} d_{\tilde{B}}(\tilde{b}) = d_C \varphi f_B(\tilde{b}) - f_Z d_{\tilde{C}} \tilde{\varphi}(\tilde{b})$$

$$= d_C f_C \tilde{\varphi}(\tilde{b}) - d_C f_C \tilde{\varphi}(\tilde{b}) = 0 .$$

Thus (5.48) is in the image of i_X. Hence there is a map

$\tau : \tilde{C} \to X$ in $\mathcal{C}$ with

$$i_X \tau \tilde{\varphi}(\tilde{b}) = d_B f_B(\tilde{b}) - (f_X p_{\tilde{X}} d_{\tilde{B}}(\tilde{b}), f_Z p_{\tilde{Z}} d_{\tilde{B}}(\tilde{b})) .$$

Setting up

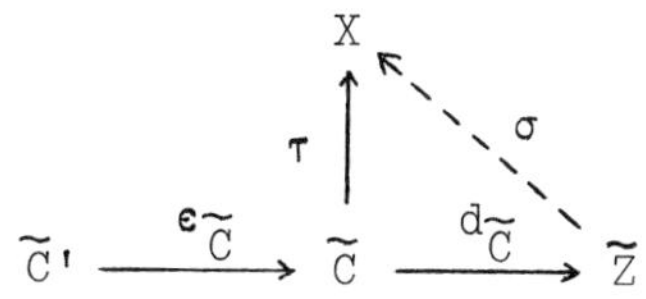

we prepare to invoke Lemma 5.5. We have

$$\begin{aligned} i_X \tau \varepsilon_{\tilde{C}}(\tilde{c}') &= i_X \tau \varepsilon_{\tilde{C}} \tilde{\varphi}'(\tilde{b}') = i_X \tau \tilde{\varphi} \varepsilon_{\tilde{B}}(\tilde{b}') \\ &= d_B f_B \varepsilon_{\tilde{B}}(\tilde{b}') - (f_X p_{\tilde{X}} d_{\tilde{B}} \varepsilon_{\tilde{B}}(\tilde{b}'), f_Z p_{\tilde{Z}} d_{\tilde{B}} \varepsilon_{\tilde{B}}(\tilde{b}')) \\ &= d_B \varepsilon_B f_{B'}(\tilde{b}') - (0,0) \qquad \text{since } d_{\tilde{B}} \varepsilon_{\tilde{B}} = 0 \\ &= 0 \qquad \text{since } d_B \varepsilon_B = 0 . \end{aligned}$$

Since i_X is one-one, $\tau \varepsilon_{\tilde{C}} = 0$ and Lemma 5.5 applies. Thus we obtain $\sigma : \tilde{Z} \to X$ with $\sigma d_{\tilde{C}} = \tau$, and σ satisfies

$$(\sigma p_{\tilde{Z}} d_{\tilde{B}}(\tilde{b}), 0) = i_X \sigma d_{\tilde{C}} \tilde{\varphi}(\tilde{b}) = d_B f_B(\tilde{b}) - (f_X p_{\tilde{X}} d_{\tilde{B}}(\tilde{b}), f_Z p_{\tilde{Z}} d_{\tilde{B}}(\tilde{b})) . \tag{5.49}$$

Define

$$f_{X \oplus Z}(\tilde{x}, \tilde{z}) = (f_X(\tilde{x}) + \sigma(\tilde{z}), f_Z(\tilde{z})) . \tag{5.50}$$

The three commutativity relations that are to be proved are

$$i_X f_X(\tilde{x}) = f_{X \oplus Z} i_{\tilde{X}}(\tilde{x}) \tag{5.51}$$

$$p_Z f_{X \oplus Z}(\tilde{x}, \tilde{z}) = f_Z p_{\tilde{Z}}(\tilde{x}, \tilde{z}) \tag{5.52}$$

$$d_B f_B(\tilde{b}) = f_{X\oplus Z} d_{\tilde{B}}(\tilde{b}) . \tag{5.53}$$

Relation (5.51) follows from (5.50) by taking $\tilde{z} = 0$, and (5.52) is just the second coordinate of (5.50). For (5.53) we use (5.50) and (5.49) to write

$$f_{X\oplus Z} d_{\tilde{B}}(\tilde{b}) = (f_X p_{\tilde{X}} d_{\tilde{B}}(\tilde{b}) + \sigma p_{\tilde{Z}} d_{\tilde{B}}(\tilde{b}), f_Z p_{\tilde{Z}} d_{\tilde{B}}(\tilde{b})) = d_B f_B(\tilde{b}) ,$$

and the result is (5.53). This proves the lemma.

Proof of Theorem 5.29. Let us start from a diagram (5.34) of compatible projective resolutions of R, S, T and a corresponding twiddle diagram of compatible projective resolutions of $\tilde{R}$, $\tilde{S}$, $\tilde{T}$ above. We put the vertical maps f_R, f_S, and f_T in place and, by Proposition 5.3, form chain maps $f_{X_0}, f_{X_1}, \ldots$ and $f_{Z_0}, f_{Z_1}, \ldots$ over f_R and f_T, respectively. To complete the diagram, we shall fill in $f_{X_0\oplus Z_0}, f_{X_1\oplus Z_1}, \ldots$ so that the remaining vertical squares commute. For $f_{X_0\oplus Z_0}$, we apply Lemma 5.30 with $A' = B' = C' = 0$, $A = R$, $B = S$, $C = T$, $X = X_0$, and $Z = Z_0$. Having constructed $f_{X_0\oplus Z_0}$, we obtain $f_{X_1\oplus Z_1}$ by applying Lemma 5.30 with $A' = R$, $B' = S$, $C' = T$, $A = X_0$, $B = X_0 \oplus Z_0$, $C = Z_0$, $X = X_1$, and $Z = Z_1$. Proceeding inductively, we obtain $f_{X_n\oplus Z_n}$ for all n. The result is that our two diagrams (5.34) are connected by compatible chain maps over f_R, f_S, and f_T.

Proceeding in the same way with Lemma 5.31, we can handle two diagrams (5.38) of injective resolutions. The result is that the two diagrams (5.38), one with twiddles and one without, are connected by compatible cochain maps over

f_R, f_S, and f_T.

Finally we apply F to one or the other of the above three-dimensional diagrams, argue that the hypotheses of Proposition 5.21 are satisfied (in the same way that we argued for Theorem 5.28 that the hypotheses of Theorem 5.20 are satisfied), and invoke Proposition 5.21 to complete the proof of Theorem 5.29.

CHAPTER VI

APPLICATION TO LIE ALGEBRAS

1. Projectives and injectives

This chapter continues the project begun in Chapter V of generalizing and extending the tools used in the example of a cohomology computation in Chapter IV. We now take up those aspects of homological algebra that are special to the situation where the ring is a universal enveloping algebra.

Let $\mathfrak{g}$ be a finite-dimensional complex Lie algebra, and let $\mathcal{C}(\mathfrak{g})$ be the good category of all left $U(\mathfrak{g})$ modules in which 1 acts as 1. In studying $\mathcal{C}(\mathfrak{g})$, we are thus studying all representations of $\mathfrak{g}$. To emphasize that the behavior of $\mathfrak{g}$ controls the behavior of $U(\mathfrak{g})$, we shall often write $\mathrm{Hom}_{\mathfrak{g}}$ and $\otimes_{\mathfrak{g}}$ in place of $\mathrm{Hom}_{U(\mathfrak{g})}$ and $\otimes_{U(\mathfrak{g})}$ when there is no possibility of confusion.

Recall that left $U(\mathfrak{g})$ modules are automatically right $U(\mathfrak{g})$ modules via the transpose map: If X is in $\mathfrak{g}$ and V is a left $U(\mathfrak{g})$ module, we can make V into a right $U(\mathfrak{g})$ module by the definition $vX = -Xv$. When we form $U \otimes_{\mathfrak{g}} V$ for two members U and V of $U(\mathfrak{g})$, this convention is in force; thus we first make U into a right $U(\mathfrak{g})$ module, and then we tensor U with V over the algebra $U(\mathfrak{g})$.

Before coming to projectives and injectives, we need two associativity identities special to $U(\mathfrak{g})$. Let A, B, and C be in $\mathcal{C}(\mathfrak{g})$. The first identity is

$$\operatorname{Hom}_{\mathfrak{g}}(A \otimes_{\mathbb{C}} B, C) \cong \operatorname{Hom}_{\mathfrak{g}}(A, \operatorname{Hom}_{\mathbb{C}}(B,C)) \tag{6.1}$$

naturally in A, B, and C. This identity is given in Corollary 4.3, and the map Φ from left to right is $\Phi(\varphi)(a)(b) = \varphi(a \otimes b)$.

The second identity is

$$(A \otimes_{\mathbb{C}} B) \otimes_{\mathfrak{g}} C \cong A \otimes_{\mathfrak{g}} (B \otimes_{\mathbb{C}} C) \tag{6.2}$$

naturally in A, B, and C, the mapping Φ from left to right satisfying $\Phi((a \otimes b) \otimes c) = a \otimes (b \otimes c)$. This identity can be derived from Proposition 2.6, but it is instructive to use Proposition 5.14. For V in $\mathcal{C}(\mathfrak{g})$ with trivial action, we have

$$\begin{aligned}
\operatorname{Hom}_{\mathbb{C}}((A \otimes_{\mathbb{C}} B) \otimes_{\mathfrak{g}} C, V) &\cong \operatorname{Hom}_{\mathfrak{g}}(A \otimes_{\mathbb{C}} B, \operatorname{Hom}_{\mathbb{C}}(C,V)) && \text{by (2.49)} \\
&\cong \operatorname{Hom}_{\mathfrak{g}}(A, \operatorname{Hom}_{\mathbb{C}}(B, \operatorname{Hom}_{\mathbb{C}}(C,V))) && \text{by (5.1)} \\
&\cong \operatorname{Hom}_{\mathfrak{g}}(A, \operatorname{Hom}_{\mathbb{C}}(B \otimes_{\mathbb{C}} C, V)) && \text{by Proposition 2.3} \\
&\cong \operatorname{Hom}_{\mathbb{C}}(A \otimes_{\mathfrak{g}} (B \otimes_{\mathbb{C}} C), V) && \text{by (2.49)},
\end{aligned}$$

with all isomorphisms natural in A, B, C, and V. Applying Proposition 5.14, we obtain (6.2), and the proposition even tells us how to read off Φ.

In Chapter V, we carried along $\operatorname{Hom}_{\mathbb{C}}(\cdot, V)$, $\operatorname{Hom}_{\mathbb{C}}(V, \cdot)$, and $V \otimes_{\mathbb{C}} (\cdot)$ as examples of functors from $\mathcal{C}(\mathfrak{g})$ to $\mathcal{C}(\mathfrak{g})$.

These functors will play a special role for us.

Proposition 6.1. Let V be in $\mathcal{C}(\mathfrak{g})$. Then

(a) the functor $W \to V \otimes_{\mathbb{C}} W$ on $\mathcal{C}(\mathfrak{g})$ is covariant and exact, and it sends projectives to projectives.

(b) the functor $W \to \operatorname{Hom}_{\mathbb{C}}(W,V)$ on $\mathcal{C}(\mathfrak{g})$ is contravariant and exact, and it sends projectives to injectives.

(c) the functor $W \to \operatorname{Hom}_{\mathbb{C}}(V,W)$ on $\mathcal{C}(\mathfrak{g})$ is covariant and exact, and it sends injectives to injectives.

Proof. The results about covariance and contravariance are obvious, and the exactness is from Corollary 5.11. For (a), let P be projective. We use (6.1). The functor

$$W \to \operatorname{Hom}_{\mathfrak{g}}(P \otimes_{\mathbb{C}} V, W) \cong \operatorname{Hom}_{\mathfrak{g}}(P, \operatorname{Hom}_{\mathbb{C}}(V,W))$$

is the composition of $W \to \operatorname{Hom}_{\mathbb{C}}(V,W)$, which is exact, and $U \to \operatorname{Hom}_{\mathfrak{g}}(P,U)$, which is exact by Proposition 5.10a. Hence it is exact, and a second application of Proposition 5.10a says $P \otimes_{\mathbb{C}} V$ is projective.

For (b), let P be projective. The functor

$$W \to \operatorname{Hom}_{\mathfrak{g}}(W, \operatorname{Hom}_{\mathbb{C}}(P,V)) \cong \operatorname{Hom}_{\mathfrak{g}}(W \otimes_{\mathbb{C}} P, V) \cong \operatorname{Hom}_{\mathfrak{g}}(P, \operatorname{Hom}_{\mathbb{C}}(W,V))$$

is the composition of $W \to \operatorname{Hom}_{\mathbb{C}}(W,V)$ and $U \to \operatorname{Hom}_{\mathfrak{g}}(P,U)$, both of which are exact. Hence $\operatorname{Hom}_{\mathbb{C}}(P,V)$ is injective by Proposition 5.10b.

For (c), let I be injective. The functor

$$W \to \operatorname{Hom}_{\mathfrak{g}}(W, \operatorname{Hom}_{\mathbb{C}}(V,I)) \cong \operatorname{Hom}_{\mathfrak{g}}(W \otimes_{\mathbb{C}} V, I)$$

is the composition of $W \to W \otimes_{\mathbb{C}} V$ and $U \to \operatorname{Hom}_{\mathfrak{g}}(U,I)$, both of which are exact. Hence $\mathrm{Hom}_{\mathbb{C}}(V, I)$ is injective by Proposition 5.10b. This completes the proof.

Let us recall the Koszul resolution (of $\mathbb{C}$) given in (4.19). The spaces are

$$X_n = U(\mathfrak{g}) \otimes_{\mathbb{C}} \wedge^n \mathfrak{g} \tag{6.3a}$$

with trivial action on $\wedge^n \mathfrak{g}$. And the resolution is

$$0 \longleftarrow \mathbb{C} \overset{\varepsilon}{\longleftarrow} X_0 \overset{\partial_0}{\longleftarrow} X_1 \overset{\partial_1}{\longleftarrow} X_2 \longleftarrow \cdots \longleftarrow X_N \longleftarrow 0 \tag{6.3b}$$

with $N = \dim \mathfrak{g}$ and with ε and ∂_n given by

$$\varepsilon(u) = \text{constant term of } u, \qquad u \in X_0 = U(\mathfrak{g}) \otimes_{\mathbb{C}} \mathbb{C} \cong U(\mathfrak{g})$$

$$\begin{aligned} \partial_{n-1}&(u \otimes X_1 \wedge \ldots \wedge X_n) \\ &= \sum_{i=1}^{n} (-1)^{i+1} (uX_i \otimes X_1 \wedge \ldots \wedge \widehat{X}_i \wedge \ldots \wedge X_n) \\ &\quad + \sum_{k<\ell} (-1)^{k+\ell} (u \otimes [X_k, X_\ell] \wedge X_1 \wedge \ldots \wedge \widehat{X}_k \wedge \ldots \wedge \widehat{X}_\ell \wedge \ldots \wedge X_n) . \end{aligned} \tag{6.3c}$$

If we apply (a) and (b) of Proposition 6.1 to the Koszul resolution, we obtain a projective resolution

$$0 \longleftarrow V \longleftarrow X_0 \otimes_{\mathbb{C}} V \longleftarrow X_1 \otimes_{\mathbb{C}} V \longleftarrow \cdots \longleftarrow X_N \otimes_{\mathbb{C}} V \longleftarrow 0 \tag{6.4}$$

and an injective resolution

$$0 \longrightarrow V \longrightarrow \mathrm{Hom}_{\mathbb{C}}(X_0, V) \longrightarrow \mathrm{Hom}_{\mathbb{C}}(X_1, V) \longrightarrow \cdots \longrightarrow \mathrm{Hom}_{\mathbb{C}}(X_N, V) \longrightarrow 0 . \tag{6.5}$$

These are called the <u>standard resolutions</u> of V. The existence of these resolutions proves that $C(\mathfrak{g})$ has enough projectives and enough injectives. (Other more general proofs are possible, but they will not concern us.)

Let $\mathfrak{q}$ be a Lie subalgebra of $\mathfrak{g}$. If V is in $C(\mathfrak{q})$, we define

$$P(V) = P_{\mathfrak{q}}^{\mathfrak{g}}(V) = U(\mathfrak{g}) \otimes_{U(\mathfrak{q})} V \tag{6.6}$$

as a left $U(\mathfrak{g})$ module under $u_1(u_2 \otimes x) = u_1u_2 \otimes x$; the convention for $\otimes_{U(\mathfrak{q})}$ is that $uq \otimes v = u \otimes qv$. Also we define

$$I(V) = I_{\mathfrak{q}}^{\mathfrak{g}}(V) = \mathrm{Hom}_{U(\mathfrak{q})}(U(\mathfrak{g}), V) \tag{6.7}$$

as a left $U(\mathfrak{g})$ module under $u_1\varphi(u_2) = \varphi(u_2u_1)$; the convention for $\mathrm{Hom}_{U(\mathfrak{q})}$ is that $\varphi(qu) = q(\varphi(u))$.

There is a second way of defining $P(V)$, namely

$$P'(V) = V \otimes_{U(\mathfrak{q})} U(\mathfrak{g}) \tag{6.8}$$

as a left $U(\mathfrak{g})$ module under $u_1(v \otimes u_2) = v \otimes u_2u_1^{tr}$; the convention for $\otimes_{U(\mathfrak{q})}$ is that $q^{tr}v \otimes u = v \otimes qu$. Both P and P' give functors; their definitions give naturally isomorphic $U(\mathfrak{g})$ modules under the map $u \otimes v \to v \otimes u^{tr}$. Thus we are free to use either definition, but we shall normally use the first definition (6.6).

Similarly there is a second way of defining $I(V)$, namely with $u_1\varphi(u_2) = \varphi(u_1^{tr}u_2)$; the convention for $\mathrm{Hom}_{U(\mathfrak{q})}$ is that $\varphi(uq) = q^{tr}(\varphi(u))$. The two definitions give naturally isomorphic $U(\mathfrak{g})$ modules, and we are free to use either definition. Except in the motivating examples below, we shall invariably use the first definition, however.

Algebraically, P and I are change-of-rings functors. (Our earlier example of a change-of-rings functor was complexification of a real vector space.) But there is an interesting analytic interpretation, particularly for I, that is worth bearing in mind. We give the interpretation by means of examples.

Example 1. The Borel-Weil Theorem of §III.7 gave a geometric construction of irreducible representations. In the discussion, we took G_0 to be $U(n)$, L_0 to be a block diagonal subgroup, and χ to be an irreducible representation of L_0 on a finite-dimensional space V satisfying certain properties. The C^∞ sections $\gamma : G_0/L_0 \to G_0 \times_{L_0} V$ corresponded to C^∞ functions $\varphi : G_0 \to V$ such that

$$\varphi(g_0 \ell_0) = \chi(\ell_0)^{-1}\varphi(g_0) \tag{6.9}$$

for all $g_0 \in G_0$ and $\ell_0 \in L_0$. We imposed complex structures on G_0/L_0 and $G_0 \times_{L_0} V$ with the aid of a block lower triangular subgroup Q of $G = GL(n,\mathbb{C})$ and found that the C^∞ section γ is holomorphic if and only if its corresponding φ satisfies $Z\varphi = 0$ for every block strictly lower triangular member of $\mathfrak{gl}(n,\mathbb{C})$.

Let $\mathfrak{g}_0 = \mathfrak{u}(n)$ be the Lie algebra of G_0, let $\mathfrak{g} = \mathfrak{g}_0^{\mathbb{C}} = \mathfrak{gl}(n,\mathbb{C})$ be the complexification, let $\mathfrak{l}_0$ be the Lie algebra of L_0, and let $\mathfrak{l} = \mathfrak{l}_0^{\mathbb{C}}$. If $\mathfrak{q}$ denotes the Lie algebra of Q, then we can write $\mathfrak{q} = \mathfrak{l} \oplus \mathfrak{u}$, where $\mathfrak{u}$ is the

Lie algebra of block strictly lower triangular members of $\mathfrak{gl}(n,\mathbb{C})$. To each smooth $\varphi : G_0 \to V$ satisfying (6.9), we can associate a member φ_0 of $\operatorname{Hom}_{\mathbb{C}}(U(\mathfrak{g}),V)$ by the definition

$$\varphi_0(u) = (u\ \varphi)(1) , \tag{6.10}$$

where $u\ \varphi$ refers to left-invariant differentiation by u. The condition (6.9) forces φ_0 to be in $\operatorname{Hom}_{U(\mathfrak{l})}(U(\mathfrak{g}),V)$, with the condition $\operatorname{Hom}_{U(\mathfrak{l})}$ given as $\varphi_0(u\ell) = \ell^{tr}(\varphi_0(u))$ and with $U(\mathfrak{l})$ acting on V by virtue of the representation χ. The condition that φ be holomorphic is that $Z\varphi = 0$ for Z in $\mathfrak{u}$. This condition forces φ_0 to be in $\operatorname{Hom}_{U(\mathfrak{q})}(U(\mathfrak{g}),V)$, where V is a left $U(\mathfrak{q})$ module by virtue of $\chi(L_0)$ and the zero action by $\mathfrak{u}$.

Theorem 3.22 identifies the representation of G_0 in the space of holomorphic sections, and what we have just seen is that (6.10) gives us a map (clearly $\mathfrak{g}$-equivariant) of this space into $I_{\mathfrak{q}}^{\mathfrak{g}}(V)$. On the space of holomorphic sections, this map is one-one, since a holomorphic function is determined by its derivatives at a point. It is not necessarily onto——for two reasons: One is that $I_{\mathfrak{q}}^{\mathfrak{g}}(V)$ includes all formal power series, not just the convergent ones. The other is that even if a power series is convergent, there is no guarantee that it satisfies the consistency conditions needed to give a globally defined φ.

Example 2. Let G be a Lie group, let Q be a closed subgroup, and let π be a unitary representation of Q on a Hilbert space V. Then the subspace V_0 of V of C^∞

vectors is a representation space for $\mathfrak{q}$. If the coset space G/Q has a G-invariant measure, the induced representation $\Pi = \mathrm{ind}_Q^G \pi$ acts in the space

$$U = \{\varphi : G \to V \mid |\varphi| \in L^2(G/Q) \text{ and } \varphi(xq) = \pi(q)^{-1}\varphi(x)\}$$

by the left regular representation

$$(\Pi(g)\varphi)(x) = \varphi(g^{-1}x) .$$

(If G/Q does not have an invariant measure, a definition is still possible, but some adjustment needs to be made.) Any smooth φ in U leads to a member of $\mathrm{Hom}_{\mathbb{C}}(U(\mathfrak{g}), V_0)$ by the definition (6.10), and we check that $\varphi \to \varphi_0$ actually carries the smooth members of U into $\mathrm{Hom}_{U(\mathfrak{q})}(U(\mathfrak{g}), V_0)$. Again $\varphi \to \varphi_0$ is one-one on the subspace of real analytic φ's, and it is not onto $\mathrm{Hom}_{U(\mathfrak{q})}(U(\mathfrak{g}), V)$ for the same reasons as in Example 1.

Thus for now we can regard $I_{\mathfrak{q}}^{\mathfrak{g}}(V)$ as an imperfect analog of an induced representation (as in Example 2) or a "holomorphically induced representation" (as in Example 1). In Chapter VII we shall see how to modify the algebraic construction so as to get an improved analogy.

Let us return to the algebraic definitions and establish properties of P and I.

Proposition 6.2. Let $\mathfrak{q}$ be a Lie subalgebra of $\mathfrak{g}$.

(a) Then $P_{\mathfrak{q}}^{\mathfrak{g}}$ and $I_{\mathfrak{q}}^{\mathfrak{g}}$ are covariant exact functors.

(b) If $\mathfrak{h}$ is a Lie subalgebra of $\mathfrak{q}$, then

$$P_{\mathfrak{h}}^{\mathfrak{g}} \cong P_{\mathfrak{q}}^{\mathfrak{g}} \circ P_{\mathfrak{h}}^{\mathfrak{q}} \quad \text{and} \quad I_{\mathfrak{h}}^{\mathfrak{g}} \cong I_{\mathfrak{q}}^{\mathfrak{g}} \circ I_{\mathfrak{h}}^{\mathfrak{q}} . \tag{6.11}$$

Proof. (a) P and I are functors, and they are covariant. Let $\mathfrak{a}$ be a vector space complement to $\mathfrak{q}$ in $\mathfrak{g}$. Then we have natural vector space isomorphisms

$$P_{\mathfrak{q}}^{\mathfrak{g}}(V) = U(\mathfrak{g}) \otimes_{U(\mathfrak{q})} V \cong (S(\mathfrak{a}) \otimes_{\mathbb{C}} U(\mathfrak{q})) \otimes_{U(\mathfrak{q})} V$$

$$\cong S(\mathfrak{a}) \otimes_{\mathbb{C}} (U(\mathfrak{q}) \otimes_{U(\mathfrak{q})} V) \cong S(\mathfrak{a}) \otimes_{\mathbb{C}} V ,$$

from which it follows that $P_{\mathfrak{q}}^{\mathfrak{g}}$ is naturally isomorphic to $S(\mathfrak{a}) \otimes_{\mathbb{C}} (\cdot)$ if we ignore the Lie algebra structure. This functor on the category of vector spaces is exact, and hence $P_{\mathfrak{q}}^{\mathfrak{g}}$ is exact. Similarly the identity

$$I_{\mathfrak{q}}^{\mathfrak{g}}(V) = \mathrm{Hom}_{U(\mathfrak{q})}(U(\mathfrak{g}),V) \cong \mathrm{Hom}_{U(\mathfrak{q})}(S(\mathfrak{a}) \otimes_{\mathbb{C}} U(\mathfrak{q}),V)$$

$$\cong \mathrm{Hom}_{\mathbb{C}}(S(\mathfrak{a}),U(\mathfrak{q}) \otimes_{U(\mathfrak{q})} V) \cong \mathrm{Hom}_{\mathbb{C}}(S(\mathfrak{a}),V)$$

shows that $I_{\mathfrak{q}}^{\mathfrak{g}}$ is exact.

(b) We have natural isomorphisms

$$P_{\mathfrak{q}}^{\mathfrak{g}} \circ P_{\mathfrak{h}}^{\mathfrak{q}}(V) = U(\mathfrak{g}) \otimes_{U(\mathfrak{q})} (U(\mathfrak{q}) \otimes_{U(\mathfrak{h})} V)$$

$$\cong (U(\mathfrak{g}) \otimes_{U(\mathfrak{q})} U(\mathfrak{q})) \otimes_{U(\mathfrak{h})} V \cong U(\mathfrak{g}) \otimes_{U(\mathfrak{h})} V = P_{\mathfrak{h}}^{\mathfrak{g}}(V)$$

and

$$I_{\mathfrak{q}}^{\mathfrak{g}} \circ I_{\mathfrak{h}}^{\mathfrak{q}}(V) = \mathrm{Hom}_{U(\mathfrak{q})}(U(\mathfrak{g}),\mathrm{Hom}_{U(\mathfrak{h})}(U(\mathfrak{q}),V))$$

$$\cong \mathrm{Hom}_{U(\mathfrak{h})}(U(\mathfrak{g}) \otimes_{U(\mathfrak{q})} U(\mathfrak{q}),V)$$

$$\cong \mathrm{Hom}_{U(\mathfrak{h})}(U(\mathfrak{g}),V) = I_{\mathfrak{h}}^{\mathfrak{g}}(V) ,$$

and the proof is complete.

The analytic analog of (b) for the I functor is the "induction-in-stages formula": induced of induced is induced.

A companion functor to $P_{\mathfrak{q}}^{\mathfrak{g}}$ and $I_{\mathfrak{q}}^{\mathfrak{g}}$ is the <u>forgetful functor</u> $\mathcal{F}_{\mathfrak{g}}^{\mathfrak{q}}$, which takes a $U(\mathfrak{g})$ module and regards it simply as a $U(\mathfrak{q})$ module by forgetting the unnecessary action. Sometimes the forgetful functor is so transparent that we omit writing it altogether. At any rate, it is obviously exact.

<u>Proposition 6.3</u>. Let $\mathfrak{q}$ be a Lie subalgebra of $\mathfrak{g}$. Then there are isomorphisms

$$\operatorname{Hom}_{\mathfrak{g}}(P_{\mathfrak{q}}^{\mathfrak{g}}(V),W) \cong \operatorname{Hom}_{\mathfrak{q}}(V,\mathcal{F}_{\mathfrak{g}}^{\mathfrak{q}}(W)) \tag{6.12}$$

$$\operatorname{Hom}_{\mathfrak{g}}(W,I_{\mathfrak{q}}^{\mathfrak{g}}(V)) \cong \operatorname{Hom}_{\mathfrak{q}}(\mathcal{F}_{\mathfrak{g}}^{\mathfrak{q}}(W),V) \tag{6.13}$$

natural for V in $\mathcal{C}(\mathfrak{q})$ and W in $\mathcal{C}(\mathfrak{g})$.

For P we have natural isomorphisms

$$\begin{aligned}\operatorname{Hom}_{\mathfrak{g}}(P_{\mathfrak{q}}^{\mathfrak{g}}(V),W) &= \operatorname{Hom}_{U(\mathfrak{g})}(U(\mathfrak{g})\otimes_{U(\mathfrak{q})} V,W)\\ &\cong \operatorname{Hom}_{U(\mathfrak{q})}(V,\operatorname{Hom}_{U(\mathfrak{g})}(U(\mathfrak{g}),W))\\ &\cong \operatorname{Hom}_{U(\mathfrak{q})}(V,W) = \operatorname{Hom}_{\mathfrak{q}}(V,\mathcal{F}_{\mathfrak{g}}^{\mathfrak{q}}(W)),\end{aligned}$$

by associativity (2.50), and for I we have natural isomorphisms

$$\begin{aligned}\operatorname{Hom}_{\mathfrak{g}}(W,I_{\mathfrak{q}}^{\mathfrak{g}}(V)) &= \operatorname{Hom}_{U(\mathfrak{g})}(W,\operatorname{Hom}_{U(\mathfrak{q})}(U(\mathfrak{g}),V))\\ &\cong \operatorname{Hom}_{U(\mathfrak{q})}(U(\mathfrak{g})\otimes_{U(\mathfrak{g})} W,V)\\ &\cong \operatorname{Hom}_{U(\mathfrak{q})}(W,V) = \operatorname{Hom}_{\mathfrak{q}}(\mathcal{F}_{\mathfrak{g}}^{\mathfrak{q}}(W),V).\end{aligned}$$

The explicit formulas implementing (6.12) and (6.13) are of interest. For P, let us say that the two isomorphisms in the proof carry $\varphi_1 \to \varphi_2 \to \varphi_3$. Then the relationships are

$$\varphi_2(v)(u) = \varphi_1(u \otimes v)$$

$$\varphi_3(v) = \varphi_2(v)(1), \qquad \varphi_2(v)(u) = u(\varphi_3(v)).$$

Thus if Φ is the isomorphism from left to right in (6.12), we have

$$\begin{aligned} \Phi(\varphi)(v) &= \varphi(1 \otimes v) \\ \Phi^{-1}(\psi)(u \otimes v) &= u(\psi(v)). \end{aligned} \tag{6.14}$$

Similarly for I, let us say that the two isomorphisms in the proof carry $\varphi_1 \to \varphi_2 \to \varphi_3$. Then the relationships are

$$\varphi_2(u \otimes w) = \varphi_1(w)(u)$$

$$\varphi_3(w) = \varphi_2(1 \otimes w), \qquad \varphi_2(u \otimes w) = \varphi_3(uw).$$

Thus if Φ is the isomorphism from left to right in (6.13), we have

$$\begin{aligned} \Phi(\varphi)(w) &= \varphi(w)(1) \\ \Phi^{-1}(\psi)(w)(u) &= \psi(uw). \end{aligned} \tag{6.15}$$

The analytic analog of (6.13) in the case of induced representations is called <u>Frobenius reciprocity</u>. Some additional hypothesis is needed: Let us suppose that G and Q are compact. Then the formula is

$$\operatorname{Hom}_G(W, \operatorname{ind}_Q^G V) \cong \operatorname{Hom}_Q(W|_Q, V), \tag{6.16}$$

with the isomorphism implemented as in (6.15) by evaluation at the identity. If we take into account Corollary 3.8, we see that (6.16) can be interpreted as saying: the number of times that an irreducible representation τ of G (on W) occurs in an induced representation equals the number of times that the inducing representation of Q (on V) occurs in $\tau|_Q$.

<u>Corollary 6.4</u>. Let $\mathfrak{q}$ be a Lie subalgebra of $\mathfrak{g}$. Then $P_\mathfrak{q}^\mathfrak{g}$ sends projectives to projectives, and $I_\mathfrak{q}^\mathfrak{g}$ sends injectives to injectives.

Proof. If P is projective in $\mathcal{C}(\mathfrak{q})$, then (6.9) says that

$$W \to \mathrm{Hom}_\mathfrak{g}(P_\mathfrak{q}^\mathfrak{g}(P), W)$$

is naturally isomorphic with

$$W \to \mathrm{Hom}_\mathfrak{q}(P, \mathcal{F}_\mathfrak{g}^\mathfrak{q}(W)),$$

which is exact, being the composition of the exact functors $\mathcal{F}_\mathfrak{g}^\mathfrak{q}$ and $\mathrm{Hom}_\mathfrak{q}(P,\cdot)$. Hence $P_\mathfrak{q}^\mathfrak{g}(P)$ is projective in $\mathcal{C}(\mathfrak{g})$.

If I is injective in $\mathcal{C}(\mathfrak{q})$, then (6.10) says that

$$W \to \mathrm{Hom}_\mathfrak{g}(W, I_\mathfrak{q}^\mathfrak{g}(I))$$

is naturally isomorphic with

$$W \to \mathrm{Hom}_\mathfrak{q}(\mathcal{F}_\mathfrak{g}^\mathfrak{q}(W), I),$$

which is exact, being the composition of the exact functors $\mathcal{F}_\mathfrak{g}^\mathfrak{q}$ and $\mathrm{Hom}_\mathfrak{q}(\cdot, I)$. Hence $I_\mathfrak{q}^\mathfrak{g}(I)$ is injective in $\mathcal{C}(\mathfrak{g})$.

We have seen five associativity formulas for $\otimes$ and Hom that have bearing on $U(\mathfrak{g})$ modules. Three of these are valid for any algebras and are given as (2.49), (2.50), and (2.51). The other two are special to the algebras $U(\mathfrak{g})$ and are given as (6.1) and (6.2). What the five have in common is that the mapping implementing the isomorphism is semi-trivial; the three variables are mapped to the three variables, with no twisting or mixing. The isomorphisms in the two propositions and corollary that follow are associativity formulas that lie deeper because the variables get twisted. We call them Mackey isomorphisms because of Mackey's early use of the analytic analog of these isomorphisms for induced representations.

Proposition 6.5. Let $\mathfrak{q}$ be a Lie subalgebra of $\mathfrak{g}$, let U be in $\mathcal{C}(\mathfrak{g})$, and let V be in $\mathcal{C}(\mathfrak{q})$. Then there exists a unique isomorphism of $U(\mathfrak{g})$ modules

$$\Phi : P_{\mathfrak{q}}^{\mathfrak{g}}(U \otimes_{\mathbb{C}} V) \to U \otimes_{\mathbb{C}} P_{\mathfrak{q}}^{\mathfrak{g}}(V) \tag{6.17a}$$

such that

$$\Phi(1 \otimes (f \otimes v)) = f \otimes (1 \otimes v) \tag{6.17b}$$

for f in U and v in V. The isomorphism is natural with respect to U and V.

Remark. Observe from (6.17c) below that Φ is not the obvious product map unless the action on U is trivial.

Proof. Uniqueness is clear since u in $U(\mathfrak{g})$ then implies

$$\Phi(u\otimes(f\otimes v)) = u(\Phi(1\otimes(f\otimes v)) = u(f\otimes(1\otimes v)). \quad (6.17c)$$

To prove existence, let X be in $\mathcal{C}(\mathfrak{g})$. Then we have natural isomorphisms

$$\begin{aligned}
\operatorname{Hom}_{\mathfrak{g}}(P_{\mathfrak{q}}^{\mathfrak{g}}(U\otimes_{\mathbb{C}} V), X) &\cong \operatorname{Hom}_{\mathfrak{q}}(U\otimes_{\mathbb{C}} V, X) && \text{by (6.12)}\\
&\cong \operatorname{Hom}_{\mathfrak{q}}(V, \operatorname{Hom}_{\mathbb{C}}(U,X)) && \text{by (6.1)}\\
&\cong \operatorname{Hom}_{\mathfrak{g}}(P_{\mathfrak{q}}^{\mathfrak{g}}(V), \operatorname{Hom}_{\mathbb{C}}(U,X)) && \text{by (6.12)}\\
&\cong \operatorname{Hom}_{\mathfrak{g}}(U\otimes_{\mathbb{C}} P_{\mathfrak{q}}^{\mathfrak{g}}(V), X) && \text{by (6.1).}
\end{aligned}$$

Thus (6.17a) follows by Proposition 5.14.

Let us compute the map that implements (6.17a). If $\Phi_1,\ldots,\Phi_5$ are corresponding elements in the above Hom's, then (6.14) gives

$$\Phi_2(f\otimes v) = \Phi_1(1\otimes(f\otimes v))$$

$$\Phi_3(v)(f) = \Phi_2(f\otimes v)$$

$$\Phi_4(1\otimes v)(f) = \Phi_3(v)(f)$$

$$\Phi_5(f\otimes(u\otimes v)) = \Phi_4(u\otimes v)(f).$$

Taking $X = U\otimes_{\mathbb{C}} P_{\mathfrak{q}}^{\mathfrak{g}}(V)$ and $\Phi_5 = 1$ gives us

$$\Phi_1(1\otimes(f\otimes v)) = f\otimes(1\otimes v),$$

and Φ_1 is the required mapping Φ (according to Proposition 5.14).

Proposition 6.6. Let $\mathfrak{q}$ be a Lie subalgebra of $\mathfrak{g}$, let U be in $\mathcal{C}(\mathfrak{g})$, and let V be in $\mathcal{C}(\mathfrak{q})$. Then there exists a unique isomorphism of $U(\mathfrak{g})$ modules

$$\Phi : \operatorname{Hom}_{\mathbb{C}}(U, I_{\mathfrak{q}}^{\mathfrak{g}}(V)) \to I_{\mathfrak{q}}^{\mathfrak{g}}(\operatorname{Hom}_{\mathbb{C}}(U,V)) \tag{6.18a}$$

such that

$$\Phi(\varphi)(1)(f) = \varphi(f)(1) \tag{6.18b}$$

for f in U and φ in $\operatorname{Hom}_{\mathbb{C}}(U, I_{\mathfrak{q}}^{\mathfrak{g}}(V))$. The isomorphism is natural with respect to U and V.

Proof. Uniqueness is clear since u in $U(\mathfrak{g})$ then implies

$$\Phi(\varphi)(u)(f) = u(\Phi(\varphi))(1)(f) = \Phi(u\varphi)(1)(f) = (u\varphi)(f)(1) . \tag{6.18c}$$

To prove existence, let X be in $\mathcal{C}(\mathfrak{g})$. Then we have natural isomorphisms

$$\begin{aligned}
\operatorname{Hom}_{\mathfrak{g}}(X, \operatorname{Hom}_{\mathbb{C}}(U, I_{\mathfrak{q}}^{\mathfrak{g}}(V))) &\cong \operatorname{Hom}_{\mathfrak{g}}(U \otimes_{\mathbb{C}} X, I_{\mathfrak{q}}^{\mathfrak{g}}(V)) && \text{by (6.1)} \\
&\cong \operatorname{Hom}_{\mathfrak{q}}(U \otimes_{\mathbb{C}} X, V) && \text{by (6.13)} \\
&\cong \operatorname{Hom}_{\mathfrak{q}}(X, \operatorname{Hom}_{\mathbb{C}}(U,V)) && \text{by (6.1)} \\
&\cong \operatorname{Hom}_{\mathfrak{g}}(X, I_{\mathfrak{q}}^{\mathfrak{g}}(\operatorname{Hom}_{\mathbb{C}}(U,V))) && \text{by (6.13).}
\end{aligned}$$

Thus (6.18a) follows by Proposition 5.14.

To verify that the implementing map satisfies (6.18b), let $\Phi_1, \ldots, \Phi_5$ be the corresponding elements in the above Hom's. Then (6.15) gives

$$\Phi_1(x)(f) = \Phi_2(f \otimes x)$$

$$\Phi_2(f \otimes x)(1) = \Phi_3(f \otimes x)$$

$$\Phi_3(f \otimes x) = \Phi_4(x)(f)$$

$$\Phi_4(x)(f) = \Phi_5(x)(1)(f) .$$

Taking $X = \operatorname{Hom}_{\mathbb{C}}(U, I_{\mathfrak{q}}^{\mathfrak{g}}(V))$ and $\Phi_1 = 1$ gives us

$$\Phi_5(\varphi)(1)(f) = \varphi(f)(1) ,$$

and Φ_5 is the required mapping Φ, according to Proposition 5.14.

<u>Corollary 6.7</u>. Let $\mathfrak{q}$ be a Lie subalgebra of $\mathfrak{g}$, let F be a finite-dimensional member of $\mathcal{C}(\mathfrak{g})$, and let V be in $\mathcal{C}(\mathfrak{q})$. Then there exists a unique isomorphism of $U(\mathfrak{g})$ modules

$$\Phi : F \otimes_{\mathbb{C}} I_{\mathfrak{q}}^{\mathfrak{g}}(V) \to I_{\mathfrak{q}}^{\mathfrak{g}}(F \otimes_{\mathbb{C}} V) \tag{6.19a}$$

such that

$$\Phi(f \otimes \varphi)(1) = f \otimes \varphi(1) \tag{6.19b}$$

for f in F and φ in $I_{\mathfrak{q}}^{\mathfrak{g}}(V)$. The isomorphism is natural with respect to F and V.

Proof. We put $U = F^*$ in Proposition 6.6. Let $\{f_j\}$ be a basis of F, and let $\{f_j^*\}$ be the dual basis of F^*. We have

$$F \otimes_{\mathbb{C}} I_{\mathfrak{q}}^{\mathfrak{g}}(V) \cong \operatorname{Hom}_{\mathbb{C}}(F^*, I_{\mathfrak{q}}^{\mathfrak{g}}(V))$$

under $f \otimes \varphi \to \psi$ with $\psi(f^*) = f^*(f)\varphi$ with inverse $\psi \to \Sigma f_j \otimes \psi(f_j^*)$. This ψ maps under (6.18) to ψ' in

$I_{\mathfrak{q}}^{\mathfrak{g}}(\mathrm{Hom}_{\mathbb{C}}(F^*, V))$ with

$$\psi'(1)(f^*) = \psi(f^*)(1) = f^*(f)\varphi(1) .$$

The member of $I_{\mathfrak{q}}^{\mathfrak{g}}(F \otimes_{\mathbb{C}} V)$ that corresponds to ψ' is $\Phi(f \otimes \varphi)$, and it has

$$\Phi(f \otimes \varphi)(1) = \Sigma\, f_j \otimes \psi'(1)(f_j^*) = \Sigma\, f_j \otimes f_j^*(f)\varphi(1) = f \otimes \varphi(1) .$$

Thus the map Φ satisfies (6.19b).

2. Lie algebra homology and cohomology

Recall that Lie algebra cohomology $H^n(\mathfrak{g}, V)$ was defined concretely by a coboundary operator (given in (4.18)), and then we matched formulas to see that $H^n(\mathfrak{g}, V)$ was given by the cohomology of the complex

$$0 \longrightarrow \mathrm{Hom}_{\mathfrak{g}}(X_0, V) \xrightarrow{\mathrm{Hom}(\partial_0, 1)} \mathrm{Hom}_{\mathfrak{g}}(X_1, V) \xrightarrow{\mathrm{Hom}(\partial_1, 1)} \cdots , \tag{6.20}$$

where $\{X_n\}$ is the Koszul resolution (see (4.19) or (6.3)).

We know that the Koszul resolution is a projective resolution of the trivial module $\mathbb{C}$, and Proposition 6.1b allowed us to conclude that

$$0 \longrightarrow V \longrightarrow \mathrm{Hom}_{\mathbb{C}}(X_0, V) \xrightarrow{\mathrm{Hom}(\partial_0, 1)} \mathrm{Hom}_{\mathbb{C}}(X_1, V) \longrightarrow \cdots \tag{6.21}$$

is an injective resolution of V. By Proposition 4.2, $\mathrm{Hom}_{\mathfrak{g}}(X_n, V)$ is the result of applying the invariants functor

$W \to W^{\mathfrak{g}}$ to $\mathrm{Hom}_{\mathbb{C}}(X_n, V)$. Since this functor is covariant and left exact, (6.20) and (6.21) allow us to recognize the nth <u>Lie algebra cohomology</u> as the nth derived functor of the invariants functor. In particular, the identity

$$H^0(\mathfrak{g}, V) \cong V^{\mathfrak{g}} \tag{6.22}$$

is a consequence of the general result Proposition 5.15.

For future reference, let us reassemble the formulas for cohomology that we obtain from (6.20). The identity

$$\begin{aligned} \mathrm{Hom}_{\mathfrak{g}}(X_n, V) &= \mathrm{Hom}_{\mathfrak{g}}(U(\mathfrak{g}) \otimes_{\mathbb{C}} \Lambda^n \mathfrak{g}, V) \\ &\cong \mathrm{Hom}_{\mathbb{C}}(\Lambda^n \mathfrak{g}, \mathrm{Hom}_{\mathfrak{g}}(U(\mathfrak{g}), V)) \\ &\cong \mathrm{Hom}_{\mathbb{C}}(\Lambda^n \mathfrak{g}, V) \end{aligned} \tag{6.23a}$$

allows us to rewrite (6.20) as

$$0 \longrightarrow \mathrm{Hom}_{\mathbb{C}}(\Lambda^0 \mathfrak{g}, V) \xrightarrow{d_0} \mathrm{Hom}_{\mathbb{C}}(\Lambda^1 \mathfrak{g}, V) \xrightarrow{d_1} \cdots, \tag{6.23b}$$

and the operator $d : \mathrm{Hom}_{\mathbb{C}}(\Lambda^n \mathfrak{g}, V) \to \mathrm{Hom}_{\mathbb{C}}(\Lambda^{n+1} \mathfrak{g}, V)$ is given by

$$\begin{aligned} &(d\omega)(Y_1 \wedge \cdots \wedge Y_{n+1}) \\ &= \sum_{\ell=1}^{n+1} (-1)^{\ell+1} Y_\ell (\omega(Y_1 \wedge \cdots \wedge \hat{Y}_\ell \wedge \cdots \wedge Y_{n+1})) \\ &\quad + \sum_{r < s} (-1)^{r+s} \omega([Y_r, Y_s] \wedge Y_1 \wedge \cdots \wedge \hat{Y}_r \wedge \cdots \wedge \hat{Y}_s \wedge \cdots \wedge Y_{n+1}) . \end{aligned} \tag{6.23c}$$

We use the derived functor construction, suitably dualized, to define the nth <u>Lie algebra homology</u> functor $H_n(\mathfrak{g}, \cdot)$ as the nth derived functor of the <u>co-invariants</u>

functor

$$V \to V/\mathfrak{g}V \cong V \otimes_{\mathfrak{g}} \mathbb{C}\,. \tag{6.24}$$

The right member of (6.24) shows that the functor is covariant and right exact, so that its derived functors are obtained from a projective resolution. In particular, we have

$$H_0(\mathfrak{g}, V) \cong V/\mathfrak{g}V \cong V \otimes_{\mathfrak{g}} \mathbb{C}\,. \tag{6.25}$$

To obtain explicit formulas, we can use the standard projective resolution (6.4). Application of the functor $(\cdot) \otimes_{\mathfrak{g}} \mathbb{C}$ leads to the terms

$$\begin{aligned} (X_n \otimes_{\mathbb{C}} V) \otimes_{\mathfrak{g}} \mathbb{C} &= (U(\mathfrak{g}) \otimes_{\mathbb{C}} (\Lambda^n \mathfrak{g} \otimes_{\mathbb{C}} V)) \otimes_{\mathfrak{g}} \mathbb{C} \\ &\cong U(\mathfrak{g}) \otimes_{\mathfrak{g}} ((\Lambda^n \mathfrak{g} \otimes_{\mathbb{C}} V) \otimes_{\mathbb{C}} \mathbb{C}) \qquad \text{by (6.2)} \\ &\cong \Lambda^n \mathfrak{g} \otimes_{\mathbb{C}} V\,. \end{aligned} \tag{6.26a}$$

Tracking down the formulas for $\partial_n \otimes 1$ and rewriting $\partial_n \otimes 1$ as ∂, we are led to the complex

$$0 \longleftarrow \Lambda^0 \mathfrak{g} \otimes_{\mathbb{C}} V \xleftarrow{\ \partial_0\ } \Lambda^1 \mathfrak{g} \otimes_{\mathbb{C}} V \xleftarrow{\ \partial_1\ } \cdots \tag{6.26b}$$

with the operator $\partial : \Lambda^n \mathfrak{g} \otimes_{\mathbb{C}} V \to \Lambda^{n-1} \mathfrak{g} \otimes_{\mathbb{C}} V$ given by

$$\begin{aligned} &\partial(Y_1 \wedge \cdots \wedge Y_n \otimes v) \\ &\quad = \sum_{\ell=1}^{n} (-1)^{\ell} (Y_1 \wedge \cdots \wedge \widehat{Y}_\ell \wedge \cdots \wedge Y_n \otimes Y_\ell v) \\ &\qquad + \sum_{r<s} (-1)^{r+s} ([Y_r, Y_s] \wedge Y_1 \wedge \cdots \wedge \widehat{Y}_r \wedge \cdots \wedge \widehat{Y}_s \wedge \cdots \wedge Y_n \otimes v)\,. \end{aligned} \tag{6.26c}$$

(Note the slight difference in sign between (6.26c) and (6.23c).)

Now let us return to the example in §IV.4, seeing where we stand. We had $\mathfrak{h} \subseteq \mathfrak{g}$, took V in $\mathcal{C}(\mathfrak{h})$, defined $U = U(\mathfrak{g}) \otimes_{\mathfrak{h}} V$, and asked about the Lie algebra cohomology of U^*. If F is the invariants functor, we have

$$H^n(\mathfrak{g}, U^*) \cong F^n(U^*)$$

$$\cong (F \circ \text{dual})^n(U) \quad \text{by Proposition 5.16b.}$$

To compute this, we are to compute the cohomology of the complex $F \circ \text{dual}(P_n)$, where P_n is a projective resolution of U in $\mathcal{C}(\mathfrak{g})$. But $U = P_{\mathfrak{h}}^{\mathfrak{g}}(V)$, and $P_{\mathfrak{h}}^{\mathfrak{g}}$ carries a projective resolution Z_n of V in $\mathcal{C}(\mathfrak{h})$ to a projective resolution of U in $\mathcal{C}(\mathfrak{g})$, by Proposition 6.2a and Corollary 6.4. Thus we can take $P_n = P_{\mathfrak{h}}^{\mathfrak{g}}(Z_n)$, where Z_n is the standard resolution of V in $\mathcal{C}(\mathfrak{h})$. Since $P_n = 0$ for $n > \dim \mathfrak{h}$, we conclude $H^n(\mathfrak{g}, U^*) = 0$ for $n > \dim \mathfrak{h}$.

These steps show how our model example is handled by the methods of Chapters V and VI. The model example is not very useful, but there are useful examples that are handled in the same way. Some of these go under the name Shapiro's Lemma, which we discuss at the end of this section.

By way of preparation, we discuss a functor Ext that generalizes the invariants functor. The invariants functor can be rewritten as $V \to \text{Hom}_{\mathfrak{g}}(\mathbb{C}, V)$, and we know its derived functors to be $V \to H^n(\mathfrak{g}, V)$. We define $\text{Ext}_{\mathfrak{g}}^n(U, \cdot)$ to be the nth derived functor of $\text{Hom}_{\mathfrak{g}}(U, \cdot)$ on the category $\mathcal{C}(\mathfrak{g})$;

here U is in $\mathcal{C}(\mathfrak{g})$.

Proposition 6.8. $\mathrm{Ext}^n_{\mathfrak{g}}(\cdot,V)$ is the nth derived functor of $\mathrm{Hom}_{\mathfrak{g}}(\cdot,V)$, and also $\mathrm{Ext}^n_{\mathfrak{g}}(U,V) \cong H^n(\mathfrak{g},\mathrm{Hom}_{\mathbb{C}}(U,V))$ naturally for U and V in $\mathcal{C}(\mathfrak{g})$.

Remark. The first statement of the proposition is valid with $U(\mathfrak{g})$ replaced by any ring, but its proof is more complicated in such generality.

Proof. Let P_n and I_n be projective and injective resolutions of U and V, respectively. The derived functors of $V \to \mathrm{Hom}_{\mathfrak{g}}(U,V)$ are computed by applying $\mathrm{Hom}_{\mathfrak{g}}(U,\cdot)$ to $\{I_n\}$, and this can be done in two steps— forming $\mathrm{Hom}_{\mathbb{C}}(U,I_n)$ and then applying the invariants functor. Proposition 6.1c shows that $\mathrm{Hom}_{\mathbb{C}}(U,I_n)$ is an injective resolution of $\mathrm{Hom}_{\mathbb{C}}(U,V)$, and thus $\mathrm{Ext}^n(U,V)$ is isomorphic to $H^n(\mathfrak{g},\mathrm{Hom}_{\mathbb{C}}(U,V))$. It is easy to see that the isomorphism is natural.

Similarly the derived fucntors of $U \to \mathrm{Hom}_{\mathfrak{g}}(U,V)$ are computed by applying $\mathrm{Hom}_{\mathfrak{g}}(\cdot,V)$ to $\{P_n\}$, and this can be done by applying the invariants functor to the injective resolution $\mathrm{Hom}_{\mathbb{C}}(P_n,V)$ of $\mathrm{Hom}_{\mathbb{C}}(U,V)$. The result follows.

Theorem 6.9 (Shapiro's Lemma). If $\mathfrak{q}$ is a Lie subalgebra of $\mathfrak{g}$, then

$$\mathrm{Ext}^n_{\mathfrak{g}}(P^{\mathfrak{g}}_{\mathfrak{q}}(V),W) \cong \mathrm{Ext}^n_{\mathfrak{q}}(V,\mathcal{F}^{\mathfrak{q}}_{\mathfrak{g}}(W)) \tag{6.27}$$

$$\mathrm{Ext}^n_{\mathfrak{g}}(W,I^{\mathfrak{g}}_{\mathfrak{q}}(V)) \cong \mathrm{Ext}^n_{\mathfrak{q}}(\mathcal{F}^{\mathfrak{q}}_{\mathfrak{g}}(W),V) \tag{6.28}$$

naturally for V in $\mathcal{C}(\mathfrak{q})$ and W in $\mathcal{C}(\mathfrak{g})$.

Remark. These should be regarded as derived functor versions of the adjoint relations for $\operatorname{Hom}_{\mathfrak{g}}$ and $\operatorname{Hom}_{\mathfrak{q}}$ in (6.12) and (6.13).

Proof. The two functors $\operatorname{Hom}_{\mathfrak{g}}(\cdot,W)\circ P_{\mathfrak{q}}^{\mathfrak{g}}$ and $\operatorname{Hom}_{\mathfrak{q}}(\cdot,\mathcal{F}_{\mathfrak{g}}^{\mathfrak{q}}(W))$ are naturally isomorphic by (6.12). Hence their derived functors are naturally isomorphic. Since $P_{\mathfrak{q}}^{\mathfrak{g}}$ is exact and sends projectives to projectives (Proposition 6.2a and Corollary 6.4), Proposition 5.17 allows us to compute the derived functors of $\operatorname{Hom}_{\mathfrak{g}}(\cdot,W)\circ P_{\mathfrak{q}}^{\mathfrak{g}}$ as $\{\operatorname{Hom}_{\mathfrak{g}}(\cdot,W)\}^n\circ P_{\mathfrak{q}}^{\mathfrak{g}}$. Thus (6.27) follows from Proposition 6.8.

Similarly (6.13) gives us a natural isomorphism of the two functors $\operatorname{Hom}_{\mathfrak{g}}(W,\cdot)\circ I_{\mathfrak{q}}^{\mathfrak{g}}$ and $\operatorname{Hom}_{\mathfrak{q}}(\mathcal{F}_{\mathfrak{g}}^{\mathfrak{q}}(W),\cdot)$. Hence their derived functors are naturally isomorphic. Since $I_{\mathfrak{q}}^{\mathfrak{g}}$ is exact and sends injectives to injectives (Proposition 6.2a and Corollary 6.4), Proposition 5.17 allows us to compute the derived functors of $\operatorname{Hom}_{\mathfrak{g}}(W,\cdot)\circ I_{\mathfrak{q}}^{\mathfrak{g}}$ as $\{\operatorname{Hom}_{\mathfrak{g}}(W,\cdot)\}^n\circ I_{\mathfrak{q}}^{\mathfrak{g}}$, and (6.28) follows.

Let us return once more to our example from Chapter IV, in which $\mathfrak{h}$ is a Lie subalgebra of $\mathfrak{g}$, V is in $\mathcal{C}(\mathfrak{h})$, and $U = U(\mathfrak{g})\otimes_{\mathfrak{h}} V = P_{\mathfrak{h}}^{\mathfrak{g}}(V)$. We were interested in seeing that $H^n(\mathfrak{g},U^*) = 0$ for $n > \dim\mathfrak{h}$. In fact, more is true. Let us apply (6.27) with $\mathfrak{q} = \mathfrak{h}$ and $W = \mathbb{C}$. Using Proposition 6.8, we obtain

$$H^n(\mathfrak{g},U^*) = H^n(\mathfrak{g},\operatorname{Hom}(P_{\mathfrak{h}}^{\mathfrak{g}}(V),\mathbb{C})) \cong \operatorname{Ext}_{\mathfrak{g}}^n(P_{\mathfrak{h}}^{\mathfrak{g}}(V),\mathbb{C})$$

$$\cong \mathrm{Ext}^n_{\mathfrak{h}}(V,\mathbb{C}) \cong H^n(\mathfrak{h},\mathrm{Hom}_{\mathbb{C}}(V,\mathbb{C})) = H^n(\mathfrak{h},V^*) .$$

Thus the cohomologies match in all dimensions and in particular vanish for $n > \dim \mathfrak{h}$.

3. Poincaré duality

We continue with $\mathfrak{g}$ as a complex Lie algebra of finite complex dimension N. When forming terms $X_n = U(\mathfrak{g}) \otimes_{\mathbb{C}} \Lambda^n\mathfrak{g}$ of the Koszul resolution, we treated $\Lambda^n\mathfrak{g}$ as having trivial action by $\mathfrak{g}$. But now we shall use $\Lambda^n\mathfrak{g}$ for a different purpose, at least when $n = N$, and we assume that $\mathfrak{g}$ operates on it as in Chapter II.

Theorem 6.10 (Poincaré duality). For $n \leq N$ and for V in $\mathcal{C}(\mathfrak{g})$, there are vector space isomorphisms

$$H^n(\mathfrak{g},V^*) \cong H_n(\mathfrak{a},V)^* \tag{6.29}$$

and

$$H^n(\mathfrak{g},V) \cong H_{N-n}(\mathfrak{g},V\otimes_{\mathbb{C}} (\Lambda^N\mathfrak{g})^*) \tag{6.30}$$

natural in V. Consequently

$$H^n(\mathfrak{g},V^*) \cong H^{N-n}(\mathfrak{g},V\otimes_{\mathbb{C}} \Lambda^n\mathfrak{g})^* \tag{6.31}$$

naturally in V.

Remarks. (1) The isomorphism (6.29) is easy, and (6.30) is hard; (6.31) comes by substituting (6.30) into (6.29).

(2) If $\mathfrak{g}$ acts trivially on $\Lambda^N\mathfrak{g}$, then $(\Lambda^N\mathfrak{g})^*$ can pull out of the homology in (6.30) and $\Lambda^N\mathfrak{g}$ can pull out of

the cohomology in (6.31).

(3) An important special case of this theorem is the formula that results for $H^N(\mathfrak{g},V)$ from (6.30) and the formula $H_0(\mathfrak{g},W) \cong W/\mathfrak{g}W \cong W \otimes_{\mathfrak{g}} \mathbb{C}$.

Proof of (6.29). Let F and G be the functors $F(W) = W^{\mathfrak{g}}$ and $G(W) = W \otimes_{\mathfrak{g}} \mathbb{C}$. Then

$$F(V^*) = \mathrm{Hom}_{\mathfrak{g}}(\mathbb{C},V^*) = \mathrm{Hom}_{\mathfrak{g}}(\mathbb{C},\mathrm{Hom}_{\mathbb{C}}(V,\mathbb{C})) \cong \mathrm{Hom}_{\mathbb{C}}(V \otimes_{\mathfrak{g}} \mathbb{C},\mathbb{C}) = G(V)^* ,$$

which we can rewrite as a natural isomorphism

$$F \circ \mathrm{dual}(V) \cong \mathrm{dual} \circ G(V) .$$

Thus

$$\begin{aligned}
(F^n \circ \mathrm{dual})(V) &\cong (F \circ \mathrm{dual})^n && \text{by Proposition 5.17, part (b2)} \\
&\cong (\mathrm{dual} \circ G)^n \\
&\cong \mathrm{dual} \circ G_n && \text{by Proposition 5.16.}
\end{aligned}$$

This isomorphism is just (6.29).

Proof of (6.30). We break the argument into three steps.

(1) Preliminaries. We calculate homology and cohomology from the explicit complexes (6.26) and (6.23). In this setting we shall exhibit a natural isomorphism on the chain-cochain level and show that it passes to the homology/cohomology. First we define ∂ on all of $\wedge^n\mathfrak{g} \otimes_{\mathbb{C}} V \otimes_{\mathbb{C}} (\wedge^N\mathfrak{g})^*$ by (6.26) and d on all of $\mathrm{Hom}_{\mathbb{C}}(\wedge^n\mathfrak{g},V)$ by (6.23). We give the isomorphism on the level of a map

$$\wedge^n\mathfrak{g} \otimes_{\mathbb{C}} V \otimes_{\mathbb{C}} (\wedge^N\mathfrak{g})^* \to \mathrm{Hom}_{\mathbb{C}}(\wedge^{N-n}\mathfrak{g},V) . \qquad (6.32)$$

Namely for $\xi \in \Lambda^n \mathfrak{g}$, $\gamma \in \Lambda^{N-n}\mathfrak{g}$, and $\varepsilon \in (\Lambda^N \mathfrak{g})^*$, we define $\xi \otimes v \otimes \varepsilon \to \lambda_{\xi\otimes v\otimes\varepsilon}$ by

$$\lambda_{\xi\otimes v\otimes\varepsilon}(\gamma) = \varepsilon(\xi \wedge \gamma)v . \tag{6.33}$$

Then $\xi \otimes v \otimes \varepsilon \to \lambda_{\xi\otimes v\otimes\varepsilon}$ certainly extends to give (6.32) as a linear isomorphism. Equation (6.30) will thus follow if we show that, under this map, ∂ corresponds to $(-1)^{|\xi|}d$, where $|\xi|$ is the degree of ξ . That is, we are to show that

$$d\lambda_{\xi\otimes v\otimes\varepsilon} = (-1)^{|\xi|}\lambda_{\partial(\xi\otimes v\otimes\varepsilon)} . \tag{6.34}$$

(2) Action of $\mathfrak{g}$ on $(\Lambda^N \mathfrak{g})^*$. Let $X_1, \ldots, X_N$ be in $\mathfrak{g}$, and put $\eta = X_1 \wedge \ldots \wedge X_N$. If Z belongs to $\mathfrak{g}$, then Z acts on η in $\Lambda^N \mathfrak{g}$ by the adjoint action:

$$(\operatorname{ad} Z)\eta = \sum_{i=1}^{N} (-1)^{i+1}([Z,X_i] \wedge X_1 \wedge \ldots \wedge \widehat{X}_i \wedge \ldots \wedge X_N) .$$

For the action of Z on the element of $(\Lambda^N \mathfrak{g})^*$, we use the contragredient action:

$$(Z\varepsilon)(\eta) = \sum_{i=1}^{N} (-1)^{i}\varepsilon([Z,X_i] \wedge X_1 \wedge \ldots \wedge \widehat{X}_i \wedge \ldots \wedge X_N). \tag{6.35}$$

(3) Verification of (6.34). Let $\xi = X_1 \wedge \ldots \wedge X_n$ and $\gamma = Y_0 \wedge \ldots \wedge Y_{N-n}$ be in $\Lambda^n \mathfrak{g}$ and $\Lambda^{N-n+1}\mathfrak{g}$, respectively. To prove (6.34), we want to prove that

$$d\lambda_{\xi\otimes v\otimes\varepsilon}(\gamma) + (-1)^{n+1}\lambda_{\partial(\xi\otimes v\otimes\varepsilon)}(\gamma) = 0 . \tag{6.36}$$

Actually we shall prove that

$$(-1)^n d_{\xi\otimes v\otimes\epsilon}(\gamma) - \lambda_{\partial(\xi\otimes v\otimes\epsilon)}(\gamma) = d(\epsilon\otimes v)(\xi\wedge\gamma)\,. \tag{6.37}$$

On the right side, $\epsilon\otimes v$ is to be regarded as a member of $\mathrm{Hom}_{\mathbb{C}}(\Lambda^N\mathfrak{g},V)$. Since $\xi\wedge\gamma$ belongs to $\Lambda^{N+1}\mathfrak{g}$, which is 0, the right side of (6.37) is 0. Thus (6.36) follows from (6.37).

To prove (6.37), we introduce

$$\begin{aligned}
\xi_i &= X_1\wedge\dots\wedge\widehat{X}_i\wedge\dots\wedge X_n && \text{in } \Lambda^{n-1}\mathfrak{g}\\
\xi_{ij} &= X_1\wedge\dots\wedge\widehat{X}_i\wedge\dots\wedge\widehat{X}_j\wedge\dots\wedge X_n && \text{in } \Lambda^{n-2}\mathfrak{g}\\
\gamma_i &= Y_0\wedge\dots\wedge\widehat{Y}_i\wedge\dots\wedge Y_{N-n} && \text{in } \Lambda^{N-n}\mathfrak{g}\\
\gamma_{ij} &= Y_0\wedge\dots\wedge\widehat{Y}_i\wedge\dots\wedge\widehat{Y}_j\wedge\dots\wedge Y_{N-n} && \text{in } \Lambda^{N-n-1}\mathfrak{g}\,.
\end{aligned} \tag{6.38}$$

We compute the left side of (6.37), using (6.23) and (6.26) to obtain

$$\begin{aligned}
&(-1)^n\sum_{i=0}^{N-n}(-1)^i\epsilon(\xi\wedge\gamma_i)Y_i v + (-1)^n\sum_{i<j}(-1)^{i+j}\epsilon(\xi\wedge[Y_i,Y_j]\wedge\gamma_{ij})v\\
&\qquad-\sum_{r=1}^{n}(-1)^r\epsilon(\xi_r\wedge\gamma)X_r v - \sum_{r=1}^{n}(-1)^r(X_r\epsilon)(\xi_r\wedge\gamma)v\\
&\qquad-\sum_{r<s}(-1)^{r+s}\epsilon([X_r,X_s]\wedge\xi_{rs}\wedge\gamma)v\,.
\end{aligned} \tag{6.39}$$

(Here the third and fourth terms incorporate the action of $\mathfrak{g}$ on $V\otimes_{\mathbb{C}}(\Lambda^N\mathfrak{g})^*$.) We use (6.35) to compute $X_r\epsilon$ and see that the fourth term of (6.39) is

$$= - \sum_{s<r} (-1)^{r+s} \varepsilon([X_r, X_s] \wedge \xi_{rs} \wedge \gamma) v$$

$$- \sum_{s>r} (-1)^{r+s-1} \varepsilon([X_r, X_s] \wedge \xi_{rs} \wedge \gamma) v$$

$$- \sum_{r=1}^{n} \sum_{i=0}^{N-n} (-1)^{r+i+n-1} \varepsilon([X_r, Y_i] \wedge \xi_r \wedge \gamma_i) v$$

$$= 2 \sum_{r<s} (-1)^{r+s} \varepsilon([X_r, X_s] \wedge \xi_{rs} \wedge \gamma) v$$

$$+ \sum_{r,i} (-1)^{r+i+n} \varepsilon([X_r, Y_i] \wedge \xi_r \wedge \gamma_i) v .$$

When we substitute this expression into (6.39), the first term of this expression adds to reverse the sign of the last term of (6.39). Thus (6.39) is

$$= \sum_{r=1}^{n} (-1)^{r+1} \varepsilon(\xi_r \wedge \gamma) X_r v + \sum_{i=0}^{N-n} (-1)^{i+n} \varepsilon(\xi \wedge \gamma_i) Y_i v$$

$$+ \sum_{r<s} (-1)^{r+s} \varepsilon([X_r, X_s] \wedge \xi_{rs} \wedge \gamma)$$

$$+ \sum_{r,i} (-1)^{r+i+n} \varepsilon([X_r, Y_i] \wedge \xi_r \wedge \gamma_i) v$$

$$+ \sum_{i<j} (-1)^{i+j} \varepsilon([Y_i, Y_j] \wedge \xi \wedge \gamma_{ij}) v ,$$

which is precisely the right side of (6.37), by (6.23). Thus (6.37) is proved, and Theorem 6.10 follows.

4. Kostant's Theorem for U(n)

When we first introduced Lie algebra cohomology, we

mentioned that its computation would usually be done without computation of cocycles and coboundaries. The actual tools that are used are the ones of the previous three sections and, most importantly, the long exact sequence of §V.5. With these tools and some additional information, we shall compute some specific cohomology spaces in the remainder of this chapter.

The setting will be $G = U(n)$, $\mathfrak{g}_0 = \mathfrak{u}(n)$, and $\mathfrak{g} = \mathfrak{u}(n)^{\mathbb{C}} = \mathfrak{gl}(n,\mathbb{C})$, with other notation as established in Chapter III. Fix a lexicographic ordering, the corresponding set Δ^+ of positive roots, and the half-sum δ of the members of Δ^+. Let $\mathfrak{n}$ be the Lie subalgebra of $\mathfrak{g}$ given by $\mathfrak{n} = \sum_{\alpha\in\Delta^+} \mathfrak{g}_\alpha$. (This is the strictly upper triangular subalgebra if Δ^+ is the standard system of positive roots.) Kostant's Theorem will compute the cohomology $H^k(\mathfrak{n},V)$ for each irreducible representation space V of G. (Here it is understood that the G action on V yields a $\mathfrak{g}$ action and therefore an $\mathfrak{n}$ action by restriction.) In turn, Kostant's Theorem becomes a fundamental tool in its own right for computing more complicated cohomology.

Actually Kostant's Theorem does something further beyond computing $H^k(\mathfrak{n},V)$. Namely we shall introduce an action of the diagonal subgroup T (and hence also of the diagonal subalgebra $\mathfrak{h}$) on $H^k(\mathfrak{n},V)$, and the theorem identifies the cohomology as a representation of T. In Chapter VII we shall reinterpret this result as a computation of certain relative Lie algebra cohomology spaces.

Let us specify the action by T and then state the theorem. Recall that $H^k(\mathfrak{n},V)$ can be computed explicitly from the complex (6.23) whose k^{th} term is $\mathrm{Hom}_{\mathbb{C}}(\Lambda^k\mathfrak{n}, V)$. Without assuming that V is finite-dimensional, assume that we have a representation of G or $\mathfrak{g}$ on V. If the action is by the group G, then T acts on $\Lambda^k\mathfrak{n}$ (since it acts on $\mathfrak{n}$ by Ad) and on V, hence on $\mathrm{Hom}_{\mathbb{C}}(\Lambda^k\mathfrak{n}, V)$ by the formula

$$(t\omega)(\xi) = t(\omega(\mathrm{Ad}(t)^{-1}\xi)) . \qquad (6.40a)$$

(See §II.4.) If the action is by the Lie algebra $\mathfrak{g}$, then $\mathfrak{h}$ acts on $\Lambda^k\mathfrak{n}$ and on V, hence on $\mathrm{Hom}_{\mathbb{C}}(\Lambda^k\mathfrak{n}, V)$ by the formula

$$(H\omega)(\xi) = H(\omega(\xi)) - \omega((\mathrm{ad}\ H)\xi) . \qquad (6.40b)$$

<u>Proposition 6.11</u>. If G or $\mathfrak{g}$ acts on V, then the corresponding representation of T or $\mathfrak{h}$ on $\mathrm{Hom}_{\mathbb{C}}(\Lambda^k\mathfrak{n}, V)$ given in (6.40) commutes with the d operator in (6.23) and therefore descends to an action of T or $\mathfrak{h}$ on the cohomology $H^k(\mathfrak{n}, V)$.

Remarks. We give the argument for an action by T. The argument with $\mathfrak{h}$ is longer, but formally it is obtained simply by differentiating all the formulas in the argument with T.

Proof. For an action by T,

$$d(t\omega)(X_1 \wedge \ldots \wedge X_k)$$

$$= \sum_{j=1}^{k} (-1)^{j+1} X_j((t\omega)(X_1 \wedge \ldots \wedge \hat{X}_j \wedge \ldots \wedge X_k))$$

$$+ \sum_{r<s} (-1)^{r+s}(t\omega)([X_r,X_s] \wedge X_1 \wedge \cdots \wedge \hat{X}_r \wedge \cdots \wedge \hat{X}_s \wedge \cdots \wedge X_k)$$

$$= \sum_{j=1}^{k} (-1)^{j+1} X_j(t(\omega(\mathrm{Ad}(t)^{-1}X_1 \wedge \cdots \wedge \widehat{\mathrm{Ad}(t)^{-1}X_j} \wedge \cdots \wedge \mathrm{Ad}(t)^{-1}X_k)))$$

$$+ \sum_{r<s} (-1)^{r+s} t(\omega(\mathrm{Ad}(t)^{-1}[X_r,X_s] \wedge \mathrm{Ad}(t)^{-1}X_1 \wedge \cdots \wedge \widehat{\mathrm{Ad}(t)^{-1}X_r} \wedge \cdots \wedge \widehat{\mathrm{Ad}(t)^{-1}X_s} \wedge \cdots \wedge \mathrm{Ad}(t)^{-1}X_k)) .$$

Meanwhile

$$(t(d\omega))(X_1 \wedge \cdots \wedge X_k) = t(d\omega(\mathrm{Ad}(t)^{-1}X_1 \wedge \cdots \wedge \mathrm{Ad}(t)^{-1}X_k))$$

$$= \sum_{j=1}^{k} (-1)^{j+1} t(\mathrm{Ad}(t)^{-1}X_j(\omega(\mathrm{Ad}(t)^{-1}X_1 \wedge \cdots \wedge \widehat{\mathrm{Ad}(t)^{-1}X_j} \wedge \cdots \wedge \mathrm{Ad}(t)^{-1}X_k)))$$

$$+ \sum_{r<s} (-1)^{r+s} t(\omega[\mathrm{Ad}(t)^{-1}X_r, \mathrm{Ad}(t)^{-1}X_s] \wedge \mathrm{Ad}(t)^{-1}X_1 \wedge \cdots \wedge \widehat{\mathrm{Ad}(t)^{-1}X_r} \wedge \cdots \wedge \widehat{\mathrm{Ad}(t)^{-1}X_s} \wedge \cdots \wedge \mathrm{Ad}(t)^{-1}X_k)) ,$$

and this matches the previous expression since

$$t(\mathrm{Ad}(t)^{-1}X_j(v)) = X_j(t(v))$$

for v in V.

Theorem 6.12 (Kostant). Let V be an irreducible finite-dimensional representation space for G with highest weight λ relative to Δ^+, and let $0 \leq k \leq \dim \mathfrak{n}$. As a representation space for T, $H^k(\mathfrak{n},V)$ is the direct sum of all the one-dimensional representations of T with weights

$$w(\lambda+\delta)-\delta\ , \qquad w\in W\ , \qquad \ell(w)=k\ ,$$

all occurring with multiplicity one.

Recall from §III.6 that δ is half the sum of the positive roots. There are two easy cases to the theorem: $k = 0$ and $k = \dim \mathfrak{n}$. For $k = 0$, $H^0(\mathfrak{n}, V)$ is just the space $V^{\mathfrak{n}}$ of invariants, acted upon by the G action restricted to T. Theorem 3.13c identifies $V^{\mathfrak{n}}$ as the highest weight space of V, which is one-dimensional with weight $\lambda = 1(\lambda+\delta)-\delta$.

For $k = \dim \mathfrak{n}$, the theorem follows from Poincaré duality. This implication is proved most clearly with the aid of Proposition 7.34 below, but we give a direct argument now. It is necessary first to establish a natural action of T on $\bigwedge^{\ell}\mathfrak{n} \otimes_{\mathbb{C}} V$ (which we do in the obvious way), to see that it descends to homology (which we do by an argument analogous to the one for Proposition 6.11), and to see that Poincaré duality respects the T actions. Recall from (6.32) and (6.33) that Poincaré duality is implemented by the map $\xi \otimes v \otimes \epsilon \to \lambda_{\xi\otimes v\otimes\epsilon}$ of

$$\textstyle\bigwedge^{\ell}\mathfrak{n} \otimes_{\mathbb{C}} V \otimes_{\mathbb{C}} (\bigwedge^{\dim \mathfrak{n}} \mathfrak{n})^* \to \operatorname{Hom}_{\mathbb{C}}(\bigwedge^{\dim \mathfrak{n} - \ell}, V)$$

given by

$$\lambda_{\xi\otimes v\otimes\epsilon}(\gamma) = \epsilon(\xi \wedge \gamma)v\ .$$

Since, with abbreviated notation,

$$(t\lambda_{\xi\otimes v\otimes\epsilon})(\gamma) = t(\lambda_{\xi\otimes v\otimes\epsilon}(t^{-1}\gamma)) = \epsilon(\xi \wedge t^{-1}\gamma)(tv)$$

$$= ((t\varepsilon)(t\xi \wedge \gamma))(tv) = \lambda_{t\xi \otimes tv \otimes t\varepsilon}(\gamma) = \lambda_{t(\xi \otimes v \otimes \varepsilon)}(\gamma),$$

the duality map commutes with T. Therefore the isomorphism (6.30), namely

$$H^k(\mathfrak{n}, V) \cong H_{\dim \mathfrak{n} - k}(\mathfrak{n}, V \otimes_{\mathbb{C}} (\Lambda^{\dim \mathfrak{n}} \mathfrak{n})^*),$$

is an equivalence of representations of T. Taking $k = \dim \mathfrak{n}$, we obtain

$$\begin{aligned} H^{\dim \mathfrak{n}}(\mathfrak{n}, V) &\cong H_0(\mathfrak{n}, V \otimes_{\mathbb{C}} \mathbb{C}_{-2\delta}) \\ &\cong (V \otimes_{\mathbb{C}} \mathbb{C}_{-2\delta}) / \mathfrak{n}(V \otimes_{\mathbb{C}} \mathbb{C}_{-2\delta}) \\ &\cong (V \otimes_{\mathbb{C}} \mathbb{C}_{-2\delta}) / (\mathfrak{n} V \otimes_{\mathbb{C}} \mathbb{C}_{-2\delta}) \\ &\cong V/\mathfrak{n} V \otimes_{\mathbb{C}} \mathbb{C}_{-2\delta} . \end{aligned}$$

Application of a Birkhoff-Witt basis from $\{\mathfrak{n}, \mathfrak{h}, \bar{\mathfrak{n}}\}$, where $\bar{\mathfrak{n}} = \sum_{\alpha \in \Delta^+} \mathfrak{g}_{-\alpha}$, to a lowest weight vector shows that $V/\mathfrak{n}V$ is isomorphic to the lowest weight space, which one can see is $\mathbb{C}_{w_0\lambda}$, where w_0 is the unique element of W carrying Δ^+ to $-\Delta^+$. (If Δ^+ is the standard positive system, then $w_0 = (1\ \ n)(2\ \ n{-}1)(3\ \ n{-}2)\cdots$.) Thus $H^{\dim \mathfrak{n}}(\mathfrak{n}, V)$ is one-dimensional, with weight $w_0\lambda - 2\delta = w_0(\lambda + \delta) - \delta$.

The proof of Theorem 6.12 in other degrees of cohomology lies deeper and divides into two steps. The first step, which is just a combinatorial calculation with weights, is to see that the weight $w(\lambda + \delta) - \delta$ occurs in $\mathrm{Hom}_{\mathbb{C}}(\Lambda^{\ell(w)}\mathfrak{n}, V)$ with multiplicity one and in all other $\mathrm{Hom}_{\mathbb{C}}(\Lambda^k \mathfrak{n}, V)$ with multiplicity zero. Then d of its occurrence in $\mathrm{Hom}_{\mathbb{C}}(\Lambda^{\ell(w)}\mathfrak{n}, V)$ must be 0, and its occurrence cannot arise from d applied to $\operatorname{Hom}_{\mathbb{C}}(\Lambda^{\ell(w)-1}\mathfrak{n}, V)$. Hence the weight passes to cohomology with multiplicity one in degree $\ell(w)$ and with multiplicity zero in other degrees.

We can even identify explicitly the one-dimensional subspace of $\operatorname{Hom}_{\mathbb{C}}(\Lambda^{\ell(w)}\mathfrak{n}, V)$ of weight $w(\lambda+\delta)-\delta$. Namely for each $\alpha \in \Delta^+$, let E_α be a nonzero root vector for α. Enumerate

$$\Delta^+(w) = \{\beta \in \Delta^+ \mid w^{-1}\beta < 0\}$$

as $\beta_1, \ldots, \beta_{\ell(w)}$, and let $v_{w\lambda}$ be a nonzero vector in V of weight $w\lambda$. Define ω in $\operatorname{Hom}_{\mathbb{C}}(\Lambda^{\ell(w)}\mathfrak{n}, V)$ by

$$\omega(E_{\alpha_1} \wedge \ldots \wedge E_{\alpha_{\ell(w)}}) = \begin{cases} v_{w\lambda} & \text{if } \alpha_1 = \beta_1, \ldots, \alpha_{\ell(w)} = \beta_{\ell(w)} \\ 0 & \text{if } \{\alpha_1, \ldots, \alpha_{\ell(w)}\} \neq \{\beta_1, \ldots, \beta_{\ell(w)}\}. \end{cases}$$

Then $\mathbb{C}\omega$ is a one-dimensional T-invariant subspace with weight $w\lambda - \Sigma_{\beta\in\Delta^+(w)}\beta = w\lambda - (\delta - w\delta) = w(\lambda+\delta)-\delta$. (See Proposition 3.19.)

We shall carry out the first step of the proof after proving Lemma 6.13 below. The second step of the proof is to show that no other weight occurs in the cohomology. This step is harder and will be carried out in §6, using the Harish-Chandra isomorphism established in the next section.

<u>Lemma 6.13</u>. Let λ be dominant, and let S be a subset of Δ^+. If w is in W, then

$$|w(\lambda+\delta) - \delta + \sum_{\alpha\in S} \alpha|^2 \geq |\lambda|^2 \tag{6.41}$$

with equality only if $S = \Delta^+(w)$.

Proof. Since $\delta - w\delta = \Sigma_{\beta\in\Delta^+(w)}\beta$ by Proposition 3.19, we have

$$|w(\lambda+\delta) - \delta + \sum_{\alpha\in S}\alpha|^2 - |w\lambda|^2 = |w\lambda + \sum_{\alpha\in S}\alpha - \sum_{\beta\in\Delta^+(w)}\beta|^2 - |w\lambda|^2$$

$$= 2\langle w\lambda , \sum_{\alpha\in S}\alpha - \sum_{\beta\in\Delta^+(w)}\beta\rangle + |\sum_{\alpha\in S}\alpha - \sum_{\beta\in\Delta^+(w)}\beta|^2 . \quad (6.42)$$

In considering $\Sigma_{\alpha\in S}\alpha - \Sigma_{\beta\in\Delta^+(w)}\beta$, let us think of cancelling the roots that match in the two sums. Let γ be a root that survives this cancellation. If γ comes from the first term, then γ is positive but γ is not in $\Delta^+(w)$. Thus $w^{-1}\gamma > 0$. Since λ and δ are dominant, we have

$$\langle w\lambda,\gamma\rangle = \langle\lambda,w^{-1}\gamma\rangle \geq 0 \quad \text{and} \quad \langle w\delta,\gamma\rangle = \langle\delta,w^{-1}\gamma\rangle > 0 . \quad (6.43)$$

If γ is some $-\beta$ with β from the second term, then β is in $\Delta^+(w)$ and satisfies $w^{-1}\beta < 0$. Thus $\gamma = -\beta$ has $w^{-1}\gamma > 0$, and the dominance of λ and δ again gives (6.43). In other words, every γ of interest satisfies (6.43).

Adding the λ inequalities of (6.43), we see that the first term on the right side of (6.42) is ≥ 0, and thus (6.41) follows. Adding the δ inequalities of (6.43), we see that $\Sigma_{\alpha\in S}\alpha - \Sigma_{\beta\in\Delta^+(w)}\beta$ is nonzero unless the roots in S coincide with those in $\Delta^+(w)$. Referring to the second term on the right side of (6.42), we see that the inequality (6.41) is strict unless $S = \Delta^+(w)$. This proves the lemma.

<u>Proof of first part of Theorem 6.12.</u> We compute the multiplicity of $w(\lambda+\delta)-\delta$ in $\mathrm{Hom}_{\mathbb{C}}(\Lambda^k\mathfrak{n},V)$, showing that it

is one if $k = \ell(w)$ and is zero otherwise.

In fact, let μ be a weight of $\mathrm{Hom}_{\mathbb{C}}(\Lambda^k\mathfrak{n}, V)$. The weights of $\Lambda^k\mathfrak{n}$ are of the form $\Sigma_{\alpha\in S}\,\alpha$, where S is a subset of k elements of Δ^+, and each such weight has multiplicity equal to the number of such S's. Thus μ is of the form

$$\mu = \gamma - \sum_{\alpha\in S} \alpha ,$$

where γ is a weight of V and S is a subset of k elements of Δ^+. Moreover, the multiplicity of μ is the sum over all such γ and S of the multiplicity of γ.

We are interested in $\mu = w(\lambda+\delta) - \delta$. Thus we consider the relation

$$w(\lambda+\delta) - \delta = \gamma - \sum_{\alpha\in S} \alpha . \tag{6.44}$$

Then Lemma 6.13 says that $|\gamma|^2 \geq |\lambda|^2$, while Proposition 3.16 says that $|\gamma|^2 \leq |\lambda|^2$. Hence $|\gamma|^2 = |\lambda|^2$, and Lemma 6.13 thus says $S = \Delta^+(w)$. Then $\Sigma_{\alpha\in S}\,\alpha = \delta - w\delta$, so that (6.44) gives $\gamma = w\lambda$.

In other words, the only possible γ is $w\lambda$ (which has multiplicity one in V) and the only possible S is $\Delta^+(w)$. Hence $w(\lambda+\delta)-\delta$ has multiplicity one if $k = \ell(w)$, and it has multiplicity zero otherwise.

The second part of the proof of Theorem 6.12 will use the Harish-Chandra isomorphism, which is the subject of the next section. We shall resume the proof of Theorem 6.12 in §6.

5. Harish-Chandra isomorphism for U(n)

For studying representations of Lie groups, particularly irreducible representations, the center of the universal enveloping algebra is a useful tool. For example, in any situation where Schur's Lemma applies to a representation of a Lie algebra, every member of the center of the universal enveloping algebra will act as a scalar, and that scalar will be an invariant of the representation. If the center is large, we should expect to learn a large amount about the representation in this way.

We continue to take $G = U(n)$ (or sometimes $SU(2)$), writing $\mathfrak{g} = \mathfrak{u}(n)^{\mathbb{C}} = \mathfrak{gl}(n,\mathbb{C})$. In this case the center $Z(\mathfrak{g})$ of $U(\mathfrak{g})$ is indeed large, as a consequence of the theorem we prove in this section.

We begin by exhibiting some members of $Z(\mathfrak{g})$. It is usually easier to test for members of $Z(\mathfrak{g})$ by using the group than by using the Lie algebra, and we begin by relating actions of the group and the Lie algebra on $U(\mathfrak{g})$. We have already defined a representation ad of $\mathfrak{g}$ on $U(\mathfrak{g})$ in §II.7 by

$$(\operatorname{ad} X)u = Xu - uX \quad \text{for } X \in \mathfrak{g}, \ u \in U(\mathfrak{g}) .$$

Let us define a representation Ad of G on $U(\mathfrak{g})$. To do so, we start with the representation Ad of G on the tensor algebra of $\mathfrak{g}$ and observe that

$$\begin{aligned} &\operatorname{Ad}(x)(X\otimes Y - Y\otimes X - [X,Y]) \\ &\qquad = \operatorname{Ad}(x)X\otimes \operatorname{Ad}(x)Y - \operatorname{Ad}(x)Y\otimes \operatorname{Ad}(x)X - [\operatorname{Ad}(x)X,\operatorname{Ad}(x)Y] . \end{aligned}$$

Therefore the action descends to $U(\mathfrak{g})$. Each $U^n(\mathfrak{g})$ is a finite-dimensional invariant subspace under $\mathrm{Ad}(G)$, and the action is therefore controlled by the differentiated version of Ad, which the proposition below identifies as ad.

Proposition 6.14. The differentiated version of Ad on $U(\mathfrak{g})$ is ad. A member z of $U(\mathfrak{g})$ is in the center $Z(\mathfrak{g})$ if and only if $\mathrm{Ad}(x)z = z$ for all $x \in G$.

Proof. If $u = X_1 \cdots X_k$ is a monomial in $U(\mathfrak{g})$ and X is in $\mathfrak{g}_0$, then

$$\begin{aligned}
&\frac{d}{dt}\,\mathrm{Ad}(\exp tX)(X_1 \cdots X_k)\Big|_{t=0} \\
&\quad = \frac{d}{dt}\{(\mathrm{Ad}(\exp tX)X_1)\cdots(\mathrm{Ad}(\exp tX)X_k)\}\Big|_{t=0} \\
&\quad = ((\mathrm{ad}\,X)X_1)X_2 \cdots X_k + \cdots + X_1 \cdots X_{k-1}((\mathrm{ad}\,X)X_k) \\
&\quad = XX_1X_2 \cdots X_k - X_1XX_2 \cdots X_k + \cdots + X_1 \cdots X_{k-1}XX_k - X_1 \cdots X_{k-1}X_kX \\
&\quad = XX_1 \cdots X_k - X_1 \cdots X_kX \\
&\quad = (\mathrm{ad}\,X)(X_1 \cdots X_k).
\end{aligned}$$

Consequently ad is the differentiated version of Ad.

It follows that $\mathrm{Ad}(\exp X) = e^{\mathrm{ad}\,X}$ for all X in $\mathfrak{g}_0$. Thus, since G is connected, the members z of $U(\mathfrak{g})$ on which all $\mathrm{Ad}(x)$ act as the identity coincide with those for which all ad X act as 0. The latter z's are the members of $Z(\mathfrak{g})$, and the proposition follows.

Let B_0 be the trace form. Choose a basis $X_1, \ldots, X_N$ of $\mathfrak{g}$, and let $g_{ij} = B_0(X_i, X_j)$. Since B_0 is nondegenerate, (g_{ij}) is nonsingular; let (g^{ij}) denote the inverse matrix $(g_{ij})^{-1}$. Put $X^j = \Sigma_i\, g^{ij}X_i$, so that $X_i = \Sigma_j\, g_{ji}X^j$. The Casimir element of $U(\mathfrak{g})$ is defined by

$$\Omega = \sum_{i,j} g_{ij}X^iX^j .$$

Proposition 6.15. The Casimir element Ω of $U(\mathfrak{g})$ is independent of the basis. It satisfies $\mathrm{Ad}(x)\Omega = \Omega$ for all x in G and hence is in the center $Z(\mathfrak{g})$ of $U(\mathfrak{g})$.

Proof. Let $Y_j = \Sigma\, a^i_jX_i$ be another basis, and let $(b^i_j) = (a^i_j)^{-1}$, so that $\Sigma_k\, a^i_kb^k_j = \delta^i_j = \Sigma_k\, a^k_jb^i_k$. Define

$$h_{ij} = B_0(Y_i,Y_j) = \sum_{k,m} a^k_ia^m_jB_0(X_k,X_m) = \sum_{k,m} a^k_ig_{km}a^m_j .$$

If we put $H^{jn} = \Sigma\, b^j_{k'}g^{k'm'}b^n_{m'}$, then

$$\sum_j h_{ij}H^{jn} = \Sigma\, a^k_ig_{km}a^m_jb^j_{k'}g^{k'm'}b^n_{m'} = \delta^n_i ,$$

so that $(H^{ij}) = (h_{ij})^{-1}$. Thus

$$Y^j = \Sigma\, h^{ij}Y_i = \Sigma\, b^i_kg^{km}b^j_mY_i = \Sigma\, b^i_kg^{km}b^j_ma^{i'}_iX_{i'}$$

$$= \Sigma\, g^{km}b^j_mX_k = \Sigma\, g^{km}b^j_mg_{rk}X^r = \Sigma\, b^j_rX^r .$$

Letting Ω_Y and Ω_X denote the two possible versions of Ω, we thus have

$$\Omega_Y = \Sigma\, h_{ij}Y^iY^j = \Sigma\, a^k_ig_{km}a^m_jb^i_rb^j_sX^rX^s = \Sigma\, g_{rs}X^rX^s = \Omega_X .$$

Hence Ω is independent of the basis.

Now if x is in G, we have

$$\mathrm{Ad}(x)\Omega = \sum_{i,j} B_0(X_i,X_j)\mathrm{Ad}(x)X^i\mathrm{Ad}(x)X^j$$

$$= \sum_{i,j} B_0(\mathrm{Ad}(x)^{-1}X_i, \mathrm{Ad}(x)^{-1}X_j)\mathrm{Ad}(x)X^i\mathrm{Ad}(x)X^j$$

by invariance of B_0

$$= \sum_{i,j} B_0(X_i, X_j)X^iX^j \qquad \text{by the above independence}$$

$$= \Omega .$$

By Proposition 6.14, Ω is in $Z(\mathfrak{g})$.

<u>Example</u>. The same considerations apply to $SU(2)$ with $\mathfrak{g} = \mathfrak{sl}(2,\mathbb{C})$. We use $\{h, e, f\}$ as basis. Then

$$B_0(h,h) = 2, \qquad B_0(e,f) = B_0(f,e) = 1,$$

and B_0 is 0 on other pairs. Thus

$$(g_{ij}) = \begin{pmatrix} 2 & 0 & 0 \\ 0 & 0 & 1 \\ 0 & 1 & 0 \end{pmatrix} \quad \text{and} \quad (g^{ij}) = \begin{pmatrix} \frac{1}{2} & 0 & 0 \\ 0 & 0 & 1 \\ 0 & 1 & 0 \end{pmatrix}.$$

(rows and columns indexed by h, e, f), so that

$$\Omega = \tfrac{1}{2}h^2 + ef + fe .$$

To get a handle on the most general element of $Z(\mathfrak{g})$, we fix Δ^+ and return to the monomial basis (3.16) of $U(\mathfrak{g})$, namely

$$E_{-\beta_1}^{i_1} \cdots E_{-\beta_m}^{i_m} H_1^{j_1} \cdots H_n^{j_n} E_{\beta_1}^{k_1} \cdots E_{\beta_m}^{k_m}, \tag{6.45}$$

and the action on a highest weight vector v of an irreducible representation of G with highest weight λ. Think of expanding $z \in Z(\mathfrak{g})$ in terms of this basis and applying the result to v. Any term with some E_{β_j} present acts as 0, and the H's act as scalars. In Lemma 6.16b, we shall see for this z that the presence of some $E_{-\beta_j}$ forces the presence of some $E_{\beta_{j'}}$. Thus only the terms with just H's present contribute to the result. They contribute a scalar, and z acts by that same scalar on the whole representation. The Harish-Chandra mapping will make this calculation universally, without reference to the highest weight λ.

Let

$$\mathfrak{n} = \sum_{\alpha\in\Delta^+} \mathfrak{g}_\alpha \quad \text{and} \quad \bar{\mathfrak{n}} = \sum_{\alpha\in\Delta^+} \mathfrak{g}_{-\alpha} .$$

If Δ^+ is the standard positive system, these are the strictly upper triangular and strictly lower triangular subalgebras of $\mathfrak{g}$. In any event, the universal enveloping algebras of $\mathfrak{n}$ and $\bar{\mathfrak{n}}$ result from the right and left ends of (6.45).

Lemma 6.16.

(a) $U(\mathfrak{g}) = U(\mathfrak{h}) \oplus (U(\mathfrak{g})\mathfrak{n} + \bar{\mathfrak{n}}U(\mathfrak{g}))$.

(b) If z is in $Z(\mathfrak{g})$, then its projection into the second factor of the right side of (a) lies in $U(\mathfrak{g})\mathfrak{n}$.

Proof. (a) We express members of $U(\mathfrak{g})$ in the basis (6.45). Then $U(\mathfrak{h})$ consists of all members with just H's present, $U(\mathfrak{g})\mathfrak{n}$ consists of all members requiring some E_{β_j} in each term, and $\bar{\mathfrak{n}}U(\mathfrak{g})$ consists of all members requiring some $E_{-\beta_j}$ in each term. Then (a) is clear.

(b) Expand z in terms of the basis. Each monomial in the basis is an eigenvector for each $\operatorname{ad} H$ with $H \in \mathfrak{h}$, and the eigenvalue is

$$\sum_j k_j \beta_j(H) - \sum_j \ell_j \beta_j(H) . \tag{6.46}$$

Also $(\operatorname{ad} H)z = 0$. Then it follows that each nonzero term in the expansion of z has (6.46) equal to 0. If some $E_{-\beta_j}$ is present in a term, it follows that some $E_{\beta_{j'}}$ is present. Thus (b) is proved.

Let

$$\gamma' : U(\mathfrak{g}) \to U(\mathfrak{h})$$

be the projection to the first factor in Lemma 6.16a. If z is in $Z(\mathfrak{g})$, then Lemma 6.16b says that

$$z - \gamma'(z) \quad \text{is in} \quad U(\mathfrak{g})\mathfrak{n} . \tag{6.47}$$

Example. For $\mathfrak{g} = \mathfrak{sl}(2,\mathbb{C})$, let us compute $\gamma'(\Omega)$. The positive root will be $\alpha = e_1 - e_2$, and then $\mathfrak{n} = \mathbb{C}e$ and $\bar{\mathfrak{n}} = \mathbb{C}f$. The procedure is to write out Ω in the Birkhoff-Witt basis $f^i h^j e^k$ and then to retain only those terms with no f's or e's. Since

$$\Omega = \tfrac{1}{2}h^2 + ef + fe = \tfrac{1}{2}h^2 + (fe + [e,f]) + fe = \tfrac{1}{2}h^2 + h + 2fe ,$$

we obtain $\gamma'(\Omega) = \tfrac{1}{2}h^2 + h$.

Let V_λ be an irreducible representation of G of highest weight λ. Then $\lambda : \mathfrak{h} \to \mathbb{C}$ is linear and so extends to a homomorphism of the symmetric algebra $\lambda : S(\mathfrak{h}) \to \mathbb{C}$ with

$\lambda(1) = 1$. Since $\mathfrak{h}$ is abelian, we have $S(\mathfrak{h}) = U(\mathfrak{h})$, and thus λ is a homomorphism

$$\lambda : U(\mathfrak{h}) \to \mathbb{C} \tag{6.48}$$

with $\lambda(1) = 1$. If z is in $Z(\mathfrak{g})$, our calculations before Lemma 6.16 show that

$$zv = \lambda(\gamma'(z))v \quad \text{for all} \quad v \in V_\lambda . \tag{6.49}$$

Harish-Chandra found that a slight adjustment of γ' leads to a more symmetric formula for the behavior of the center. Before giving the general construction, let us see it in $\mathfrak{sl}(2,\mathbb{C})$.

Example. For $\mathfrak{g} = \mathfrak{sl}(2,\mathbb{C})$, we extend $h \to h-1$ linearly to a map of $\mathfrak{h}$ into $U(\mathfrak{h})$. This map then extends (multiplicatively) to an isomorphism of $U(\mathfrak{h})$ onto itself, and we shall apply this isomorphism to $\gamma'(\Omega)$. Our map sends $\frac{1}{2}h^2 + h$ into

$$\tfrac{1}{2}(h-1)^2 + (h-1) = \tfrac{1}{2}h^2 - h + \tfrac{1}{2} + h - 1 = \tfrac{1}{2}(h^2 - 1) .$$

The significance of this expression comes from the action of the Weyl group W. The nontrivial element of the 2-element W sends $\begin{pmatrix}1 & 0\\ 0 & -1\end{pmatrix}$ into $\begin{pmatrix}-1 & 0\\ 0 & 1\end{pmatrix}$, i.e., h into $-h$. The element $\frac{1}{2}(h^2-1)$ that we have obtained is thus invariant under the Weyl group. We shall see that this phenomenon generalizes.

As in §III.6, we let $\delta = \frac{1}{2}\sum_{\alpha\in\Delta^+}\alpha$. If Δ^+ is the standard positive system, then δ is given by

$$\delta = \tfrac{1}{2} \sum_{\alpha \in \Delta^+} \alpha = \tfrac{1}{2}(n-1)e_1 + \tfrac{1}{2}(n-3)e_2 + \ldots + (-\tfrac{1}{2})(n-1)e_n .$$

(In the case of $\mathfrak{sl}(2,\mathbb{C})$ with $e_1 - e_2$ as positive root, note that $\delta = \frac{1}{2}(e_1 - e_2)$ acts on $h = \begin{pmatrix} 1 & 0 \\ 0 & -1 \end{pmatrix}$ as 1.) Define $\sigma : \mathfrak{h} \to U(\mathfrak{h})$ by

$$\sigma(H) = H - \delta(H)1 ,$$

and extend σ to an algebra automorphism of $U(\mathfrak{h})$. The <u>Harish-Chandra map</u> γ is defined by

$$\gamma = \sigma \circ \gamma' \tag{6.50}$$

as a mapping of $Z(\mathfrak{g})$ into $U(\mathfrak{h})$. The maps γ and γ' are related by

$$(\lambda + \delta)(\gamma(z)) = \lambda(\gamma'(z)) . \tag{6.51}$$

Thus on the irreducible representation V_λ, members z of $Z(\mathfrak{g})$ act by

$$zv = (\lambda + \delta)(\gamma(z))v \quad \text{for all} \quad v \in V_\lambda . \tag{6.52}$$

The symmetry property of γ appears in the theorem that follows.

<u>Theorem 6.17</u> (Harish-Chandra). The map $\gamma : Z(\mathfrak{g}) \to U(\mathfrak{h})$ is an algebra isomorphism of $Z(\mathfrak{g})$ onto the algebra

$$U(\mathfrak{h})^W = \{\text{members of } U(\mathfrak{h}) \text{ fixed by the Weyl group } W\} .$$

Remark. Members of the Weyl group have representatives in G normalizing $\mathfrak{h}$ under the adjoint action, and the action of the Weyl group comes from the extension of that adjoint action to $U(\mathfrak{h})$. Recall that $H_1 = E_{11}, \ldots, H_n = E_{nn}$ is a basis of $\mathfrak{h}$ (in the case $G = U(n)$). Then $U(\mathfrak{h})$ consists of all polynomials in the n variables $H_1, \ldots, H_n$, and W acts by permuting the variables. Thus $U(\mathfrak{h})^W$ is just the subalgebra of symmetric polynomials.

The step that requires new ideas is the proof that image $\gamma \subseteq U(\mathfrak{h})^W$. We shall use the P functor of §VI.1 in a new way. Let $\mathfrak{b}$ be the Lie subalgebra of $\mathfrak{g}$ given by

$$\mathfrak{b} = \mathfrak{h} \oplus \mathfrak{n} .$$

In case Δ^+ is the standard positive system, $\mathfrak{b}$ is the full upper triangular subalgebra of $\mathfrak{g}$. Fix λ in $\mathfrak{h}^*$ and define a one-dimensional representation of $\mathfrak{b}$ on a space $\mathbb{C}_{\lambda-\delta}$ by

$$Xz = \begin{cases} (\lambda-\delta)(X)z & \text{for } X \in \mathfrak{h} \\ 0 & \text{for } X \in \mathfrak{n} . \end{cases}$$

(Here z is in $\mathbb{C}$.) Then $\mathbb{C}_{\lambda-\delta}$ is a $U(\mathfrak{b})$ module, and we let

$$V(\lambda) = P_{\mathfrak{b}}^{\mathfrak{g}}(\mathbb{C}_{\lambda-\delta}) = U(\mathfrak{g}) \otimes_{U(\mathfrak{b})} \mathbb{C}_{\lambda-\delta} .$$

The $U(\mathfrak{g})$ module $V(\lambda)$ is called a <u>Verma module</u>. We shall derive some of its properties.

First we have $\mathfrak{g} = \bar{\mathfrak{n}} \oplus \mathfrak{b}$ as vector spaces and hence

$$U(\mathfrak{g}) \cong U(\bar{\mathfrak{n}}) \otimes_{\mathbb{C}} U(\mathfrak{b})$$

as a left $U(\bar{\mathfrak{n}})$ module and right $U(\mathfrak{b})$ module. Consequently we have a $U(\bar{\mathfrak{n}})$ isomorphism

$$V(\lambda) \cong (U(\bar{\mathfrak{n}}) \otimes_{\mathbb{C}} U(\mathfrak{b})) \otimes_{U(\mathfrak{b})} \mathbb{C}_{\lambda-\delta}$$

$$\cong U(\bar{\mathfrak{n}}) \otimes_{\mathbb{C}} (U(\mathfrak{b}) \otimes_{U(\mathfrak{b})} \mathbb{C}_{\lambda-\delta}) \cong U(\bar{\mathfrak{n}}) \otimes_{\mathbb{C}} \mathbb{C}_{\lambda-\delta} . \tag{6.53}$$

Next let us compute the effect of $z \in Z(\mathfrak{g})$ on the element $1 \otimes 1$. For $X \in \mathfrak{n}$, we have

$$X(1 \otimes 1) = X \otimes 1 = 1 \otimes X(1) = 0 , \tag{6.54a}$$

so that $U(\mathfrak{g})\mathfrak{n}$ acts as 0. For $X \in \mathfrak{h}$, we have

$$X(1 \otimes 1) = 1 \otimes X(1) = (\lambda-\delta)(X)(1 \otimes 1) , \tag{6.54b}$$

so that $\gamma'(z)$ acts as $(\lambda-\delta)(\gamma'(z))$. By (6.47), z acts as $(\lambda-\delta)(\gamma'(z))$ on $1 \otimes 1$, and (6.51) thus says that z acts as $\lambda(\gamma(z))$ on $1 \otimes 1$. Now the most general element of $V(\lambda)$ is $u \otimes 1$ with $u \in U(\mathfrak{g})$, and z acts on this by

$$z(u \otimes 1) = zu(1 \otimes 1) = uz(1 \otimes 1) = \lambda(\gamma(z))u(1 \otimes 1) = \lambda(\gamma(z))(u \otimes 1) .$$

That is,

$$z \text{ acts in } V(\lambda) \text{ by the scalar } \lambda(\gamma(z)) . \tag{6.55}$$

<u>Lemma 6.18</u>. Let λ be in $\mathfrak{h}^*$, let α be a simple root, and suppose that $m = 2\langle\lambda,\alpha\rangle/|\alpha|^2$ is an integer ≥ 0. Then the $U(\mathfrak{g})$ submodule $M = U(\mathfrak{g})(E_{-\alpha})^m(1 \otimes 1)$ of $V(\lambda)$ is isomorphic to $V(s_\alpha\lambda)$.

Proof. We may assume $m > 0$. The vector $v = (E_{-\alpha})^m(1 \otimes 1)$ is not 0, as a consequence of (6.53). Let H be in $\mathfrak{h}$, and apply H to v. Commuting H past each factor of $(E_{-\alpha})^m$ one step at a time, we see that H acts on

v by the scalar

$$-m\alpha(H) + (\lambda-\delta)(H) = \left(\lambda - \frac{2\langle\lambda,\alpha\rangle}{|\alpha|^2}\alpha\right)(H) - \delta(H) = (s_\alpha\lambda-\delta)(H).$$

Therefore the linear extension of $1 \to v$ maps $\mathbb{C}_{s_\alpha\lambda-\delta}$ into M while respecting the $\mathfrak{h}$ action. Let us see that it respects the full $\mathfrak{b}$ action. We are to see that each E_β with $\beta \in \Delta^+$ maps v to 0. Since a general E_β with $\beta \in \Delta^+$ is an iterated bracket of E_β's with β simple, it is enough to prove that $E_\beta v = 0$ for β simple. For $\beta \neq \alpha$, $[E_\beta, E_{-\alpha}] = 0$ since $\beta - \alpha$ is not a root and is not 0. Thus E_β commutes with $(E_{-\alpha})^m$. Since $E_\beta(1\otimes 1) = 0$ by (6.54a), $E_\beta v = 0$.

Let us now show that $E_\alpha v = 0$. By (6.54a), we have

$$\begin{aligned} E_\alpha v &= E_\alpha(E_{-\alpha}^m)(1\otimes 1) \\ &= [E_\alpha, E_{-\alpha}^m](1\otimes 1) + E_{-\alpha}^m E_\alpha(1\otimes 1) = [E_\alpha, E_{-\alpha}^m](1\otimes 1). \end{aligned}$$

Apart from normalizations, we can now make the computation in $\mathfrak{sl}(2,\mathbb{C})$. We write

$$[e, f^m] = (Rf)^m e - (Lf)^m e,$$

where Rf is right multiplication by f and Lf is left multiplication by f. Now $Rf = Lf - \operatorname{ad} f$, and the terms on the right commute. By the binomial theorem,

$$\begin{aligned} (Rf)^m e &= \sum_{j=0}^{m} \binom{m}{j} (Lf)^{m-j}(-\operatorname{ad} f)^j e \\ &= (Lf)^m e + m(Lf)^{m-1}(-\operatorname{ad} f)e + \frac{m(m-1)}{2}(Lf)^{m-2}(-\operatorname{ad} f)^2 e, \end{aligned}$$

with the higher order terms vanishing on the right since $(\text{ad } f)^3 e = 0$. The above expression is

$$= (Lf)^m e + m(Lf)^{m-1} h + \frac{m(m-1)}{2} f^{m-2}(-2f)$$

$$= (Lf)^m e + m f^{m-1}(h - (m-1)1) .$$

Thus

$$[e, f^m] = (Rf)^m e - (Lf)^m e = m f^{m-1}(h - (m-1)1) . \qquad (6.56)$$

Back in $U(\mathfrak{g})$, let us use the normalized root vectors E'_α and $E'_{-\alpha}$ with $[E'_\alpha, E'_{-\alpha}] = 2|\alpha|^{-2} H_\alpha$. Then (6.56) and (6.54b) show that

$$[E'_\alpha, (E'_{-\alpha})^m](1 \otimes 1) = m(E'_{-\alpha})^{m-1}(2|\alpha|^{-2} H_\alpha - (m-1))(1 \otimes 1)$$

$$= m\left(\frac{2\langle \lambda - \delta, \alpha \rangle}{|\alpha|^2} - (m-1) \right) (E'_{-\alpha})^{m-1}(1 \otimes 1)$$

$$= 0 ,$$

the last step holding since $2\langle \delta, \alpha \rangle / |\alpha|^2 = 1$ by inspection. We conclude that $E_\alpha v = 0$ and thus that $1 \to v$ gives a $U(\mathfrak{b})$ map of $\mathbb{C}_{s_\alpha \lambda - \delta}$ into $V(\lambda)$.

By Frobenius reciprocity (6.12), we have an explicit isomorphism

$$\text{Hom}_{\mathfrak{b}}(\mathbb{C}_{s_\alpha \lambda - \delta}, M) \cong \text{Hom}_{\mathfrak{g}}(P_{\mathfrak{b}}^{\mathfrak{g}}(\mathbb{C}_{s_\alpha \lambda - \delta}), M) ,$$

and thus we obtain a $U(\mathfrak{g})$ map of $V(s_\alpha \lambda)$ into M carrying $1 \otimes 1$ into v. Then

$$V(s_\alpha \lambda) = U(\mathfrak{g})(1 \otimes 1) \quad \text{maps to} \quad U(\mathfrak{g})v = M ,$$

and our map is onto. A general element of $V(s_\alpha \lambda)$ is of the form $u \otimes 1$ with u in $U(\bar{\mathfrak{n}})$, by (6.53), and this maps to $uv = u(E_{-\alpha})^m(1 \otimes 1)$. If $u \neq 0$, then $u(E_{-\alpha})^m$ is a nonzero member of $U(\bar{\mathfrak{n}})$, and this element is nonzero, again by (6.53). Thus our map is one-one, and the lemma is proved.

Proof that $\text{image}(\gamma) \subseteq U(\mathfrak{h})^W$. We are to prove that every w in the Weyl group has

$$(w\lambda)(\gamma(z)) = \lambda(\gamma(z)), \tag{6.57}$$

and it is enough to handle w equal to a simple reflection. Moreover each side for fixed z is a polynomial in λ, and thus it is enough to prove (6.57) for λ dominant integral.

Form the Verma module $V(\lambda)$. We know from (6.55) that z acts in $V(\lambda)$ by the scalar $\lambda(\gamma(z))$. Also z acts in $V(s_\alpha\lambda)$ by the scalar $(s_\alpha\lambda)(\gamma(z))$. But Lemma 6.18 says that $V(s_\alpha\lambda)$ is a (clearly nonzero) $U(\mathfrak{g})$ submodule of $V(\lambda)$. Thus the two scalars must match, and (6.57) is proved.

Proof that γ is multiplicative. Since σ is an algebra isomorphism, we need to see that

$$\gamma'(z_1z_2) = \gamma'(z_1)\gamma'(z_2). \tag{6.58}$$

We have

$$z_1z_2 - \gamma'(z_1)\gamma'(z_2) = z_1(z_2 - \gamma'(z_2)) + \gamma'(z_2)(z_1 - \gamma'(z_1)),$$

which is in $U(\mathfrak{g})\mathfrak{n}$, and therefore (6.58) follows.

Proof that γ is one-one. If $\gamma(z) = 0$, then $\gamma'(z) = 0$ also, and hence $\lambda(\gamma'(z)) = 0$ for every highest weight λ. This is the scalar by which z acts in the irreducible representation of highest weight λ. Since every finite-dimensional representation of G is the direct sum of irreducibles, we have $\pi(z) = 0$ for every finite-dimensional representation π of G.

Let $(\pi(g)u,v)$ be a matrix coefficient of π. If X is in $\mathfrak{g}_0$ and we regard X as acting by left-invariant differentiation, we have

$$X(\pi(g)u,v) = \frac{d}{dt}\,(\pi(g\ \exp\ tX)u,v)\Big|_{t=0}$$

$$= \frac{d}{dt}\,(\pi(\exp\ tX)u,\pi(g)^*v)\Big|_{t=0} = (\pi(g)\pi(X)u,v)\ .$$

Iterating this equality, we obtain

$$D(\pi(g)u,v) = (\pi(g)\pi(D)u,v)$$

for every D in $U(\mathfrak{g})$. Taking $D = z$, we see that z, as a left-invariant differential operator, annihilates every matrix coefficient of every finite-dimensional representation of $\mathfrak{g}$. This holds for the k-fold tensor product of the standard representation of G with itself, for all k, and in this case the matrix coefficients at x are the kth degree monomials in the entries of the matrix x. It holds also for the conjugate of any representation. Thus z, as a differential operator, annihilates all monomials in the real and imaginary parts of the coordinates, and $z = 0$ as a differential operator. Then it follows that $z = 0$ in $U(\mathfrak{g})$. Hence γ is one-one.

<u>Proof that γ is onto for $\mathfrak{g} = \mathfrak{gl}(n,\mathbb{C})$</u>. Let $\{H_i\}$ be the basis of $\mathfrak{h}$ with $H_i = E_{ii}$. Then $U(\mathfrak{h})^W$ is the algebra of

symmetric polynomials in the H_i and is well known to be given by

$$U(\mathfrak{h})^W = \mathbb{C}[p_1, \ldots, p_n],$$

where

$$p_1 = \sum_i H_i$$

$$p_2 = \sum_{i<j} H_i H_j$$

$$p_3 = \sum_{i<j<k} H_i H_j H_k$$

$$\vdots$$

$$p_n = H_1 H_2 \cdots H_n .$$

The polynomials $p_1, \ldots, p_n$ are the elementary symmetric polynomials and are given also by the formula

$$(X-H_1)\cdots(X-H_n) = X^n - p_1 X^{n-1} + p_2 X^{n-2} - \ldots + (-1)^n p_n, \quad (6.59)$$

where X is an indeterminate.

Let X_i run through the basis of vectors $E_{-\alpha}, H_j, E_\alpha$ of $\mathfrak{g}$ (α positive). A calculation like that in Proposition 6.15 shows that

$$Z_k = \sum_{i_1,\ldots,i_k} \operatorname{Tr}(X_{i_1}\cdots X_{i_k}) X^{i_1}\cdots X^{i_k}$$

is in $Z(\mathfrak{g})$, where X^j is defined by the relations $g_{ij} = \operatorname{Tr}(X_i X_j)$, $(g^{ij}) = (g_{ij})^{-1}$, and $X^j = \sum g^{ij} X_i$. It is clear that

$$\gamma'(Z_k) \equiv \sum_{i_1,\ldots,i_k} \operatorname{Tr}(H_{i_1}\cdots H_{i_k}) H^{i_1}\cdots H^{i_k} \mod U^{k-1}(\mathfrak{g})$$

and then that

$$\gamma(Z_k) \equiv \sum_{i_1,\dots,i_k} \operatorname{Tr}(H_{i_1}\cdots H_{i_k}) H^{i_1}\cdots H^{i_k} \mod U^{k-1}(\mathfrak{g}) .$$

If we have chosen each E_α and $E_{-\alpha}$ to be 1 in the nonzero entry, then g_{ij} is given in obvious notation by

$$(g_{ij}) = \begin{array}{c} \begin{matrix} E_\alpha & H & E_{-\alpha} \end{matrix} \\ \begin{pmatrix} 0 & 0 & I \\ 0 & I & 0 \\ I & 0 & 0 \end{pmatrix} \end{array} \begin{matrix} E_\alpha \\ H \\ E_{-\alpha} \end{matrix} = (g^{ij})$$

Hence $H^j = H_j$, and evaluation of the traces gives

$$\gamma(Z_k) \equiv \sum_i (H_i)^k \mod U^{k-1}(\mathfrak{g}) .$$

Let $1 \leq k \leq n$. Since $\gamma(Z_k)$ and $\Sigma_i (H_i)^k$ are both symmetric polynomials, their difference is a polynomial in $p_1, \dots, p_{k-1}$. Thus we see inductively that p_k will be in the image of γ if $\Sigma (H_i)^k$ requires p_k for its expansion in terms of elementary symmetric polynomials.

To see that $\Sigma (H_i)^k$ requires p_k, we let $\zeta_1, \dots, \zeta_k$ be the kth roots of unity, and we introduce the homomorphism

$$H_j \to \begin{cases} \zeta_j & \text{for } 1 \leq j \leq k \\ 0 & \text{for } k+1 \leq j \leq n . \end{cases}$$

Under this homomorphism, the left side of (6.59) becomes $X^n - X^{n-k}$, and thus (6.59) gives

$$p_j(\zeta_1,\ldots,\zeta_k,0,\ldots,0) = \begin{cases} 0 & \text{for } j \neq k \\ (-1)^{k+1} & \text{for } j = k. \end{cases} \tag{6.60}$$

The homomorphism sends $\Sigma\,(H_i)^k$ into n, and (6.60) thus implies that $\Sigma\,(H_i)^k$ cannot be expressed in terms of $p_1,\ldots,p_{k-1}$. This completes the proof.

6. Casselman-Osborne Theorem

In this section we take up the second part of the proof of Kostant's Theorem (Theorem 6.12). We continue with the setting $G = U(n)$, $\mathfrak{g}_0 = \mathfrak{u}(n)$, $\mathfrak{g} = \mathfrak{u}(n)^{\mathbb{C}} = \mathfrak{gl}(n,\mathbb{C})$; other notation is as in §§4-5.

We shall bring to bear the Harish-Chandra isomorphism. Let V be a $U(\mathfrak{g})$ module, not necessarily finite-dimensional. Proposition 6.11 still gives us a $U(\mathfrak{h})$ module structure on $H^k(\mathfrak{n},V)$, even if this structure does not come from a representation of T.

We define an action of $Z(\mathfrak{g})$ on $H^k(\mathfrak{n},V)$ as follows: If z is in $Z(\mathfrak{g})$, we make z act on $\mathrm{Hom}_{\mathbb{C}}(\Lambda^k\mathfrak{n},V)$ by the formula

$$(z\omega)(\xi) = z(\omega(\xi))\,;$$

this is meaningful since $Z(\mathfrak{g}) \subseteq U(\mathfrak{g})$ acts on V. Moreover it is clear from (6.23) that this action commutes with the d operator and thus descends to $H^k(\mathfrak{n},V)$. The $Z(\mathfrak{g})$ and $U(\mathfrak{h})$ actions on $H^k(\mathfrak{n},V)$ are related as in Theorem 6.19 below. We shall state the theorem and then immediately show how it finishes the proof of Kostant's Theorem.

Theorem 6.19 (Casselman-Osborne). Let V be a $U(\mathfrak{g})$ module. Then the $Z(\mathfrak{g})$ and $U(\mathfrak{h})$ actions on $H^k(\mathfrak{n}, V)$ are related by the formula

$$z\omega = \gamma'(z)\omega$$

for $z \in Z(\mathfrak{g})$ and $\omega \in H^k(\mathfrak{n}, V)$.

Proof of second part of Theorem 6.12. If V is irreducible with highest weight λ, then (6.52) shows that each $z \in Z(\mathfrak{g})$ acts on V by the scalar $(\lambda+\delta)(\gamma(z))$. Hence also z acts on $H^k(\mathfrak{n}, V)$ by $(\lambda+\delta)(\gamma(z))$.

On the other hand, if $\omega \neq 0$ is a weight vector under T (hence under $\mathfrak{h}$) with weight μ, then $\gamma'(z)\omega = \mu(\gamma'(z))\omega$. From Theorem 6.19 we can therefore conclude

$$\mu(\gamma'(z)) = (\lambda+\delta)(\gamma(z))$$

for all z in $Z(\mathfrak{g})$. By (6.51) we thus have

$$(\mu+\delta)(\gamma(z)) = (\lambda+\delta)(\gamma(z)) \tag{6.61}$$

for all z in $Z(\mathfrak{g})$.

It is helpful to think of (6.61) as expressing the equality of the values of a polynomial $\gamma(z)$ at two points $\mu+\delta$ and $\lambda+\delta$. We claim that $\mu+\delta = w(\lambda+\delta)$ for some $w \in W$. Assuming the contrary, we can choose a polynomial p in $H_1, \ldots, H_n$ that is 1 at every $w(\mu+\delta)$ and is 0 at every $w(\lambda+\delta)$. Then $q = (n!)^{-1} \sum_{w \in W} wp$ is a W-invariant polynomial that is 1 at $\mu+\delta$ and is 0 at $\lambda+\delta$. According to the Harish-Chandra theorem (Theorem 6.17), q is $\gamma(z)$ for some z in $Z(\mathfrak{g})$. For this z, the respective sides of (6.61) are 1 and 0, and we have a contradiction.

We conclude that $\mu+\delta = w(\lambda+\delta)$ for some $w \in W$. Thus the only weights that occur in $H^k(\mathfrak{n}, V)$ are of the form $\mu = w(\lambda+\delta) - \delta$, and these have been handled by the first part of the proof. Thus Theorem 6.19 implies Theorem 6.12.

Our objective in the remainder of this section will be to prove Theorem 6.19 for $G = U(n)$ and $\mathfrak{g} = \mathfrak{gl}(n, \mathbb{C})$. The proof will use a cohomological technique known as dimension shifting. Namely we shall make use of a long exact cohomology sequence in which every third term is 0, so that we obtain isomorphisms of cohomology that cross from one dimension to the next.

The long exact sequence in question is the following. First suppose that we have an exact sequence

$$0 \longrightarrow V \xrightarrow{\psi} U \xrightarrow{\varphi} V' \longrightarrow 0 \tag{6.62}$$

in the category $\mathcal{C}(\mathfrak{n})$. Applying Theorem 5.28c to the invariants functor $(\cdot)^{\mathfrak{n}}$, we obtain a long exact sequence

$$\begin{aligned} 0 \longrightarrow H^0(\mathfrak{n}, V) &\longrightarrow H^0(\mathfrak{n}, U) \longrightarrow H^0(\mathfrak{n}, V') \\ \longrightarrow H^1(\mathfrak{n}, V) &\longrightarrow H^1(\mathfrak{n}, U) \longrightarrow H^1(\mathfrak{n}, V') \\ \longrightarrow H^2(\mathfrak{n}, V) &\longrightarrow H^2(\mathfrak{n}, U) \longrightarrow H^2(\mathfrak{n}, V') \\ \longrightarrow \cdots \longrightarrow H^{\dim \mathfrak{n}}(\mathfrak{n}, U) &\longrightarrow H^{\dim \mathfrak{n}}(\mathfrak{n}, V') \longrightarrow 0, \end{aligned} \tag{6.63}$$

because Lie algebra cohomology is a derived functor of the invariants functor. If (6.62) is actually an exact sequence in $\mathcal{C}(\mathfrak{g})$, then each cohomology space in (6.63) possesses actions by $U(\mathfrak{h})$ and $Z(\mathfrak{g})$ from §4. In order to make use of

(6.63), we need to know that the maps in (6.63) respect these actions.

Lemma 6.20. If the sequence (6.62) is exact in the category $\mathcal{C}(\mathfrak{g})$, then the maps in the long exact sequence (6.63) commute with the actions of $U(\mathfrak{h})$ and $Z(\mathfrak{g})$.

Proof. Let us set up matters as in §§4-5 of Chapter V. We are to begin with compatible injective resolutions of (6.62), and these we take as the rows of

$$\begin{array}{ccccc}
 & 0 & & 0 & \\
 & \downarrow & & \downarrow & \\
\longrightarrow & \mathrm{Hom}_{\mathbb{C}}(U(\mathfrak{n})\otimes_{\mathbb{C}}\Lambda^{k-1}\mathfrak{n},V) & \xrightarrow{\mathrm{Hom}(\partial,1)} & \mathrm{Hom}_{\mathbb{C}}(U(\mathfrak{n})\otimes_{\mathbb{C}}\Lambda^{k}\otimes,V) & \longrightarrow \\
 & \downarrow{\scriptstyle \mathrm{Hom}(1,\psi)} & & \downarrow{\scriptstyle \mathrm{Hom}(1,\psi)} & \\
\longrightarrow & \mathrm{Hom}_{\mathbb{C}}(U(\mathfrak{n})\otimes_{\mathbb{C}}\Lambda^{k-1}\mathfrak{n},U) & \xrightarrow{\mathrm{Hom}(\partial,1)} & \mathrm{Hom}_{\mathbb{C}}(U(\mathfrak{n})\otimes_{\mathbb{C}}\Lambda^{k}\mathfrak{n},U) & \longrightarrow \\
 & \downarrow{\scriptstyle \mathrm{Hom}(1,\varphi)} & & \downarrow{\scriptstyle \mathrm{Hom}(1,\varphi)} & \\
\longrightarrow & \mathrm{Hom}_{\mathbb{C}}(U(\mathfrak{n})\otimes_{\mathbb{C}}\Lambda^{k-1}\mathfrak{n},V') & \xrightarrow{\mathrm{Hom}(\partial,1)} & \mathrm{Hom}_{\mathbb{C}}(U(\mathfrak{n})\otimes_{\mathbb{C}}\Lambda^{k}\mathfrak{n},V') & \longrightarrow \\
 & \downarrow & & \downarrow & \\
 & 0 & & 0 &
\end{array}$$

In fact, the columns are exact by Proposition 6.1c, and the squares commute because the horizontal and vertical maps operate in different coordinates. Applying the invariants functor $(\cdot)^{\mathfrak{n}}$, we are led to a commutative diagram

$$\begin{array}{ccccccc}
 & & 0 & & 0 & & \\
 & & \downarrow & & \downarrow & & \\
\longrightarrow & & \mathrm{Hom}_{\mathbb{C}}(\Lambda^{k-1}\mathfrak{n}, V) & \xrightarrow{d} & \mathrm{Hom}_{\mathbb{C}}(\Lambda^{k}\mathfrak{n}, V) & \longrightarrow & \\
 & & \Big\downarrow \mathrm{Hom}(1,\psi) & & \Big\downarrow \mathrm{Hom}(1,\psi) & & \\
\longrightarrow & & \mathrm{Hom}_{\mathbb{C}}(\Lambda^{k-1}\mathfrak{n}, U) & \xrightarrow{d} & \mathrm{Hom}_{\mathbb{C}}(\Lambda^{k}\mathfrak{n}, U) & \longrightarrow & \\
 & & \Big\downarrow \mathrm{Hom}(1,\varphi) & & \Big\downarrow \mathrm{Hom}(1,\varphi) & & \\
\longrightarrow & & \mathrm{Hom}_{\mathbb{C}}(\Lambda^{k-1}\mathfrak{n}, V') & \xrightarrow{d} & \mathrm{Hom}_{\mathbb{C}}(\Lambda^{k}\mathfrak{n}, V') & \longrightarrow & \\
 & & \downarrow & & \downarrow & & \\
 & & 0 & & 0 & &
\end{array} \qquad (6.64)$$

with exact columns (cf. Lemma 5.22). Section V.5 tells us how to compute the maps of the long exact sequence from this diagram. The ones that preserve degree in (6.63) are quotients of $\mathrm{Hom}(1,\psi)$ and $\mathrm{Hom}(1,\varphi)$, while the ones that increase degree come from the connecting homomorphism for (6.64).

Now let $H \in \mathfrak{h}$ be given. For $\omega \in \mathrm{Hom}_{\mathbb{C}}(\Lambda^k\mathfrak{n}, V)$ and $\xi \in \Lambda^k\mathfrak{n}$, we have

$$\begin{aligned}
(H(\mathrm{Hom}(1,\psi)\omega))(\xi) &= H(\mathrm{Hom}(1,\psi)\omega(\xi)) - \mathrm{Hom}(1,\psi)\omega(H\xi) \\
&= H(\psi(\omega(\xi))) - \psi(\omega(H\xi)) \\
&= \psi(H(\omega(\xi))) - \psi(\omega(H\xi)) \\
&= \psi(H\omega(\xi)) = (\mathrm{Hom}(1,\psi)(H\omega))(\xi) .
\end{aligned}$$

Thus $\mathrm{Hom}(1,\psi)$ is $\mathfrak{h}$-linear and descends to an $\mathfrak{h}$-linear map on cohomology. Similar remarks apply to $\mathrm{Hom}(1,\varphi)$.

Recall from Lemma 5.18 how the connecting homomorphism ρ is defined. We start with ω in $\mathrm{Hom}_{\mathbb{C}}(\Lambda^{k-1}\mathfrak{n}, V')$ such that $d\omega = 0$. We choose ξ in $\mathrm{Hom}_{\mathbb{C}}(\Lambda^{k-1}\mathfrak{n}, U)$ with $\mathrm{Hom}(1,\varphi)\xi = \omega$. Then $d\xi$ is of the form $\mathrm{Hom}(1,\psi)\eta$ for some η in $\mathrm{Hom}_{\mathbb{C}}(\Lambda^{k}\mathfrak{n}, V)$, and we define $\rho(\omega)$ to be the class of η in $H^k(\mathfrak{n}, V)$.

Now let us consider $H\omega$. We have $d(H\omega) = H(d\omega) = 0$. The element $H\xi$ has $\mathrm{Hom}(1,\varphi)(H\xi) = H(\mathrm{Hom}(1,\varphi)\xi) = H\omega$. Moreover

$$\mathrm{Hom}(1,\psi)(H\eta) = H(\mathrm{Hom}(1,\psi)\eta) = H(d\xi) = d(H\xi).$$

Thus $\rho(H\omega)$ is the class of $H\eta$ in $H^k(\mathfrak{n}, V)$. This statement says that ρ commutes with the action of $\mathfrak{h}$.

The action of $Z(\mathfrak{g})$ is formally the same, except that the action is trivial on the alternating tensors. Thus the above arguments go through, suitably interpreted, and the lemma follows.

<u>Proof of Theorem 6.19</u>. We show by induction on k that the identity

$$z\omega = \gamma'(z)\omega, \qquad z \in Z(\mathfrak{g}),$$

holds on $H^k(\mathfrak{n}, V)$ for all $U(\mathfrak{q})$ modules V.

For $k = 0$, $H^0(\mathfrak{n}, V) \cong V^{\mathfrak{n}}$ as a $U(\mathfrak{h})$ module. Both z and $\gamma'(z)$ act by the restriction of their actions on V. Since (6.47) says that $z - \gamma'(z)$ is in $U(\mathfrak{g})\mathfrak{n}$, we see that $(z - \gamma'(z))\omega = 0$ for ω in $H^0(\mathfrak{n}, V)$. This is the desired conclusion for $k = 0$.

Now assume the result for degree $k-1$. The beginning of the standard injective resolution of V (in the category $\mathcal{C}(\mathfrak{g})$) gives us an exact sequence

$$0 \longrightarrow V \longrightarrow \operatorname{Hom}_{\mathbb{C}}(U(\mathfrak{g}),V) \longrightarrow Q \longrightarrow 0, \quad (6.65)$$

where Q is a suitable quotient $U(\mathfrak{g})$ module. Regarding (6.65) as an exact sequence in $\mathcal{C}(\mathfrak{n})$, we obtain from (6.63) an exact sequence

$$H^{k-1}(\mathfrak{n},Q) \xrightarrow{\ \rho\ } H^k(\mathfrak{n},V) \longrightarrow H^k(\mathfrak{n},\operatorname{Hom}_{\mathbb{C}}(U(\mathfrak{g}),V)), \quad (6.66)$$

and Lemma 6.20 says that this exact sequence respects the $U(\mathfrak{h})$ and $Z(\mathfrak{g})$ actions.

As vector spaces, $\mathfrak{g} = (\bar{\mathfrak{n}} + \mathfrak{h}) \oplus \mathfrak{n}$. Thus as $\mathfrak{n}$ modules,

$$\begin{aligned}\operatorname{Hom}_{\mathbb{C}}(U(\mathfrak{g}),V) &\cong \operatorname{Hom}_{\mathbb{C}}(S(\bar{\mathfrak{n}}+\mathfrak{h}) \otimes_{\mathbb{C}} U(\mathfrak{n}),V) \\ &\cong \operatorname{Hom}_{\mathbb{C}}(U(\mathfrak{n}),\operatorname{Hom}_{\mathbb{C}}(S(\bar{\mathfrak{n}}+\mathfrak{h}),V)),\end{aligned}$$

where $\mathfrak{n}$ acts trivially on the symmetric algebra $S(\bar{\mathfrak{n}}+\mathfrak{h})$. The right side here is injective in $\mathcal{C}(\mathfrak{n})$, by Proposition 6.1b, and therefore $I = \operatorname{Hom}_{\mathbb{C}}(U(\mathfrak{g}),V)$ is injective in $\mathcal{C}(\mathfrak{n})$.

Since I is injective, $0 \to I \to I \to 0$ is an injective resolution of I, and it follows that $H^{\ell}(\mathfrak{n},I) = 0$ for $\ell > 0$. In particular, $H^k(\mathfrak{n},I) = 0$, and the exactness of (6.66) therefore says that the connecting homomorphism ρ is onto $H^k(\mathfrak{n},V)$. Given ω in $H^k(\mathfrak{n},V)$, choose $\tilde{\omega}$ in $H^{k-1}(\mathfrak{n},Q)$ with $\rho(\tilde{\omega}) = \omega$. By inductive hypothesis, $z\tilde{\omega} = \gamma'(z)\tilde{\omega}$. Applying ρ to both sides and using Lemma 6.20, we obtain

$$z\omega = z\rho(\tilde{\omega}) = \rho(z\tilde{\omega}) = \rho(\gamma'(z)\tilde{\omega}) = \gamma'(z)\rho(\tilde{\omega}) = \gamma'(z)\omega .$$

This completes the induction and the proof.

CHAPTER VII

RELATIVE LIE ALGEBRA COHOMOLOGY

1. Motivation for how to construct representations

Cohomological constructions in representation theory have been most useful in the case of semisimple groups or the slightly wider class of reductive groups. For our purposes we can think of a reductive group as a closed linear group G (of real or complex matrices, as in Chapter I) that is closed under the operation of conjugate transpose. Sometimes one requires that G not be too badly disconnected, but we shall not do so yet. The group is semisimple if the center of its identity component is 0-dimensional (and then it will be finite).

For example, the compact groups $U(n)$, $SU(n)$, and $SO(n)$, as defined in §I.3, are reductive, and $SU(n)$ and $SO(n)$ are semisimple if $n \geq 2$ and $n \geq 3$, respectively. Two examples of noncompact reductive groups from §I.3 are $SL(n,\mathbb{R})$ and $SL(n,\mathbb{C})$. Further examples are the following:

$SO(n,\mathbb{C})$	Linear isometry group with $\det = 1$ preserving complex quadratic form $z_1^2 + \cdots + z_n^2$ in $\mathbb{C}^n$
$SO(m,n)$	Linear isometry group with $\det = 1$ preserving $x_1^2 + \cdots + x_m^2 - x_{m+1}^2 - \cdots - x_{m+n}^2$ in $\mathbb{R}^{m+n}$

$U(m,n)$	Linear isometry group in $\mathbb{C}^{m+n}$ preserving Hermitian quadratic form given by $\|z_1\|^2 + \ldots + \|z_m\|^2 - \|z_{m+1}\|^2 - \ldots - \|z_{m+n}\|^2$
$SU(m,n)$	Subgroup of $U(m,n)$ with $\det = 1$
$Sp(n,\mathbb{C})$	$\{g \in SL(2n,\mathbb{C}) \mid g^{tr}Jg = J \text{ for } J = \begin{pmatrix} 0 & I_n \\ -I_n & 0 \end{pmatrix}\}$
$Sp(n,\mathbb{R})$	$Sp(n,\mathbb{C}) \cap GL(2n,\mathbb{R})$

Each example is reductive, and each example except $U(m,n)$ is semisimple except when m and n are so small that the example degenerates.

If G is given as a subgroup of $GL(n,\mathbb{C})$, then its intersection with the unitary group $U(n)$ turns out to be a maximal compact subgroup K. Moreover K captures all of the disconnectedness of G in that $g = K \exp \mathfrak{p}_0$ for a suitable subspace $\mathfrak{p}_0$ of the Lie algebra $\mathfrak{g}_0$ of G.

Our definition of "representation" of G assumes initially that the underlying vector space is a Hilbert space. For all practical purposes when G is reductive, there is no serious loss of generality in assuming also that the restriction of the representation to K is unitary. Then it follows from the Peter-Weyl Theorem that the representation of K splits into the orthogonal (infinite) sum of irreducible finite-dimensional subspaces. In particular, the vector subspace of K-finite vectors, those vectors whose K translates span a finite-dimensional space, is dense.

In many nice situations (such as when the representation of G is irreducible), every K-finite vector is C^∞ in the sense of §I.8: The map $g \to \Phi(g)v$ is C^∞, where Φ is the representation. In such cases we get a representation of the Lie algebra $\mathfrak{g}_0$ of G on the space of K-finite vectors. (And in every case we at least get a representation of $\mathfrak{g}_0$ on the subspace of C^∞ K-finite vectors.)

Thus from a representation Φ of G we get two representations, one of $\mathfrak{g}_0$ and one of K. These satisfy certain compatibility conditions, namely

(i) the differentiated version of the representation $\Phi|_K$ is the restriction of $\Phi|_{\mathfrak{g}_0}$ to the Lie algebra $\mathfrak{k}_0$ of K, and

(ii) $\Phi(k)\Phi(X)\Phi(k^{-1})v = \Phi(\mathrm{Ad}(k)X)v$ for all $X \in \mathfrak{g}_0$ and $k \in K$.

(7.1)

We shall study such pairs of representations in this chapter. We can regard the presence of the K representation as imposing a condition for the $\mathfrak{g}_0$ representation to lift to our group G; in other words, we are discarding many Lie algebra representations that are of no interest to us.

To proceed abstractly we shall work with a "pair" $(\mathfrak{g}, K)$ satisfying some compatibility conditions. Here we should regard $\mathfrak{g}$ as the complexification of $\mathfrak{g}_0$ above. Our representations will be "$(\mathfrak{g}, K)$ modules," representations of both $\mathfrak{g}$ and K having every vector K-finite and satisfying (i) and (ii) above. Just as we can study representations of $\mathfrak{g}$ by studying $U(\mathfrak{g})$ modules that are <u>unital</u> in the sense that 1 acts as 1, so it will turn out that we can study

$(\mathfrak{g}, K)$ modules by studying unital $R(\mathfrak{g}, K)$ modules for a suitable algebra $R(\mathfrak{g}, K)$.

A qualitatively different situation where we shall use $(\mathfrak{g}, K)$ modules is in connection with Kostant's Theorem. The Lie algebra $\mathfrak{g}$ for a pair will be $\mathfrak{b} = \mathfrak{h} \oplus \mathfrak{n}$ within the complexified Lie algebra of the unitary group $U(n)$, and the group K will be the torus T. Our $(\mathfrak{b}, T)$ module will be an irreducible representation of $U(n)$ with its action suitably restricted. Notice with this example that $\mathfrak{b}$ does not arise (as we construct it) as the complexification of a real Lie algebra.

Our interest will be in constructing representations, and it will be helpful to have some sample constructions in mind. First recall from Example 2 in §VI.1 that the functor $I_{\mathfrak{q}}^{\mathfrak{g}}$ from $U(\mathfrak{q})$ modules to $U(\mathfrak{g})$ modules is related to the construction of induced representations $\operatorname{ind}_Q^G \pi$ in the case that $\mathfrak{q}$ is the complexified Lie algebra of a closed subgroup Q of G. Let us be more precise now. We change notation so that G acts by right translation, in order to eliminate inverses and transposes in the formulas.

Suppose G is $SL(n, \mathbb{R})$ and Q is the upper triangular subgroup. Let $\mathfrak{g} = \mathfrak{sl}(n, \mathbb{R})^{\mathbb{C}} = \mathfrak{sl}(n, \mathbb{C})$, and let $\mathfrak{q}$ be the upper triangular subalgebra of $\mathfrak{g}$. Let π be a representation of Q on V. The induced representation $\operatorname{ind}_Q^G \pi$ acts (modulo some adjustment of parameters because $Q \backslash G$ has no invariant measure) in the space of reasonable $f : G \to V$ such that

$$f(qx) = \pi(q)f(x) \quad \text{for} \quad q \in Q, \quad x \in G. \tag{7.2}$$

As we saw in §VI.1, any smooth such f with values in the C^∞ subspace V_0 of V leads to a member φ_f of

$$I_{\mathfrak{q}}^{\mathfrak{g}}(V_0) = \operatorname{Hom}_{U(\mathfrak{q})}(U(\mathfrak{g}), V_0)$$

by the definition $\varphi_f(u) = (uf)(1)$ for $u \in U(\mathfrak{g})$. We should think of φ_f as giving the Taylor coefficients of f at the identity. On the subspace of real analytic f's, the Taylor coefficient map $f \to \varphi_f$ is one-one, and we would like some way to ensure that it is onto.

To fix things, we shall make use of the subgroup $K = SO(n)$. Roughly speaking, we shall work only with members of $I_{\mathfrak{q}}^{\mathfrak{g}}(V_0)$ such that K acts on the systems of Taylor coefficients compatibly with the right regular representation on $\operatorname{ind}_Q^G \pi$. If a member of $I_{\mathfrak{q}}^{\mathfrak{g}}(V_0)$ has this property, then we shall see that there is a function on K with the desired coefficients. Then we shall observe that $G = QK$ and that the function therefore extends to G. The key operation for us, therefore, will be (approximately) extraction of the subspace of $I_{\mathfrak{q}}^{\mathfrak{g}}(V_0)$ on which the representation of the Lie algebra of K globalizes to K. In more detail we proceed in four steps.

1) The intersection $M = K \cap Q$ is just the (finite) diagonal subgroup with diagonal entries ± 1, and we can regard V_0 as a $(\mathfrak{q}, M)$ module instead of just a $\mathfrak{q}$ module. Let us see that $I_{\mathfrak{q}}^{\mathfrak{g}}(V_0)$ becomes a $(\mathfrak{g}, M)$ module. To get at the M action, first suppose that f is in the induced space

and k is in K. Then the function kf (the right translate of f by k) has

$$\varphi_{kf}(u) = (ukf)(1) = (k(\mathrm{Ad}(k)^{-1}u)f)(1) = ((\mathrm{Ad}(k)^{-1}u)f)(k) .$$

For m in M, (7.2) shows that this formula becomes

$$\varphi_{mf}(u) = \pi(m)(\varphi_f(\mathrm{Ad}(m)^{-1}u)) . \tag{7.3}$$

Motivated by this formula, we define

$$(m\varphi)(u) = \pi(m)(\varphi(\mathrm{Ad}(m)^{-1}u)) \tag{7.4}$$

for $\varphi \in I_{\mathfrak{q}}^{\mathfrak{g}}(V_0)$, $m \in M$, and $u \in U(\mathfrak{g})$, and we readily check that $I_{\mathfrak{q}}^{\mathfrak{g}}(V_0)$ becomes a $(\mathfrak{g}, M)$ module in such a way that the map $f \to \varphi_f$ respects the $\mathfrak{g}$ and M actions.

2) Suppose we are given a K-finite function in the induced space. Let us see that we are led to a finite-dimensional representation of K in a subspace of $I_{\mathfrak{q}}^{\mathfrak{g}}(V_0)$ satisfying certain properties. To begin with, there are linearly independent functions $f_1, \ldots, f_r$ with the given function in their linear span such that

$$kf_j = \sum_i \rho_{ij}(k) f_i \tag{7.5}$$

for all k in K. (Here kf_j refers to the right regular representation, and each $\rho_{ij}(k)$ is a constant.) We define operators $\rho(k)$ on the vector space $\sum \mathbb{C}\varphi_{f_j}$ by

$$\rho(k)\varphi_{f_j} = \sum_i \rho_{ij}(k)\varphi_{f_i} , \tag{7.6}$$

and then ρ is a finite-dimensional representation of K.

Writing ρ also for the corresponding representation of $U(\mathfrak{k})$, we shall show that

$$\rho(u)\varphi_{f_j} = u\varphi_{f_j} \quad \text{for } u \in U(\mathfrak{k}) \tag{7.7a}$$

$$\rho(m)\varphi_{f_j} = m\varphi_{f_j} \quad \text{for } m \in M, \tag{7.7b}$$

with the right side of (7.7) referring to the action in Step 1. For (7.7a) it is enough to handle $u = X$ in $\mathfrak{k}$. Differentiating (7.5) and (7.6) and combining them, we have

$$(\rho(X)\varphi_{f_j})(u') = \varphi_{Xf_j}(u') = u'Xf_j(1) = \varphi_{f_j}(u'X) = (X\varphi_{f_j})(u')$$

for $u' \in U(\mathfrak{g})$, and (7.7a) follows. For (7.7b) we take $m \in M$ and write

$$\begin{aligned} \rho(m)\varphi_{f_j} &= \varphi_{mf_j} && \text{by (7.5) and (7.6)} \\ &= m\varphi_{f_j} && \text{by (7.3) and (7.4);} \end{aligned}$$

then (7.7b) follows.

3) Conversely suppose we are given a representation ρ of K on a finite-dimensional subspace of $I_{\mathfrak{q}}^{\mathfrak{g}}(V_0)$ with basis $\varphi_1, \ldots, \varphi_r$, and suppose that the corresponding $(\mathfrak{k},M)$ action on this subspace coincides with the restriction to $(\mathfrak{k},M)$ of the $(\mathfrak{g},M)$ action on $I_{\mathfrak{q}}^{\mathfrak{g}}(V_0)$. Then we define smooth functions $f_j : K \to V_0$ by

$$f_j(k) = (\rho(k)\varphi_j)(1). \tag{7.8}$$

For m in M we have

$$f_j(mk) = (\rho(mk)\varphi_j)(1) = (\rho(m)(\rho(k)\varphi_j))(1)$$

$$= \pi(m)((\rho(k)\varphi_j)(1)) \qquad \text{by (7.4)}$$

$$= \pi(m)(f_j(k)) \qquad \text{by (7.8).}$$

Hence every f in the span of the f_j's satisfies

$$f(mk) = \pi(m)(f(k)) \quad \text{for } m \in M, \ k \in K. \tag{7.9}$$

As in (3.25), where the global group is $GL(m,\mathbb{C})$, one can show that $G = QK$ as a consequence of the Gram-Schmidt orthogonalization process for $\mathbb{R}^n$. If we attempt to extend a function $f : K \to V_0$ to G by the definition

$$f(qk) = \pi(q)(f(k)), \tag{7.10}$$

then (7.9) is exactly the consistency condition that allows us to make the extension. Hence each f in the span of the f_j's extends to G by means of (7.10) and then satisfies (7.2). In short, each of the f's is in our induced space.

4) In fact, Step 3 provides a two-sided inverse to Step 2. To see this, we need to see that the function f_j in (7.8) satisfies $\varphi_{f_j} = \varphi_j$. If u is in $U(\mathfrak{k})$, we have

$$\varphi_{f_j}(u) = uf_j(1) = (\rho(u)\varphi_j)(1) \qquad \text{by (7.8)}$$

$$= (u\varphi_j)(1) = \varphi_j(u).$$

Thus φ_{f_j} and φ_j agree on $U(\mathfrak{k})$. The claim is that they therefore agree on $U(\mathfrak{g})$. In fact, we have $\mathfrak{g} = \mathfrak{q} \oplus \mathfrak{k}$ and thus $U(\mathfrak{g}) = U(\mathfrak{q}) \otimes_{\mathbb{C}} U(\mathfrak{k})$ by (2.36). If $u' \in U(\mathfrak{q})$ and $u \in U(\mathfrak{k})$, then we have

$$\varphi_{f_j}(u'u) = u'(\varphi_{f_j}(u)) = u'(\varphi_j(u)) = \varphi_j(u'u) ,$$

and φ_{f_j} indeed agrees with φ_j on $U(\mathfrak{k})$.

The four steps above show the precise sense in which the map $f \to \varphi_f$ carries the K-finite subspace of $\text{ind}_Q^G \pi$ onto the subspace of $I_{\mathfrak{q}}^{\mathfrak{g}}(V_0)$ for which the representation globalizes to K.

For another example, let us consider Example 1 of §VI.1, rewritten with new notation. Suppose G is $U(n)$, T is the diagonal subgroup, $G^{\mathbb{C}}$ is the complexification $GL(n,\mathbb{C})$, and $\overline{B}$ is the lower triangular subgroup of $G^{\mathbb{C}}$. Then $T\backslash G$ maps into $\overline{B}\backslash G^{\mathbb{C}}$ by $Tx \to \overline{B}x$, and this mapping is a diffeomorphism onto. Since $\overline{B}\backslash G^{\mathbb{C}}$ is a complex manifold in a natural way, $T\backslash G$ acquires the structure of a complex manifold. Now let π be a one-dimensional representation of T on $\mathbb{C}$. As above, the Taylor coefficient mapping carries the smooth f's in the induced space $\text{ind}_T^G \pi$ into $\text{Hom}_{U(\mathfrak{h})}(U(\mathfrak{g}),\mathbb{C})$, where $\mathfrak{g}$ and $\mathfrak{h}$ are the complexified Lie algebras of G and T, respectively. Under this map the holomorphic f's in the induced space go into $\text{Hom}_{U(\overline{\mathfrak{b}})}(U(\mathfrak{g}),\mathbb{C})$, where $\overline{\mathfrak{b}} = \mathfrak{h} \oplus \overline{\mathfrak{n}}$ is the lower triangular subalgebra and $\overline{\mathfrak{n}}$ acts as 0 on $\mathbb{C}$. The holomorphic f's map in one-one fashion into this space, and again we would like to arrange that the map is onto. Again what we should do is look only at the subspace of $\text{Hom}_{U(\overline{\mathfrak{b}})}(U(\mathfrak{g}),\mathbb{C})$ on which the representation of $\mathfrak{g}_0 = \mathfrak{u}(n)$ globalizes to the compact group G.

Putting aside these examples, let us return to our

general pair $(\mathfrak{g},K)$ but assume that K is connected. The operation Γ of passing from a member of $\mathcal{C}(\mathfrak{g})$ to the subspace on which K operates satisfies

$$\operatorname{Hom}_{\mathfrak{g},K}(U,\Gamma(V)) \cong \operatorname{Hom}_{\mathfrak{g}}(U,V)$$

for every $(\mathfrak{g},K)$ module U, since a $\mathfrak{g}$ map from U into V will necessarily carry vectors of U that lie in a K-finite subspace (namely all vectors of U) into vectors of the same kind in V. This property of Γ is similar to the property (6.13) of the I functor, and we shall see that Γ can be defined with our still-to-be-defined algebras $R(\mathfrak{g},K)$ in much the same way that I was defined with the algebras $U(\mathfrak{g})$.

But there is one major difference between Γ and our old I: Γ need not be exact. Consequently the derived functors of Γ may be interesting. In fact, in our second example above, use of the derived functors of Γ corresponds to replacing holomorphic sections over $T\backslash G$ by differential-form sections of type $(0,q)$ over $T\backslash G$. Later in this chapter, we shall see that we can construct all the irreducible representations of $U(n)$ in our algebraic analog of this setting with differential forms; the relevant theorem is an algebraic analog of an analytic result known as the Borel-Weil-Bott Theorem.

2. $(\mathfrak{g},K)$ modules

Let $\mathfrak{g}$ be a finite-dimensional complex Lie algebra, and let K be a compact Lie group (possibly disconnected). We assume that $\mathfrak{g}$ and K satisfy the following compatibility conditions:

(i) the complexified Lie algebra $\mathfrak{k}$ of K is a Lie subalgebra of $\mathfrak{g}$

(ii) K acts on $\mathfrak{g}$ by automorphisms (called $\mathrm{Ad}(k)$, $k \in K$)

(iii) the differentiated version of $\mathrm{Ad}(K)$ is $\mathrm{ad}\,\mathfrak{k} \subseteq \mathrm{ad}\,\mathfrak{g}$.

In this case we refer to $(\mathfrak{g}, K)$ as a <u>pair</u>. If G is a linear reductive group, then §1 shows how G leads to a pair.

Fix a pair $(\mathfrak{g}, K)$. A <u>$(\mathfrak{g}, K)$ module</u> is a complex vector space V carrying representations of $\mathfrak{g}$ and of K such that

(i) the K representation splits as the (possibly infinite) algebraic direct sum of finite-dimensional irreducible representations

(ii) the differentiated version of the K action is the restriction to $\mathfrak{k}$ of the $\mathfrak{g}$ action

(iii) $(\mathrm{Ad}(k)u)x = k(u(k^{-1}x))$ for $k \in K$, $u \in U(\mathfrak{g})$, and $x \in V$.

If V and W are two $(\mathfrak{g}, K)$ modules, we denote by $\mathrm{Hom}_{\mathfrak{g},K}(V, W)$ the vector space of linear maps from V into W that respect both the $\mathfrak{g}$ representations and the K representations.

In our definition we tacitly assume that the K representation is continuous, hence smooth, on each of the finite-dimensional irreducible subspaces. Then we make no assumption about having a topology on V as a whole.

The class of $(\mathfrak{g}, K)$ modules is closed under passage to submodules, quotients, and finite direct sums, as a consequence of the following lemma.

Lemma 7.1. Let π be a representation of the compact group K on a vector space V such that every vector lies in a finite-dimensional K-invariant subspace on which π operates continuously. Then V is the (possibly infinite) algebraic direct sum of finite dimensional K-invariant subspaces on which π operates irreducibly.

Proof. Let us say that one direct sum decomposition (into irreducibles) of an invariant subspace V_1 is included in another direct sum decomposition for an invariant subspace V_2 if each summand for V_1 is one of the summands for V_2. Then Zorn's Lemma allows us to choose a maximal such direct sum, say $V' = \Sigma \oplus V_\alpha$. Arguing by contradiction, suppose there is some v in the whole space V that is not in the subspace V'. Let V_0 be a finite-dimensional K-invariant subspace containing v, and let $v_1, \ldots, v_n$ be a basis of $V_0 \cap V'$. Each v_i is in a finite sum of V_α's, and hence we can write $V_0 \cap V' \subseteq \Sigma_{j=1}^N \oplus V_{\alpha_j}$. Then it follows that

$$V_0 + V' = \{V_0 + (\sum_j \oplus V_{\alpha_j})\} \oplus \{\sum_{\alpha \neq \alpha_j} \oplus V_\alpha\}$$

The first factor in braces on the right is finite-dimensional. Thus $\Sigma_j \oplus V_{\alpha_j}$ has an invariant complement in $V_0 + \Sigma_j \oplus V_{\alpha_j}$ by Proposition 3.1, and this splits into irreducibles by Corollary 3.2, say as $\Sigma_{i=1}^M V_{\beta_i}$. Then

$$V_0 + V' = \{\sum_{i=1}^{M} \oplus V_{\beta_i}\} \oplus \{\sum_{\text{all } \alpha} \oplus V_\alpha\}$$

is a direct sum decomposition that contradicts the maximality, and the lemma follows.

Next we shall exhibit closure properties for $(\mathfrak{g},K)$ modules under $\otimes_{\mathbb{C}}$ and $\mathrm{Hom}_{\mathbb{C}}$. If V and W are $(\mathfrak{g},K)$ modules, then we obtain representations of $\mathfrak{g}$ and K on $V \otimes_{\mathbb{C}} W$ in the usual way:

$$\begin{aligned} X(v\otimes w) &= Xv\otimes w + v\otimes Xw\,, && X\in\mathfrak{g} \\ k(v\otimes w) &= kv\otimes kw\,, && k\in K\,. \end{aligned} \tag{7.11}$$

It is clear that every vector is K-finite, and Lemma 7.1 then shows that axiom (i) for a $(\mathfrak{g},K)$ module is satisfied. The other two axioms are fairly clear, and thus $V\otimes_{\mathbb{C}} W$ is a $(\mathfrak{g},K)$ module.

For $\mathrm{Hom}_{\mathbb{C}}$, the situation is more complicated. If V and W are $(\mathfrak{g},K)$ modules, then we obtain representations of $\mathfrak{g}$ and K on $\mathrm{Hom}_{\mathbb{C}}(V,W)$ as usual by

$$\begin{aligned} (X\varphi)(v) &= X(\varphi(v)) - \varphi(Xv)\,, && X\in\mathfrak{g} \\ (k\varphi)(v) &= k(\varphi(k^{-1}v))\,, && k\in K\,. \end{aligned} \tag{7.12}$$

But there is no reason why every φ in $\mathrm{Hom}_{\mathbb{C}}(V,W)$ should be K-finite.

<u>Example</u>. Let $K = \{e^{i\psi}\}$ be the circle, and let $\mathfrak{g} = \mathfrak{k}$. Let V be all trigonometric polynomials

$$P(\theta) = \sum_{\text{finite}} c_n e^{in\theta}$$

with $(e^{i\psi}P)(\theta) = P(\theta-\psi)$. Let $W=\mathbb{C}$ with trivial action, and let $\varphi \in \mathrm{Hom}_{\mathbb{C}}(V,W)$ be given by $\varphi(e^{in\theta}) = 1$ for all n. Then φ is not K-finite, as a consequence of the following lemma.

Lemma 7.2. Let V and W be $(\mathfrak{k},K)$ modules, and suppose the action on W is trivial. Then the K-finite members of $\mathrm{Hom}_{\mathbb{C}}(V,W)$ are those that vanish except on finitely many K-isotypic subspaces of V.

Remark. If τ is an irreducible representation of K, the K-isotypic subspace for the K type τ is the sum of all irreducible subspaces equivalent with τ.

Proof. Let φ be K-finite, and let V_0 be an irreducible K-stable subspace of V on which φ is not identically 0. Let V_0^* be the contragredient representation, and let $W_0 = \varphi(V_0)$. Then $\varphi|_{V_0}$ is in $\mathrm{Hom}_{\mathbb{C}}(V_0,W_0) \cong V_0^* \otimes_{\mathbb{C}} W_0$, which is the sum of a finite number of copies of V_0^*. Hence $\varphi|_{V_0}$ transforms under K according to V_0^*. Since φ lies in a finite-dimensional K-stable subspace, V_0^* is one of the finitely many K types occurring in $\mathrm{span}\{K\varphi\}$. So φ vanishes on the K-isotypic subspaces for all K types whose contragredients do not occur in $\mathrm{span}\{K\varphi\}$.

In the converse direction it is enough to show that if just one K type occurs in V, then every φ in $\mathrm{Hom}_{\mathbb{C}}(V,W)$ is K-finite. Thus write $V = \sum \oplus V_\alpha$ with all V_α K-isomorphic and irreducible. Single out some V_{α_0}, and let $\psi_\alpha : V_{\alpha_0} \to V_\alpha$ be a K isomorphism. By irreducibility, choose a basis of $\mathrm{End}_{\mathbb{C}} V_{\alpha_0}$ of the form $k_j^{-1}|_{V_{\alpha_0}}$, $1 \leq j \leq n$.

Here $\{k_j\}$ refers to a subset of the actions by K. If k is in K, then there exist $c_j(k)$, $1 \le j \le n$, such that

$$k^{-1}x = \sum c_j(k)k_j^{-1}x \quad \text{for all} \quad x \in V_{\alpha_0}.$$

Now $x \in V_{\alpha_0}$ implies

$$(k\varphi)(\psi_\alpha(x)) = \varphi(k^{-1}(\psi_\alpha(x))) = \varphi(\psi_\alpha(k^{-1}x))$$
$$= \varphi(\psi_\alpha(\sum c_j(k)k_j^{-1}x)) = \sum c_j(k)(k_j\varphi)(\psi_\alpha(x)).$$

Hence $k\varphi = \sum c_j(k)(k_j\varphi)$, and φ is K-finite. This proves the lemma.

Returning to the general case, let V and W be two $(\mathfrak{g},K)$ modules, and let $\mathrm{Hom}_{\mathbb{C}}(V,W)_K$ be the subspace of K-finite members of $\mathrm{Hom}_{\mathbb{C}}(V,W)$.

Proposition 7.3. If V and W are $(\mathfrak{g},K)$ modules, then $\mathrm{Hom}_{\mathbb{C}}(V,W)_K$ is a $\mathfrak{g}$-invariant subspace of $\mathrm{Hom}_{\mathbb{C}}(V,W)$ and is therefore a $(\mathfrak{g},K)$ module.

Proof. For $k \in K$, $X \in \mathfrak{g}$, and $\varphi \in \mathrm{Hom}_{\mathbb{C}}(V,W)$, an easy computation gives

$$kX\varphi = (\mathrm{Ad}(k)X)(k\varphi) \in \mathfrak{g}(K\varphi), \tag{7.13}$$

and this lies in a finite-dimensional space if φ is K-finite. Thus $\mathrm{Hom}_{\mathbb{C}}(V,W)_K$ is $\mathfrak{g}$-invariant. Axiom (i) for a $(\mathfrak{g},K)$ module follows from Lemma 7.1, axiom (ii) is trivial, and axiom (iii) is (7.13). Thus $\mathrm{Hom}_{\mathbb{C}}(V,W)_K$ is a $(\mathfrak{g},K)$ module.

Remark. In the special case that $W = \mathbb{C}$, we denote $\mathrm{Hom}_{\mathbb{C}}(V,\mathbb{C})_K$ by V^c and call V^c the K-finite contragredient of V.

3. The algebra $R(\mathfrak{g},K)$

We shall introduce an algebra $R(\mathfrak{g},K)$ such that the $(\mathfrak{g},K)$ modules are exactly the "unital" $R(\mathfrak{g},K)$ modules and such that $\mathrm{Hom}_{\mathfrak{g},K}(V,W)$ coincides with all $R(\mathfrak{g},K)$ maps from V into W. Then our homological constructions from earlier chapters will be available, and we will be prepared to construct new representations.

Before treating general $(\mathfrak{g},K)$, let us consider two extreme cases. One case is that $K = \{1\}$. In this case the $(\mathfrak{g},K)$ modules are just the members of $\mathcal{C}(\mathfrak{g})$, and $U(\mathfrak{g})$ will serve as the desired algebra $R(\mathfrak{g},K)$.

The opposite extreme is that of $(\mathfrak{k},K)$, where $\mathfrak{k}$ is the Lie algebra of K. In this case we let C_K be the algebra of all linear combinations of matrix coefficients of finite-dimensional representations of K, with convolution as multiplication. Unless K is a finite group, the algebra C_K does not have an identity. But it does have an "approximate identity." Namely for each irreducible representation τ of K, let χ_τ be the product of the degree of τ by the character of τ. This is in C_K. For each finite set F of equivalence classes of irreducible representations of K, put $\chi_F = \sum_{\tau\in F}\chi_\tau$. As F varies through such finite sets, the functions χ_F comprise the approximate identity (with the F's partially ordered by inclusion). To be precise, if c is in C_K, then $\chi_F * c = c = c * \chi_F$ for all sufficiently large F. In fact, it is enough to take F large enough so that it includes all irreducibles contributing to the linear combination c of matrix coefficients; the formula $\chi_F * c = c$ $= c * \chi_F$ then follows from Schur orthogonality. A left C_K module V will be called unital if, for each x in V, $\chi_F x = x$ for all χ_F sufficiently far out in the approximate identity. In this case $C_K x$ is finite-dimensional for all x in V (since $C_K * \chi_F$ is finite-dimensional). The following lemma shows that C_K will serve as the desired algebra $R(\mathfrak{k},K)$.

Lemma 7.4. A $(\mathfrak{k},K)$ module in a natural way is a unital left C_K module. Conversely any unital C_K module is a $(\mathfrak{k},K)$ module in a natural way inverse to the first process. For any two such modules V and W, $\mathrm{Hom}_{\mathfrak{k},K}(V,W)$ coincides with $\mathrm{Hom}_{C_K}(V,W)$.

Proof. If π is a representation of K on V that splits as the algebraic direct sum of finite-dimensional irreducible representations, then we can define

$$cx = \int_K c(k)\pi(k)x\,dk \qquad \text{for } c \in C_K,\ x \in V,$$

and the result will be meaningful since $\{\pi(k)x\}$ lies in a finite-dimensional space as k varies. Moreover it is routine to check that we get a unital C_K module in this way.

Conversely suppose V is a unital C_K module. For

$x \in V$, choose a finite set $F(x)$ of irreducible representations of K such that $\chi_{F(x)}x = x$, and define

$$\pi(k)x = (\ell(k)\chi_{F(x)})x \quad \text{for } k \in K,$$

where ℓ is the left regular representation of K on C_K: $\ell(k_0)f(k) = f(k_0^{-1}k)$. If $F \supseteq F(x)$, then $\chi_F x = x$ and

$$\begin{aligned}\ell(k_0)\chi_{F(x)}(k) &= \ell(k_0)(\chi_F * \chi_{F(x)})(k) = \chi_F * \chi_{F(x)}(k_0^{-1}k)\\ &= \int_K \chi_F(k_0^{-1}km^{-1})\chi_{F(x)}(m)\,dm\\ &= (\ell(k_0)\chi_F) * \chi_{F(x)}(k),\end{aligned}$$

so that

$$\begin{aligned}(\ell(k_0)\chi_{F(x)})x &= ((\ell(k_0)\chi_F) * \chi_{F(x)})(x)\\ &= (\ell(k_0)\chi_F)(\chi_{F(x)}x) = (\ell(k_0)\chi_F)x\end{aligned}$$

and $\pi(k)x$ is well defined. Since we are allowed to increase F, it follows that $\pi(k)$ is linear.

Next we have

$$\begin{aligned}\ell(k_1)\chi_F * \ell(k_2)\chi_F(k_0) &= \int_K \chi_F(k_1^{-1}k_0m^{-1})\chi_F(k_2^{-1}m)\,dm\\ &= \int_K \chi_F(k_1^{-1}k_0m^{-1}k_2^{-1})\chi_F(m)\,dm\\ &= \int_K \chi_F(k_2^{-1}k_1^{-1}k_0m^{-1})\chi_F(m)\,dm\\ &\qquad \text{since } \chi_\tau(ab) = \chi_\tau(ba) \text{ for all } \tau\\ &= \chi_F(k_2^{-1}k_1^{-1}k_0)\\ &= \ell(k_1k_2)\chi_F(k_0),\end{aligned}$$

and it follows readily that

$$\pi(k_1k_2) = \pi(k_1)\pi(k_2) .$$

Also $\pi(1) = 1$, so that π is a homomorphism of K into invertible operators. Since

$$\operatorname*{span}_{k\in K} \{\ell(k)\chi_{F(x)}\}$$

is finite-dimensional in C_K, $\text{span}_{k\in K}\{\pi(k)x\}$ is finite-dimensional in V. Thus every vector is K-finite.

Fix a finite-dimensional K-invariant subspace V_0 of V. We check that $k \to \pi(k)\big|_{V_0}$ is continuous. Choose F so that $\chi_F x = x$ for all x in V_0. The map $k \to \ell(k)\chi_F$ is continuous into a finite-dimensional subspace S of C_K (with the supremum norm, say), the map $c \in S \to cx \in V_0$ is linear between finite-dimensional spaces and hence is continuous, and the map $k \to \pi(k)x$ is the composition. Thus $k \to \pi(k)x$ is continuous for each x in V_0, and $k \to \pi(k)\big|_{V_0}$ is continuous.

Thus π is a representation. If $c \in C_K$ and $x \in V$ are given, choose a finite-dimensional K-invariant subspace containing x, and let F be large enough so that χ_F acts as the identity on this space and $c * \chi_F = c$. Then we recover our original C_K module structure from π because

$$\begin{aligned} \int_K c(k)\pi(k)x\,dk &= \int_K c(k)(\ell(k)\chi_F)x\,dk \\ &= \int_K c(k)\chi_F(k^{-1}\cdot)x\,dk \\ &= (c * \chi_F)x = cx . \end{aligned}$$

Finally if V and W are $(\mathfrak{k},K)$ modules, we show that

$$\mathrm{Hom}_{\mathfrak{k},K}(V,W) = \mathrm{Hom}_{C_K}(V,W) .$$

Let π and π' be the representations of K on V and W, respectively. If φ is in $\mathrm{Hom}_{C_K}(V,W)$ and $\chi_F x = x$ and $\chi_F(\varphi(x)) = \varphi(x)$, then

$$\varphi(\pi(k_0)x) = \varphi((\ell(k_0)\chi_F)x) = (\ell(k_0)\chi_F)\varphi(x) = \pi'(k_0)\varphi(x) ,$$

so that φ is in $\mathrm{Hom}_{\mathfrak{k},K}(V,W)$. Conversely if φ is in $\mathrm{Hom}_{\mathfrak{k},K}(V,W)$, then

$$\pi'(k)\varphi(x) = \varphi(\pi(k)x) \quad \text{for} \quad k \in K .$$

Multiplying by $c(k)$ and integrating, we obtain

$$c\cdot\varphi(x) = \int_K c(k)\pi'(k)\varphi(x)\,dk = \int_K c(k)\varphi(\pi(k)x)\,dk$$

$$= \varphi\Big(\int_K c(k)\pi(k)x\,dk\Big) = \varphi(c\cdot x) ,$$

the next-to-last equality holding since φ, being linear, is automatically continuous between the relevant finite-dimensional subspaces. This completes the proof.

For a general pair $(\mathfrak{g},K)$, we need to construct $R(\mathfrak{g},K)$. Motivation for the definition comes from the following considerations: Suppose $(\mathfrak{g},K)$ comes from a group G. Then $U(\mathfrak{g})$ may be identified with the left-invariant differential operators on G, while C_K may be identified with the bi-K-finite signed measures on $K \subseteq G$. Both $U(\mathfrak{g})$ and C_K may thus be regarded as sets of distributions on G

(acting on $C^\infty_{com}(G)$). For $R(\mathfrak{g},K)$ we take all bi-K-finite distributions on G whose support is in K, using convolution as multiplication. We omit the verification that this definition of $R(\mathfrak{g},K)$ is actually equivalent with the one we give below.

To define $R(\mathfrak{g},K)$, we recall the left and right regular representations on C_K and $U(\mathfrak{g})$. The ones for C_K come from differentiating the usual group actions:

$$\begin{aligned} \ell(X)c(k) &= \frac{d}{dt}\, c((\exp tX)^{-1}k)\Big|_{t=0}, \qquad X\in\mathfrak{k},\quad c\in C_K \\ r(X)c(k) &= \frac{d}{dt}\, c(k\exp tX)\Big|_{t=0}, \qquad X\in\mathfrak{k},\quad c\in C_K . \end{aligned} \tag{7.14}$$

The ones for $U(\mathfrak{g})$ are given by multiplication:

$$\begin{aligned} \ell(X)u &= Xu, \qquad X\in\mathfrak{g},\quad u\in U(\mathfrak{g}) \\ r(X)u &= -uX, \qquad X\in\mathfrak{g},\quad u\in U(\mathfrak{g}) . \end{aligned} \tag{7.15}$$

We define $R(\mathfrak{g},K)$ as a set to be

$$R(\mathfrak{g},K) = C_K \otimes_{\mathfrak{k}} U(\mathfrak{g}) . \tag{7.16a}$$

Here $U(\mathfrak{g})$ is a left $\mathfrak{k}$ module under ℓ, and C_K is a right $\mathfrak{k}$ module under the right action. Since r in (7.14) is a left $\mathfrak{k}$ module action, the identification from $\otimes_{\mathfrak{k}}$ is given by

$$0 = c\otimes Xu - cX\otimes u = c\otimes \ell(X)u + r(X)c\otimes u, \qquad X\in\mathfrak{k} . \tag{7.16b}$$

Under this definition, $R(\mathfrak{g},K)$ has surviving actions by C_K on the left and $U(\mathfrak{g})$ on the right, but these do not immediately make $R(\mathfrak{g},K)$ into an algebra. We need an explicit definition.

The "definition" of multiplication that is usually handy for calculations is

$$(c\otimes u)(c'\otimes u') = c * (c'(\cdot)\mathrm{Ad}(\cdot)^{-1}u)\times u' .$$

To make sense out of this, we compute the indicated convolution with values in $U(\mathfrak{g})$, write the result in terms of a basis of $U(\mathfrak{g})$, and separate the coefficient functions into the C_K factor and the $U(\mathfrak{g})$ elements into the $U(\mathfrak{g})$ factor, where they get multiplied by u' on the right. More precisely, choose an $\mathrm{Ad}(K)$-invariant inner product on $U^n(\mathfrak{g})$ for sufficiently large n, and let $\{u_i\}$ be an orthonormal basis of $U^n(\mathfrak{g})$. Then

$$(c\otimes u)(c'\otimes u') = \sum_i c * c'(\cdot)(\mathrm{Ad}(\cdot)u,u_i)\otimes u_i u' . \qquad (7.16c)$$

Initially this product is defined within $C_K\otimes_{\mathbb{C}} U(\mathfrak{g})$. Certainly it is independent of the basis u_i. We must check that it descends to $R(\mathfrak{g},K)$ and makes $R(\mathfrak{g},K)$ into a complex associative algebra. To prove it is well defined, we use the following identities valid for $X\in\mathfrak{k}$, $c\in C_K$, $c'\in C_K$, and $k\in K$:

$$(\ell(X)c)(k) = -\sum (\mathrm{Ad}(k)^{-1}X,X_j)(r(X_j)c)(k) \qquad (7.17a)$$

$$(r(X)c)(k) = -\sum (\mathrm{Ad}(k)X,X_j)(\ell(X_j)c)(k) \qquad (7.17b)$$

$$(r(X)c) * c' = -\,c * (\ell(X)c') \qquad (7.17c)$$

$$r(X)(c * c') = c * (r(X)c') \qquad (7.17d)$$

$$\ell(X)(c * c') = (\ell(X)c) * c' . \qquad (7.17e)$$

In (a) and (b), it is to be understood that X_j is an orthonormal basis of $\mathfrak{k}$. The last three identities are elementary, and (a) and (b) are similar to each other. Let us check (a). With k fixed, we change from the orthonormal basis X_j to the orthonormal basis $Y_j = Ad(k)^{-1}X_j$. Then

$$-\sum (Ad(k)^{-1}X,X_j)(r(X_j)c)(k) = -\sum (Ad(k)^{-1}X,Y_j)(r(Y_j)c)(k)$$

$$= -\sum (Ad(k)^{-1}X,Ad(k)^{-1}X_j) \frac{d}{dt} c(k \exp tY_j)\Big|_{t=0}$$

$$= -\sum (X,X_j) \frac{d}{dt} c(k \exp\{t\ Ad(k)^{-1}X_j\})\Big|_{t=0}$$

$$= -\sum (X,X_j) \frac{d}{dt} c(kk^{-1}\{\exp tX_j\}k)\Big|_{t=0}$$

$$= \sum (X,X_j)(\ell(X_j)c)(k)$$

$$= (\ell(X)c)(k) .$$

Proposition 7.5. The multiplication (7.16c) descends to $R(\mathfrak{g},K)$ and makes $R(\mathfrak{g},K)$ into a complex associative algebra.

Proof. To check that multiplication is well defined in the first factor, we have to check, in view of (7.16b), that

$$(c\otimes \ell(X)u)(c'\otimes u') + (r(X)c\otimes u)(c'\otimes u')$$

in $C_K\otimes_{\mathbb{C}} U(\mathfrak{g})$ descends to 0 in $C_K\otimes_{\mathfrak{k}} U(\mathfrak{g})$, for $X\in\mathfrak{k}$. We have

$$(c\otimes \ell(X)u)(c'\otimes u') = \sum_i c*c'(\cdot)(Ad(\cdot)^{-1}(Xu),u_i)\otimes u_iu'$$

$$= \sum_{i,j,k} c*c'(\cdot)(Ad(\cdot)^{-1}X,X_j)(Ad(\cdot)^{-1}u,u_k)(X_ju_k,u_i)\otimes u_iu'$$

$$= \sum_{j,k} c*c'(\cdot)(Ad(\cdot)^{-1}X,X_j)(Ad(\cdot)^{-1}u,u_k)\otimes X_ju_ku'$$

$$\equiv \sum_{j,k} -r(X_j)\{c * c'(\cdot)(\mathrm{Ad}(\cdot)^{-1}X, X_j)(\mathrm{Ad}(\cdot)^{-1}u, u_k)\} \otimes u_k u' \qquad \text{by (7.16b)}$$

$$= \sum_{j,k} -c * r(X_j)\{c'(\cdot)(\mathrm{Ad}(\cdot)^{-1}X, X_j)(\mathrm{Ad}(\cdot)^{-1}u, u_k)\} \otimes u_k u' \qquad \text{by (7.17d)}$$

$$= \sum_k c * [\, -\sum_j (\mathrm{Ad}(\cdot)^{-1}X, X_j) r(X_j)\{c'(\cdot)(\mathrm{Ad}(\cdot)^{-1}u, u_k)\}] \otimes u_k u' \qquad \text{since } r(X_j)(\mathrm{Ad}(\cdot)^{-1}X, X_j) = 0$$

$$= \sum_k c * \ell(X)\{c'(\cdot)(\mathrm{Ad}(\cdot)^{-1}u, u_k)\} \otimes u_k u' \qquad \text{by (7.17a)}$$

$$= \sum_k -\, r(X)c * c'(\cdot)(\mathrm{Ad}(\cdot)^{-1}u, u_k) \otimes u_k u' \qquad \text{by (7.17c)}$$

$$= -(r(X)c \otimes u)(c' \otimes u') \qquad \text{by (7.16c),}$$

and thus multiplication is well defined in the first factor. A similar computation shows that it is well defined in the second factor.

Thus we obtain a multiplication, clearly bilinear, and we have to prove it is associative. Let us show that

$$((c \otimes u)(c' \otimes u'))(c'' \otimes u'') = (c \otimes u)((c' \otimes u')(c'' \otimes u'')) . \qquad (7.18)$$

For f and g in C_K we have

$$\sum_i f(\cdot)(\mathrm{Ad}(\cdot)^{-1}u, u_i) * g(\cdot)(\mathrm{Ad}(\cdot)^{-1}u_i, u_m)$$

$$= \sum_i \int_K f(\cdot k^{-1})(\mathrm{Ad}(\cdot k^{-1})^{-1}u, u_i) g(k)(\mathrm{Ad}(k)^{-1}u_i, u_m)\, dk$$

$$= \sum_i \int_K f(\cdot k^{-1})(\mathrm{Ad}(\cdot)^{-1}u, \mathrm{Ad}(k)^{-1}u_i) g(k)(\mathrm{Ad}(k)^{-1}u_i, u_m)\, dk$$

$$= \int_K f(\cdot k^{-1})(\mathrm{Ad}(\cdot)^{-1}u, u_m) g(k)\, dk$$

$$= (f * g)(\cdot)(\mathrm{Ad}(\cdot)^{-1}u, u_m) \, . \tag{7.19}$$

Taking $f = c'$ and $g = c''(\cdot)(u_m \mathrm{Ad}(\cdot)^{-1}u', u_j)$, we obtain

$$\sum_m (c' * c''(\cdot)(u_m \mathrm{Ad}(\cdot)^{-1}u', u_j))(\mathrm{Ad}(\cdot)^{-1}u, u_m)$$

$$= \sum_{i,m} c'(\cdot)(\mathrm{Ad}(\cdot)^{-1}u, u_i) * c''(\cdot)(u_m \mathrm{Ad}(\cdot)^{-1}u', u_j)(\mathrm{Ad}(\cdot)^{-1}u_i, u_m)$$

$$= \sum_i c'(\cdot)(\mathrm{Ad}(\cdot)^{-1}u, u_i) * c''(\cdot)(\mathrm{Ad}(\cdot)^{-1}(u_i u'), u_j) \, . \tag{7.20}$$

Therefore

$$((c \otimes u)(c' \otimes u'))(c'' \otimes u'')$$

$$= \sum_{i,j} (c * c'(\cdot)(\mathrm{Ad}(\cdot)^{-1}u, u_i)) * c''(\cdot)(\mathrm{Ad}(\cdot)^{-1}(u_i u'), u_j) \otimes u_j u''$$

$$= \sum_{i,j} c * (c'(\cdot)(\mathrm{Ad}(\cdot)^{-1}u, u_i) * c''(\cdot)(\mathrm{Ad}(\cdot)^{-1}(u_i u'), u_j)) \otimes u_j u''$$

since convolution is associative in C_K

$$= \sum_{j,m} c * (c' * c''(\cdot)(u_m \mathrm{Ad}(\cdot)^{-1}u', u_j))(\mathrm{Ad}(\cdot)^{-1}u, u_m) \otimes u_j u''$$

by (7.20)

$$= \sum_{j,k,m} c * (c' * c''(\cdot)(u_m u_k, u_j)(\mathrm{Ad}(\cdot)^{-1}u', u_k))(\mathrm{Ad}(\cdot)^{-1}u, u_m) \otimes u_j u''$$

$$= \sum_{k,m} c * (c' * c''(\cdot)(\mathrm{Ad}(\cdot)^{-1}u', u_k))(\mathrm{Ad}(\cdot)^{-1}u, u_m) \otimes u_m u_k u''$$

$$= (c \otimes u)((c' \otimes u')(c'' \otimes u'')) \, ,$$

and the proposition follows.

There is a second way of defining $R(\mathfrak{g},K)$, which will turn out to be isomorphic to the first. For the set, we take

$$R(\mathfrak{g},K) = U(\mathfrak{g}) \otimes_{\mathfrak{k}} C_K, \tag{7.21a}$$

the identification under $\otimes_{\mathfrak{k}}$ being

$$0 = u \otimes Xc - uX \otimes c = u \otimes \ell(X)c + r(X)u \otimes c, \quad X \in \mathfrak{k}. \tag{7.21b}$$

Multiplication is given formally by

$$(u \otimes c)(u' \otimes c') = u \times (c(\cdot)\mathrm{Ad}(\cdot)u') * c'$$

and more precisely by

$$(u \otimes c)(u' \otimes c') = \sum_i uu_i \otimes (\mathrm{Ad}(\cdot)u', u_i)c(\cdot) * c', \tag{7.21c}$$

and one checks as in Proposition 7.5 that multiplication descends to $R(\mathfrak{g},K)$ and makes $R(\mathfrak{g},K)$ into a complex associative algebra.

Proposition 7.6. The two definitions (7.16) and (7.21) of $R(\mathfrak{g},K)$ are isomorphic, the correspondence being

$$c \otimes u \longrightarrow \sum u_i \otimes c(\cdot)(\mathrm{Ad}(\cdot)u, u_i) \tag{7.22a}$$

with inverse

$$u \otimes c \longrightarrow \sum_j c(\cdot)(\mathrm{Ad}(\cdot)^{-1}u, u_j) \otimes u_j. \tag{7.22b}$$

Remark. In particular $c \otimes 1 \to 1 \otimes c$, and vice versa.

Proof. First we show that (7.22a) descends to the quotient $R(\mathfrak{g},K)$. For $X \in \mathfrak{k}$, we have

$$c \otimes \ell(X)u \to \sum_i u_i \otimes c(\cdot)(\mathrm{Ad}(\cdot)(Xu), u_i)$$

$$= \sum_i u_i \otimes c(\cdot)(\mathrm{Ad}(\cdot)(uX),u_i) + \sum_i u_i \otimes c(\cdot)(\mathrm{Ad}(\cdot)(\mathrm{ad}\ X)u,u_i)$$

$$= \sum_{i,j,k} u_i \otimes c(\cdot)(\mathrm{Ad}(\cdot)u,u_k)(\mathrm{Ad}(\cdot)X,X_j)(u_kX_j,u_i)$$

$$+ \sum_i u_i \otimes c(\cdot)r(X)\{(\mathrm{Ad}(\cdot)u,u_i)\}$$

$$= \mathrm{I} + \mathrm{II},$$

say. Then

$$\mathrm{I} = \sum_{j,k} u_kX_j \otimes c(\cdot)(\mathrm{Ad}(\cdot)u,u_k)(\mathrm{Ad}(\cdot)X,X_j)$$

$$\equiv \sum_{j,k} u_k \otimes \ell(X_j)\{c(\cdot)(\mathrm{Ad}(\cdot)u,u_k)(\mathrm{Ad}(\cdot)X,X_j)\}.$$

Now

$$\ell(X_j)\{(\mathrm{Ad}(\cdot)X,X_j)\} = (-(\mathrm{ad}\ X_j)\mathrm{Ad}(\cdot)X,X_j) = (\mathrm{Ad}(\cdot)X,[X_j,X_j]) = 0.$$

Hence

$$\mathrm{I} \equiv \sum_{j,k} u_k \otimes \ell(X_j)\{c(\cdot)(\mathrm{Ad}(\cdot)u,u_k)\}(\mathrm{Ad}(\cdot)X,X_j)$$

$$= -\sum_k u_k \otimes r(X)\{c(\cdot)(\mathrm{Ad}(\cdot)u,u_k)\} \quad \text{by (7.17b)}.$$

Adding I and II, we see that

$$c \otimes \ell(X)u \to -\sum_i u_i \otimes r(X)\{c(\cdot)\}(\mathrm{Ad}(\cdot)u,u_i),$$

which is minus the image of $r(X)c \otimes u$. So (7.22a) descends to the quotient, and a similar argument shows that (7.22b) descends to the quotient.

Formulas (7.22a) and (7.22b) are two-sided inverses. In fact, the result of applying (7.22a) and then (7.22b) is

$$c \otimes u \to \sum_i u_i \otimes c(\cdot)(\mathrm{Ad}(\cdot)u, u_i)$$

$$\to \sum_{i,j} c(\cdot)(\mathrm{Ad}(\cdot)u, u_i)(\mathrm{Ad}(\cdot)^{-1}u_i, u_j) \otimes u_j$$

$$= \sum_j c(\cdot)(\mathrm{Ad}(\cdot)u, \mathrm{Ad}(\cdot)u_j) \otimes u_j$$

$$= \sum_j c(\cdot)(u, u_j) \otimes u_j = c \otimes u .$$

Thus the composition in one order is the identity, and similarly the composition in the other order is the identity.

Finally let us see that (7.22a) respects multiplication. We have

$$(c \otimes u)(c' \otimes u') = \sum_i c * c'(\cdot)(\mathrm{Ad}(\cdot)^{-1}u, u_i) \otimes u_i u'$$

$$\to \sum_{i,m} u_m \otimes (c * c'(\cdot)(\mathrm{Ad}(\cdot)^{-1}u, u_i))(\mathrm{Ad}(\cdot)(u_i u'), u_m) . \tag{7.23}$$

The same kind of argument as for (7.19) shows that

$$(f * g)(\cdot)(\mathrm{Ad}(\cdot)(u_i u'), u_m) =$$

$$= \sum_j f(\cdot)(\mathrm{Ad}(\cdot)u_j, u_m) * g(\cdot)(\mathrm{Ad}(\cdot)(u_i u', u_j) .$$

Taking $f = c$ and $g = c'(\cdot)(\mathrm{Ad}(\cdot)^{-1}u, u_i)$, we see that (7.23) is

$$= \sum_{i,j,m} u_m \otimes c(\cdot)(\mathrm{Ad}(\cdot)u_j, u_m) * c'(\cdot)(\mathrm{Ad}(\cdot)^{-1}u, u_i)(\mathrm{Ad}(\cdot)(u_i u'), u_j)$$

$$= \sum_{i,j,k,m} u_m \otimes c(\cdot)(\mathrm{Ad}(\cdot)u_j, u_m)$$
$$* c'(\cdot)(\mathrm{Ad}(\cdot)u', u_k)(\mathrm{Ad}(\cdot)^{-1}u, u_i)((\mathrm{Ad}(\cdot)u_i)u_k, u_j)$$

$$= \sum_{j,k,m} u_m \otimes c(\cdot)(\mathrm{Ad}(\cdot)u_j, u_m) * c'(\cdot)(\mathrm{Ad}(\cdot)u', u_k)(uu_k, u_j)$$

$$= \sum_{k,m} u_m \otimes c(\cdot)(\mathrm{Ad}(\cdot)(uu_k),u_m) * c'(\cdot)(\mathrm{Ad}(\cdot)u',u_k)$$

$$= \sum_{i,j,k,m} u_m \otimes c(\cdot)(\mathrm{Ad}(\cdot)u,u_i)(\mathrm{Ad}(\cdot)u_k,u_j)(u_iu_j,u_m) * c'(\cdot)(\mathrm{Ad}(\cdot)u',u_k)$$

$$= \sum_{i,j,k} u_iu_j \otimes c(\cdot)(\mathrm{Ad}(\cdot)u,u_i)(\mathrm{Ad}(\cdot)u_k,u_j) * c'(\cdot)(\mathrm{Ad}(\cdot)u',u_k)$$

$$= \Big(\sum_i u_i \otimes c(\cdot)(\mathrm{Ad}(\cdot)u,u_i)\Big)\Big(\sum_k u_k \otimes c'(\cdot)(\mathrm{Ad}(\cdot)u',u_k)\Big),$$

and this is the product of the images of $c \otimes u$ and $c' \otimes u'$. The proposition follows.

The algebra $R(\mathfrak{g},K)$ has a special antiautomorphism

$$(c \otimes u)^{tr} = u^{tr} \otimes c^{\sim}$$
$$(u \otimes c)^{tr} = c^{\sim} \otimes u^{tr},$$

where $c^{\sim}(k) = c(k^{-1})$. We omit the easy verifications that are involved.

The algebra $R(\mathfrak{g},K)$ does not have an identity unless K is a finite group, but, like C_K, it does have an approximate identity. As F varies through finite sets of K types, we put

$$e_F = \chi_F \otimes 1 = 1 \otimes \chi_F,$$

where $\chi_F \in C_K$ is defined as before Lemma 7.4; here 1 denotes the identity of $U(\mathfrak{g})$. We notice from (7.16c) that $e_F(c \otimes u) = c \otimes u$ if F is sufficiently large, and we notice from (7.21c) that $(u \otimes c)e_F = u \otimes c$ if F is sufficiently large. Therefore any $r \in R$ has $e_F r = r = re_F$ if F is sufficiently large.

A left $R(\mathfrak{g},K)$ module V will be called unital if, for each x in V, $e_F x = x$ for all e_F sufficiently far out in the approximate identity. In this case $C_K x$ is finite-dimensional for all x in V, since $C_K * \chi_F$ is finite-dimensional.

Proposition 7.7. A $(\mathfrak{g},K)$ module in a natural way is a unital left $R(\mathfrak{g},K)$ module. Conversely any unital left $R(\mathfrak{g},K)$ module is a $(\mathfrak{g},K)$ module in a natural way inverse to the first process. For any two such modules V and W, $\mathrm{Hom}_{\mathfrak{g},K}(V,W)$ coincides with $\mathrm{Hom}_{R(\mathfrak{g},K)}(V,W)$.

Remark. The correspondence between the $R(\mathfrak{g},K)$ action and the $(\mathfrak{g},K)$ action will be given by (7.24) and (7.25) in the course of the proof.

Proof. Let (π,V) be a $(\mathfrak{g},K)$ module, and define

$$(c \otimes u)x = \pi(c)\pi(u)x . \tag{7.24}$$

To see that we have an $R(\mathfrak{g},K)$ module, we need only check multiplicative associativity, the other properties being obvious. We have

$$((c \otimes u)(c' \otimes v))x = (c(\cdot) * c'(\cdot)(\mathrm{Ad}(\cdot)^{-1}u)v)x \qquad \text{by (7.16c)}$$

$$= \sum_i \pi(c(\cdot) * c'(\cdot)(\mathrm{Ad}(\cdot)^{-1}u,u_i))\pi(u_i v)x \qquad \text{by (7.24)}$$

$$= \sum_i \pi(c)\pi(c'(\cdot)(\mathrm{Ad}(\cdot)^{-1}u,u_i))\pi(u_i)\pi(v)x$$

$$= \sum_i \pi(c) \int_K c'(k)(\mathrm{Ad}(k)^{-1}u, u_i)\pi(k)\pi(u_i)\pi(v)x\, dk$$

$$= \pi(c) \int_K c'(k)\pi(k)\pi(\mathrm{Ad}(k)^{-1}u)\pi(v)x\, dk$$

$$= \pi(c) \int_K c'(k)\pi(u)\pi(k)\pi(v)x\, dk \qquad \text{by (7.1)}$$

$$= \pi(c)\pi(u)\pi(c')\pi(v)x$$

$$= (c \otimes u)((c' \otimes v)x) \qquad \text{by (7.24).}$$

To see V is unital, we write $e_F x = \pi(\chi_F \otimes 1)x = \pi(\chi_F)x$, and this is x for F large since x is K-finite.

Parenthetically let us record the formula that expresses how an element $u \otimes c$ of $R(\mathfrak{g},K)$ acts:

$$(u \otimes c)x = \pi(u)\pi(c)x. \tag{7.25}$$

In fact, (7.22) gives

$$(u \otimes c)x = \sum_i \pi(c(\cdot)(\mathrm{Ad}(\cdot)^{-1}u, u_i))\pi(u_i)x$$

$$= \sum_i \int_K c(k)(\mathrm{Ad}(k)^{-1}u, u_i)\pi(k)\pi(u_i)x\, dk$$

$$= \int_K c(k)\pi(k)\pi(\mathrm{Ad}(k)^{-1}u)x\, dk$$

$$= \int_K c(k)\pi(u)\pi(k)x\, dk \qquad \text{by (7.1)}$$

$$= \pi(u)\pi(c)x.$$

Conversely let V be a unital $R(\mathfrak{g},K)$ module, and let $u \in U(\mathfrak{g})$ and $x \in V$ be given. Choose F so that $e_F x = x$, and let

$$\pi(u)x = (u \otimes \chi_F)x\,. \tag{7.26}$$

To see this is well defined, let $F' \supseteq F$, so that $\chi_{F'} * \chi_F = \chi_F$. Then

$$\begin{aligned}(u \otimes \chi_{F'})x &= (u \otimes \chi_{F'})e_F x = (u \otimes \chi_{F'})(1 \otimes \chi_F)x \\ &= (u \otimes (\chi_{F'} * \chi_F))x = (u \otimes \chi_F)x\,.\end{aligned}$$

So $\pi(u)x$ is well defined.

Clearly $\pi(1) = 1$. Before checking that $\pi(uv)$ equals $\pi(u)\pi(v)$, let us define π on K. The action of $R(\mathfrak{g}, K)$ gives us an action of the subring C_K by restriction, and $C_K x$ is finite-dimensional for all x since V is unital. By Lemma 7.4, we obtain a $(\mathfrak{k}, K)$ action of K defined compatibly with the action of C_K:

$$(c \otimes 1)x = (1 \otimes c)x = \int_K c(k)\pi(k)x\,dk\,.$$

The claim is that $e_F x = x$ implies

$$(u \otimes c)x = \pi(u)\pi(c)x \tag{7.27}$$

for u in $U(\mathfrak{g})$ and c in C_K. (In view of (7.25), our definitions are then inverse to those in the first part of the proof.) In fact, we have

$$\begin{aligned}e_F\pi(c)x &= \pi(\chi_F)\pi(c)x = \pi(\chi_F * c)x \\ &= \pi(c * \chi_F)x \qquad \text{since } \chi_F \text{ is central in } C_K \\ &= \pi(c)\pi(\chi_F)x = \pi(c)e_F x = \pi(c)x\,.\end{aligned} \tag{7.28}$$

Thus (7.26) says

$$\begin{aligned}\pi(u)\pi(c)x &= (u\otimes\chi_F)\pi(c)x = (u\otimes\chi_F)(1\otimes c)x = (u\otimes(\chi_F * c))x \\ &= (u\otimes(c*\chi_F))x = (u\otimes c)(1\otimes\chi_F)x \\ &= (u\otimes c)e_F x = (u\otimes c)x\,,\end{aligned}$$

and (7.27) follows.

Now we show that $\pi(uv) = \pi(u)\pi(v)$. If $e_F x = x$ and $e_{F'}\pi(v)x = \pi(v)x$, then

$$\begin{aligned}\pi(u)\pi(v)x &= (u\otimes\chi_{F'})(v\otimes\chi_F)x \\ &= ((u\ \mathrm{Ad}(\cdot)v)\chi_{F'}(\cdot) * \chi_F)x \\ &= \Big(\sum_j uv_j\otimes\Big\{\int_K (\mathrm{Ad}(\cdot\ell^{-1})v, v_j)\chi_{F'}(\cdot\ell^{-1})\chi_F(\ell)\,d\ell\Big\}\Big)x \\ &= \sum_j \pi(uv_j)\pi\Big(\Big\{\int_K (\mathrm{Ad}(\cdot\ell^{-1})v, v_j)\chi_{F'}(\cdot\ell^{-1})\chi_F(\ell)\,d\ell\Big\}\Big)x \\ &\qquad\text{by (7.27)} \\ &= \sum_j \pi(uv_j)\iint_{K\times K}(\mathrm{Ad}(k\ell^{-1})v, v_j)\chi_{F'}(k\ell^{-1})\chi_F(\ell)\pi(k)x\,d\ell\,dk \\ &= \sum_j \pi(uv_j)\iint_{K\times K}(\mathrm{Ad}(\ell^{-1})v, \mathrm{Ad}(k^{-1})v_j)\chi_{F'}(k\ell^{-1})\chi_F(\ell)\pi(k)x\,d\ell\,dk \\ &= \sum \pi(uv_j)\int_K (\mathrm{Ad}(k^{-1})v, \mathrm{Ad}(k^{-1})v_j)\chi_F(k)\pi(k)x\,dk \\ &\qquad\text{if } F' \text{ is large enough} \\ &= \pi(uv)\pi(\chi_F)x \\ &= \pi(uv)x \qquad\text{since } e_F x = x\,.\end{aligned}$$

We have to verify axioms (ii) and (iii). For axiom (ii), let $X \in \mathfrak{k}$ be given. We are to show that $\pi(X)x$ equals $\frac{d}{dt}\pi(\exp tX)x\Big|_{t=0}$. Choose F so that $e_F x = x$. If $F' \supseteq F$, then

$$\begin{aligned}
\pi(\chi_{F'}) \frac{d}{dt} \pi(\exp tX)x\Big|_{t=0} &= \frac{d}{dt} \int_K \chi_{F'}(k)\pi(k \exp tX)x\, dk \Big|_{t=0} \\
&= \frac{d}{dt} \int_K \chi_{F'}(k(\exp tX)^{-1})\pi(k)x\, dk \Big|_{t=0} \\
&= \frac{d}{dt} \int_K \chi_{F'}((\exp tX)^{-1}k)\pi(k)x\, dk \Big|_{t=0} \\
&\qquad \text{since } \chi_{F'} \text{ is central in } C_K \\
&= \int_K (\ell(X)\chi_{F'})(k)\pi(k)x\, dk \\
&= \pi(\ell(X)\chi_{F'})x \\
&= (1 \otimes \ell(X)\chi_{F'})x \qquad \text{by (7.27)} \\
&= (X \otimes \chi_{F'})x \qquad \text{by (7.21b)} \\
&= \pi(X)x \qquad \text{by (7.26)}.
\end{aligned}$$

Since F' can be arbitrarily large, axiom (ii) follows.

For axiom (iii) we are to show that

$$\pi(\mathrm{Ad}(k_0)u)\pi(k_0)x = \pi(k_0)\pi(u)x$$

for u in $U(\mathfrak{g})$ and k_0 in K. Choose F with $e_F x = x$. Then

$$\pi(k_0)x = \pi(k_0)\pi(\chi_F)x = \int_K \chi_F(k)\pi(k_0)\pi(k)x\, dk = \pi(\ell(k_0)\chi_F)x. \tag{7.29}$$

Applying (7.27), we obtain

$$\begin{aligned}
\pi(\mathrm{Ad}(k_0)u)\pi(k_0)x &= \pi(\mathrm{Ad}(k_0)u)\pi(\ell(k_0)\chi_F)x \\
&= (\mathrm{Ad}(k_0)u \otimes \ell(k_0)\chi_F)x. \qquad (7.30)
\end{aligned}$$

If $F' \supseteq F$ is large enough so that $e_{F'}\pi(u)x = \pi(u)x$, then

$$\pi(k_0)\pi(u)x = \pi(\ell(k_0)\chi_{F'})\pi(u)x = (\ell(k_0)\chi_{F'} \otimes 1)(u \otimes \chi_F)x$$

$$= (\ell(k_0)\chi_{F'} \otimes 1)(\chi_F \otimes \mathrm{Ad}(\cdot)^{-1}u)x$$

$$= (\ell(k_0)\chi_{F'} * \chi_F(\cdot)(\mathrm{Ad}(\cdot)^{-1}u))x$$

$$= \left(\int_K \chi_{F'}(k_0^{-1}\cdot\ell^{-1})\chi_F(\ell)\mathrm{Ad}(\ell)^{-1}u\, d\ell\right)x$$

$$= (\chi_F(k_0^{-1}\cdot)\mathrm{Ad}(k_0^{-1}\cdot)^{-1}u)x$$

if F' is increased sufficiently

$$= (\mathrm{Ad}(k_0)u \otimes \ell(k_0)\chi_F)x \quad \text{by (7.22).}$$

Comparison with (7.30) completes the proof of axiom (iii).

Finally let us show that $\mathrm{Hom}_{R(\mathfrak{g},K)}(V, W) = \mathrm{Hom}_{\mathfrak{g},K}(V, W)$. Let π act on V and π' act on W. If φ is in $\mathrm{Hom}_{R(\mathfrak{g},K)}(V, W)$ and F is chosen with $e_F x = x$, then (7.29) applied twice gives

$$\varphi(\pi(k_0)x) = \varphi((\ell(k_0)\chi_F \otimes 1)x)$$

$$= (\ell(k_0)\chi_F \otimes 1)\varphi(x) = \pi'(k_0)\pi'(\chi_F)\varphi(x).$$

For large F the right side is $\pi'(k_0)\varphi(x)$, as required. Also u in $U(\mathfrak{g})$ implies

$$\varphi(\pi(u)x) = \varphi((u \otimes \chi_F)x) = (u \otimes \chi_F)\varphi(x) = \pi'(u)\varphi(x)$$

if F is large enough. So φ is in $\mathrm{Hom}_{\mathfrak{g},K}(V, W)$.

Conversely let φ be in $\mathrm{Hom}_{\mathfrak{g},K}(V, W)$. Then $\varphi(\pi(u)x) = \pi'(u)\varphi(x)$ for u in $U(\mathfrak{g})$ and $\varphi(\pi(k_0)x)$ equals $\pi'(k_0)\varphi(x)$ for k_0 in K. Hence also $\varphi(\pi(c)x) = \pi'(c)\varphi(x)$

for c in C_K. Consequently (7.24) gives

$$\varphi((c\otimes u)x) = \varphi(\pi(c)\pi(u)x) = \pi'(c)\pi'(u)\varphi(x) = (c\otimes u)\varphi(x),$$

and φ is in $\mathrm{Hom}_{R(\mathfrak{g},K)}(V,W)$. This completes the proof.

4. The category $\mathcal{C}(\mathfrak{g},K)$

Let $(\mathfrak{g},K)$ be a pair. Then it follows from Proposition 7.7 and Lemma 7.1 that all $(\mathfrak{g},K)$ modules and $(\mathfrak{g},K)$ maps can be regarded as a good category of $R(\mathfrak{g},K)$ modules, namely the category of all unital left $R(\mathfrak{g},K)$ modules. We denote this category by $\mathcal{C}(\mathfrak{g},K)$. In the special case $K=\{1\}$, $\mathcal{C}(\mathfrak{g},K)$ reduces to $\mathcal{C}(\mathfrak{g})$.

We should bear in mind that left and right $R(\mathfrak{g},K)$ modules are interchangeable via the transpose map: Right modules become left modules under the definition $rx = xr^{tr}$ for $r\in R(\mathfrak{g},K)$. This fact will be of use in applying the operations $\otimes_{R(\mathfrak{g},K)}$ and $\mathrm{Hom}_{R(\mathfrak{g},K)}$ shortly.

We have seen that $\mathcal{C}(\mathfrak{g},K)$ is closed under appropriate tensor product and Hom operations. Specifically if V and W are in $\mathcal{C}(\mathfrak{g},K)$, then so are $V\otimes_{\mathbb{C}}W$ and $\mathrm{Hom}_{\mathbb{C}}(V,W)_K$. We can make these operations into functors from $\mathcal{C}(\mathfrak{g},K)$ into $\mathcal{C}(\mathfrak{g},K)$ as follows:

(1) With $F(V) = V\otimes_{\mathbb{C}}W$, define $F(\varphi) = \varphi\otimes 1$. Then it is routine to see that F is a covariant functor. Similar remarks apply to $F(W) = V\otimes_{\mathbb{C}}W$.

(2) With $F(V) = \mathrm{Hom}_{\mathbb{C}}(V,W)_K$, define $F(\varphi) = \mathrm{Hom}(\varphi,1)$.

In seeing that F is a contravariant functor, we need to check that if φ is a $(\mathfrak{g},K)$ map from V into V', then $\mathrm{Hom}(\varphi,1)$ is a $(\mathfrak{g},K)$ map from $\mathrm{Hom}_{\mathbb{C}}(V',W)_K$ into $\mathrm{Hom}_{\mathbb{C}}(V,W)_K$. Certainly $\mathrm{Hom}(\varphi,1)$ is a $(\mathfrak{g},K)$ map from $\mathrm{Hom}_{\mathbb{C}}(V',W)$ into $\mathrm{Hom}_{\mathbb{C}}(V,W)$ and therefore from $\mathrm{Hom}_{\mathbb{C}}(V',W)_K$ into $\mathrm{Hom}_{\mathbb{C}}(V,W)$. Since $\mathrm{Hom}(\varphi,1)$ is a K map, it carries K-finite elements to K-finite elements. Therefore the image must be in $\mathrm{Hom}_{\mathbb{C}}(V,W)_K$. The other properties of F are standard, and thus F is a contravariant functor.

(3) With $F(W) = \mathrm{Hom}_{\mathbb{C}}(V,W)_K$, define $F(\varphi) = \mathrm{Hom}(1,\varphi)$. The same kind of argument as in (2) shows that F is a covariant functor.

One instance of (2) is worth special note. The functor $V \to \mathrm{Hom}_{\mathbb{C}}(V,\mathbb{C})_K$ is called <u>contragredient</u>, to be consistent with our earlier name for $V^c = \mathrm{Hom}_{\mathbb{C}}(V,\mathbb{C})_K$.

In the category $\mathcal{C}(\mathfrak{g},K)$, there are analogs of the associativity formulas (6.1) and (6.2). The first one is

$$\mathrm{Hom}_{\mathfrak{g},K}(A \otimes_{\mathbb{C}} B, C) \cong \mathrm{Hom}_{\mathfrak{g},K}(A, \mathrm{Hom}_{\mathbb{C}}(B,C)_K) \tag{7.31}$$

naturally in A, B, and C. To see this, we start from the isomorphism

$$\mathrm{Hom}_{\mathbb{C}}(A \otimes_{\mathbb{C}} B, C) \cong \mathrm{Hom}_{\mathbb{C}}(A, \mathrm{Hom}_{\mathbb{C}}(B,C)),$$

which we know respects the $\mathfrak{g}$ action and which is easily seen to respect the K action. Then the invariants on the two sides must correspond, and we obtain

$$\mathrm{Hom}_{\mathfrak{g},K}(A \otimes_{\mathbb{C}} B, C) \cong \mathrm{Hom}_{\mathfrak{g},K}(A, \mathrm{Hom}_{\mathbb{C}}(B,C)).$$

On the right side, a $(\mathfrak{g}, K)$ map from A into $\text{Hom}_{\mathbb{C}}(B, C)$ must carry K-finite vectors to K-finite vectors; since every vector in A is K-finite, the image must be in $\text{Hom}_{\mathbb{C}}(B, C)_K$. Then (7.31) follows.

The second identity is

$$(A \otimes_{\mathbb{C}} B) \otimes_{R(\mathfrak{g},K)} C \cong A \otimes_{R(\mathfrak{g},K)} (B \otimes_{\mathbb{C}} C)\,, \tag{7.32}$$

which follows by the same kind of calculation as for (6.2):

$$\begin{aligned}
&\text{Hom}_{\mathbb{C}}((A \otimes_{\mathbb{C}} B) \otimes_{R(\mathfrak{g},K)} C, V) \\
&\qquad \cong \text{Hom}_{R(\mathfrak{g},K)}(A \otimes_{\mathbb{C}} B, \text{Hom}_{\mathbb{C}}(C,V)) && \text{by (2.49)} \\
&\qquad \cong \text{Hom}_{R(\mathfrak{g},K)}(A, \text{Hom}_{\mathbb{C}}(B, \text{Hom}_{\mathbb{C}}(C,V))_K) && \text{by (7.31)} \\
&\qquad = \text{Hom}_{\mathfrak{g},K}(A, \text{Hom}_{\mathbb{C}}(B, \text{Hom}_{\mathbb{C}}(C,V))) \\
&\qquad \cong \text{Hom}_{\mathfrak{g},K}(A, \text{Hom}_{\mathbb{C}}(B \otimes_{\mathbb{C}} C, V)) && \text{by Proposition 2.3} \\
&\qquad = \text{Hom}_{R(\mathfrak{g},K)}A, \text{Hom}_{\mathbb{C}}(B \otimes_{\mathbb{C}} C, V)) \\
&\qquad \cong \text{Hom}_{\mathbb{C}}(A \otimes_{R(\mathfrak{g},K)} (B \otimes_{\mathbb{C}} C), V) && \text{by (2.49).}
\end{aligned}$$

To obtain (7.32) from this, we apply Proposition 5.14.

<u>Lemma 7.8</u>. If V is in $\mathcal{C}(\mathfrak{k}, K)$ and W is a $(\mathfrak{k}, K)$ submodule of V, then there exists a $(\mathfrak{k}, K)$ submodule W' of V such that $V = W \oplus W'$. Consequently every module in $\mathcal{C}(\mathfrak{k}, K)$ is projective and injective.

Proof. By Zorn's Lemma, let W' be a maximal $(\mathfrak{k}, K)$ submodule of V with $W \cap W' = 0$. Write $W = \sum \oplus W_\alpha$ and $W' = \sum \oplus W'_\beta$ by Lemma 7.1, with each W_α and W'_β irreducible finite-dimensional. Arguing by contradiction, suppose there is some v in V that is not in $W+W'$. Let V_0 be a finite-dimensional K-invariant subspace containing v, and let $v_1, \ldots, v_n$ be a basis of $V_0 \cap (W+W')$. Each v_i is in a finite sum of W_α's and W'_β's, and hence we can write

$$V_0 \cap (W+W') \subseteq \Big(\sum_{j=1}^{M} \oplus\, W_{\alpha_j}\Big) \oplus \Big(\sum_{j=1}^{N} \oplus\, W'_{\beta_j}\Big).$$

Then it follows that

$$V_0 + (W+W') = \{V_0 + (\sum_j \oplus\, W_{\alpha_j}) + (\sum_j \oplus\, W'_{\beta_j})\} \oplus \{\sum_{\alpha \neq \alpha_j} \oplus\, W_\alpha\} \oplus \{\sum_{\beta \neq \beta_j} \oplus\, W'_\beta\}.$$

The first factor in braces on the right is finite-dimensional. By Proposition 3.1, $(\Sigma_j \oplus W_{\alpha_j}) + (\Sigma_j \oplus W'_{\beta_j})$ has an invariant complement in it, and this invariant complement is not 0 since V_0 is not contained in $W+W'$. This invariant complement splits into irreducibles by Corollary 3.2. If W_0 is one such irreducible, then $W_0 + W'$ contains W' strictly and $(W_0 + W') \cap W = 0$, contradiction. We conclude that $V = W + W'$ and hence that $V = W \oplus W'$.

The first example of injective after Lemma 5.5 shows how to deduce from this fact that every module in the category is injective. A similar argument shows that every module in the category is projective.

Proposition 7.9. Let V be in $\mathcal{C}(\mathfrak{g},K)$. Then

(a) the covariant functor $W \to V \otimes_{\mathbb{C}} W$ on $\mathcal{C}(\mathfrak{g},K)$ is exact, and it sends projectives to projectives.

(b) the contravariant functor $W \to \mathrm{Hom}_{\mathbb{C}}(W,V)_K$ on $\mathcal{C}(\mathfrak{g},K)$ is exact, and it sends projectives to injectives.

(c) the covariant functor $W \to \operatorname{Hom}_{\mathbb{C}}(V,W)_K$ on $\mathcal{C}(\mathfrak{g},K)$ is exact, and it sends injectives to injectives.

Proof. The exactness in (a) is from Corollary 5.11, but the exactness in (b) and (c) needs proof. Let us prove exactness for (c), omitting the analogous argument for (b). Let

$$0 \longrightarrow A \xrightarrow{\psi} B \xrightarrow{\varphi} C \longrightarrow 0 \qquad (7.33)$$

be exact in $\mathcal{C}(\mathfrak{g},K)$, and form the exact sequence of vector spaces

$$0 \longrightarrow \operatorname{Hom}_{\mathbb{C}}(V,A) \xrightarrow{\operatorname{Hom}(1,\psi)} \operatorname{Hom}_{\mathbb{C}}(V,B) \xrightarrow{\operatorname{Hom}(1,\varphi)} \operatorname{Hom}_{\mathbb{C}}(V,C) \longrightarrow 0 \qquad (7.34a)$$

and the sequence to be investigated:

$$0 \longrightarrow \operatorname{Hom}_{\mathbb{C}}(V,A)_K \xrightarrow{\operatorname{Hom}(1,\psi)} \operatorname{Hom}_{\mathbb{C}}(V,B)_K \xrightarrow{\operatorname{Hom}(1,\varphi)} \operatorname{Hom}_{\mathbb{C}}(V,C)_K \longrightarrow 0 \qquad (7.34b)$$

Since $\operatorname{Hom}(1,\psi)$ is one-one in (7.34a), its restriction in (7.34b) is one-one also.

Now regard (7.33) as an exact sequence in $\mathcal{C}(\mathfrak{k},K)$. By Lemmas 7.8 and 5.22, this sequence splits, so that there is another exact sequence in $\mathcal{C}(\mathfrak{k},K)$ of the form

$$0 \longleftarrow A \xleftarrow{\psi'} B \xleftarrow{\varphi'} C \longleftarrow 0$$

with $\psi'\psi = 1_A$ and $\varphi\varphi' = 1_C$. Let η be in $\ker(\operatorname{Hom}(1,\varphi))$ in (7.34b). Then η is in $\ker(\operatorname{Hom}(1,\varphi)) = \operatorname{image}(\operatorname{Hom}(1,\psi))$ in (7.34a), and we can write $\eta = \operatorname{Hom}(1,\psi)\xi$ for some ξ in $\operatorname{Hom}_{\mathbb{C}}(V,A)$. Now

$$\xi = \mathrm{Hom}(1,1_A)\xi = \mathrm{Hom}(1,\psi')\mathrm{Hom}(1,\psi)\xi = \mathrm{Hom}(1,\psi')\eta$$

exhibits ξ as the image under a $(\mathfrak{k},K)$ map of something K-finite. Thus ξ is K-finite, and $\ker(\mathrm{Hom}(1,\varphi))$ equals $\mathrm{image}(\mathrm{Hom}(1,\psi))$ in (7.34b).

Finally let η be in $\mathrm{Hom}_{\mathbb{C}}(V,C)_K$, and put $\xi = \mathrm{Hom}(1,\varphi')\eta$. Then ξ is in $\mathrm{Hom}_{\mathbb{C}}(V,B)_K$, and

$$\mathrm{Hom}(1,\varphi)\xi = \mathrm{Hom}(1,\varphi)\mathrm{Hom}(1,\varphi')\eta = \mathrm{Hom}(1,1_C)\eta = \eta .$$

Hence $\mathrm{Hom}(1,\varphi)$ is onto $\mathrm{Hom}_{\mathbb{C}}(V,C)_K$ in (7.34b), and (7.34b) is exact.

As we said, the exactness in (b) is proved similarly. Then the remainder of the proof proceeds as with Proposition 6.1, using (7.31) in place of (6.1).

5. The functors P and I

Let $(\mathfrak{g},K)$ and $(\mathfrak{q},H)$ be two pairs with $\mathfrak{g} \supseteq \mathfrak{q}$ and $K \supseteq H$ compatibly. (By "compatibly" we mean that the composite inclusions $\mathfrak{g} \supseteq \mathfrak{q} \supseteq \mathfrak{h}$ and $\mathfrak{g} \supseteq \mathfrak{k} \supseteq \mathfrak{h}$ are the same and that the Ad_H action of H on $\mathfrak{q}$ coincides with the restriction to $\mathfrak{q}$ of the Ad_K action of H.)

Except in special circumstances, $R(\mathfrak{q},H)$ is not naturally a subalgebra of $R(\mathfrak{g},K)$. Nevertheless $R(\mathfrak{g},K)$ is still an $R(\mathfrak{q},H)$ module. To see this without computation, we recall that $R(\mathfrak{g},K)$ is a unital left $R(\mathfrak{g},K)$ module, hence a $(\mathfrak{g},K)$ module, hence a $(\mathfrak{q},H)$ module, hence a unital left $R(\mathfrak{q},H)$ module. By means of the transpose map for $R(\mathfrak{q},H)$,

$R(\mathfrak{g},K)$ has the structure of a right $R(\mathfrak{q},H)$ module as well. If $c\otimes u$ and $u'\otimes c'$ are in $R(\mathfrak{g},K)$ and $a\otimes q$ and $q'\otimes a'$ are in $R(\mathfrak{q},H)$, then the actions work out to be

$$(c\otimes u)(a\otimes q) = c *_H a(\cdot)(\mathrm{Ad}(\cdot)^{-1}u)q \tag{7.35a}$$

with

$$(c *_H b)(k) = \int_H c(kh^{-1})b(h)\,dh\,,$$

and

$$(q'\otimes a')(u'\otimes c') = q'(\mathrm{Ad}(\cdot)u')a'(\cdot) *_H c' \tag{7.35b}$$

with

$$(b' *_H c')(k) = \int_H b'(h)c'(h^{-1}k)\,dh\,.$$

If V is in $\mathcal{C}(\mathfrak{q},H)$, we define

$$P(V) = P_{\mathfrak{q},H}^{\mathfrak{g},K}(V) = R(\mathfrak{g},K)\otimes_{R(\mathfrak{q},H)} V \tag{7.36}$$

as a left $R(\mathfrak{g},K)$ module under $r(s\otimes x) = rs\otimes x$; the convention for $\otimes_{R(\mathfrak{q},H)}$ is that $rq\otimes x = r\otimes qx$. Also we make

$$\mathrm{Hom}_{R(\mathfrak{q},H)}(R(\mathfrak{g},K),V) \tag{7.37}$$

into a left $R(\mathfrak{g},K)$ module by letting $r\varphi(s) = \varphi(sr)$; the convention for $\mathrm{Hom}_{R(\mathfrak{q},H)}$ is that $\varphi(qr) = q(\varphi(r))$. We define a subset $I(V)$ to be the subset of K-finite members:

$$I(V) = I_{\mathfrak{q},H}^{\mathfrak{g},K}(V) = \mathrm{Hom}_{R(\mathfrak{q},H)}(R(\mathfrak{g},K),V)_K\,. \tag{7.38}$$

There is a second way of defining $P(V)$, namely

$$P'(V) = V\otimes_{R(\mathfrak{q},H)} R(\mathfrak{g},K)$$

as a left $R(\mathfrak{g},K)$ module under $r(x\otimes s) = x\otimes sr^{tr}$; the convention for $\otimes_{R(\mathfrak{q},H)}$ is that $q^{tr}x\otimes r = x\otimes qr$. These definitions give naturally isomorphic $R(\mathfrak{g},K)$ modules under the map $r\otimes x \to x\otimes r^{tr}$. Thus we are free to use either definition, but normally we shall use the first definition.

Similarly there is a second way of defining $I(V)$, namely with $r\varphi(s) = \varphi(r^{tr}s)$; the convention for $\mathrm{Hom}_{R(\mathfrak{q},H)}$ is that $\varphi(rq) = q^{tr}(\varphi(r))$. The two definitions give naturally isomorphic $R(\mathfrak{g},K)$ modules, and we are free to use either definition. However, we shall invariably use the first definition.

<u>Proposition 7.10</u>. For V in $\mathcal{C}(\mathfrak{q},H)$ and notation as above, $P(V)$ and $I(V)$ are in $\mathcal{C}(\mathfrak{g},K)$. Both P and I are covariant functors, P is right exact, and I is left exact.

Remark. The choice of notation P and I is again to be a reminder of facts to be proved about preservation of projectives and injectives. When $K = H = \{1\}$, P and I here reduce to the P and I functors in Chapter VI.

Proof. We know that the left $R(\mathfrak{g},K)$ module $R(\mathfrak{g},K)$ is unital and that every member is K-finite. Hence the same is true of $P(V)$, and $P(V)$ is in $\mathcal{C}(\mathfrak{g},K)$ by Proposition 7.7. If $\varphi: V\to W$ is a map, then $P(\varphi)$ is the map $1\otimes\varphi$. Then P is a covariant functor, and it is right exact by Proposition 5.9c.

For $I(V)$, we regard $R(\mathfrak{g},K)$ as a right $R(\mathfrak{g},K)$ module and V as a trivial $R(\mathfrak{g},K)$ module, and we apply Lemma 7.1 to $\mathrm{Hom}_{\mathbb{C}}(R(\mathfrak{g},K),V)_K$. Members of this set are characterized by vanishing except on finitely many K types of $R(\mathfrak{g},K)$, and hence members of the subset (7.38) of (7.37) are characterized by this same property. This property is preserved under the left $R(\mathfrak{g},K)$ action on (7.37), and thus $I(V)$ is a left $R(\mathfrak{g},K)$ module. Using an element e_F that fixes finitely many K types of $R(\mathfrak{g},K)$, we see that $I(V)$ is a unital left $R(\mathfrak{g},K)$ module. The members $I(V)$ are K-finite by definition, and Proposition 7.7 thus shows that $I(V)$ is in $\mathcal{C}(\mathfrak{g},K)$.

To see that I is a functor, let $\varphi : V \to W$ be a map in $\mathcal{C}(\mathfrak{q},H)$. Then $I(\varphi) = \varphi\circ(\cdot)$ carries $I(V)$ into $\mathrm{Hom}_{R(\mathfrak{q},H)}(R(\mathfrak{g},K),W)$, and the problem is to show that the image is in the subset $I(W)$. If ψ is in $I(V)$, then ψ vanishes on all but finitely many K types of $R(\mathfrak{g},K)$, by Lemma 7.2. Hence so does $\varphi\circ\psi$. Then $\varphi\circ\psi$ is in $I(W)$ by Lemma 7.2, and it follows that I is a covariant functor.

To prove that I is left exact, let

$$0 \longrightarrow A \xrightarrow{\ \psi\ } B \xrightarrow{\ \varphi\ } C \longrightarrow 0$$

be exact in $\mathcal{C}(\mathfrak{q},H)$. Then

$$\begin{aligned} 0 &\longrightarrow \mathrm{Hom}_{R(\mathfrak{q},H)}(R(\mathfrak{g},K),A) \\ &\xrightarrow{\mathrm{Hom}(1,\psi)} \mathrm{Hom}_{R(\mathfrak{q},H)}(R(\mathfrak{q},K),B) \xrightarrow{\mathrm{Hom}(1,\varphi)} \mathrm{Hom}_{R(\mathfrak{q},H)}(R(\mathfrak{g},K),C) \end{aligned} \tag{7.39}$$

is exact in the good category of all left $R(\mathfrak{g},K)$ modules, by

Proposition 5.9b. Consider the complex

$$0 \longrightarrow \mathrm{Hom}_{R(\mathfrak{q},H)}(R(\mathfrak{g},K),A)_K$$

$$\xrightarrow{\mathrm{Hom}(1,\psi)} \mathrm{Hom}_{R(\mathfrak{q},H)}(R(\mathfrak{g},K),B)_K \xrightarrow{\mathrm{Hom}(1,\varphi)} \mathrm{Hom}_{R(\mathfrak{q},H)}(R(\mathfrak{g},K),C)_K \tag{7.40}$$

in the category $\mathcal{C}(\mathfrak{g},K)$. The map $\mathrm{Hom}(1,\psi)$ in (7.40) is a restriction of $\mathrm{Hom}(1,\psi)$ in (7.39) and remains one-one. Let η be in $\ker(\mathrm{Hom}(1,\varphi))$ in (7.40). The exactness of (7.39) allows us to find ξ in $\mathrm{Hom}_{R(\mathfrak{q},H)}(R(\mathfrak{g},K),A)$, necessarily unique, such that $\eta = \mathrm{Hom}(1,\psi)\xi$. We are to prove ξ is K-finite. Since $\eta = \psi\xi$ is K-finite, Lemma 7.2 says η vanishes on all but finitely many K types of $R(\mathfrak{g},K)$. Since ψ is one-one, ξ must vanish on all but finitely many K types of $R(\mathfrak{g},K)$. Therefore ξ is K-finite by Lemma 7.2, and (7.40) is exact. This completes the proof.

Remarks. Proposition 7.10 is the best that we have for a generalization of Proposition 6.2a. In Chapter VI, $P_{\mathfrak{q}}^{\mathfrak{g}}$ and $I_{\mathfrak{q}}^{\mathfrak{g}}$ were exact functors, not just one-sided exact. In the new context of our two compatible pairs $(\mathfrak{g},K)$ and $(\mathfrak{q},H)$, P and I need not be exact, as we shall see in §6 after proving a version of Frobenius reciprocity.

To proceed further, we need further facts about $R(\mathfrak{g},K)$. The first two propositions say that P and I are the identity functor if $(\mathfrak{g},K) = (\mathfrak{q},H)$.

Proposition 7.11. If V is in $\mathcal{C}(\mathfrak{g},K)$, then

$$R(\mathfrak{g},K)\otimes_{R(\mathfrak{g},K)} V \cong V \tag{7.41}$$

as a natural isomorphism of left $R(\mathfrak{g},K)$ modules.

Proof. The map $r\otimes v \to rv$ is an $R(\mathfrak{g},K)$ map of the left side of (7.41) into the right side. It is onto because of the presence of approximate identities. Let us prove it is one-one. If $\Sigma(r_i\otimes v_i)$ maps to 0, then $\Sigma\, r_i v_i = 0$. Choose F so that $e_F r_i = r_i$ for all i. Then

$$\begin{aligned}\Sigma(r_i\otimes v_i) = \Sigma(e_F r_i\otimes v_i) &= \Sigma(e_F r_i\otimes v_i) - \Sigma\, e_F\otimes r_i v_i\\ &= \Sigma(e_F r_i\otimes v_i - e_F\otimes r_i v_i)\end{aligned}$$

shows $\Sigma(r_i\otimes v_i) = 0$ on the left side of (7.41). So the map is one-one. Clearly the isomorphism is natural in V.

Proposition 7.12. If W is in $\mathcal{C}(\mathfrak{g},K)$, then

$$\operatorname{Hom}_{R(\mathfrak{g},K)}(R(\mathfrak{g},K),W)_K \cong W \tag{7.42}$$

as a natural isomorphism of left $R(\mathfrak{g},K)$ modules.

Proof. Let $F(W)$ be the left side of (7.42), and let V be in $\mathcal{C}(\mathfrak{g},K)$. Then we have natural isomorphisms

$$\begin{aligned}\operatorname{Hom}_{R(\mathfrak{g},K)}(V,F(W)) &= \operatorname{Hom}_{R(\mathfrak{g},K)}(V,\operatorname{Hom}_{R(\mathfrak{g},K)}(R(\mathfrak{g},K),W))\\ &\cong \operatorname{Hom}_{R(\mathfrak{g},K)}(R(\mathfrak{g},K)\otimes_{R(\mathfrak{g},K)} V,\ W) \qquad \text{by (2.50)}\\ &\cong \operatorname{Hom}_{R(\mathfrak{g},K)}(V,W) \qquad \text{by Proposition 7.11,}\end{aligned}$$

and it follows from Proposition 5.14 that F is naturally equivalent with the identity functor.

Let $S(\mathfrak{g})$ be the symmetric algebra of $\mathfrak{g}$. This is a $(\mathfrak{g},K)$ module under $(\mathrm{ad},\mathrm{Ad})$, and the symmetrization map $\mathcal{S} : S(\mathfrak{g}) \to U(\mathfrak{g})$ commutes with ad and Ad. (See Proposition 2.25 for the assertion about ad; the assertion about Ad is easier.) Since K is compact, we can choose an $\mathrm{Ad}(K)$-invariant subspace $\mathfrak{p}$ of $\mathfrak{g}$ such that $\mathfrak{g} = \mathfrak{k} \oplus \mathfrak{p}$. Let $S(\mathfrak{p}) \subseteq S(\mathfrak{g})$ be the symmetric algebra of $\mathfrak{p}$. This is a $(\mathfrak{k},K)$ module under $(\mathrm{ad},\mathrm{Ad})$.

We denote various left and right regular representations on $R(\mathfrak{g},K)$ by L and R.

<u>Proposition 7.13</u>. Let $(\mathfrak{g},K)$ and $(\mathfrak{g},H)$ be two pairs with $K \supseteq H$ compatibly, and write $\mathfrak{g} = \mathfrak{k} \oplus \mathfrak{p}$ with $\mathfrak{p}$ stable under $\mathrm{Ad}(K)$. Then

$$R(\mathfrak{k},H) \otimes_{\mathbb{C}} S(\mathfrak{p}) \cong R(\mathfrak{g},H) \tag{7.43}$$

under $(c \otimes u) \otimes p \to c \otimes u\mathcal{S}(p)$. The isomorphism carries the left $R(\mathfrak{k},H)$ module structure $L \otimes 1$ on the left side to L on the right side, and it carries the left $R(\mathfrak{k},H)$ structure $R \otimes \mathrm{Ad}$ on the left side to R on the right side.

Proof. The chain of vector space isomorphisms

$$\begin{aligned} R(\mathfrak{k},H) \otimes_{\mathbb{C}} S(\mathfrak{p}) &\cong (C_H \otimes_{U(\mathfrak{h})} U(\mathfrak{k})) \otimes_{\mathbb{C}} S(\mathfrak{p}) \\ &\cong C_H \otimes_{U(\mathfrak{h})} (U(\mathfrak{k}) \otimes_{\mathbb{C}} S(\mathfrak{p})) \\ &\cong C_H \otimes_{U(\mathfrak{h})} U(\mathfrak{g}) \cong R(\mathfrak{g},H) \end{aligned}$$

shows that our map is one-one onto. Clearly our map respects the action of H under $L \otimes 1$ and the action of $\mathfrak{k}$ under $R \otimes \mathrm{ad}$. It respects the action of $\mathfrak{k}$ under $L \otimes 1$ because Y in $\mathfrak{k}$ implies

$$\begin{aligned}(L\otimes 1)(Y)((c\otimes u)\otimes p) &= (c\otimes(\mathrm{Ad}(\cdot)^{-1}Y)u)\otimes p \\ &\to c\otimes(\mathrm{Ad}(\cdot)^{-1}Y)u\mathfrak{s}(p) \\ &= (L\otimes 1)(Y)(c\otimes u\mathfrak{s}(p)).\end{aligned}$$

It respects the action of H under $R\otimes \mathrm{Ad}$ because h_0 in H implies

$$\begin{aligned}(R\otimes \mathrm{Ad})(h_0)((c\otimes u)\otimes p) &= R(h_0)(\mathrm{Ad}(\cdot)u\otimes c)\otimes \mathrm{Ad}(h_0)p \\ &= (R(h_0)c\otimes \mathrm{Ad}(h_0)u)\otimes \mathrm{Ad}(h_0)p \\ &\to R(h_0)c\otimes(\mathrm{Ad}(h_0)u)\mathfrak{s}(\mathrm{Ad}(h_0)p) \\ &= R(h_0)c\otimes(\mathrm{Ad}(h_0)u)\mathrm{Ad}(h_0)\mathfrak{s}(p) \\ &= (R\otimes \mathrm{Ad})(h_0)(c\otimes u\mathfrak{s}(p)).\end{aligned}$$

Proposition 7.14. Let $(\mathfrak{g}, K)$ and $(\mathfrak{g}, H)$ be two pairs with $K \supseteq H$ compatibly. Then

$$C_K \otimes_{R(\mathfrak{k},H)} R(\mathfrak{g},H) \cong R(\mathfrak{g},K) \tag{7.44}$$

under convolution: $c_1 \otimes (c_2 \otimes u) \to (c_1 *_H c_2)\otimes u$. The isomorphism carries the left K action to the left regular representation of K on $R(\mathfrak{g},K)$, and it carries the $R(\mathfrak{g},H)$ module structure $1\otimes R$ to the right regular representation on $R(\mathfrak{g},K)$.

Proof. Proposition 7.13 gives

$$C_K \otimes_{R(\mathfrak{k},H)} R(\mathfrak{g},H) \cong C_K \otimes_{R(\mathfrak{k},H)} (R(\mathfrak{k},H) \otimes_{\mathbb{C}} S(\mathfrak{p}))$$

with the left $\mathfrak{k}$ action on $R(\mathfrak{g},H)$ going into $L \otimes 1$. The right side here by (2.51) is

$$\begin{aligned} &\cong (C_K \otimes_{R(\mathfrak{k},H)} R(\mathfrak{k},H)) \otimes_{\mathbb{C}} S(\mathfrak{p}) \\ &\cong C_K \otimes_{\mathbb{C}} S(\mathfrak{p}) \qquad \text{by Proposition 7.11} \\ &\cong R(\mathfrak{k},K) \otimes_{\mathbb{C}} S(\mathfrak{p}) \\ &\cong R(\mathfrak{g},K) \qquad \text{by Proposition 7.13.} \end{aligned}$$

We readily check that the actions correspond as indicated.

6. Projectives and injectives

We saw in Proposition 7.10 that P and I are one-sided exact. The full analog of Proposition 6.2a is available when the compact group does not change.

Proposition 7.15. Let $(\mathfrak{g},H)$ and $(\mathfrak{q},H)$ be pairs with $\mathfrak{g} \supseteq \mathfrak{q}$ compatibly. Then the functors $P^{\mathfrak{g},H}_{\mathfrak{q},H}$ and $I^{\mathfrak{g},H}_{\mathfrak{q},H}$ are exact.

Proof. Let $\mathfrak{p}'$ be a vector space complement to $\mathfrak{q}$ in $\mathfrak{g}$. Then $U(\mathfrak{g}) \cong S(\mathfrak{p}') \otimes_{\mathbb{C}} U(\mathfrak{q})$ as a right $\mathfrak{q}$ module (with trivial action on $S(\mathfrak{p}')$), and we have vector space isomorphisms

$$\begin{aligned} P^{\mathfrak{g},H}_{\mathfrak{q},H}(V) &= R(\mathfrak{g},H) \otimes_{R(\mathfrak{q},H)} V \cong U(\mathfrak{g}) \otimes_{U(\mathfrak{h})} C_H \otimes_{R(\mathfrak{q},H)} V \\ &\cong S(\mathfrak{p}') \otimes_{\mathbb{C}} U(\mathfrak{q}) \otimes_{U(\mathfrak{h})} C_H \otimes_{R(\mathfrak{q},H)} V \end{aligned}$$

$$\cong S(\mathfrak{p}') \otimes_{\mathbb{C}} R(\mathfrak{q},H) \otimes_{R(\mathfrak{q},H)} V \cong S(\mathfrak{p}') \otimes_{\mathbb{C}} V,$$

the last step holding by Proposition 7.11. Thus $P_{\mathfrak{q},H}^{\mathfrak{g},H}$ is exact.

For $I_{\mathfrak{q},H}^{\mathfrak{g},H}$, the same isomorphism $R(\mathfrak{g},H) \cong S(\mathfrak{p}') \otimes_{\mathbb{C}} R(\mathfrak{q},H)$ as above gives a vector space isomorphism

$$\begin{aligned} I_{\mathfrak{q},H}^{\mathfrak{g},H}(V) &= \mathrm{Hom}_{R(\mathfrak{q},H)}(R(\mathfrak{g},H),V)_H \cong \mathrm{Hom}_{R(\mathfrak{q},H)}(S(\mathfrak{p}') \otimes_{\mathbb{C}} R(\mathfrak{q},H),V)_H \\ &\cong \mathrm{Hom}_{\mathbb{C}}(S(\mathfrak{p}'),\mathrm{Hom}_{R(\mathfrak{q},H)}(R(\mathfrak{q},H),V)_H)_H \quad \text{by (2.49)} \\ &\cong \mathrm{Hom}_{\mathbb{C}}(S(\mathfrak{p}'),V)_H \quad \text{by Proposition 7.12.} \end{aligned}$$

Thus $I_{\mathfrak{q},H}^{\mathfrak{g},H}$ is exact.

Application of P or I in stages works just as in Proposition 6.2b.

<u>Proposition 7.16</u>. Let $(\mathfrak{g},K)$, $(\mathfrak{q},H)$, and $(\mathfrak{q}_1,B)$ be three pairs with $\mathfrak{g} \supseteq \mathfrak{q} \supseteq \mathfrak{q}_1$ and $K \supseteq H \supseteq B$ compatibly. For V in $\mathcal{C}(\mathfrak{q}_1,B)$, there are natural isomorphisms

$$P_{\mathfrak{q},H}^{\mathfrak{g},K} \circ P_{\mathfrak{q}_1,B}^{\mathfrak{q},H}(V) \cong P_{\mathfrak{q}_1,B}^{\mathfrak{g},K}(V)$$

and

$$I_{\mathfrak{q},H}^{\mathfrak{g},K} \circ I_{\mathfrak{q}_1,B}^{\mathfrak{q},H}(V) \cong I_{\mathfrak{q}_1,B}^{\mathfrak{g},K}(V).$$

Proof. Formula (2.51) and Proposition 7.11 give

$$\begin{aligned} P_{\mathfrak{q},H}^{\mathfrak{g},K} \circ P_{\mathfrak{q}_1,B}^{\mathfrak{q},H}(V) &= R(\mathfrak{g},K) \otimes_{R(\mathfrak{q},H)} R(\mathfrak{q},H) \otimes_{R(\mathfrak{q}_1,B)} V \\ &\cong R(\mathfrak{g},K) \otimes_{R(\mathfrak{q}_1,B)} V = P_{\mathfrak{q}_1,B}^{\mathfrak{g},K}(V), \end{aligned}$$

while (2.49) and Proposition 7.11 give

$$I^{\mathfrak{g},K}_{\mathfrak{q},H}\circ I^{\mathfrak{q},H}_{\mathfrak{q}_1,B}(V) = \mathrm{Hom}_{R(\mathfrak{q},H)}(R(\mathfrak{g},K),\mathrm{Hom}_{R(\mathfrak{q}_1,B)}(R(\mathfrak{q},H),V)_H)_K$$

$$= \mathrm{Hom}_{R(\mathfrak{q},H)}(R(\mathfrak{g},K),\mathrm{Hom}_{R(\mathfrak{q}_1,B)}(R(\mathfrak{q},H),V))_K$$

$$\cong \mathrm{Hom}_{R(\mathfrak{q}_1,B)}(R(\mathfrak{g},K)\otimes_{R(\mathfrak{q},H)} R(\mathfrak{q},H),V)_K$$

$$\cong \mathrm{Hom}_{R(\mathfrak{q}_1,B)}(R(\mathfrak{g},K),V)_K = I^{\mathfrak{g},K}_{\mathfrak{q}_1,B}(V)\ .$$

Proposition 6.3 dealt with Frobenius reciprocity, showing that $P^{\mathfrak{g}}_{\mathfrak{q}}$ and $I^{\mathfrak{g}}_{\mathfrak{q}}$ are adjoint to forgetful functors. The situation is a little more complicated when the compact groups are in place.

Let $(\mathfrak{g},K)$ and $(\mathfrak{q},H)$ denote two pairs with $\mathfrak{g}\supseteq\mathfrak{q}$ and $K\supseteq H$ compatibly. If V is a $(\mathfrak{g},K)$ module, then we can regard V as a $(\mathfrak{q},H)$ module by restricting the action. We denote the resulting forgetful functor $\mathcal{F}$ by

$$\mathcal{F}^{\mathfrak{q},H}_{\mathfrak{g},K} : \mathcal{C}(\mathfrak{g},K)\to\mathcal{C}(\mathfrak{q},H)\ .$$

Clearly $\mathcal{F}$ is covariant and exact.

In the context of $R(\mathfrak{g},K)$ and $R(\mathfrak{q},H)$, the definition of $\mathcal{F}$ is less transparent, since $R(\mathfrak{q},H)$ need not be a subring of $R(\mathfrak{g},K)$. But we can define the action this way: Let $q\in\mathfrak{q}$ and $h\in H$. If $x\in V$ is given, choose e_F in $R(\mathfrak{g},K)$ with $e_F x = x$. Then qx and hx are given as $(qe_F)x$ and $(he_F)x$, where qe_F and he_F are computed with $R(\mathfrak{g},K)$ considered as a left $R(\mathfrak{q},H)$ module. (See (7.35).)

Let W be a $(\mathfrak{g},K)$ module. We define a pseudo-forgetful functor $\mathcal{F}^\vee$ with

$$(\mathcal{F}^\vee)^{\mathfrak{q},H}_{\mathfrak{g},K} : \mathcal{C}(\mathfrak{g},K)\to\mathcal{C}(\mathfrak{q},H)$$

by

$$\mathcal{F}^\vee(W) = \mathrm{Hom}_{R(\mathfrak{g},K)}(R(\mathfrak{g},K),W)_H .$$

The fact that $\mathcal{F}^\vee(W)$ is a $(\mathfrak{q},H)$ module and $\mathcal{F}^\vee$ is a functor is proved by the same arguments as in the second half of Proposition 7.10.

<u>Lemma 7.17</u>. The covariant functor $(\mathcal{F}^\vee)^{\mathfrak{q},H}_{\mathfrak{g},K}$ is exact.

Proof. Let

$$0 \longrightarrow W' \xrightarrow{\ \psi\ } W \xrightarrow{\ \varphi\ } W'' \longrightarrow 0$$

be an exact sequence in $\mathcal{C}(\mathfrak{g},K)$. If η is in $\mathcal{F}^\vee(W')$ and $\psi\circ\eta = 0$, then $\psi(\eta(r)) = 0$ for all r, and $\eta(r) = 0$ for all r. Hence $\eta = 0$. Thus $\mathcal{F}^\vee(\psi)$ is one-one.

If η is in $\mathcal{F}^\vee(W)$ and $\varphi\circ\eta = 0$, then $\varphi(\eta(r)) = 0$ for all r, $\eta(r)$ is in $\ker\varphi = \text{image}\ \psi$, and $\eta(r) = \psi(w')$. Define $\xi(r) =$ this w', necessarily unique. Then ξ is in $\mathrm{Hom}_{R(\mathfrak{g},K)}(R(\mathfrak{g},K),W')$. Since η vanishes off finitely many H types of $R(\mathfrak{g},K)$, so does ξ. Thus ξ is in $\mathcal{F}^\vee(W')$. Then $\psi(\xi(r)) = \psi(w') = \eta(r)$ shows $\ker\mathcal{F}^\vee(\varphi) = \text{image}\ \mathcal{F}^\vee(\psi)$.

Finally let η be in $\mathcal{F}^\vee(W'')$. We seek ξ in $\mathcal{F}^\vee(W)$ with $\varphi\circ\xi = \eta$. For each irreducible representation τ of K, the element $\eta(e_{\{\tau\}})$ is in W'' and is thus of the form $\varphi(w_{\{\tau\}})$, $w_{\{\tau\}}\in W$. Now $R(\mathfrak{g},K)$ is a direct sum of left ideals $R(\mathfrak{g},K) = \Sigma_\tau\ R(\mathfrak{g},K)e_{\{\tau\}}$, as is apparent from (7.21c). Thus we can define $\xi_0 \in \mathrm{Hom}_{R(\mathfrak{g},K)}(R(\mathfrak{g},K),W)$ consistently by

$$\xi_0(\Sigma\ r_\tau e_{\{\tau\}}) = \Sigma\ r_\tau w_{\{\tau\}} .$$

Then

$$\varphi(\xi_0(\Sigma\, r_\tau e_{\{\tau\}})) = \varphi\ (\Sigma\, r_\tau w_{\{\tau\}}) = \Sigma\, r_\tau \varphi(w_{\{\tau\}})$$
$$= \Sigma\, r_\tau \eta(e_{\{\tau\}}) = \eta(\Sigma\, r_\tau e_{\{\tau\}})$$

says $\varphi \circ \xi_0 = \eta$. Now $\eta = 0$ except on finitely many H types of $R(\mathfrak{g},K)$ under the right regular representation. Thus let us define ξ to be ξ_0 on the finitely many H types and to be 0 on the remaining H types. Then ξ respects the left action of $R(\mathfrak{g},K)$, is in $\mathcal{F}^\vee(W)$, and has $\varphi\circ\xi = \eta$. Hence ξ has the required properties.

Proposition 7.18 (Frobenius reciprocity). Let $(\mathfrak{g},K)$ and $(\mathfrak{q},H)$ be two pairs with $\mathfrak{g} \supseteq \mathfrak{q}$ and $K \supseteq H$ compatibly. Then there are natural isomorphisms

$$\operatorname{Hom}_{R(\mathfrak{g},K)}(P_{\mathfrak{q},H}^{\mathfrak{g},K}(V),W) \cong \operatorname{Hom}_{R(\mathfrak{q},H)}(V,(\mathcal{F}^\vee)_{\mathfrak{g},K}^{\mathfrak{q},H}(W)) \qquad (7.45)$$

$$\operatorname{Hom}_{R(\mathfrak{g},K)}(W,I_{\mathfrak{q},H}^{\mathfrak{g},K}(V)) \cong \operatorname{Hom}_{R(\mathfrak{q},H)}(\mathcal{F}_{\mathfrak{g},K}^{\mathfrak{q},H}(W),V) \qquad (7.46)$$

for V in $\mathcal{C}(\mathfrak{q},H)$ and W in $\mathcal{C}(\mathfrak{g},K)$.

Proof of (7.45).

$$\operatorname{Hom}_{R(\mathfrak{g},K)}(P_{\mathfrak{q},H}^{\mathfrak{g},K}(V),W) = \operatorname{Hom}_{R(\mathfrak{g},K)}(R(\mathfrak{g},K)\otimes_{R(\mathfrak{q},H)} V,W)$$
$$\cong \operatorname{Hom}_{R(\mathfrak{q},H)}(V,\operatorname{Hom}_{R(\mathfrak{g},K)}(R(\mathfrak{g},K),W)), \quad \text{by (2.50)}$$
$$= \operatorname{Hom}_{R(\mathfrak{q},H)}(V,\operatorname{Hom}_{R(\mathfrak{g},K)}(R(\mathfrak{g},K),W)_H)$$
$$= \operatorname{Hom}_{R(\mathfrak{q},H)}(V,(\mathcal{F}^\vee)_{\mathfrak{g},K}^{\mathfrak{q},H}(W)).$$

Proof of (7.46).

$$\begin{aligned}\mathrm{Hom}_{R(\mathfrak{g},K)}(W, I_{\mathfrak{q},H}^{\mathfrak{g},K}(V)) &= \mathrm{Hom}_{R(\mathfrak{g},K)}(W, \mathrm{Hom}_{R(\mathfrak{q},H)}(R(\mathfrak{g},K),V)_K) \\ &= \mathrm{Hom}_{R(\mathfrak{g},K)}(W, \mathrm{Hom}_{R(\mathfrak{q},H)}(R(\mathfrak{g},K),V)) \\ &\cong \mathrm{Hom}_{R(\mathfrak{q},H)}(R(\mathfrak{g},K) \otimes_{R(\mathfrak{g},K)} W, V) \quad \text{by (2.50)} \\ &\cong \mathrm{Hom}_{R(\mathfrak{q},H)}(\mathcal{F}_{\mathfrak{g},K}^{\mathfrak{q},H}(W),V) \end{aligned}$$

by Proposition 7.11.

It is worth tracking down the explicit formulas for (7.45) and (7.46). If we write (7.45) as $\Phi \longleftrightarrow \Phi'$, then the relationship is

$$\Phi(r \otimes v) = \Phi'(v)(r) . \tag{7.47}$$

If we write (7.46) as $\Phi \longleftrightarrow \Phi'$, then the relationship is

$$\Phi(w)(r) = \Phi'(rw) . \tag{7.48}$$

Proposition 7.18 allows us to give an alternate formula for $I_{\mathfrak{g},H}^{\mathfrak{g},K}$ when $K \supseteq H$ compatibly and K is connected. Namely for V in $\mathcal{C}(\mathfrak{g},H)$, let $\Gamma(V)$ be the subspace of vectors lying in a finite-dimensional space where the action of $\mathfrak{k}$ exponentiates to K compatibly with that on H. It is easy to see that $\Gamma(V)$ is a $(\mathfrak{g},K)$ module. If W is in $\mathcal{C}(\mathfrak{g},K)$ and φ is a $(\mathfrak{g},H)$ map into V, then we shall check below that the image of φ lies in $\Gamma(V)$. Consequently

$$\mathrm{Hom}_{R(\mathfrak{g},K)}(W,\Gamma(V)) \cong \mathrm{Hom}_{R(\mathfrak{g},H)}(\mathcal{F}_{\mathfrak{g},K}^{\mathfrak{g},H}(W),V)$$

naturally in V and W. Combining (7.46) and Proposition 5.14, we see that $I_{\mathfrak{g},H}^{\mathfrak{g},K}(V) \cong \Gamma(V)$ naturally in V. In the situations of our two examples in §1, what we showed is that $\Gamma(V)$ corresponds in each case to the representation of interest; what we now see is that the representations of interest can be reinterpreted as coming from suitable functors $I_{\mathfrak{g},H}^{\mathfrak{g},K}$.

Let us verify the assertion above that if $\varphi : W \to V$ is a $(\mathfrak{g},H)$ map and W is a $(\mathfrak{g},K)$ module, then image $\varphi \subseteq \Gamma(V)$. Let ρ be the representation of $(\mathfrak{k},K)$ on W, and let ρ' be the representation of $(\mathfrak{k},H)$ on V. We are to show on image φ that ρ' extends to a representation of $(\mathfrak{k},K)$. If $v = \varphi(w)$ is given, we define

$$\rho'(k)v = \varphi(\rho(k)w) \quad \text{for } k \in K.$$

To see that $\rho'(k)$ is well defined, let $v = \varphi(w')$ also. We are to check that $\varphi(\rho(k)w') = \varphi(\rho(k)w)$. That is, we want $\varphi(w-w') = 0$ to imply $\varphi(\rho(k)(w-w')) = 0$. Thus we are to check that $\ker \varphi$ is K-invariant. Now $\ker \varphi$ is the direct sum of finite-dimensional $\mathfrak{k}$-invariant subspaces, and each of these is K-invariant; hence $\ker \varphi$ is K-invariant. We conclude that $\rho'(k)$ is well defined. Then it is a simple exercise to check that $\rho'|_K$ is a representation, that $\rho'|_K$ differentiates to give the original $\rho'|_{\mathfrak{k}}$, and that $\rho'|_K$ restricts to give the original $\rho'|_H$. Hence image $\varphi \subseteq \Gamma(V)$ as asserted.

Example. Let $(\mathfrak{k},K)$ be obtained from the circle group $\{e^{i\theta}\}$. Then we can arrange for a two-dimensional representation of $\mathfrak{k} \cong \mathbb{C}$ on $\mathbb{C}^2$ that does not split into the sum of one-dimensional representations such that $\mathbb{C}^2$ has a one-dimensional subrepresentation for K and a one-dimensional quotient for K. Then the sequence

$$0 \longrightarrow \mathbb{C} \longrightarrow \mathbb{C}^2 \longrightarrow \mathbb{C} \longrightarrow 0$$

in $\mathcal{C}(\mathfrak{t},\{1\})$ is exact. Applying Γ as above, we obtain the complex

$$0 \longrightarrow \mathbb{C} \longrightarrow \mathbb{C} \longrightarrow \mathbb{C} \longrightarrow 0,$$

in $\mathcal{C}(\mathfrak{t},K)$, and this is not exact. Therefore $I_{\mathfrak{t},\{1\}}^{\mathfrak{t},K}$ is not an exact functor.

Corollary 7.19. If $(\mathfrak{g},K)$ and $(\mathfrak{q},H)$ are pairs with $\mathfrak{g} \supseteq \mathfrak{q}$ and $K \supseteq H$ compatibly, then $P_{\mathfrak{q},H}^{\mathfrak{g},K}$ carries projectives to projectives and $I_{\mathfrak{q},H}^{\mathfrak{g},K}$ carries injectives to injectives.

Proof. If P is projective in $\mathcal{C}(\mathfrak{q},H)$, then (7.45) says that

$$W \to \operatorname{Hom}_{R(\mathfrak{g},K)}(P_{\mathfrak{q},H}^{\mathfrak{g},K}(P), W)$$

is equivalent with

$$W \to \operatorname{Hom}_{R(\mathfrak{q},H)}(P, (\mathcal{F}^\vee)_{\mathfrak{g},K}^{\mathfrak{q},H}(W)),$$

which is exact, being the composition of the exact functors $(\mathcal{F}^\vee)_{\mathfrak{g},K}^{\mathfrak{q},H}$ and $\operatorname{Hom}_{R(\mathfrak{q},H)}(P,\cdot)$. (See Lemma 7.17.) Hence $P_{\mathfrak{q},H}^{\mathfrak{g},K}(P)$ is projective.

If I is injective in $\mathcal{C}(\mathfrak{q},H)$, then (7.46) says that

$$W \to \operatorname{Hom}_{R(\mathfrak{g},K)}(W, I_{\mathfrak{q},H}^{\mathfrak{g},K}(I))$$

is equivalent with

$$W \to \mathrm{Hom}_{R(\mathfrak{q},H)}(\mathfrak{F}^{\mathfrak{q},H}_{\mathfrak{g},K}(W), I),$$

which is exact, being the composition of the exact functors $\mathfrak{F}^{\mathfrak{q},H}_{\mathfrak{g},K}$ and $\mathrm{Hom}_{R(\mathfrak{q},H)}(\cdot, I)$. Hence $I^{\mathfrak{g},K}_{\mathfrak{q},H}(I)$ is injective.

Corollary 7.20. If $(\mathfrak{g},H)$ is a pair, if $\mathfrak{h}$ is the complexified Lie algebra of H, and if V is in $\mathcal{C}(\mathfrak{h},H)$, then $P^{\mathfrak{g},H}_{\mathfrak{h},H}(V)$ is projective and $I^{\mathfrak{g},H}_{\mathfrak{h},H}$ is injective.

Proof. We combine Lemma 7.8 and Corollary 7.19.

Remark. The projectives and injectives in this corollary are called standard projectives and standard injectives in the category $\mathcal{C}(\mathfrak{g},H)$.

Corollary 7.21. Let $(\mathfrak{g},H)$ be a pair, and let W be in $\mathcal{C}(\mathfrak{g},H)$.

(a) Then W is the quotient of a standard projective, and W is a submodule of a standard injective. Consequently the category $\mathcal{C}(\mathfrak{g},H)$ has enough projectives and enough injectives.

(b) If W is projective, then W is a direct summand of a standard projective. If W is injective, then W is a direct summand of a standard injective.

Proof for projectives. Let P be the standard projective $P = P^{\mathfrak{g},H}_{\mathfrak{h},H}\mathfrak{F}^{\mathfrak{h},H}_{\mathfrak{g},H}(W)$. By (7.45) and Proposition 7.12 we have

$$\mathrm{Hom}_{\mathfrak{g},H}(P,W) \cong \mathrm{Hom}_{\mathfrak{h},H}(\mathfrak{F}^{\mathfrak{h},H}_{\mathfrak{g},H}W, (\mathfrak{F}^{\vee})^{\mathfrak{h},H}_{\mathfrak{g},H}(W)) \cong \mathrm{Hom}_{\mathfrak{h},H}(\mathfrak{F}^{\mathfrak{h},H}_{\mathfrak{g},H}W, \mathfrak{F}^{\mathfrak{h},H}_{\mathfrak{g},H}(W)).$$

Moreover (7.47) shows that if Φ, Φ', and Φ'' are corresponding members of these Hom's, then

$$\Phi(r \otimes w) = \Phi'(w)(r) = r\Phi''(w).$$

Let φ be the member of $\mathrm{Hom}_{\mathfrak{g},H}(P,W)$ that corresponds to $\Phi'' = 1$. Then $\varphi(r \otimes w) = rw$, and $\varphi : P \to W$ is onto since W is unital. This proves (a) for projectives. For (b) we use P obtained from (a) and set up the exact sequence

$$0 \longrightarrow Q \longrightarrow P \longrightarrow W \longrightarrow 0$$

for some Q. Since W is projective, Lemma 5.22 says this sequence splits. This proves (b) for projectives.

Proof for injectives. Let I be the standard injective $I = I_{\mathfrak{h},H}^{\mathfrak{g},H}\mathcal{F}_{\mathfrak{g},H}^{\mathfrak{h},H}(W)$. By (7.46) we have

$$\mathrm{Hom}_{\mathfrak{g},H}(W,I) \cong \mathrm{Hom}_{\mathfrak{h},H}(\mathcal{F}_{\mathfrak{g},H}^{\mathfrak{h},H}W, \mathcal{F}_{\mathfrak{g},H}^{\mathfrak{h},H}W).$$

Moreover (7.48) shows that if Φ and Φ' are corresponding members of these Hom's, then

$$\Phi(w)(r) = \Phi'(rw).$$

Let φ be the member of $\mathrm{Hom}_{\mathfrak{g},H}(W,I)$ that corresponds to $\Phi' = 1$. Then $\varphi(w)(r) = rw$. If $\varphi(w) = 0$, then $w = 0$ since W is unital. Thus $\varphi : W \to I$ is one-one. This proves (a) for injectives. For (b) we use I obtained from (a) and set up the exact sequence

$$0 \longrightarrow W \longrightarrow I \longrightarrow I/W \longrightarrow 0.$$

Since W is injective, Lemma 5.22 says this sequence splits. This proves (b) for injectives.

There is a certain asymmetry in Proposition 7.18 because $I_{\mathfrak{q},H}^{\mathfrak{g},K}$ is adjoint to the genuine forgetful functor $\mathcal{F}_{\mathfrak{g},K}^{\mathfrak{q},H}$ whereas $P_{\mathfrak{q},H}^{\mathfrak{g},K}$ is adjoint merely to the pseudo-forgetful functor $(\mathcal{F}^{\vee})_{\mathfrak{g},K}^{\mathfrak{q},H}$. There are two notable cases where $\mathcal{F}^{\vee}$ reduces to a forgetful functor. One is that $H = K$; then $\mathcal{F}^{\vee}$ is naturally isomorphic to $\mathcal{F}$ as a consequence of Proposition 7.12. The other is the subject of the next proposition.

<u>Proposition 7.22</u>. Let $(\mathfrak{k},K)$ and $(\mathfrak{k},H)$ be two pairs with $K \supseteq H$ compatibly. Then

$$(\mathcal{F}^{\vee})_{\mathfrak{k},K}^{\mathfrak{k},H}(W) \cong W$$

for every finite-dimensional $(\mathfrak{k},K)$ module W, and the isomorphism is natural on the good category of finite-dimensional $(\mathfrak{k},K)$ modules.

Proof. We can write

$$(\mathcal{F}^{\vee})_{\mathfrak{k},K}^{\mathfrak{k},H}(W) = \mathrm{Hom}_{C_K}(C_K, W)_H .$$

Since W is finite-dimensional, we can choose a member χ_F of the approximate identity in C_K such that $\chi_F w = w$ for all w in W. Define $\Phi : \mathrm{Hom}_{C_K}(C_K, W)_H \to W$ by $\Phi(\varphi) = \varphi(\chi_F)$. Then Φ is C_K-linear because

$$\Phi(c\varphi) = (c\varphi)(\chi_F) = \varphi(\chi_F * c) = \varphi(c * \chi_F) = c(\varphi(\chi_F)) = c(\Phi(\varphi)) .$$

The map Φ is onto because we can take any w in W, define φ in $\mathcal{F}^{\vee}(W)$ by $\varphi(c) = cw$, and get

$$\Phi(\varphi) = \varphi(\chi_F) = \chi_F w = w .$$

And Φ is one-one because $\Phi(\varphi) = 0$ for some φ forces $\varphi(\chi_F) = 0$ and then

$$\varphi(c) = \chi_F(\varphi(c)) = \varphi(\chi_F * c) = \varphi(c * \chi_F) = c(\varphi(\chi_F)) = 0$$

for all c.

7. Homology, cohomology, and Ext

Fix a pair $(\mathfrak{g}, K)$. The functor $V \to V \otimes_{R(\mathfrak{g},K)} \mathbb{C}$ is covariant and right exact from $\mathcal{C}(\mathfrak{g}, K)$ into the category $\mathcal{C}(0,\{1\})$ of all vector spaces. Its derived functors, defined by a projective resolution of V, are defined to be the relative homology functors $V \to H_n(\mathfrak{g}, K; V)$.

Similarly, for V in $\mathcal{C}(\mathfrak{g}, K)$, let

$$V^{R(\mathfrak{g},K)} \cong \mathrm{Hom}_{R(\mathfrak{g},K)}(\mathbb{C}, V)$$

be the subspace of $R(\mathfrak{g}, K)$ invariants; these are the $\mathfrak{g}$ invariants that are fixed by K. The functor $V \to V^{R(\mathfrak{g},K)}$ is covariant and left exact from $\mathcal{C}(\mathfrak{g}, K)$ into $\mathcal{C}(0,\{1\})$. Its derived functors, defined by an injective resolution of V, are defined to be the relative cohomology functors $V \to H^n(\mathfrak{g}, K; V)$.

Since $V \to V \otimes_{R(\mathfrak{g},K)} \mathbb{C}$ is right exact, we have a natural isomorphism

$$H_0(\mathfrak{g}, K; V) \cong V \otimes_{R(\mathfrak{g},K)} \mathbb{C}. \tag{7.49}$$

Since $V \to V^{R(\mathfrak{g},K)}$ is left exact, we have a natural isomorphism

$$H^0(\mathfrak{g},K;V) \cong V^{R(\mathfrak{g},K)} . \tag{7.50}$$

As in §VI.2, we can generalize cohomology to an Ext functor. The cohomology functors are the derived functors of $\mathrm{Hom}_{\mathfrak{g},K}(\mathbb{C},V)$. We define $\mathrm{Ext}^n_{\mathfrak{g},K}(U,\cdot)$ to be the nth derived functor of $\mathrm{Hom}_{\mathfrak{g},K}(U,\cdot)$ on the category $\mathcal{C}(\mathfrak{g},K)$. Here U is in $\mathcal{C}(\mathfrak{g},K)$, and the derived functors are computed from an injective resolution.

Proposition 7.23. $\mathrm{Ext}^n_{\mathfrak{g},K}(\cdot,V)$ is the nth derived functor of $\mathrm{Hom}_{\mathfrak{g},K}(\cdot,V)$, and also $\mathrm{Ext}^n_{\mathfrak{g},K}(U,V)$ is isomorphic to $H^n(\mathfrak{g},K;\mathrm{Hom}_{\mathbb{C}}(U,V)_K)$ naturally for U and V in $\mathcal{C}(\mathfrak{g},K)$.

Proof. The argument is the same as for Proposition 6.8; the three versions of Ext, one in the definition and two in the proposition, are calculated by a process that eventually involves taking the invariants of an injective resolution of $\mathrm{Hom}_{\mathbb{C}}(U,V)_K$.

Theorem 7.24 (Shapiro's Lemma). If $(\mathfrak{g},H)$ and $(\mathfrak{q},H)$ are pairs with $\mathfrak{g} \supseteq \mathfrak{q}$ compatibly, then

$$\mathrm{Ext}^n_{\mathfrak{g},H}(P^{\mathfrak{g},H}_{\mathfrak{q},H}(V),W) \cong \mathrm{Ext}^n_{\mathfrak{q},H}(V,\mathcal{F}^{\mathfrak{q},H}_{\mathfrak{g},H}(W)) \tag{7.51}$$

$$\mathrm{Ext}^n_{\mathfrak{g},H}(W,I^{\mathfrak{g},H}_{\mathfrak{q},H}(V)) \cong \mathrm{Ext}^n_{\mathfrak{q},H}(\mathcal{F}^{\mathfrak{q},H}_{\mathfrak{g},H}(W),V) \tag{7.52}$$

naturally for V in $\mathcal{C}(\mathfrak{q},H)$ and W in $\mathcal{C}(\mathfrak{g},H)$.

Remark. Notice that the two compact groups are the same in the statement of the theorem.

Proof. Since the two compact groups are the same, Proposition 7.12 shows that Frobenius reciprocity (7.45) and (7.46) reduces to

$$\operatorname{Hom}_{\mathfrak{g},H}(P^{\mathfrak{g},H}_{\mathfrak{q},H}(V),W) \cong \operatorname{Hom}_{\mathfrak{q},H}(V,\mathcal{F}^{\mathfrak{q},H}_{\mathfrak{g},H}(W))$$

$$\operatorname{Hom}_{\mathfrak{g},H}(W,I^{\mathfrak{g},H}_{\mathfrak{q},H}(V)) \cong \operatorname{Hom}_{\mathfrak{q},H}(\mathcal{F}^{\mathfrak{q},H}_{\mathfrak{g},H}(W),V) .$$

Then the proof proceeds as in Theorem 6.9, using the exactness of $P^{\mathfrak{g},H}_{\mathfrak{q},H}$ and $I^{\mathfrak{g},H}_{\mathfrak{q},H}$ proved in Proposition 7.15.

8. Standard resolutions

Fix a pair $(\mathfrak{g},K)$. Then $\operatorname{Ad}(K)$ operates on $\mathfrak{g}/\mathfrak{k}$ and hence on $\Lambda^n(\mathfrak{g}/\mathfrak{k})$. Thus $\Lambda^n(\mathfrak{g}/\mathfrak{k})$ becomes a $(\mathfrak{k},K)$ module. By Corollary 7.20,

$$X_n = P^{\mathfrak{g},K}_{\mathfrak{k},K}(\Lambda^n(\mathfrak{g}/\mathfrak{k})) = R(\mathfrak{g},K)\otimes_{C_K}\Lambda^n(\mathfrak{g}/\mathfrak{k}) \qquad (7.53a)$$

is a projective in $\mathcal{C}(\mathfrak{g},K)$. If $\mathfrak{p}$ denotes an $\operatorname{Ad}(K)$-invariant complement to $\mathfrak{k}$ in $\mathfrak{g}$, then we can write X_n also as

$$X_n = R(\mathfrak{g},K)\otimes_{C_K}\Lambda^n\mathfrak{p} . \qquad (7.53b)$$

Since $R(\mathfrak{g},K)$ is more difficult than $U(\mathfrak{g})$ to use in calculations, it will be helpful to have a realization of X_n that involves $U(\mathfrak{g})$ in place of $R(\mathfrak{g},K)$. In fact, we have a $U(\mathfrak{g})$ isomorphism

$$\begin{aligned} R(\mathfrak{g},K)\otimes_{C_K}\Lambda^n\mathfrak{p} &\cong (U(\mathfrak{g})\otimes_{\mathfrak{k}} C_K)\otimes_{C_K}\Lambda^n\mathfrak{p} \\ &\cong U(\mathfrak{g})\otimes_{\mathfrak{k}}(C_K\otimes_{C_K}\Lambda^n\mathfrak{p}) \cong U(\mathfrak{g})\otimes_{\mathfrak{k}}\Lambda^n\mathfrak{p} . \end{aligned}$$

Thus we can write

$$X_n = U(\mathfrak{g}) \otimes_{\mathfrak{k}} \Lambda^n(\mathfrak{g}/\mathfrak{k}) \tag{7.53c}$$

or

$$X_n = U(\mathfrak{g}) \otimes_{\mathfrak{k}} \Lambda^n \mathfrak{p} \tag{7.53d}$$

as a $\mathfrak{g}$ module. We shall identify the action by K in these realizations.

Lemma 7.25. If $R(\mathfrak{g},K)$, which is a unital left $R(\mathfrak{g},K)$ module under the left regular action, is considered as a $(\mathfrak{g},K)$ module, then the action by $k \in K$ on $u \otimes c$ in $U(\mathfrak{g}) \otimes_{\mathfrak{k}} C_K$ is given by

$$L(k)(u \otimes c) = \mathrm{Ad}(k)u \otimes \ell(k)c .$$

Proof.
$$\begin{aligned} L(k)(u \otimes c) &= L(k)\Big(\sum_j c(\cdot)(\mathrm{Ad}(\cdot)^{-1}u,u_j) \otimes u_j\Big) \qquad \text{by (7.22b)} \\ &= \sum_j \ell(k)\{c(\cdot)(\mathrm{Ad}(\cdot)^{-1}u,u_j)\} \otimes u_j \\ &= \sum_j (\ell(k)c)(\cdot)(\mathrm{Ad}(k^{-1}\cdot)^{-1}u,u_j) \otimes u_j \\ &= \sum_j (\ell(k)c)(\cdot)(\mathrm{Ad}(\cdot)^{-1}\mathrm{Ad}(k)u,u_j) \otimes u_j \\ &= \mathrm{Ad}(k)u \otimes \ell(k)c \qquad \text{by (7.22b).} \end{aligned}$$

Lemma 7.26. In (7.53c) and (7.53d), the action of K on the $(\mathfrak{g},K)$ module X_n is given by $\mathrm{Ad} \otimes \mathrm{Ad}$.

Proof. Under the isomorphism that proves (7.53c) and (7.53d), we have

$$(u \otimes c) \otimes \omega \to u \otimes (c \otimes \omega) \to u \otimes c\omega . \tag{7.54}$$

Applying Lemma 7.25, we see that

$$k((u \otimes c) \otimes \omega) = L(k)(u \otimes c) \otimes \omega = (\mathrm{Ad}(k)u \otimes \ell(k)c) \otimes \omega$$

$$\to \mathrm{Ad}(k)u \otimes (\ell(k)c)\omega .$$

In (7.54) the elements $(u \otimes c) \otimes \omega$ span X_n even when c ranges only through the elements χ_F with F large. For sufficiently large F, the proof of Lemma 7.4 shows that $(\ell(k)\chi_F)\omega = \mathrm{Ad}(k)\omega$. For such χ_F, we have

$$k((u \otimes \chi_F) \otimes \omega) = \mathrm{Ad}(k)u \otimes \mathrm{Ad}(k)\omega ,$$

and thus the lemma follows.

Let us realize X_n as in (7.53c). The <u>Koszul resolution</u> in $\mathcal{C}(\mathfrak{g},K)$ is the sequence

$$0 \longleftarrow \mathbb{C} \overset{\epsilon}{\longleftarrow} X_0 \overset{\partial_0}{\longleftarrow} X_1 \overset{\partial_1}{\longleftarrow} X_2$$

$$\cdots \longleftarrow X_{\dim(\mathfrak{g}/\mathfrak{k})} \longleftarrow 0 \tag{7.55a}$$

with $\epsilon : U(\mathfrak{g}) \otimes_{\mathfrak{k}} \mathbb{C}$ given as

$$\epsilon = \text{projection to constant term of } U(\mathfrak{g}) \tag{7.55b}$$

and with $\partial_{n-1} : X_n \to X_{n-1}$ given by

$$\partial_{n-1}(u \otimes Y_1 \wedge \dots \wedge Y_n)$$

$$= \sum_{\ell=1}^{n} (-1)^{\ell+1}(u\widetilde{Y}_\ell \otimes Y_1 \wedge \dots \wedge \widehat{Y}_\ell \wedge \dots \wedge Y_n) \tag{7.55c}$$

$$+ \sum_{r<s} (-1)^{r+s}(u \otimes ([\widetilde{Y}_r,\widetilde{Y}_s]+\mathfrak{k}) \wedge Y_1 \wedge \dots \wedge \widehat{Y}_r \wedge \dots \wedge \widehat{Y}_s \wedge \dots \wedge Y_n) ,$$

where $Y_1, \ldots, Y_n$ are in $\mathfrak{g}/\mathfrak{k}$ and $\tilde{Y}_1, \ldots, \tilde{Y}_n$ are representatives in $\mathfrak{g}$.

<u>Lemma 7.27</u>. The maps ϵ and ∂_{n-1} are well defined, relative to the choice of the representatives $\tilde{Y}_1, \ldots, \tilde{Y}_n$ and the passage to the tensor product $\otimes_{\mathfrak{k}}$, and they are $(\mathfrak{g}, K)$ maps.

We omit the routine but lengthy proof. Once (7.55c) is well defined, we may want to make a particular choice of representatives. Let $\mathfrak{g} = \mathfrak{k} \oplus \mathfrak{p}$ as above, and let ρ be the projection of $\mathfrak{g}$ onto $\mathfrak{p}$ along $\mathfrak{k}$. If we realize X_n as in (7.53d), then the formula for ∂_{n-1} becomes

$$\partial_{n-1}(u \otimes Y_1 \wedge \ldots \wedge Y_n)$$
$$= \sum_{\ell=1}^{n} (-1)^{\ell+1}(uY_\ell \otimes Y_1 \wedge \ldots \wedge \hat{Y}_\ell \wedge \ldots \wedge Y_n)$$
$$+ \sum_{r<s} (-1)^{r+s}(u \otimes \rho[Y_r, Y_s] \wedge Y_1 \wedge \ldots \wedge \hat{Y}_r \wedge \ldots \wedge \hat{Y}_s \wedge \ldots \wedge Y_n), \tag{7.55d}$$

where $Y_1, \ldots, Y_n$ are in $\mathfrak{p}$.

<u>Lemma 7.28</u>. The sequence (7.55a) is a complex.

This is another routine but lengthy computation that we omit.

<u>Theorem 7.29</u>. The sequence (7.55a) is exact.

Proof. For $0 \leq n \leq k$, let $X_n^{(k)}$ be the image of $U^{k-n}(\mathfrak{g}) \otimes_{\mathbb{C}} \Lambda^n\mathfrak{p}$ in $U(\mathfrak{g}) \otimes_{\mathfrak{k}} \Lambda^n\mathfrak{p}$. The formulas (7.55b) and

(7.55d) for ϵ and ∂_{n-1} make sense on $U^{k-n}(\mathfrak{g}) \otimes_{\mathbb{C}} \Lambda^n \mathfrak{p}$ (although we do not necessarily have a complex on this level), and it is thus clear that ∂ maps $X_n^{(k)}$ into $X_{n-1}^{(k)}$. Hence we have a complex of vector spaces

$$0 \longleftarrow \mathbb{C} \overset{\epsilon}{\longleftarrow} X_0^{(k)} \overset{\partial_0}{\longleftarrow} X_1^{(k)} \longleftarrow \cdots \longleftarrow X_k^{(k)} \longleftarrow 0 . \tag{7.56}$$

We prove exactness of (7.56) for $k \geq 1$ by induction on k. To see that exactness of (7.55a) follows, let x be in $\ker \partial_n \subseteq X_{n+1}$. Then x is in $X_{n+1}^{(k)}$ for large enough k, and $x = \partial_{n+1}\tilde{x}$ for some $\tilde{x}$ in $X_{n+2}^{(k)} \subseteq X_{n+2}$. So exactness of (7.55a) will follow from Lemma 7.28 and exactness of (7.56).

Let $N = \dim \mathfrak{p}$, let $x_1, \ldots, x_N$ be a basis of $\mathfrak{p}$, and let $\{\omega_r\}$ be a basis of $\Lambda^n \mathfrak{p}$. We shall show that the images in the quotient of the elements

$$x^J \otimes \omega_r = x_1^{j_1} \cdots x_N^{j_N} \otimes \omega_r , \qquad \Sigma\, j_i \leq k-n ,$$

are a basis of $X_n^{(k)}$. It is clear from the Poincaré-Birkhoff-Witt Theorem that they span. Let us see that they are linearly independent. If a linear combination $\Sigma_{j,r}\, c_{jr} x^J \otimes \omega_r$ of such elements descends to 0, it is a linear combination of elements

$$x^{\ell} u \otimes \omega' - x^{\ell} \otimes u\omega'$$

with $u \in U(\mathfrak{k})$ of constant term 0, $\omega' \in \Lambda^n \mathfrak{p}$, $\ell = (\ell_1, \ldots, \ell_N)$, and $\Sigma\, \ell_i \leq k-n$. Thus we have an equality of the form

$$\sum_{j,r} c_{jr} x^j \otimes \omega_r = \sum_m c_m (x^{\ell_m} u_m \otimes \omega_m' - x^{\ell_m} \otimes u_m \omega_m') \qquad (7.57)$$

with $\{x^{\ell_m} \otimes \omega_m'\}$ linearly independent in $U(\mathfrak{g}) \otimes_{\mathbb{C}} \Lambda^n \mathfrak{p}$ as m varies. Since each u_m has constant term 0, the Poincaré-Birkhoff-Witt Theorem implies that

$$(\sum_m \mathbb{C} x^{\ell_m} u_m) \cap (\sum_\ell \mathbb{C} x^\ell) = 0 .$$

Therefore (7.57) implies that

$$\sum c_m (x^{\ell_m} u_m \otimes \omega_m') = 0 ,$$

and we deduce that all $c_m = 0$. Hence the images of the elements $x^j \otimes \omega_r$ with $\Sigma j_i \leq k-n$ are a basis of $X_n^{(k)}$.

Returning to (7.56), fix $k \geq 1$. We have just seen that the elements $x^j \otimes 1$ with $\Sigma j_i \leq k$ give a basis of $X_0^{(k)}$. Since ϵ annihilates all of these but $1 \otimes 1$, the elements $x^j \otimes 1$ with $1 \leq \Sigma j_i \leq k$ give a basis of $\ker \epsilon \subseteq X_0^{(k)}$. Meanwhile the elements $x^{j'} \otimes x_m$ with $\Sigma j_i' \leq k-1$ give a basis of $X_1^{(k)}$. If M is the last index i with $j_i' > 0$, then

$$\partial_0 (x_1^{j_1'} \cdots x_M^{j_M'-1} \otimes x_M) = x_1^{j_1'} \cdots x_M^{j_M'} = x^{j'}$$

shows that image $\partial_0 |_{X_1^{(k)}} = \ker \epsilon |_{X_0^{(k)}}$. Thus the exactness of (7.56) is equivalent with the exactness of

$$0 \longleftarrow \sum_{1 \leq \Sigma j_i \leq k} \mathbb{C} x^j \otimes 1 \xleftarrow{\partial_0} X_1^{(k)} \longleftarrow \cdots \longleftarrow X_k^{(k)} \longleftarrow 0 . \qquad (7.58)$$

For $k = 1$, this complex is

$$0 \longleftarrow \mathfrak{p} \otimes_{\mathbb{C}} \mathbb{C} \xleftarrow{\ \partial_0\ } \mathbb{C} \otimes_{\mathbb{C}} \mathfrak{p} \longleftarrow 0$$

with $\partial_0(1 \otimes x) = x \otimes 1$, and the complex is exact.

Proceeding inductively, suppose (7.58) is exact for some $k-1 > 0$, and consider (7.58) for k. We set up the diagram below, in which the columns are exact by construction, the top row of horizontal maps is exact by inductive hypothesis, the middle row of maps is a complex by Lemma 7.28, the top row of squares commutes by construction, and the maps $\bar{\partial}_{n-1}$ are obtained from Lemma 4.17 so that the bottom row of squares commutes. Certainly the bottom row of maps is a complex. We shall prove that this row is exact, and then the exactness of the middle row will follow by Lemma 4.16.

$$\begin{array}{ccccccccccccc}
 & & 0 & & 0 & & & & 0 & & 0 & & \\
 & & \downarrow & & \downarrow & & & & \downarrow & & \downarrow & & \\
0 & \longleftarrow & \sum\limits_{1 \le \Sigma j_i \le k-1} \mathbb{C}x^j \otimes 1 & \xleftarrow{\partial_0} & X_1^{(k-1)} & \longleftarrow & \cdots & \longleftarrow & X_{k-1}^{(k-1)} & \longleftarrow & 0 & \longleftarrow & 0 \\
 & & \downarrow & & \downarrow & & & & \downarrow & & \downarrow & & \\
0 & \longleftarrow & \sum\limits_{1 \le \Sigma j_i \le k} \mathbb{C}x^j \otimes 1 & \xleftarrow{\partial_0} & X_1^{(k)} & \longleftarrow & \cdots & \longleftarrow & X_{k-1}^{(k)} & \longleftarrow & X_k^{(k)} & \longleftarrow & 0 \\
 & & \downarrow & & \downarrow & & & & \downarrow & & \downarrow & & \\
0 & \longleftarrow & \sum\limits_{\Sigma j_i = k} \mathbb{C}x^j \otimes 1 & \xleftarrow{\bar{\partial}_0} & X_1^{(k)}/X_1^{(k-1)} & \longleftarrow & \cdots & \longleftarrow & X_{k-1}^{(k)}/X_{k-1}^{(k-1)} & \longleftarrow & X_k^{(k)} & \longleftarrow & 0 \\
 & & \downarrow & & \downarrow & & & & \downarrow & & \downarrow & & \\
 & & 0 & & 0 & & & & 0 & & 0 & &
\end{array}$$

(7.59)

From our bases of $X_n^{(k)}$ and $X_n^{(k-1)}$, we see that a basis of $X_n^{(k)}/X_n^{(k-1)}$ is the image of all $x^j \otimes \omega_r$ with

$\Sigma\, j_i = k-n$. Hence

$$X_n^{(k)}/X_n^{(k-1)} \cong S^{k-n}(\mathfrak{p}) \otimes_{\mathbb{C}} \Lambda^n \mathfrak{p}$$

$$X_k^{(k)} \cong \mathbb{C} \otimes_{\mathbb{C}} \Lambda^k \mathfrak{p} .$$

Under these identifications, we want to identify $\bar\partial_{n-1}$. So let $\Sigma\, j_i = k-n$. Then

$$\bar\partial_{n-1}(x^j \otimes x_{i_1} \wedge \ldots \wedge x_{i_n})$$

$$= \sum_{\ell=1}^{n} (-1)^{\ell+1}(x^j x_{i_\ell} \otimes x_{i_1} \wedge \ldots \wedge \hat{x}_{i_\ell} \wedge \ldots \wedge x_{i_n})$$

$$+ \sum_{r<s} (-1)^{r+s}(x^j \otimes [x_{i_r}, x_{i_s}] \wedge x_{i_1} \wedge \ldots \wedge \hat{x}_{i_r} \wedge \ldots \wedge \hat{x}_{i_s} \wedge \ldots \wedge x_{i_n})$$

$$\equiv \sum_{\ell=1}^{n} (-1)^{\ell+1}(x^j x_{i_\ell} \otimes x_{i_1} \wedge \ldots \wedge \hat{x}_{i_\ell} \wedge \ldots \wedge x_{i_n}) \mod X_{n-1}^{(k-1)} .$$

Thus the bottom row of (7.59) is isomorphic to that part of the standard complex for $\mathfrak{p}$ (made into an abelian Lie algebra) in which the $U(\mathfrak{p}) = S(\mathfrak{p})$ coefficients are homogeneous of degree $k-n$. This sequence is exact by Theorem 4.6 (in the special case of an abelian Lie algebra), and the induction is complete.

Let V be in $\mathcal{C}(\mathfrak{g}, K)$, and let $N = \dim(\mathfrak{g}/\mathfrak{k}) = \dim \mathfrak{p}$. Applying (a) and (b) of Proposition 7.9 to the Koszul resolution, we obtain a projective resolution

$$0 \longleftarrow V \longleftarrow X_0 \otimes_{\mathbb{C}} V \longleftarrow X_1 \otimes_{\mathbb{C}} V \longleftarrow \cdots \longleftarrow X_N \otimes_{\mathbb{C}} V \longleftarrow 0 \tag{7.60}$$

and an injective resolution

$$0 \longrightarrow V \longrightarrow \mathrm{Hom}_{\mathbb{C}}(X_0, V)_K \longrightarrow \mathrm{Hom}_{\mathbb{C}}(X_1, V)_K \longrightarrow$$
$$\cdots \longrightarrow \mathrm{Hom}_{\mathbb{C}}(X_N, V)_K \longrightarrow 0 . \tag{7.61}$$

These are called the <u>standard resolutions</u> of V.

Let us now make homology more explicit. We consider the standard projective resolution (7.60) of V. A term is $X_n \otimes_{\mathbb{C}} V$, and we write $X_n = R(\mathfrak{g}, K) \otimes_{R(\mathfrak{k}, K)} \wedge^n \mathfrak{p}$ as in (7.53b). Applying the functor $V \to V \otimes_{R(\mathfrak{g}, K)} \mathbb{C}$ to the resolution and using (7.32), we have

$$\begin{aligned}
(X_n \otimes_{\mathbb{C}} V) \otimes_{R(\mathfrak{g}, K)} \mathbb{C} &\cong X_n \otimes_{R(\mathfrak{g}, K)} V \\
&= (\wedge^n \mathfrak{p} \otimes_{R(\mathfrak{k}, K)} R(\mathfrak{g}, K)) \otimes_{R(\mathfrak{g}, K)} V \\
&\cong \wedge^n \mathfrak{p} \otimes_{R(\mathfrak{k}, K)} (R(\mathfrak{g}, K) \otimes_{R(\mathfrak{g}, K)} V) && \text{by (2.51)} \\
&\cong \wedge^n \mathfrak{p} \otimes_{R(\mathfrak{k}, K)} V && \text{by Proposition 7.11} \\
&= \wedge^n \mathfrak{p} \otimes_K V .
\end{aligned}$$

Thus the complex from which we obtain homology is

$$0 \longleftarrow \wedge^0 \mathfrak{p} \otimes_K V \xleftarrow{\partial} \wedge^1 \mathfrak{p} \otimes_K V \xleftarrow{\partial} \cdots \xleftarrow{\partial} \wedge^N \mathfrak{p} \otimes_K V \longleftarrow 0 , \tag{7.62a}$$

and we readily check that the operator ∂ is given on $\wedge^n \mathfrak{p} \otimes_K V$ by

$$\partial(Y_1 \wedge \cdots \wedge Y_n \otimes v)$$

$$= \sum_{\ell=1}^{n} (-1)^{\ell} (Y_1 \wedge \cdots \wedge \hat{Y}_\ell \wedge \cdots \wedge Y_n \otimes Y_\ell v)$$

$$+ \sum_{r<s} (-1)^{r+s} (\rho[Y_r, Y_s] \wedge Y_1 \wedge \cdots \wedge \hat{Y}_r \wedge \cdots \wedge \hat{Y}_s \wedge \cdots \wedge Y_n \otimes v) . \tag{7.62b}$$

Next let us make cohomology more explicit. We consider the standard injective resolution (7.61) of V. A term is $\operatorname{Hom}_{\mathbb{C}}(X_n, V)_K$, and we again write X_n as in (7.53b). Applying the functor $V \to V^{R(\mathfrak{g},K)}$ to the resolution, we have

$$\begin{aligned}
(\operatorname{Hom}_{\mathbb{C}}(X_n, V)_K)^{R(\mathfrak{g},K)} &= \operatorname{Hom}_{R(\mathfrak{g},K)}(X_n, V) \\
&= \operatorname{Hom}_{R(\mathfrak{g},K)}(R(\mathfrak{g},K) \otimes_{R(\mathfrak{k},K)} \wedge^n \mathfrak{p}, V) \\
&\cong \operatorname{Hom}_{R(\mathfrak{k},K)}(\wedge^n \mathfrak{p}, \operatorname{Hom}_{R(\mathfrak{g},K)}(R(\mathfrak{g},V), V)) \\
&\qquad \text{by (2.50)} \\
&= \operatorname{Hom}_{R(\mathfrak{k},K)}(\wedge^n \mathfrak{p}, \operatorname{Hom}_{R(\mathfrak{g},K)}(R(\mathfrak{g},K), V)_K) \\
&\cong \operatorname{Hom}_{R(\mathfrak{k},K)}(\wedge^n \mathfrak{p}, V) \quad \text{by Proposition 7.12} \\
&= \operatorname{Hom}_K(\wedge^n \mathfrak{p}, V) .
\end{aligned}$$

Thus the complex from which we obtain cohomology is

$$0 \longrightarrow \operatorname{Hom}_K(\wedge^0 \mathfrak{p}, V) \xrightarrow{d} \operatorname{Hom}_K(\wedge^1 \mathfrak{p}, V) \xrightarrow{d} \cdots \xrightarrow{d} \operatorname{Hom}_K(\wedge^N \mathfrak{p}, V) \longrightarrow 0 , \tag{7.63a}$$

and we readily check that the operator d is given on $\operatorname{Hom}_K(\wedge^{k-1}\mathfrak{p}, V)$ by

$$d\lambda(Y_1 \wedge \dots \wedge Y_k)$$

$$= \sum_{\ell=1}^{k} (-1)^{\ell+1} Y_\ell (\lambda(Y_1 \wedge \dots \wedge \hat{Y}_\ell \wedge \dots \wedge Y_k))$$

$$+ \sum_{r<s} (-1)^{r+s} \lambda(\mathcal{P}[Y_r, Y_s] \wedge Y_1 \wedge \dots \wedge \hat{Y}_r \wedge \dots \wedge \hat{Y}_s \wedge \dots \wedge Y_k) . \tag{7.63b}$$

9. Poincaré duality

For the pair $(\mathfrak{g}, K)$, let us continue to write $\mathfrak{g} = \mathfrak{k} \oplus \mathfrak{p}$ with $\mathfrak{p}$ stable under $\mathrm{Ad}(K)$. As in §8, we let $N = \dim_{\mathbb{C}}(\mathfrak{g}/\mathfrak{k}) = \dim_{\mathbb{C}} \mathfrak{p}$. Although Chapter II supplies us with an adjoint action of $\mathfrak{g}$ on $\Lambda^n \mathfrak{g}$, we do not get an action of $\mathfrak{g}$ on each $\Lambda^n(\mathfrak{g}/\mathfrak{k})$ since $\mathfrak{k}$ need not be an ideal in $\mathfrak{g}$. Nevertheless, $\mathfrak{g}$ does act on the top exterior power $\Lambda^N(\mathfrak{g}/\mathfrak{k}) \cong \Lambda^N \mathfrak{p}$.

Namely fix any nonzero element τ in $\Lambda^{\dim \mathfrak{k}} \mathfrak{k}$. Then right multiplication by τ gives us a vector space isomorphism $R_\tau : \Lambda^N \mathfrak{p} \to \Lambda^{\dim \mathfrak{g}} \mathfrak{g}$, and we define

$$Z \cdot \eta = R_\tau^{-1}(Z \cdot (R_\tau \eta)) \quad \text{for } \eta \in \Lambda^N \mathfrak{p}, \ Z \in \mathfrak{g} . \tag{7.64}$$

Since any two choices of τ are multiples of one another, this action does not depend on the choice of τ.

Lemma 7.30. On $\Lambda^N \mathfrak{p}$, the $\mathfrak{g}$ action given by (7.64) and the K action given by Ad make $\Lambda^N \mathfrak{p}$ into a $(\mathfrak{g}, K)$ module.

Proof. If $\mathfrak{k}_0$ denotes the Lie algebra of K, we may regard $\mathfrak{k}$ as the complexification of $\mathfrak{k}_0$ and $\Lambda^{\dim \mathfrak{k}} \mathfrak{k}$ as the complexification of $\Lambda^{\dim \mathfrak{k}_0} \mathfrak{k}_0$. We may as well assume

τ is in $\bigwedge^{\dim \mathfrak{k}_0} \mathfrak{k}_0$. For $k \in K$, $\mathrm{Ad}(k)\tau$ must be in $\bigwedge^{\dim \mathfrak{k}_0} \mathfrak{k}_0$ and hence must be $\chi(k)\tau$, where $\chi : K \to \{\pm 1\}$ is a continuous homomorphism. Differentiating, we obtain

$$(\mathrm{ad}\ X)\tau = 0 \quad \text{for all} \quad X \in \mathfrak{k}. \tag{7.65}$$

Using (7.64) with $X \in \mathfrak{k}$, we then have

$$\begin{aligned} X \cdot \eta = R_\tau^{-1}((\mathrm{ad}\ X)(\eta \wedge \tau)) &= R_\tau^{-1}((\mathrm{ad}\ X)\eta \wedge \tau + \eta \wedge (\mathrm{ad}\ X)\tau) \\ &= R_\tau^{-1}((\mathrm{ad}\ X)\eta \wedge \tau) \quad \text{by (7.65)} \\ &= (\mathrm{ad}\ X)\eta . \end{aligned}$$

Hence the action (7.64) from $\mathfrak{k}$ is the derivative of the K action.

Next we have

$$\begin{aligned} \mathrm{Ad}(k)X \cdot k\eta &= R_\tau^{-1}(\mathrm{ad}(\mathrm{Ad}(k)X)(\mathrm{Ad}(k)\eta \wedge \tau)) \\ &= R_\tau^{-1}(\mathrm{Ad}(k)(\mathrm{ad}\ X)\eta \wedge \tau) + R_\tau^{-1}(\mathrm{Ad}(k)\eta \wedge \mathrm{ad}(\mathrm{Ad}(k)X)\tau) \end{aligned} \tag{7.66}$$

and

$$\begin{aligned} k \cdot (X \cdot \eta) &= \mathrm{Ad}(k) R_\tau^{-1}((\mathrm{ad}\ X)(\eta \wedge \tau)) \\ &= \mathrm{Ad}(k) R_\tau^{-1}((\mathrm{ad}\ X)\eta \wedge \tau) + \mathrm{Ad}(k) R_\tau^{-1}(\eta \wedge (\mathrm{ad}\ X)\tau) . \end{aligned} \tag{7.67}$$

For $\eta' \in \bigwedge^N \mathfrak{p}$, we have

$$\mathrm{Ad}(k) R_\tau^{-1}(\eta' \wedge \tau) = \mathrm{Ad}(k)\eta' = R_\tau^{-1}(\mathrm{Ad}(k)\eta' \wedge \tau) ,$$

and it follows that the respective first terms on the right sides of (7.66) and (7.67) are equal. We claim that the respective second terms are 0. If X is in $\mathfrak{k}$, this assertion is a consequence of (7.65), while if X is in $\mathfrak{p}$, it is a consequence of the bracket relation $[\mathfrak{k},\mathfrak{p}] \subseteq \mathfrak{p}$ and the fact that η and $\mathrm{Ad}(k)\eta$ are in $\Lambda^N \mathfrak{p}$. We conclude that the left sides of (7.66) and (7.67) are equal, and thus $\Lambda^N \mathfrak{p}$ is a $(\mathfrak{g},K)$ module. This proves the lemma.

Recall that if V is a $(\mathfrak{g},K)$ module, then V^c denotes the $(\mathfrak{g},K)$ module $\mathrm{Hom}_{\mathbb{C}}(V,\mathbb{C})_K$.

Theorem 7.31 (Poincaré duality). For $n \leq N$ and for V in $\mathcal{C}(\mathfrak{g},K)$, there are vector space isomorphisms

$$H^i(\mathfrak{g},K;V^c) \cong H_i(\mathfrak{g},K;V)^* \tag{7.68}$$

and

$$H^i(\mathfrak{g},K;V) \cong H_{N-i}(\mathfrak{g},K;V\otimes_{\mathbb{C}} (\Lambda^N \mathfrak{p})^*) \tag{7.69}$$

natural in V. Consequently

$$H^i(\mathfrak{g},K;V^c) \cong H^{N-i}(\mathfrak{g},K;V\otimes_{\mathbb{C}} \Lambda^N \mathfrak{p})^* \tag{7.70}$$

naturally in V.

Remarks. (1) This theorem generalizes Theorem 6.10. The isomorphism (7.68) is easy, and (7.69) is hard; (7.70) comes by substituting (7.69) into (7.68).

(2) In (7.69) and (7.70), $\Lambda^N \mathfrak{p}$ is regarded as a $(\mathfrak{g},K)$ module as in Lemma 7.30. If K acts trivially on $\Lambda^N \mathfrak{p}$ (as is the case if K is connected) and if $\mathfrak{g}$ acts trivially on

$\bigwedge^{\dim \mathfrak{g}} \mathfrak{g}$, then $(\bigwedge^N \mathfrak{p})^*$ can pull out of the homology in (7.69), and $\bigwedge^N \mathfrak{p}$ can pull out of the cohomology in (7.70).

(3) An important special case of this theorem is the formula that results for $H^N(\mathfrak{g},K;V)$ from (7.69) and the formula $H_0(\mathfrak{g},K;W) \cong W \otimes_{R(\mathfrak{g},K)} \mathbb{C}$.

Proof of (7.68). Let $F(W) = W^{R(\mathfrak{g},K)}$ and $G(W) = W \otimes_{R(\mathfrak{g},K)} \mathbb{C}$. Then

$$\begin{aligned} F(V^c) &= \mathrm{Hom}_{R(\mathfrak{g},K)}(\mathbb{C},V^c) = \mathrm{Hom}_{R(\mathfrak{g},K)}(\mathbb{C},\mathrm{Hom}_{\mathbb{C}}(V,\mathbb{C})_K) \\ &\cong \mathrm{Hom}_{\mathbb{C}}(V \otimes_{R(\mathfrak{g},K)} \mathbb{C},\mathbb{C}) = G(V)^* . \end{aligned}$$

Letting "dual" refer to $(\cdot)^c$ in $\mathcal{C}(\mathfrak{g},K)$ and to $(\cdot)^*$ in $\mathcal{C}(0,\{1\})$, we can rewrite this relation as a natural isomorphism

$$F \circ \mathrm{dual}(V) \cong \mathrm{dual} \circ G(V) . \tag{7.71}$$

Thus

$$\begin{aligned} (F^n \circ \mathrm{dual})(V) &\cong (F \circ \mathrm{dual})^n(V) && \text{since dual is exact and carries projectives to injectives} \\ &\cong (\mathrm{dual} \circ G)^n(V) && \text{by (7.71)} \\ &\cong (\mathrm{dual} \circ G_n)(V) && \text{since dual is exact and contravariant.} \end{aligned}$$

This isomorphism is just (7.68).

Proof of (7.69). Imitating the proof of Theorem 6.10, we break the argument into three steps.

(1) Preliminaries. We calculate homology and cohomology from the explicit complexes (7.62) and (7.63). In this

setting, we shall exhibit a natural isomorphism on the chain/cochain level and show that it passes to the homology/cohomology. First we define ∂ on all of $\Lambda^n\mathfrak{p}\otimes_{\mathbb{C}} V\otimes_{\mathbb{C}}(\Lambda^N\mathfrak{p})^*$ by (7.62b) and d on all of $\mathrm{Hom}_{\mathbb{C}}(\Lambda^{k-1}\mathfrak{p},V)$ by (7.63b). It is trivial that ∂ commutes with the action of K on $\Lambda^n\mathfrak{p}\otimes_{\mathbb{C}} V\otimes_{\mathbb{C}}(\Lambda^N\mathfrak{p})^*$ and thus descends to the K invariants $\Lambda^n\mathfrak{p}\otimes_K(V\otimes_{\mathbb{C}}(\Lambda^N\mathfrak{p})^*)$. And, of course, our new d is an extension of the old one on $\mathrm{Hom}_K(\Lambda^{k-1}\mathfrak{p},V)$. We give the isomorphism on the level of a map

$$\Lambda^n\mathfrak{p}\otimes_{\mathbb{C}} V\otimes_{\mathbb{C}}(\Lambda^N\mathfrak{p})^* \longrightarrow \mathrm{Hom}_{\mathbb{C}}(\Lambda^{N-n}\mathfrak{p},V). \tag{7.72}$$

Namely for $\xi\in\Lambda^n\mathfrak{p}$, $\gamma\in\Lambda^{N-n}\mathfrak{p}$, and $\epsilon\in(\Lambda^N\mathfrak{p})^*$, we define $\xi\otimes v\otimes\epsilon\to\lambda_{\xi\otimes v\otimes\epsilon}$ by

$$\lambda_{\xi\otimes v\otimes\epsilon}(\gamma) = \epsilon(\xi\wedge\gamma)v. \tag{7.73}$$

Then $\xi\otimes v\otimes\epsilon\to\lambda_{\xi\otimes v\otimes\epsilon}$ extends to give the linear isomorphism (7.72). This map respects the K actions

$$k(\xi\otimes v\otimes\epsilon) = k\xi\otimes kv\otimes k\epsilon$$

$$(k\lambda)(\gamma) = k(\lambda(k^{-1}\gamma))$$

because

$$\begin{aligned}\lambda_{k(\xi\otimes v\otimes\epsilon)}(\gamma) &= \lambda_{k\xi\otimes kv\otimes k\epsilon}(\gamma) = (k\epsilon)(k\xi\wedge\gamma)kv \\ &= \epsilon(\xi\wedge k^{-1}\gamma)kv = k(\epsilon(\xi\wedge k^{-1}\gamma)v) \\ &= k(\lambda_{\xi\otimes v\otimes\epsilon}(k^{-1}\gamma)) = (k\lambda_{\xi\otimes v\otimes\epsilon})(\gamma).\end{aligned}$$

Restricted to the K invariants, our map thus gives a linear isomorphism

$$\Lambda^n \mathfrak{p} \otimes_K (V \otimes_{\mathbb{C}} (\Lambda^N \mathfrak{p})^*) \longrightarrow \mathrm{Hom}_K(\Lambda^{N-n}\mathfrak{p}, V) .$$

Equation (7.69) will thus follow if we show that, under our mapping $\xi \otimes v \otimes \varepsilon \to \lambda_{\xi\otimes v\otimes\varepsilon}$, ∂ corresponds to $(-1)^{|\xi|}d$, where $|\xi|$ is the degree of ξ. That is, we are to show that

$$d\lambda_{\xi\otimes v\otimes\varepsilon} = (-1)^{|\xi|}\lambda_{\partial(\xi\otimes v\otimes\varepsilon)} . \tag{7.74}$$

(2) Action of $\mathfrak{g}$ on $(\Lambda^N\mathfrak{p})^*$. Let $X_1, \ldots, X_N$ be in $\mathfrak{p}$, and put $\eta = X_1 \wedge \ldots \wedge X_N$. If Z belongs to $\mathfrak{g}$, then Z acts on η in $\Lambda^N\mathfrak{p}$ according to (7.64):

$$Z\cdot\eta = R_\tau^{-1}((\mathrm{ad}\ Z)(\eta\wedge\tau)) = R_\tau^{-1}((\mathrm{ad}\ Z)\eta\wedge\tau) + R_\tau^{-1}(\eta\wedge(\mathrm{ad}\ Z)\tau) .$$

The second term on the right side here is 0; if X is in $\mathfrak{k}$, this assertion follows from (7.65), while if X is in $\mathfrak{p}$, it follows from the relation $[\mathfrak{k},\mathfrak{p}] \subseteq \mathfrak{p}$ and the fact that η is in $\Lambda^N\mathfrak{p}$. Thus

$$Z\cdot\eta = R_\tau^{-1}((\mathrm{ad}\ Z)\eta\wedge\tau) .$$

Now

$$\begin{aligned}(\mathrm{ad}\ Z)\eta\wedge\tau &= \sum_{i=1}^{N}(-1)^{i+1}([Z,X_i]\wedge X_1\wedge\ldots\wedge\hat{X}_i\wedge\ldots\wedge X_N\wedge\tau)\\ &= \sum_{i=1}^{N}(-1)^{i+1}(\mathcal{P}[Z,X_i]\wedge X_1\wedge\ldots\wedge\hat{X}_i\wedge\ldots\wedge X_N\wedge\tau) ,\end{aligned}$$

where $\mathcal{P}$ is the projection of $\mathfrak{g}$ on $\mathfrak{p}$ along $\mathfrak{k}$. Consequently

$$Z\cdot\eta = \sum_{i=1}^{N}(-1)^{i+1}(\mathcal{P}[Z,X_i]\wedge X_1\wedge\ldots\wedge\hat{X}_i\wedge\ldots\wedge X_N) .$$

For the action of Z on the element ϵ of $(\Lambda^N\mathfrak{p})^*$, we use the contragredient action:

$$(Z\epsilon)(\eta) = \sum_{i=1}^{N} (-1)^i \epsilon(P[Z,X_i] \wedge X_1 \wedge \ldots \wedge \hat{X}_i \wedge \ldots \wedge X_N). \qquad (7.75)$$

(3) Verification of (7.74). The argument is the same as for Step 3 of the proof of Theorem 6.10, except that we change $\mathfrak{g}$ to $\mathfrak{p}$ everywhere and we use (7.75) in place of (6.35). Once (7.74) is verified, the proof of (7.69) is complete.

10. Revised setting for Kostant's Theorem

As we hinted at the beginning of this chapter, Kostant's Theorem (Theorem 6.12) can be viewed as a statement about objects in the setting of relative Lie algebra cohomology. Such a point of view has the advantage that we can bring to bear all the machinery of this chapter when applying the theorem. The next section will give a fundamental illustration of the resulting technique.

As in §VI.4, the setting will involve $G = U(n)$, $\mathfrak{g}_0 = \mathfrak{u}(n)$, $\mathfrak{g} = \mathfrak{u}(n)^{\mathbb{C}} = \mathfrak{gl}(n,\mathbb{C})$, T = diagonal subgroup of G, and $\mathfrak{h}$ = diagonal subalgebra of $\mathfrak{g}$. Other notation for $U(n)$ is as in Chapters III and VI. Fix a lexicographic ordering, the corresponding set Δ^+ of positive roots, and the half sum δ of the members of Δ^+. Let $\mathfrak{n}$ be the Lie subalgebra of $\mathfrak{g}$ given by $\mathfrak{n} = \sum_{\alpha\in\Delta^+} \mathfrak{g}_\alpha$, and let $\mathfrak{b} = \mathfrak{h} \oplus \mathfrak{n}$.

For V in $\mathcal{C}(\mathfrak{b},T)$, we consider the vector spaces

$$V \otimes_{\mathfrak{n}} \mathbb{C} \quad \text{and} \quad \mathrm{Hom}_{\mathfrak{n}}(\mathbb{C},V).$$

These vector spaces carry $(\mathfrak{h}, T)$ module structures under the definitions

$$X(v\otimes 1) = Xv\otimes 1, \quad X\in\mathfrak{h}, \quad \text{and} \quad t(v\otimes 1) = tv\otimes 1, \quad t\in T$$

$$(X\varphi)(z) = X(\varphi(z)), \quad X\in\mathfrak{h}, \quad \text{and} \quad (t\varphi)(z) = t(\varphi(z)), \quad t\in T.$$

[For example, to see that $t\in T$ acts on $V\otimes_{\mathfrak{n}}\mathbb{C}$, we have to check that the kernel of $V\to V\otimes_{\mathfrak{n}}\mathbb{C}$ is T-invariant. This kernel is $\mathfrak{n}V$, and

$$t(Yv) = \mathrm{Ad}(t)Y\cdot tv \quad \text{is in } \mathfrak{n}V$$

if $t\in T$ and $Y\in\mathfrak{n}$.] Then it is a simple matter to see that

$$V\to V\otimes_{\mathfrak{n}}\mathbb{C}$$

and

$$V\to \mathrm{Hom}_{\mathfrak{n}}(\mathbb{C}, V)$$

are covariant functors from $\mathcal{C}(\mathfrak{b}, T)$ to $\mathcal{C}(\mathfrak{h}, T)$. We shall relate the corresponding derived functors to the discussion of Kostant's Theorem in §VI.4.

Lemma 7.32. The forgetful functor $\mathcal{F}^{\mathfrak{n},1}_{\mathfrak{b},T}$ sends projectives to projectives and injectives to injectives.

Proof. If P is projective, then Corollary 7.21b shows that there is a standard projective $S = P^{\mathfrak{b},T}_{\mathfrak{h},T}(W)$ such that $S = P\oplus P'$. Then $\mathcal{F} = \mathcal{F}^{\mathfrak{n},1}_{\mathfrak{b},T}$ satisfies $\mathcal{F}(S)\cong\mathcal{F}(P)\oplus\mathcal{F}(P')$, and $\mathcal{F}(S)$ projective will imply $\mathcal{F}(P)$ projective. Thus it is enough to consider standard projectives. For S as above, we have $U(\mathfrak{n})$ isomorphisms

$$\mathcal{F}(S) \cong \mathcal{F}(((U(\mathfrak{n}) \otimes_{\mathbb{C}} U(\mathfrak{h})) \otimes_{U(\mathfrak{h})} C_T) \otimes_{C_T} W) \cong U(\mathfrak{n}) \otimes_{\mathbb{C}} W,$$

and the right side is projective in $\mathcal{C}(\mathfrak{n},\{1\})$. An analogous argument handles injectives.

Proposition 7.33.

(a) Let $F : \mathcal{C}(\mathfrak{b},T) \to \mathcal{C}(\mathfrak{h},T)$ be the functor $F(V) = V \otimes_{\mathfrak{n}} \mathbb{C}$. Then the derived functors of F satisfy

$$\mathcal{F}^{0,1}_{\mathfrak{h},T} \circ F_k \cong H_k(\mathfrak{n}, \mathcal{F}^{\mathfrak{n},1}_{\mathfrak{b},T}(\cdot)) .$$

That is, $F_k(V)$ may be computed, apart from the T action, as $H_k(\mathfrak{n},V)$.

(b) Let $G : \mathcal{C}(\mathfrak{b},T) \to \mathcal{C}(\mathfrak{h},T)$ be the functor $G(V) = \mathrm{Hom}_{\mathfrak{n}}(\mathbb{C},V)$. Then the derived functors of G satisfy

$$\mathcal{F}^{0,1}_{\mathfrak{h},T} \circ G^k \cong H^k(\mathfrak{n}, \mathcal{F}^{\mathfrak{n},1}_{\mathfrak{b},T}(\cdot)) .$$

That is, $G^k(V)$ may be computed, apart from the T action, as $H^k(\mathfrak{n},V)$. If V happens to be a representation space for the group G, then the T action on the member $H^k(\mathfrak{n},V)$ of $\mathcal{C}(\mathfrak{h},T)$ coincides with the T action in Proposition 6.11.

Proof. For (a) let $\widetilde{F} : \mathcal{C}(\mathfrak{n},\{1\}) \to \mathcal{C}(0,\{1\})$ be the functor $\widetilde{F}(V) = V \otimes_{\mathfrak{n}} \mathbb{C}$. Then

$$\mathcal{F}^{0,1}_{\mathfrak{h},T} \circ F = \widetilde{F} \circ \mathcal{F}^{\mathfrak{n},1}_{\mathfrak{b},T} .$$

Both forgetful functors are exact, and the one on the right sends projectives to projectives, by Lemma 7.32. Propositions 5.16a and 5.17a1 thus give

$$\mathcal{F}_{\mathfrak{h},T}^{0,1} \circ F_k \cong (\mathcal{F}_{\mathfrak{h},T}^{0,1} \circ F)_k \cong (\widetilde{F} \circ \mathcal{F}_{\mathfrak{b},T}^{\mathfrak{n},1})_k \cong \widetilde{F}_k \circ \mathcal{F}_{\mathfrak{b},T}^{\mathfrak{n},1},$$

and (a) follows.

For (b) let $\widetilde{G} : \mathcal{C}(\mathfrak{n},\{1\}) \to \mathcal{C}(0,\{1\})$ be the functor $\widetilde{G}(V) = \mathrm{Hom}_{\mathfrak{n}}(\mathbb{C}, V)$. Then

$$\mathcal{F}_{\mathfrak{h},T}^{0,1} \circ G = \widetilde{G} \circ \mathcal{F}_{\mathfrak{b},T}^{\mathfrak{n},1}.$$

Both forgetful functors are exact, and the one on the right sends injectives to injectives, by Lemma 7.32. Propositions 5.16a and 5.17a2 thus give

$$\mathcal{F}_{\mathfrak{h},T}^{0,1} \circ G^k \cong (\mathcal{F}_{\mathfrak{h},T}^{0,1} \circ G)^k \cong (\widetilde{G} \circ \mathcal{F}_{\mathfrak{b},T}^{\mathfrak{n},1})^k \cong \widetilde{G}^k \circ \mathcal{F}_{\mathfrak{b},T}^{\mathfrak{n},1},$$

and the formula in (b) follows.

To see that the T action on $H^k(\mathfrak{n}, V)$ matches that in Proposition 6.11, let X_k be the Koszul resolution in $\mathcal{C}(\mathfrak{b}, T)$, written as in (7.53d) as

$$X_k = U(\mathfrak{b}) \otimes_{\mathfrak{h}} \Lambda^k \mathfrak{n}.$$

Here T acts by $\mathrm{Ad} \otimes \mathrm{Ad}$, according to Lemma 7.26. A term of the standard injective resolution (7.61) of V is then

$$\mathrm{Hom}_{\mathbb{C}}(U(\mathfrak{b}) \otimes_{\mathfrak{h}} \Lambda^k \mathfrak{n}, V)_T,$$

and T acts by $\mathrm{Hom}(\mathrm{Ad} \otimes \mathrm{Ad}, \nu)$, where ν gives the action on V. The complex for computing $G^k(V)$ then has terms

$$\mathrm{Hom}_{\mathfrak{n}}(U(\mathfrak{b}) \otimes_{\mathfrak{h}} \Lambda^k \mathfrak{n}, V)_T, \tag{7.76}$$

still with T acting by $\mathrm{Hom}(\mathrm{Ad} \otimes \mathrm{Ad}, \nu)$. Expression (7.76) is

$$\cong \mathrm{Hom}_{\mathfrak{n}}((U(\mathfrak{n})\otimes_{\mathbb{C}} U(\mathfrak{h}))\otimes_{\mathfrak{h}} \Lambda^k\mathfrak{n}, V)_T$$

$$\cong \mathrm{Hom}_{\mathfrak{n}}(U(\mathfrak{n})\otimes_{\mathbb{C}} (U(\mathfrak{h})\otimes_{\mathfrak{h}} \Lambda^k\mathfrak{n}), V)_T \quad \text{by (2.51)}$$

$$\cong \mathrm{Hom}_{\mathfrak{n}}(U(\mathfrak{n})\otimes_{\mathbb{C}} \Lambda^k\mathfrak{n}, V)_T$$

$$\cong \mathrm{Hom}_{\mathfrak{n}}(U(\mathfrak{n}), \mathrm{Hom}_{\mathbb{C}}(\Lambda^k\mathfrak{n}, V))_T \quad \text{by (6.1)}$$

$$\cong \mathrm{Hom}_{\mathbb{C}}(\Lambda^k\mathfrak{n}, V)_T, \tag{7.77}$$

and we can drop the last subscript T since $\Lambda^k\mathfrak{n}$ is finite-dimensional. In these lines the respective T actions are given by $\mathrm{Hom}(\mathrm{Ad}\otimes 1\otimes \mathrm{Ad},\nu)$, $\mathrm{Hom}(\mathrm{Ad}\otimes 1\otimes \mathrm{Ad},\nu)$, $\mathrm{Hom}(\mathrm{Ad}\otimes \mathrm{Ad},\nu)$, $\mathrm{Hom}(\mathrm{Ad},\mathrm{Hom}(\mathrm{Ad},\nu))$, and $\mathrm{Hom}(\mathrm{Ad},\nu)$. The last of these matches (6.40a), and the proof of Proposition 7.33 is complete.

Kostant's Theorem computes $H^k(\mathfrak{n},V)$ as a T module when V is an irreducible representation of G. Proposition 7.33b reinterprets the theorem as a calculation on the module $\mathcal{F}^{\mathfrak{b},T}_{\mathfrak{g},G}(V)$ of the derived functors of the functor $\mathcal{G}:\mathcal{C}(\mathfrak{b},T)\to\mathcal{C}(\mathfrak{h},T)$ given by $\mathcal{G}(\cdot) = \mathrm{Hom}_{\mathfrak{n}}(\mathbb{C},\cdot)$.

It will be handy to know that the Poincaré duality isomorphism (6.30) in the category $\mathcal{C}(\mathfrak{n},\{1\})$ is compatible with the T actions that we have imposed on homology and cohomology by means of our functors from $\mathcal{C}(\mathfrak{b},T)$ to $\mathcal{C}(\mathfrak{h},T)$. We give such a result in the next proposition.

<u>Proposition 7.34</u>. Put $S = \dim_{\mathbb{C}}\mathfrak{n}$, and regard $(\Lambda^S\mathfrak{n})^*$ as a $(\mathfrak{b},T)$ module as in Lemma 7.30. For $k \leq S$ and V in $\mathcal{C}(\mathfrak{b},T)$, the Poincaré duality isomorphism

$$H^{S-k}(\mathfrak{n}, V) \cong H_k(\mathfrak{n}, V \otimes_{\mathbb{C}} (\Lambda^S \mathfrak{n})^*)$$

is an isomorphism of $(\mathfrak{h}, T)$ modules.

Proof. Duality is given on the level of complexes by the map

$$\Lambda^k \mathfrak{n} \otimes_{\mathbb{C}} V \otimes_{\mathbb{C}} (\Lambda^S \mathfrak{n})^* \to \mathrm{Hom}_{\mathbb{C}}(\Lambda^{S-k}\mathfrak{n}, V) \tag{7.78}$$

with

$$\xi \otimes v \otimes \epsilon \to \lambda_{\xi \otimes v \otimes \epsilon}$$

and

$$\lambda_{\xi \otimes v \otimes \epsilon}(\gamma) = \epsilon(\xi \wedge \gamma) v .$$

As in (7.77), the complexes in (7.78) arise by applying our functors to standard resolutions, and we can track down the two actions of T to see that they are

$$\mathrm{Ad} \otimes \nu \otimes \mathrm{Ad}^* \quad \text{and} \quad \mathrm{Hom}(\mathrm{Ad}, \nu) .$$

Hence

$$\begin{aligned}(t\lambda_{\xi \otimes v \otimes \epsilon})(\gamma) &= \nu(t)(\lambda_{\xi \otimes v \otimes \epsilon}(\mathrm{Ad}(t)^{-1}\gamma)) \\ &= \nu(t)(\epsilon(\xi \wedge \mathrm{Ad}(t)^{-1}\gamma)v) \\ &= (\mathrm{Ad}^*(t)\epsilon)(\mathrm{Ad}(t)\xi \wedge \gamma)\nu(t)v \\ &= \lambda_{\mathrm{Ad}(t)\xi \otimes \nu(t) v \otimes \mathrm{Ad}^*(t)\epsilon}(\gamma) \\ &= \lambda_{t(\xi \otimes v \otimes \epsilon)}(\gamma) ,\end{aligned}$$

and the result follows.

11. Borel-Weil-Bott Theorem for U(n)

We shall apply our results from this chapter to give realizations of the irreducible representations of the unitary groups $U(n)$ in homology and cohomology spaces. We continue with notation for $G = U(n)$ as in §10. In particular we take

$$\mathfrak{n} = \sum_{\alpha\in\Delta^+} \mathfrak{g}_\alpha \quad \text{and} \quad \mathfrak{b} = \mathfrak{h} \oplus \mathfrak{n} .$$

Also we let

$$\bar{\mathfrak{n}} = \sum_{\alpha\in\Delta^+} \mathfrak{g}_{-\alpha} \quad \text{and} \quad \bar{\mathfrak{b}} = \mathfrak{h} \oplus \bar{\mathfrak{n}} .$$

As we suggested in §1, if we start from a one-dimensional representation of T (say with weight λ on the space $\mathbb{C}_\lambda$), then we can extend λ to $\bar{\mathfrak{b}}$ by making it be 0 on the subalgebra $\bar{\mathfrak{n}}$, and $I^{\mathfrak{g},T}_{\bar{\mathfrak{b}},T}(\mathbb{C}_\lambda)$ will be related to the Taylor coefficients of locally defined holomorphic functions satisfying the transformation law given by λ. The remarks following Proposition 7.18 suggest that $I^{\mathfrak{g},G}_{\mathfrak{g},T}$ extracts global holomorphic functions from these Taylor coefficients. Let us write Γ in place of $I^{\mathfrak{g},G}_{\mathfrak{g},T}$ to emphasize that we are obtaining sections. The functor Γ is not exact and has nonzero higher derived functors Γ^k (obtained from an injective resolution), and these may be regarded as extracting global $(0,k)$ forms. The Borel-Weil-Bott Theorem locates the irreducible representations of $G = U(n)$ in such spaces of forms.

The proof of such a result in outline form is so

informative that we give it before stating a theorem, postponing the justification of the steps to later in this section. Let V be an irreducible representation of G. With $S = \dim_{\mathbb{C}} \mathfrak{n}$, we have

$$\operatorname{Hom}_{\mathfrak{g},G}(V,\Gamma^k(I^{\mathfrak{g},T}_{\mathfrak{b},T}(\mathbb{C}_\lambda)))$$

$$\cong \operatorname{Ext}^k_{\mathfrak{g},T}(V, I^{\mathfrak{g},T}_{\mathfrak{b},T}(\mathbb{C}_\lambda)) \tag{7.79a}$$

$$\cong \operatorname{Ext}^k_{\bar{\mathfrak{b}},T}(V,\mathbb{C}_\lambda) \tag{7.79b}$$

$$\cong \operatorname{Hom}_{\mathfrak{h},T}(H_k(\bar{\mathfrak{n}},V),\mathbb{C}_\lambda) \tag{7.79c}$$

$$\cong \operatorname{Hom}_{\mathfrak{h},T}(H^{S-k}(\bar{\mathfrak{n}},V\otimes_{\mathbb{C}}\textstyle\bigwedge^S\bar{\mathfrak{n}}),\mathbb{C}_\lambda) \tag{7.79d}$$

$$\cong \operatorname{Hom}_{\mathfrak{h},T}(H^{S-k}(\bar{\mathfrak{n}},V)\otimes_{\mathbb{C}}\textstyle\bigwedge^S\bar{\mathfrak{n}},\mathbb{C}_\lambda)\ . \tag{7.79e}$$

The right side here has dimension 0 or 1, according to Kostant's Theorem, and we know precisely when it is 0 and when it is 1. By Corollary 3.8 the left side is the multiplicity of V in the representation under study. Therefore we know exactly how our representation splits into irreducible pieces.

Theorem 7.35 (Borel-Weil-Bott). Let $G = U(n)$, and let $\mathbb{C}_\lambda$ be a one-dimensional representation space for T.

(a) If $\lambda+\delta$ is orthogonal to some root, then $\Gamma^k(I^{\mathfrak{g},T}_{\mathfrak{b},T}(\mathbb{C}_\lambda)) = 0$ for all $k \geq 0$.

(b) If $\lambda+\delta$ is orthogonal to no root, put

$$q = \#\{\alpha\in\Delta^+ \mid \langle\lambda+\delta,\alpha\rangle < 0\}\ . \tag{7.80}$$

The representation $\Gamma^k(I^{\mathfrak{g},T}_{\mathfrak{b},T}(\mathbb{C}_\lambda))$ is 0 if $k \neq q$. Moreover there exists a unique element $w \in W$ such that $w(\lambda+\delta)$ is dominant, w has length q, and $\Gamma^q(I_{\mathfrak{b},T}^{\mathfrak{g},T}(\mathbb{C}_\lambda))$ is an irreducible representation of G with highest weight $\mu = w(\lambda+\delta)-\delta$.

Remarks. If λ is dominant relative to Δ^+, then $q=0$ in (7.80). In this case the theorem says that $\Gamma^0(I_{\mathfrak{b},T}^{\mathfrak{g},T}(\mathbb{C}_\lambda))$ is an irreducible representation of G with highest weight λ. Since $\Gamma^0 \cong \Gamma$, this algebraic statement is an analog of the analytic statement that the representation of G on the global holomorphic sections of the line bundle $\mathbb{C}_\lambda \times_{\overline{B}} G^{\mathbb{C}} \to \overline{B}\backslash G^{\mathbb{C}}$ is irreducible with highest weight λ. The analytic statement is true and is the case of Theorem 3.22 known as the Borel-Weil Theorem.

If $q \neq 0$, we get 0 as the space analogous to the space of global holomorphic sections. In this case, we instead get a nontrivial representation in a space that is an analog of the space of $(0,q)$-form sections of the line bundle. The properly formulated analytic statement is known as the Borel-Weil-Bott Theorem.

Proof that (7.79) and Kostant's Theorem imply Theorem 7.35. Let V have highest weight μ relative to Δ^+. The set $-\Delta^+$ is another positive system, the $\mathfrak{n}$ that corresponds to it for Kostant's Theorem is our present $\overline{\mathfrak{n}}$, and the δ for it is our present $-\delta$. Find by Proposition 3.14 the unique element w_0 in W such that $w_0\Delta^+ = -\Delta^+$. The highest weight of V relative to $-\Delta^+$ is then $w_0\mu$. Applying Kostant's Theorem to $\overline{\mathfrak{n}}$, we see that (7.79e) has dimension equal to the number of w in W of length $S-k$ such that

$$(w(w_0\mu-\delta)+\delta)-2\delta = \lambda ,$$

hence such that

$$\mu+\delta = w_0^{-1}w^{-1}(\lambda+\delta) . \tag{7.81}$$

Suppose w satisfies (7.81). Since $\langle\mu,\alpha\rangle \geq 0$ and $\langle\delta,\alpha\rangle > 0$ for $\alpha\in\Delta^+$, $\langle\mu+\delta,\alpha\rangle \neq 0$ for all roots α, and $\langle\lambda+\delta,\alpha\rangle$ cannot be 0. This proves (a) of the theorem.

If w satisfies (7.81) and α is in Δ^+, then

$$\langle\lambda+\delta,\alpha\rangle = \langle ww_0(\mu+\delta),\alpha\rangle = \langle\mu+\delta,w_0^{-1}w^{-1}\alpha\rangle ,$$

and the right side is < 0 exactly for α in $\Delta^+(ww_0)$. Referring to (7.80), we see that $q = \ell(ww_0) = S-\ell(w)$. We get 0 unless $\ell(w) = S-k$, i.e., unless $k = q$.

Next let us see that there is at most one pair (w,μ) such that μ is dominant and (7.81) holds. In fact, if there are two such pairs (w_1,μ_1) and (w_2,μ_2), then $w_0^{-1}w_1^{-1}(\lambda+\delta)$ and $w_0^{-1}w_2^{-1}(\lambda+\delta)$ are both dominant, and the element $w_0^{-1}w_2^{-1}w_1w_0$ carries the first of these dominant elements to the second. If β is in $\Delta^+(w_0^{-1}w_2^{-1}w_1w_0)$, then $\beta > 0$ and $(w_0^{-1}w_2^{-1}w_1w_0)^{-1}\beta < 0$ say

$$0 < \langle w_0^{-1}w_2^{-1}(\lambda+\delta),\beta\rangle = \langle w_0^{-1}w_1^{-1}(\lambda+\delta),(w_0^{-1}w_2^{-1}w_1w_0)^{-1}\beta\rangle ,$$

which is impossible. Thus $\Delta^+(w_0^{-1}w_2^{-1}w_1w_0)$ is empty, and Proposition 3.18 says $w_0^{-1}w_2^{-1}w_1w_0 = 1$. So $w_1 = w_2$, and then also $\mu_1 = \mu_2$.

Finally choose w by Lemma 3.15 so that $w^{-1}(\lambda+\delta)$ is dominant for $-\Delta^+$, and put $\mu = w_0^{-1}w^{-1}(\lambda+\delta)-\delta$. We shall show that μ is dominant integral. Then (7.81) will be satisfied, an irreducible representation τ_μ will exist, and the result of (7.79) will be that

$$[\Gamma^q(I_{\mathfrak{b},T}^{\mathfrak{g},T}(\mathbb{C}_\lambda)) : \tau_\nu] = \begin{cases} 1 & \text{if } \nu = \mu \\ 0 & \text{if } \nu \neq \mu . \end{cases}$$

In other words, $\Gamma^q(I_{\mathfrak{b},T}^{\mathfrak{g},T}(\mathbb{C}_\lambda)) \cong \tau_\mu$.

Now μ is certainly integral, being the sum of integral terms

$$\mu = w_0^{-1}w^{-1}\lambda + (w_0^{-1}w^{-1}\delta - \delta) .$$

(See Proposition 3.19.) Let β be a simple root in Δ^+. Then $\langle w^{-1}(\lambda+\delta),\beta\rangle < 0$ implies $\langle w_0^{-1}w^{-1}(\lambda+\delta),\beta\rangle > 0$, and hence

$$\frac{2\langle w_0^{-1}w^{-1}(\lambda+\delta),\beta\rangle}{|\beta|^2} \geq 1 .$$

We can easily check that β simple implies $2\langle\delta,\beta\rangle/|\beta|^2 = 1$, and thus Corollary 3.12 implies

$$\frac{2\langle w_0^{-1}w^{-1}(\lambda+\delta)-\delta,\beta\rangle}{|\beta|^2} \geq 0 .$$

That is, $\langle\mu,\beta\rangle \geq 0$. Summing over suitable β's, we see that μ is dominant. This completes the proof that (7.79) and Kostant's Theorem imply Theorem 7.35.

Preliminary remarks about (7.79). The functor Γ^k is 0 when k exceeds $\dim_{\mathbb{C}}(\mathfrak{g}/\mathfrak{h}) = 2S$, as a result of the existence of the standard resolution (7.61). Isomorphisms (7.79a) and (7.79b) imply that $\Gamma^k(I_{\mathfrak{b},T}^{\mathfrak{g},T}(\mathbb{C}_\lambda))$ is 0 when k exceeds S, since $\mathrm{Ext}_{\mathfrak{b},T}^k$ is the 0 functor for $k > S$. Beyond those steps in (7.79), we assume $k \leq S$.

Proof of (7.79). Part (b) is Shapiro's Lemma (7.52), and part (d) is Poincaré duality, given as (6.30) and Proposition 7.34. Part (e) follows because $\mathrm{ad}\,\mathfrak{n}$ acts as 0 on $\Lambda^S\mathfrak{n}$. (To see this, we write the one-element basis of $\Lambda^S\mathfrak{n}$ in terms of root vectors as

$$E_{\alpha_1} \wedge \ldots \wedge E_{\alpha_S}$$

with the first $n-1$ roots simple, the next roots the sum of two simple roots, the next ones the sum of three simple roots, and so on. If $\alpha > 0$, then

$$(\mathrm{ad}\, E_\alpha)(E_{\alpha_1} \wedge \ldots \wedge E_{\alpha_S})$$

is a sum of terms with one bracket in place, and the resulting bracket duplicates a term farther along in the tensor. So each term is 0.)

The isomorphism (7.79a) is proved as follows: The two functors $\mathrm{Hom}_{\mathfrak{g},G}(V,\cdot)\circ I_{\mathfrak{g},T}^{\mathfrak{g},G}$ and $\mathrm{Hom}_{\mathfrak{g},T}(V,\cdot)$ are naturally isomorphic by (7.46). Hence their derived functors are naturally isomorphic. Since $\mathrm{Hom}_{\mathfrak{g},G}(V,\cdot)$ is exact (Lemma 7.8 and Proposition 5.10a), Proposition 5.16a gives

$$\operatorname{Hom}_{\mathfrak{g},G}(V,\cdot)\circ(I^{\mathfrak{g},G}_{\mathfrak{g},T})^k \cong (\operatorname{Hom}_{\mathfrak{g},G}(V,\cdot)\circ I^{\mathfrak{g},G}_{\mathfrak{g},T})^k$$

$$\cong (\operatorname{Hom}_{\mathfrak{g},T}(V,\cdot))^k \cong \operatorname{Ext}^k_{\mathfrak{g},T}(V,\cdot)\ ,$$

and (7.79a) follows.

Finally let us consider (7.79c). For V in $\mathcal{C}(\mathfrak{b},T)$, we have a natural isomorphism

$$\operatorname{Hom}_{\mathfrak{h},T}(V\otimes_{\bar{\mathfrak{n}}}\mathbb{C},\mathbb{C}_\lambda) \cong \operatorname{Hom}_{\bar{\mathfrak{b}},T}(V,\mathbb{C}_\lambda)\ , \tag{7.82}$$

the map Φ from left to right being given by $\Phi(\varphi)(v) = \varphi(v\otimes 1)$. If $G:\mathcal{C}(\bar{\mathfrak{b}},T)\to\mathcal{C}(\mathfrak{h},T)$ is the functor $G(V) = V\otimes_{\bar{\mathfrak{n}}}\mathbb{C}$, then (7.82) says that $\operatorname{Hom}_{\mathfrak{h},T}(\cdot,\mathbb{C}_\lambda)\circ G$ and $\operatorname{Hom}_{\bar{\mathfrak{b}},T}(\cdot,\mathbb{C}_\lambda)$ are naturally isomorphic. Since $\operatorname{Hom}_{\mathfrak{h},T}(\cdot,\mathbb{C}_\lambda)$ is exact (Lemma 7.8 and Proposition 5.10b), Proposition 5.16b gives

$$\operatorname{Hom}_{\mathfrak{h},T}(\cdot,\mathbb{C}_\lambda)\circ G_k \cong (\operatorname{Hom}_{\mathfrak{h},T}(\cdot,\mathbb{C}_\lambda)\circ G)^k$$

$$\cong (\operatorname{Hom}_{\bar{\mathfrak{b}},T}(\cdot,\mathbb{C}_\lambda))^k \cong \operatorname{Ext}^k_{\bar{\mathfrak{b}},T}(\cdot,\mathbb{C}_\lambda)\ .$$

Finally Proposition 7.33a shows that $G_k \cong H_k(\bar{\mathfrak{n}},\cdot)$ as a functor from $\mathcal{C}(\bar{\mathfrak{b}},T)$ to $\mathcal{C}(\mathfrak{h},T)$, and (7.54c) follows. This completes the proof.

In the next chapter we shall apply relative Lie algebra cohomology to the representation theory of certain noncompact groups. The results that we shall generalize will be <u>variants</u> of (7.79) and Theorem 7.35, but not of (7.79) and Theorem 7.35 as we have written them. We give those variants now.

Initially $\mathbb{C}_\lambda$ refers to a $\mathcal{C}(\mathfrak{h},T)$ module, and we extended $\mathbb{C}_\lambda$ to a $(\bar{\mathfrak{b}},T)$ module by having $\bar{\mathfrak{n}}$ act trivially.

To be clear, let us refer to this extension as $\mathbb{C}_{\lambda,\bar{\mathfrak{b}}}$. Another choice is to make $\mathbb{C}_\lambda$ into a $(\mathfrak{b},T)$ module by having $\mathfrak{n}$ act trivially; we refer to this extension as $\mathbb{C}_{\lambda,\mathfrak{b}}$.

Let λ be integral and introduce Δ^+ so that λ is dominant for Δ^+. Then define δ, $\mathfrak{n}$, and $\bar{\mathfrak{n}}$ as usual relative to Δ^+. Let V be an irreducible representation of G. Our first variant of (7.79) is

$$\mathrm{Hom}_{\mathfrak{g},G}(V,\Gamma^k(I_{\mathfrak{b},T}^{\mathfrak{g},T}(\mathbb{C}_{\lambda+2\delta,\mathfrak{b}})))$$

$$\cong \mathrm{Ext}^k_{\mathfrak{g},T}(V,I_{\mathfrak{b},T}^{\mathfrak{g},T}(\mathbb{C}_{\lambda+2\delta,\mathfrak{b}})) \tag{7.82a}$$

$$\cong \mathrm{Ext}^k_{\mathfrak{b},T}(V,\mathbb{C}_{\lambda+2\delta,\mathfrak{b}}) \tag{7.82b}$$

$$\cong \mathrm{Hom}_{\mathfrak{h},T}(H_k(\mathfrak{n},V),\mathbb{C}_{\lambda+2\delta}) \tag{7.82c}$$

$$\cong \mathrm{Hom}_{\mathfrak{h},T}(H^{S-k}(\mathfrak{n},V\otimes_{\mathbb{C}}\textstyle\bigwedge^S\mathfrak{n}),\mathbb{C}_{\lambda+2\delta}) \tag{7.82d}$$

$$\cong \mathrm{Hom}_{\mathfrak{h},T}(H^{S-k}(\mathfrak{n},V)\otimes_{\mathbb{C}}\textstyle\bigwedge^S\mathfrak{n},\mathbb{C}_{\lambda+2\delta})\,. \tag{7.82e}$$

The proof is the same as for (7.79). Since λ is dominant for Δ^+, the same proof as for Theorem 7.35 allows us to conclude

$$\Gamma^k(I_{\mathfrak{b},T}^{\mathfrak{g},T}(\mathbb{C}_{\lambda+2\delta,\mathfrak{b}})) \cong \begin{cases} \tau_\lambda & \text{if } k = S \\ 0 & \text{if } k \neq S\,. \end{cases} \tag{7.83}$$

The second variant will use the P functor in place of the I functor. Let us write Π in place of $P_{\mathfrak{g},T}^{\mathfrak{g},G}$. The functor Π does not have as nice an analytic interpretation as Γ. But it should be viewed as "currents" or "integrations over suitable subsets," in a fashion dual to taking sections. The variant of (7.79) is

$$\operatorname{Hom}_{\mathfrak{g},G}(\Pi_k(P_{\mathfrak{b},T}^{\mathfrak{g},T}(\mathbb{C}_{\lambda+2\delta,\bar{\mathfrak{b}}})),V)$$

$$\cong \operatorname{Ext}^k_{\mathfrak{g},T}(P_{\mathfrak{b},T}^{\mathfrak{g},T}(\mathbb{C}_{\lambda+2\delta,\bar{\mathfrak{b}}}),(\mathcal{F}^\vee)^{\mathfrak{g},T}_{\mathfrak{g},G}(V)) \qquad (7.84a)$$

$$\cong \operatorname{Ext}^k_{\mathfrak{g},T}(P_{\mathfrak{b},T}^{\mathfrak{g},T}(\mathbb{C}_{\lambda+2\delta,\bar{\mathfrak{b}}}),V) \qquad (7.84b)$$

$$\cong \operatorname{Ext}^k_{\bar{\mathfrak{b}},T}(\mathbb{C}_{\lambda+2\delta,\bar{\mathfrak{b}}},V) \qquad (7.84c)$$

$$\cong \operatorname{Hom}_{\mathfrak{h},T}(\mathbb{C}_{\lambda+2\delta},H^k(\bar{\mathfrak{n}},V)) . \qquad (7.84d)$$

Part (a) follows from Frobenius reciprocity in the same way as for (7.79a). Part (b) is by Proposition 7.22, and part (c) is by Shapiro's Lemma. Part (d) follows by applying to the functor $\operatorname{Hom}_{\bar{\mathfrak{n}}}(\mathbb{C},\cdot)$ the same kind of argument as for (7.79c). This proves (7.84). Since λ is dominant for Δ^+, the same proof as for Theorem 7.35 allows us to conclude

$$\Pi_k(P_{\mathfrak{b},T}^{\mathfrak{g},T}(\mathbb{C}_{\lambda+2\delta,\bar{\mathfrak{b}}})) \cong \begin{cases} \tau_\lambda & \text{if } k = S \\ 0 & \text{if } k \neq S . \end{cases} \qquad (7.85)$$

CHAPTER VIII

REPRESENTATIONS OF NONCOMPACT GROUPS

1. Structure theory for U(m,n)

In Chapter VII we introduced two functors, the tensor-type functor P and the Hom-type functor I , and we saw that their derived functors could be used to construct $(\mathfrak{g},K)$ modules for a pair $(\mathfrak{g},K)$. Whether or not it is true in complete generality, we should regard these $(\mathfrak{g},K)$ modules essentially as coming from group representations. In particular, we saw explicitly at the end of Chapter VII how these functors lead to all the irreducible finite-dimensional representations of the unitary groups U(n) .

The unitary groups are compact connected Lie groups, and our explicit results about U(n) obtained in §§3.3-3.6 and 7.10-7.11 can be extended easily to all compact connected Lie groups by first extending the theory of roots and weights and then checking that the later proofs apply——usually with only minor changes. But the result of all this effort is merely a second construction of previously understood representations ——understood because of the Theorem of the Highest Weight and the Borel-Weil Theorem.

In this chapter we shall turn our attention to noncompact

groups. We shall find that the construction in Chapter VII leads to new and interesting representations, and we shall be able to derive a number of illuminating properties of these representations. One context in which this theory can be carried out is the context of all connected reductive groups in the sense of §7.1. We shall work, however, in the narrower context of a single class of groups $U(m,n)$, where roots and the Weyl group will already be familiar objects.

The matrix group $U(m,n)$ is the group of all $(m+n)$-by-$(m+n)$ complex matrices that preserve the Hermitian quadratic form

$$|z_1|^2 + \ldots + |z_m|^2 - |z_{m+1}|^2 - \ldots - |z_{m+n}|^2$$

when interpreted as linear maps. More concretely let $I_{m,n}$ be the diagonal matrix whose first m diagonal entries are 1's and whose last n diagonal entries are (-1)'s. Then

$$U(m,n) = \{g \in GL(m+n, \mathbb{C}) \mid g^* I_{m,n} g = I_{m,n}\} .$$

We take $U(m,n)$ to be our group G. This group is closed under the conjugate transpose operation $(^*)$ since $I_{m,n}$ is Hermitian. If $m > 0$ and $n > 0$, the group G is noncompact; if $m = 0$, it reduces to the (compact) unitary group $U(n)$.

Let K be the subgroup

$$K = U(m,n) \cap U(m+n) = \left\{\begin{pmatrix} * & 0 \\ 0 & * \end{pmatrix}\right\} \cong U(m) \times U(n) .$$

This subgroup is compact, and one can check that it is maximal compact in G.

The Lie algebra $\mathfrak{g}_0$ of G is

$$\mathfrak{g}_0 = \left\{ \begin{pmatrix} a & b \\ b^* & d \end{pmatrix} \;\middle|\; a^* = -a\,,\ d^* = -d \right\},$$

and its complexification $\mathfrak{g} = \mathfrak{g}_0^{\mathbb{C}}$ is isomorphic to $\mathfrak{gl}(m+n,\mathbb{C})$. In the terminology of Chapter VII, $(\mathfrak{g},K)$ is a pair.

The Lie algebra $\mathfrak{g}_0$ is stable under the map θ equal to negative conjugate transpose, which is a Lie algebra isomorphism of order 2. The eigenspace decomposition under θ is called the <u>Cartan decomposition</u> of $\mathfrak{g}_0$ and is given by

$$\mathfrak{g}_0 = \mathfrak{k}_0 \oplus \mathfrak{p}_0 = \left\{ \begin{pmatrix} a & 0 \\ 0 & d \end{pmatrix} \;\middle|\; a^* = -a\,,\ d^* = -d \right\} \oplus \left\{ \begin{pmatrix} 0 & b \\ b^* & 0 \end{pmatrix} \right\}.$$

The $+1$ eigenspace $\mathfrak{k}_0$ is just the Lie algebra of K. Since θ respects brackets, the -1 eigenspace $\mathfrak{p}_0$ satisfies $[\mathfrak{k}_0,\mathfrak{p}_0] \subseteq \mathfrak{p}_0$ and $[\mathfrak{p}_0,\mathfrak{p}_0] \subseteq \mathfrak{k}_0$.

Complexifying θ and the Cartan decomposition, we obtain

$$\mathfrak{g} = \mathfrak{k} \oplus \mathfrak{p} = \left\{ \begin{pmatrix} a & 0 \\ 0 & d \end{pmatrix} \right\} \oplus \left\{ \begin{pmatrix} 0 & b \\ c & 0 \end{pmatrix} \right\}, \qquad (8.1)$$

where a, b, c, and d are arbitrary matrices of the appropriate sizes. Here we identify $\mathfrak{k} = \mathfrak{k}_0^{\mathbb{C}}$ and $\mathfrak{p} = \mathfrak{p}_0^{\mathbb{C}}$.

The diagonal matrices in $\mathfrak{g}_0$ are the diagonal matrices with imaginary entries. Since $\mathfrak{g}$ is the same as the complexification of $\mathfrak{u}(m+n)$, the theory of roots for the unitary group $U(m+n)$ given in §3.3 is applicable here. We let

$$\mathfrak{h}_0 = \text{diagonal matrices in } \mathfrak{g}_0$$

$$\mathfrak{h}_{\mathbb{R}} = i\mathfrak{h}_0$$

$$\mathfrak{h} = \mathfrak{h}_0 \oplus i\mathfrak{h}_0 = \mathfrak{h}_{\mathbb{R}} \oplus i\mathfrak{h}_{\mathbb{R}}$$

$$T = \text{diagonal matrices in } G .$$

We say T is a (compact) Cartan subgroup of G, and $\mathfrak{h}_0$ is a (compact) Cartan subalgebra of $\mathfrak{g}_0$. If $e_i \in \mathfrak{h}^*$ is evaluation of the ith diagonal entry, then we have the root-space decomposition of $\mathfrak{g}$ with respect to $\mathfrak{h}$:

$$\mathfrak{g} = \mathfrak{h} \oplus \sum_{i \neq j} \mathfrak{g}_{e_i - e_j} , \tag{8.2}$$

where

$$\mathfrak{g}_{e_i - e_j} = \{X \in \mathfrak{g} \mid (\text{ad } H)X = (e_i - e_j)(H)X \text{ for all } H \in \mathfrak{h}\} . \tag{8.3}$$

The set

$$\Delta = \{e_i - e_j \mid i \neq j\} \tag{8.4}$$

is the set of roots of $\mathfrak{g}$ with respect to $\mathfrak{h}$.

All the conclusions of §§3.3 and 3.4 that depend only on $\mathfrak{g}$, $\mathfrak{h}$, and T are valid here, since $\mathfrak{g}$, $\mathfrak{h}$, and T are the same as for $U(m+n)$. Thus the bracket relations (3.8) for the root spaces $\mathfrak{g}_\alpha$ are valid, as are the results of Proposition 3.11 and Corollary 3.12 concerning analytically integral linear functionals.

All the roots are real on $\mathfrak{h}_{\mathbb{R}}$. As in §3.3 we can introduce a lexicographic ordering and define positivity for certain linear functionals on $\mathfrak{h}_{\mathbb{R}}$. If we have some particular lexicographic ordering in mind, we let Δ^+ be the set of positive roots.

In defining the Weyl group W, we generalize from the

algebraic formulation related to (3.21) rather than the analytic formulation given at the beginning of §3.6: For $\alpha \in \Delta$ we define the <u>root reflection</u> s_α on $\mathfrak{h}_{\mathbb{R}}^*$ by

$$s_\alpha \lambda = \lambda - \frac{2\langle \lambda, \alpha \rangle}{|\alpha|^2} \alpha , \qquad (8.5)$$

and we let the <u>Weyl group</u> W be the group generated by these reflections. As in §3.6, W leaves Δ stable and can be identified with the permutation group on $m+n$ letters; it acts by permuting the e_i's. Under this definition of W, Proposition 3.14 remains valid: Any two positive systems Δ^+ (resulting from different lexicographic orderings) are related by one and only one member of W. Also Lemma 3.15 remains valid: Any member of $\mathfrak{h}^*$ that is real on $\mathfrak{h}_{\mathbb{R}}$ is conjugate via W to a dominant element.

If Δ^+ is given, we can introduce simple roots, the length function $\ell(w)$ on W, and the half sum δ of the positive roots as in §3.6, and their properties extend to the present situation.

Another construction that depends only on $\mathfrak{g}$ and $\mathfrak{h}$ (and uses Δ^+ and W) is the Harish-Chandra isomorphism of §6.5. If we fix a choice of Δ^+, we let

$$\mathfrak{n} = \sum_{\alpha \in \Delta^+} \mathfrak{g}_\alpha \quad \text{and} \quad \bar{\mathfrak{n}} = \sum_{\alpha \in \Delta^+} \mathfrak{g}_{-\alpha} . \qquad (8.6)$$

Lemma 6.16a gives us a vector-space direct sum decomposition

$$U(\mathfrak{g}) = U(\mathfrak{h}) \oplus (U(\mathfrak{g})\mathfrak{n} + \bar{\mathfrak{n}} U(\mathfrak{g})) ,$$

and we let $\gamma'_{\mathfrak{n}}$ be the projection to the $U(\mathfrak{h})$ factor. Define $\sigma_{\mathfrak{n}} : \mathfrak{h} \to U(\mathfrak{h})$ by

$$\sigma_{\mathfrak{n}}(H) = H - \delta(H)1 ,$$

and extend $\sigma_{\mathfrak{n}}$ to an algebra automorphism of $U(\mathfrak{h})$ with $\sigma_{\mathfrak{n}}(1) = 1$. The <u>Harish-Chandra map</u> γ is defined by $\gamma = \sigma_{\mathfrak{n}} \circ \gamma'_{\mathfrak{n}}$, and Theorem 6.17 says that γ is an algebra isomorphism of the center $Z(\mathfrak{g})$ onto the Weyl group invariants $U(\mathfrak{h})^W$ of $U(\mathfrak{h})$. Since the image of γ is in the Weyl group invariants, it is easy to see that γ is independent of the choice of Δ^+ (and hence of $\mathfrak{n}$).

If V is a $(\mathfrak{g}, K)$ module, it can happen that $Z(\mathfrak{g})$ acts by scalars in V. For example, let λ be in $\mathfrak{h}^*$ and extend the linear map $\lambda : \mathfrak{h} \to \mathbb{C}$ to an algebra homomorphism $\lambda : U(\mathfrak{h}) \to \mathbb{C}$ with $\lambda(1) = 1$. If $Z(\mathfrak{g})$ acts in V by $zv = \lambda(\gamma(z))v$, then we say that V has <u>infinitesimal character</u> λ; in this case the parameter λ is determined up to a member of W. [Example: When $n = 0$, the irreducible representation of $U(m)$ of highest weight λ has infinitesimal character $\lambda+\delta$, according to (6.52).]

Referring to (8.1), we see that each matrix E_{ij} is in $\mathfrak{k}$ or in $\mathfrak{p}$, and we call the root $e_i - e_j$ <u>compact</u> or <u>noncompact</u> accordingly. The sets of compact and noncompact roots are denoted Δ_K and Δ_n, respectively, and we have

$$\mathfrak{k} = \mathfrak{h} \oplus \sum_{\alpha \in \Delta_K} \mathfrak{g}_\alpha \tag{8.7a}$$

$$\mathfrak{p} = \sum_{\alpha \in \Delta_n} \mathfrak{g}_\alpha \tag{8.7b}$$

$$\Delta_K = \{e_i - e_j \mid i \text{ and } j \text{ both} \leq m \text{ or both} \geq m+1\} . \quad (8.7c)$$

The group K is just the product $U(m) \times U(n)$, and there is no difficulty in forming roots of $\mathfrak{k}$ with respect to $\mathfrak{h}$ (since we already understand roots for $U(m)$ and $U(n)$ separately); the resulting root-space decomposition is just (8.7a). Similarly K has a Weyl group W_K, which is just the product of the Weyl groups of $U(m)$ and $U(n)$. So W_K is the subgroup of W of all permutations that leave stable the two sets of indices $\{1,\ldots,m\}$ and $\{m+1,\ldots,m+n\}$. If Δ^+ is specified, we let Δ_K^+ and Δ_n^+ be the intersections with Δ^+ of Δ_K and Δ_n, and we let δ_K and δ_n be the half sums of the members of Δ_K^+ and Δ_n^+.

We shall refer to any subalgebra $\mathfrak{b} = \mathfrak{h} \oplus \mathfrak{n}$ of $\mathfrak{g}$ (with $\mathfrak{n}$ as in (8.6)) as a Borel subalgebra of $\mathfrak{g}$. There is one Borel subslgebra for each choice of Δ^+. Recall the involution θ of $\mathfrak{g}$ whose eigenspaces are $\mathfrak{k}$ and $\mathfrak{p}$. Since $\mathfrak{h}$ is in $\mathfrak{k}$ and since each root space is in $\mathfrak{k}$ or $\mathfrak{p}$, each of our Borel subalgebras is stable under θ.

We shall refer to any subalgebra $\mathfrak{q}$ of $\mathfrak{g}$ that contains a Borel subalgebra as a parabolic subalgebra. Let us classify these subalgebras.

Proposition 8.1. Let Δ^+ be a positive system, let $\{\alpha_1, \ldots, \alpha_r\}$ be a subset of simple roots, and let Δ_L be the subset of Δ of roots that are linear combinations of $\alpha_1, \ldots, \alpha_r$. Put

$$\mathfrak{l} = \mathfrak{h} \oplus \sum_{\alpha \in \Delta_L} \mathfrak{g}_\alpha$$

$$\mathfrak{u} = \sum_{\alpha\in\Delta^+-\Delta_L} \mathfrak{g}_\alpha .$$

Then $\mathfrak{l}$ and $\mathfrak{u}$ are subalgebras, $[\mathfrak{l},\mathfrak{u}]$ is contained in $\mathfrak{u}$, and $\mathfrak{q} = \mathfrak{l} \oplus \mathfrak{u}$ is a parabolic subalgebra. Conversely every parabolic subalgebra arises in this way for some choice of Δ^+ and $\{\alpha_1, \ldots, \alpha_r\}$.

Proof. It follows from the bracket relations (3.8) that $\mathfrak{l}$ and $\mathfrak{u}$ are subalgebras, $[\mathfrak{l},\mathfrak{u}]$ is contained in $\mathfrak{u}$, and $\mathfrak{q} = \mathfrak{l} \oplus \mathfrak{u}$ is a subalgebra. Since $\mathfrak{q}$ contains the Borel subalgebra corresponding to Δ^+, $\mathfrak{q}$ is parabolic.

Conversely let $\mathfrak{q}$ be given, and let $\mathfrak{b}$ be a Borel subalgebra such that $\mathfrak{q} \supseteq \mathfrak{b}$. Let Δ^+ be the positive system corresponding to $\mathfrak{b}$. Since $\mathfrak{h} \subseteq \mathfrak{q}$ and since each eigenspace of $\mathfrak{h}$ other than $\mathfrak{h}$ itself is one-dimensional, we have

$$\mathfrak{q} = \mathfrak{h} \oplus \sum_{\alpha\in\Phi} \mathfrak{g}_\alpha \tag{8.8a}$$

for some subset $\Phi \subseteq \Delta$. Let $\Delta_L = \Phi \cap (-\Phi)$, and let $\Pi_L = \{\alpha_1, \ldots, \alpha_r\}$ be the subset of simple roots in Δ_L. Since $\Delta^+ \subseteq \Phi$, we have

$$\Phi = \Delta_L \cup (\Delta^+ - \Delta_L) \tag{8.8b}$$

disjointly. Thus the proof will be complete if we show that Δ_L consists of all members of Δ that are linear combinations of the members of Π_L.

First let us observe from (3.8) and the closure of $\mathfrak{q}$ under brackets that Φ has the following property: Whenever β_1 and β_2 are in Φ and $\beta_1+\beta_2$ is in Δ, then $\beta_1+\beta_2$ is in Φ. Then it is clear by induction (on the sum of coefficients) that all linear combinations from Π_L that are in Δ are actually in Δ_L.

Conversely we show that every member of Δ_L is such a linear combination. If $\pm\alpha$ are in Δ_L with $\alpha < 0$, we expand $-\alpha = \sum n_i \alpha_i$ in terms of simple roots, not necessarily from Π_L. We proceed by induction on $\sum n_i$ to prove our assertion, the case $\sum n_i = 1$ being trivial. Choose α_j simple such that $-\alpha-\alpha_j$ is in Δ. This is possible since we are beyond the trivial case of the induction, so that $-\alpha$ is positive but not simple. Then $-\alpha-\alpha_j$ is in Φ since $-\alpha-\alpha_j$ is positive. Since α and $-\alpha-\alpha_j$ are in Φ and their sum $-\alpha_j$ is in Δ, $-\alpha_j$ is in Φ. Thus the simple root α_j is in Π_L. This conclusion reduces the proof of our assertion to the case of $\pm(\alpha+\alpha_j)$, and the sum of coefficients for $-(\alpha+\alpha_j)$ is a positive integer smaller than that for $-\alpha$. Hence our result follows by induction.

In the decomposition $\mathfrak{q} = \mathfrak{l} \oplus \mathfrak{u}$ of Proposition 8.1, we can characterize $\mathfrak{l}$ and $\mathfrak{u}$ in terms of $\mathfrak{q}$ as the proof suggests: If we define $\Phi \subseteq \Delta$ by (8.8a), then the roots α that contribute to $\mathfrak{l}$ must have both α and $-\alpha$ in Φ, while the roots that contribute to $\mathfrak{u}$ must not have this property. So $\mathfrak{l}$ and $\mathfrak{u}$ are uniquely defined. Accordingly we define

$$\Delta(\mathfrak{u}) = \Delta^+ - \Delta_L, \qquad \Delta(\mathfrak{u} \cap \mathfrak{k}) = \Delta(\mathfrak{u}) \cap \Delta_K, \qquad \Delta(\mathfrak{u} \cap \mathfrak{p}) = \Delta(\mathfrak{u}) \cap \Delta_n$$

and

$$\Delta(\mathfrak{l} \cap \mathfrak{k}) = \Delta_L \cap \Delta_K, \quad \Delta(\mathfrak{l} \cap \mathfrak{p}) = \Delta_L \cap \Delta_n.$$

Within $G = U(m,n)$, there is a connected closed subgroup L whose Lie algebra $\mathfrak{l}_0$ has $\mathfrak{l}_0^{\mathbb{C}} = \mathfrak{l}$. We can describe L explicitly as follows: Call two indices i and j between 1 and $m+n$ equivalent for L if $e_i - e_j$ is in Δ_L. The result is an equivalence relation (transitivity following since β_1 and β_2 in Δ_L and $\beta_1+\beta_2$ in Δ implies $\beta_1+\beta_2$ is in Δ_L). If we permute the indices from 1 to $m+n$ so that each equivalence class consists of consecutive integers, then L consists of the block diagonal matrices whose blocks correspond to the sets of consecutive integers. The theory developed in the remainder of the chapter will essentially use representations of L to construct representations of G.

<u>Proposition 8.2</u>. If $\mathfrak{q}$ is a parabolic subalgebra, then there exists a member X_0 of $\mathfrak{h}_{\mathbb{R}}$ such that

$$\mathfrak{q} = \text{sum of eigenspaces in } \mathfrak{g} \text{ for } \operatorname{ad} X_0 \text{ with eigenvalue} \geq 0.$$

Moreover the decomposition $\mathfrak{q} = \mathfrak{l} \oplus \mathfrak{u}$ of Proposition 8.1 is such that

$$\mathfrak{u} = \text{sum of eigenspaces in } \mathfrak{g} \text{ for } \operatorname{ad} X_0 \text{ with eigenvalue} > 0$$

$$\mathfrak{l} = \text{eigenspace in } \mathfrak{g} \text{ for } \operatorname{ad} X_0 \text{ for eigenvalue } 0.$$

Conversely any X_0 in $\mathfrak{h}_{\mathbb{R}}$ determines a parabolic subalgebra $\mathfrak{q} = \mathfrak{l} \oplus \mathfrak{u}$ in this way.

Remark. Proposition 8.1 constructs $\mathfrak{l}$ as containing $\mathfrak{h}$. Thus X_0 is in $\mathfrak{l}$. Since $\operatorname{ad} X_0$ is 0 on $\mathfrak{l}$, X_0 is in the center of $\mathfrak{l}$.

Proof. If $\mathfrak{q}$ is parabolic, choose Δ^+ and $\{\alpha_1, \ldots, \alpha_r\}$ as in Proposition 8.1, and let $\alpha_{r+1}, \ldots, \alpha_{m+n-1}$ be the other simple roots for Δ^+. Since the α_i's are linearly independent and are real-valued on $\mathfrak{h}_{\mathbb{R}}$, we can choose X_0 in $\mathfrak{h}_{\mathbb{R}}$ with

$$\alpha_i(X_0) = \begin{cases} 0 & \text{if } i \leq r \\ 1 & \text{if } i > r. \end{cases}$$

If $\alpha = \Sigma\, n_i \alpha_i$ is in Δ, then ad X_0 acts on $\mathfrak{g}_\alpha$ by the scalar

$$\alpha(X_0) = \sum_i n_i \alpha_i(X_0) = \sum_{i>r} n_i .$$

This number is 0 if α is in Δ_L, it is > 0 if α is in $\Delta(\mathfrak{u})$, and it is < 0 otherwise. The direct part of the corollary follows.

Conversely let X_0 in $\mathfrak{h}_{\mathbb{R}}$ be given, and extend X_0 to a basis $X_0, X_2, \ldots, X_{m+n}$ of $\mathfrak{h}_{\mathbb{R}}$. Using the lexicographic ordering of $\mathfrak{h}_{\mathbb{R}}^*$ obtained from this basis, we obtain a positive system Δ^+. Let $\mathfrak{b}$ be the Borel subalgebra corresponding to Δ^+, and let $\mathfrak{q}$ be the sum of the eigenspaces in $\mathfrak{g}$ for ad X_0 with eigenvalue ≥ 0. It is enough to prove that $\mathfrak{q}$ is a subalgebra and that $\mathfrak{q} \supseteq \mathfrak{b}$.

If Y_1 and Y_2 are eigenvectors for ad X_0 with eigenvalues $c_1 \geq 0$ and $c_2 \geq 0$, then

$$(\text{ad } X_0)[Y_1,Y_2] = [(\text{ad } X_0)Y_1,Y_2] + [Y_1,(\text{ad } X_0)Y_2] = (c_1+c_2)[Y_1,Y_2]$$

shows $[Y_1, Y_2]$ is an eigenvector with eigenvalue $c_1+c_2 \geq 0$. Hence $\mathfrak{q}$ is a subalgebra.

Since $\mathfrak{h}$ is in the 0 eigenspace for $\operatorname{ad} X_0$, $\mathfrak{h}$ is contained in $\mathfrak{q}$. Suppose α is a positive root. Then $\operatorname{ad} X_0$ acts on $\mathfrak{g}_\alpha$ by the scalar $\alpha(X_0)$, which must be ≥ 0 since α is positive and X_0 is the first member of the basis. Hence $\mathfrak{q} \supseteq \mathfrak{g}_\alpha$, and we conclude $\mathfrak{q} \supseteq \mathfrak{b}$.

Each of our parabolic subalgebras is stable under θ since Proposition 8.1 shows that such an algebra is the sum of some root spaces and the Cartan subalgebra $\mathfrak{h}$, each of which is θ-stable. It is customary to refer to these parabolic subalgebras as <u>θ-stable parabolic subalgebras</u>, to distinguish them from subalgebras constructed in analogous fashion from nondiagonal Cartan subalgebras that are not compact. This property of stability under θ will be critical for the main results of this chapter.

2. Cohomological induction

We continue with $\mathfrak{g}$ as in §1. Fix a θ-stable parabolic subalgebra $\mathfrak{q}$, and write $\mathfrak{q} = \mathfrak{l} \oplus \mathfrak{u}$ as in Proposition 8.1. In terms of root spaces, $\mathfrak{u}$ decomposes as $\mathfrak{u} = \Sigma_{\alpha\in\Delta(\mathfrak{u})} \mathfrak{g}_\alpha$, and we define

$$\bar{\mathfrak{u}} = \sum_{\alpha\in\Delta(\mathfrak{u})} \mathfrak{g}_{-\alpha} \quad \text{and} \quad \bar{\mathfrak{q}} = \mathfrak{l} \oplus \bar{\mathfrak{u}} . \tag{8.9}$$

Then $\bar{\mathfrak{q}}$ is another θ-stable parabolic subalgebra, and $\mathfrak{q} \cap \bar{\mathfrak{q}} = \mathfrak{l}$.

The situation in §7.11 is a special case. The underlying Lie algebra $\mathfrak{g}_0$ was $\mathfrak{u}(n)$, and we now consider the more general $\mathfrak{u}(m,n)$. What was $\mathfrak{h}$ in that section is now playing two roles, those of $\mathfrak{h}$ and $\mathfrak{l}$. Then $\mathfrak{n}$, $\bar{\mathfrak{n}}$, $\mathfrak{b}$, and $\bar{\mathfrak{b}}$ have been generalized to $\mathfrak{u}$, $\bar{\mathfrak{u}}$, $\mathfrak{q}$, and $\bar{\mathfrak{q}}$.

In our new setting, the following pairs will be of interest to us: $(\mathfrak{g},K)$, $(\mathfrak{g},L\cap K)$, $(\mathfrak{q},L\cap K)$, $(\bar{\mathfrak{q}},L\cap K)$, $(\mathfrak{l},L\cap K)$, and each of these five pairs with its algebra intersected with $\mathfrak{k}$. Let

$$R = \dim_{\mathbb{C}}(\mathfrak{u}\cap\mathfrak{p}) \quad \text{and} \quad S = \dim_{\mathbb{C}}(\mathfrak{u}\cap\mathfrak{k}).$$

Since $[\mathfrak{l},\mathfrak{u}] \subseteq \mathfrak{u}$, $\mathfrak{u}$ is an $(\mathfrak{l},L\cap K)$ module. Hence $\bigwedge^{R+S}\mathfrak{u}$ is a one-dimensional $(\mathfrak{l},L\cap K)$ module. Similarly $\bigwedge^{R}(\mathfrak{u}\cap\mathfrak{p})$ and $\bigwedge^{S}(\mathfrak{u}\cap\mathfrak{k})$ are one-dimensional $(\mathfrak{l}\cap\mathfrak{k},L\cap K)$ modules.

Our construction will proceed in analogy with the end of §7.11, and we hope for generalizations of the formulas (7.83) and (7.85) that identify the results of the construction. We start with Z in $\mathcal{C}(\mathfrak{l},L\cap K)$; this replaces the $\mathcal{C}(\mathfrak{h},T)$ module $\mathbb{C}_\lambda$ in §7.11. Let

$$Z^{\#} = Z\otimes_{\mathbb{C}}\bigwedge^{R+S}\mathfrak{u}.$$

This is the member of $\mathcal{C}(\mathfrak{l},L\cap K)$ that replaces $\mathbb{C}_{\lambda+2\delta}$. Define $Z^{\#}_{\mathfrak{q}}$ to be $Z^{\#}$ as a member of $\mathcal{C}(\mathfrak{q},L\cap K)$ with $\mathfrak{u}$ acting by 0, and define $Z^{\#}_{\bar{\mathfrak{q}}}$ to be $Z^{\#}$ as a member of $\mathcal{C}(\bar{\mathfrak{q}},L\cap K)$ with $\bar{\mathfrak{u}}$ acting by 0.

We write Π in place of $P^{\mathfrak{g},K}_{\mathfrak{g},L\cap K}$ and let Π_i be its ith derived functor (obtained from a projective resolution). We write Γ in place of $I_{\mathfrak{g},L\cap K}^{\mathfrak{g},K}$ and let Γ^i be its ith derived functor (obtained from an injective resolution). Let $\mathcal{L}_i(Z)$ and $\mathcal{R}^i(Z)$ be the cohomologically induced modules given as the following members of $\mathcal{C}(\mathfrak{g},K)$:

$$\mathcal{L}_i(Z) = \Pi_i(P_{\mathfrak{q},L\cap K}^{\mathfrak{g},L\cap K}(Z_{\mathfrak{q}}^{\#})) = (P_{\mathfrak{q},L\cap K}^{\mathfrak{g},K})_i(Z_{\mathfrak{q}}^{\#}) \tag{8.10a}$$

$$\mathcal{R}^i(Z) = \Gamma^i(I_{\mathfrak{q},L\cap K}^{\mathfrak{g},L\cap K}(Z_{\mathfrak{q}}^{\#})) = (I_{\mathfrak{q},L\cap K}^{\mathfrak{g},K})^i(Z_{\mathfrak{q}}^{\#}) . \tag{8.10b}$$

(The second equality in each of (8.10a) and (8.10b) follows from Propositions 5.17 and 7.16; the hypotheses of Proposition 5.17 are satisfied because of Proposition 7.15 and Corollary 7.19.)

Let us look ahead to see where we are going. Shortly we shall see that $\mathcal{L}_i(Z)$ and $\mathcal{R}^i(Z)$ are 0 for $i > 2S$. This is an elementary fact that follows from properties of Π and Γ alone and does not depend upon the composite nature of $\mathcal{L}_i$ and $\mathcal{R}^i$. In §3 we shall see that $\mathcal{L}_i(Z)$ and $\mathcal{R}^i(Z)$ are 0 for $i > S$. This is a deeper result that does depend on the composite nature of $\mathcal{L}_i$ and $\mathcal{R}^i$. Finally in §§4-6 we shall see that $\mathcal{L}_i(Z)$ and $\mathcal{R}^i(Z)$ are 0 for $i < S$ if suitable positivity conditions are satisfied for certain parameters. This is the deepest result, and the positivity conditions are crucial; the techniques of §VII.11 show that a vanishing theorem for $i < S$ need not hold without the positivity. Anyway our ultimate interest is consequently in the case that the positivity conditions hold and that $i = S$. Since the difficult range to study is $0 \leq i \leq 2S$, our ultimate interest is in the middle dimension. In §§7-8 we shall derive properties of these modules that come close to characterizing the modules uniquely.

Our procedure will be to try to imitate the theory of §VII.11 as much as possible, using (7.82) for $\mathcal{R}^i$ and (7.84) for $\mathcal{L}_i$. Unfortunately these arguments do not apply in our new wider setting; they merely provide a frame of reference. Initially we shall see that we can get some information by working with $\mathfrak{k}$ rather than $\mathfrak{g}$. This philosophy can be carried out in several ways, and we give two such ways in this section. Let

$$\Pi^K = P^{\mathfrak{k},K}_{\mathfrak{k},L\cap K} \quad \text{and} \quad \Gamma_K = I^{\mathfrak{k},K}_{\mathfrak{k},L\cap K},$$

and let Π^K_i and Γ^i_K be their respective derived functors.

Lemma 8.3. The forgetful functor $\mathcal{F}^{\mathfrak{k},L\cap K}_{\mathfrak{g},L\cap K}$ carries projectives to projectives and injectives to injectives.

Proof. We prove the assertion about projectives, the proof for injectives being similar. If P is projective in $\mathcal{C}(\mathfrak{g},L\cap K)$, then Corollary 7.21b says that P is a direct summand in a standard projective $\widetilde{P}$. It is enough to prove that $\mathcal{F}^{\mathfrak{k},L\cap K}_{\mathfrak{g},L\cap K}(\widetilde{P})$. Dropping forgetful functors judiciously, we have

$$\begin{aligned}
\mathcal{F}^{\mathfrak{k},L\cap K}_{\mathfrak{g},L\cap K}(\widetilde{P}) &= R(\mathfrak{g},L\cap K)\otimes_{C_{L\cap K}} V \\
&\cong (S(\mathfrak{p})\otimes_{\mathbb{C}} R(\mathfrak{k},L\cap K))\otimes_{C_{L\cap K}} V \quad \text{by Proposition 7.13} \\
&\cong S(\mathfrak{p})\otimes_{\mathbb{C}} (R(\mathfrak{k},L\cap K)\otimes_{C_{L\cap K}} V) .
\end{aligned}$$

The last isomorphism here is $(s \otimes r) \otimes v \to s \otimes (r \otimes v)$ since $C_{L \cap K}$ acts trivially on $S(\mathfrak{p})$ from the right. The resulting expression is $S(\mathfrak{p}) \otimes_{\mathbb{C}} P^{\mathfrak{k}, L\cap K}_{\mathfrak{l}\cap\mathfrak{k}, L\cap K}(V)$, and this is projective by Corollary 7.20 and Proposition 7.9a.

Proposition 8.4. For each $i \geq 0$, there are natural isomorphisms

$$\Pi^K_i \circ \mathcal{F}^{\mathfrak{k}, L\cap K}_{\mathfrak{g}, L\cap K} \cong \mathcal{F}^{\mathfrak{k}, K}_{\mathfrak{g}, K} \circ \Pi_i \tag{8.11a}$$

$$\Gamma^i_K \circ \mathcal{F}^{\mathfrak{k}, L\cap K}_{\mathfrak{g}, L\cap K} \cong \mathcal{F}^{\mathfrak{k}, K}_{\mathfrak{g}, K} \circ \Gamma^i \tag{8.11b}$$

Proof. Dropping the $\mathcal{F}$'s, we use Propositions 7.14 and 7.11 to write

$$\begin{aligned} P^{\mathfrak{g}, K}_{\mathfrak{g}, L\cap K}(V) &= R(\mathfrak{g}, K) \otimes_{R(\mathfrak{g}, L\cap K)} V \\ &\cong (C_K \otimes_{R(\mathfrak{k}, L\cap K)} R(\mathfrak{g}, L\cap K)) \otimes_{R(\mathfrak{g}, L\cap K)} V \\ &\cong C_K \otimes_{R(\mathfrak{k}, L\cap K)} (R(\mathfrak{g}, L\cap K) \otimes_{R(\mathfrak{g}, L\cap K)} V) \\ &\cong C_K \otimes_{R(\mathfrak{k}, L\cap K)} V = P^{\mathfrak{k}, K}_{\mathfrak{k}, L\cap K}(V) . \end{aligned}$$

More precisely this says

$$\Pi^K \circ \mathcal{F}^{\mathfrak{k}, L\cap K}_{\mathfrak{g}, L\cap K} \cong \mathcal{F}^{\mathfrak{k}, K}_{\mathfrak{g}, K} \circ \Pi .$$

We pass to derived functors on both sides, using Propositions 5.17(a1) and 5.16a to handle the compositions on each side. These propositions are applicable since both forgetful functors are exact and since $\mathcal{F}^{\mathfrak{k}, L\cap K}_{\mathfrak{g}, L\cap K}$ carries projectives to projectives, by Lemma 8.3. This proves (8.11a), and the proof of (8.11b) is similar.

<u>Corollary 8.5</u>. Π_i and Γ^i are 0 functors for $i > 2S$. In particular, $\mathcal{L}_i(Z)$ and $\mathcal{R}^i(Z)$ are 0 for $i > 2S$.

Proof. The functors Π^K and Γ_K are defined on $\mathcal{C}(\mathfrak{k}, L\cap K)$, and the standard resolutions (7.60) and (7.61) in this category are 0 beyond degree

$$\dim_{\mathbb{C}}(\mathfrak{k}/(\mathfrak{l}\cap\mathfrak{k})) = \dim_{\mathbb{C}}(\mathfrak{u}\cap\mathfrak{k}\oplus\bar{\mathfrak{u}}\cap\mathfrak{k}) = 2S\,.$$

Hence Π_i^K and Γ_K^i are 0 for $i > 2S$. Applying (8.11), we see that Π_i and Γ^i are 0 for $i > 2S$.

Recall that our procedure in handling $\mathcal{L}_i(Z)$ and $\mathcal{R}^i(Z)$ is to imitate the theory of §VII.11 as much as possible. Let us write out prospective analogs of (7.84) and (7.82) in our new setting, taking V to be a $(\mathfrak{g},K)$ module. The analog of (7.84) for $\mathcal{L}_i(Z)$ is

$$\begin{aligned}
\operatorname{Hom}_{\mathfrak{g},K}(\mathcal{L}_i(Z),V) &= \operatorname{Hom}_{\mathfrak{g},K}(\Pi_i(P_{\mathfrak{q},L\cap K}^{\mathfrak{g},L\cap K}(Z^{\#}_{\mathfrak{q}})),V) \\
&\overset{?}{\cong} \operatorname{Ext}^i_{\mathfrak{g},L\cap K}(P_{\mathfrak{q},L\cap K}^{\mathfrak{g},L\cap K}(Z^{\#}_{\mathfrak{q}}),(\mathcal{F}^{\vee})_{\mathfrak{g},K}^{\mathfrak{g},L\cap K}(V)) && (8.12a)\\
&\overset{?}{\cong} \operatorname{Ext}^i_{\mathfrak{g},L\cap K}(P_{\mathfrak{q},L\cap K}^{\mathfrak{g},L\cap K}(Z^{\#}_{\mathfrak{q}}),V) && (8.12b)\\
&\overset{?}{\cong} \operatorname{Ext}^i_{\mathfrak{q},L\cap K}(Z^{\#}_{\mathfrak{q}},V) && (8.12c)\\
&\overset{?}{\cong} \operatorname{Hom}_{\mathfrak{l},L\cap K}(Z^{\#},H^i(\bar{\mathfrak{u}},V))\,, && (8.12d)
\end{aligned}$$

and the analog of (7.82) for $\mathcal{R}^i(Z)$ is

$$\operatorname{Hom}_{\mathfrak{g},K}(V,\mathcal{R}^i(Z)) = \operatorname{Hom}_{\mathfrak{g},K}(V,\Gamma^i(I_{\mathfrak{q},L\cap K}^{\mathfrak{g},L\cap K}(Z_{\mathfrak{q}}^{\#})))$$

$$\overset{?}{\cong} \operatorname{Ext}^i_{\mathfrak{g},L\cap K}(V, I_{\mathfrak{q},L\cap K}^{\mathfrak{g},L\cap K}(Z_{\mathfrak{q}}^{\#})) \tag{8.13a}$$

$$\overset{?}{\cong} \operatorname{Ext}^i_{\mathfrak{q},L\cap K}(V, Z_{\mathfrak{q}}^{\#}) \tag{8.13b}$$

$$\overset{?}{\cong} \operatorname{Hom}_{\mathfrak{l},L\cap K}(H_i(\mathfrak{u},V),Z^{\#}) . \tag{8.13c}$$

Although they may seem superfluous at first, we can push through analogs of (8.12) and (8.13) on the $\mathfrak{k}$ level. Let

$$\mathcal{L}_i^K(Z) = \Pi_i^K(P_{\mathfrak{q}\cap\mathfrak{k},L\cap K}^{\mathfrak{k},L\cap K}(Z_{\overline{\mathfrak{q}}}^{\#})) \tag{8.14a}$$

$$\mathcal{R}_K^i(Z) = \Gamma_K^i(I_{\mathfrak{q}\cap\mathfrak{k},L\cap K}^{\mathfrak{k},L\cap K}(Z_{\mathfrak{q}}^{\#})) . \tag{8.14b}$$

Proposition 8.6. For $i > S$, $\mathcal{L}_i^K(Z) = 0$ and $\mathcal{R}_K^i(Z) = 0$. For $0 \leq i \leq S$ and for any finite-dimensional $(\mathfrak{k},K)$ module V,

$$\operatorname{Hom}_K(\mathcal{L}_i^K(Z),V) \cong \operatorname{Hom}_{L\cap K}(Z^{\#},H^i(\overline{\mathfrak{u}}\cap\mathfrak{k},V))$$

$$\operatorname{Hom}_K(V,\mathcal{R}_K^i(Z)) \cong \operatorname{Hom}_{L\cap K}(H_i(\mathfrak{u}\cap\mathfrak{k},V),Z^{\#}) .$$

Proof. Let us write down analogs of (8.12) and (8.13) and then justify the steps. In fact, the justifications will be essentially the same as for (7.84) and (7.82). The analog of (8.12) is

$$\operatorname{Hom}_K(\mathcal{L}_i^K(Z),V) = \operatorname{Hom}_{\mathfrak{k},K}(\Pi_i^K(P_{\overline{\mathfrak{q}}\cap\mathfrak{k},L\cap K}^{\mathfrak{k},L\cap K}(Z_{\overline{\mathfrak{q}}}^{\#})),V)$$

$$\cong \operatorname{Ext}^i_{\mathfrak{k},L\cap K}(P_{\overline{\mathfrak{q}}\cap\mathfrak{k},L\cap K}^{\mathfrak{k},L\cap K}(Z_{\overline{\mathfrak{q}}}^{\#}),(\mathcal{F}^\vee)_{\mathfrak{k},K}^{\mathfrak{k},L\cap K}(V)) \tag{8.15a}$$

$$\cong \operatorname{Ext}^i_{\mathfrak{k}, L\cap K}(P^{\mathfrak{k}, L\cap K}_{\mathfrak{q}\cap\mathfrak{k}, L\cap K}(Z^{\#}_{\bar{\mathfrak{q}}}), V) \tag{8.15b}$$

$$\cong \operatorname{Ext}^i_{\bar{\mathfrak{q}}\cap\mathfrak{k}, L\cap K}(Z^{\#}_{\bar{\mathfrak{q}}}, V) \tag{8.15c}$$

$$\cong \operatorname{Hom}_{\mathfrak{l}\cap\mathfrak{k}, L\cap K}(Z^{\#}, H^i(\bar{\mathfrak{u}}\cap\mathfrak{k}, V)), \tag{8.15d}$$

and the analog of (8.13) is

$$\operatorname{Hom}_K(V, \mathcal{R}^i_K(Z)) = \operatorname{Hom}_{\mathfrak{k},K}(V, \Gamma^i_K(I^{\mathfrak{k}, L\cap K}_{\mathfrak{q}\cap\mathfrak{k}, L\cap K}(Z^{\#}_{\mathfrak{q}})))$$

$$\cong \operatorname{Ext}^i_{\mathfrak{k}, L\cap K}(V, I^{\mathfrak{k}, L\cap K}_{\mathfrak{q}\cap\mathfrak{k}, L\cap K}(Z^{\#}_{\mathfrak{q}})) \tag{8.16a}$$

$$\cong \operatorname{Ext}^i_{\mathfrak{q}\cap\mathfrak{k}, L\cap K}(V, Z^{\#}_{\mathfrak{q}}) \tag{8.16b}$$

$$\cong \operatorname{Hom}_{\mathfrak{l}\cap\mathfrak{k}, L\cap K}(H_i(\mathfrak{u}\cap\mathfrak{k}, V), Z^{\#}). \tag{8.16c}$$

In proving (8.15), we use Proposition 7.22 for (8.15a) and Shapiro's Lemma (7.51) for (8.15c). For (8.15b) we start from the natural isomorphism

$$\operatorname{Hom}_{\mathfrak{k},K}(\cdot, V)\circ\Pi^K \cong \operatorname{Hom}_{\mathfrak{k}, L\cap K}(\cdot, V)$$

given in Frobenius reciprocity (7.45), and we take derived functors of both sides. Since $\operatorname{Hom}_{\mathfrak{k},K}(\cdot, V)$ is exact (Lemma 7.8 and Proposition 5.10b), we can use Proposition 5.16a to handle the derived functors of the composition, and (8.15a) results.

For (8.15d) we start from the natural isomorphism for V in $\mathcal{C}(\bar{\mathfrak{q}}\cap\mathfrak{k}, L\cap K)$ given by

$$\operatorname{Hom}_{\bar{\mathfrak{q}}\cap\mathfrak{k}, L\cap K}(Z^{\#}_{\bar{\mathfrak{q}}}, V) \cong \operatorname{Hom}_{\mathfrak{l}\cap\mathfrak{k}, L\cap K}(Z^{\#}_{\bar{\mathfrak{q}}}, \operatorname{Hom}_{\bar{\mathfrak{u}}\cap\mathfrak{k}}(\mathbb{C}, V)), \tag{8.17}$$

the map from left to right being $\varphi \to \Phi$ with $\Phi(z)(c) = c\varphi(z)$. If $G : \mathcal{C}(\bar{\mathfrak{q}} \cap \mathfrak{k}, L \cap K) \to \mathcal{C}(\mathfrak{l} \cap \mathfrak{k}, L \cap K)$ is the functor $G(V) = \mathrm{Hom}_{\bar{\mathfrak{u}} \cap \mathfrak{k}}(\mathbb{C}, V)$, then (8.17) says

$$\mathrm{Hom}_{\bar{\mathfrak{q}} \cap \mathfrak{k}, L \cap K}(Z_{\mathfrak{q}}^{\#}, \cdot) \cong \mathrm{Hom}_{\mathfrak{l} \cap \mathfrak{k}, L \cap K}(Z^{\#}, \cdot) \circ G .$$

We take derived functors of both sides and use Proposition 5.16a to handle the derived functors of the composition ($\mathrm{Hom}_{\mathfrak{l} \cap \mathfrak{k}, L \cap K}(Z^{\#}, \cdot)$ being exact by Lemma 7.8 and Proposition 5.10b). The derived functors of G are naturally isomorphic with $H^i(\bar{\mathfrak{u}} \cap \mathfrak{k}, V)$, by the same proof as for Proposition 7.33b, and (8.15d) follows.

In proving (8.16), we proceed similarly. For (8.16b) we use Shapiro's Lemma (7.52). For (8.16a) we start from

$$\mathrm{Hom}_{\mathfrak{k}, K}(V, \cdot) \circ \Gamma_K \cong \mathrm{Hom}_{\mathfrak{k}, L \cap K}(V, \cdot)$$

given in Frobenius reciprocity (7.46) and pass to derived functors, using the exactness of $\mathrm{Hom}_{\mathfrak{k}, K}(V, \cdot)$. For (8.16c) we start from

$$\mathrm{Hom}_{\mathfrak{l} \cap \mathfrak{k}, L \cap K}(V \otimes_{\mathfrak{u} \cap \mathfrak{k}} \mathbb{C}, Z^{\#}) \cong \mathrm{Hom}_{\mathfrak{q} \cap \mathfrak{k}, L \cap K}(V, Z_{\mathfrak{q}}^{\#})$$

and rewrite the isomorphism on the functor level as

$$\mathrm{Hom}_{\mathfrak{l} \cap \mathfrak{k}, L \cap K}(\cdot, Z^{\#}) \circ G' \cong \mathrm{Hom}_{\mathfrak{q} \cap \mathfrak{k}, L \cap K}(\cdot, Z_{\mathfrak{q}}^{\#})$$

with

$$G'(V) = V \otimes_{\mathfrak{u} \cap \mathfrak{k}} \mathbb{C} .$$

We pass to derived functors, using the natural isomorphism

$G_i^{!}(V) \cong H_i(\mathfrak{u} \cap \mathfrak{k}, V)$, which is essentially proved in Proposition 7.33a, and the exactness of $\operatorname{Hom}_{\mathfrak{l} \cap \mathfrak{k}, L \cap K}(\cdot, Z^{\#})$, and (8.16c) follows.

This proves the isomorphisms in the proposition. Since $H^i(\bar{\mathfrak{u}} \cap \mathfrak{k}, \cdot)$ and $H_i(\mathfrak{u} \cap \mathfrak{k}, \cdot)$ are zero functors for $i > S$, we obtain

$$\operatorname{Hom}_K(\mathcal{L}_i^K(Z), V) = 0 \quad \text{and} \quad \operatorname{Hom}_K(V, \mathcal{R}_K^i(Z)) = 0$$

for $i > S$ and all finite-dimensional V. Taking V irreducible of type τ, we see that the K-isotypic subspaces of type τ in $\mathcal{L}_i^K(Z)$ and $\mathcal{R}_K^i(Z)$ are 0. Since τ is arbitrary, $\mathcal{L}_i^K(Z) = 0$ and $\mathcal{R}_K^i(Z) = 0$.

3. Vanishing above the middle dimension

Although there is no reason to expect (8.12) and (8.13) to be valid at every step, the proof of Proposition 8.6 suggests that we may make some progress by settling for Hom_K rather than $\operatorname{Hom}_{\mathfrak{g},K}$. In fact, we shall see in this section that we can estimate Hom_K without computing it exactly, and the estimate will be good enough to prove $\mathcal{L}_i(Z) = 0$ for $i > S$ and $\mathcal{R}^i(Z) = 0$ for $i > S$. The precise result is as follows.

Theorem 8.7.

(a) For $i > S$, $\mathcal{L}_i(Z) = 0$. For $0 \leq i \leq S$ and for any finite-dimensional $(\mathfrak{k}, K)$ module V,

$$\dim \operatorname{Hom}_K(\mathcal{L}_i(Z),V) \leq \sum_{n=0}^{\infty} \dim \operatorname{Hom}_{L\cap K}(S^n(\mathfrak{u}\cap\mathfrak{p})\otimes_{\mathbb{C}} Z^{\#}, H^i(\overline{\mathfrak{u}}\cap\mathfrak{k},V)) . \tag{8.18a}$$

Here S^n refers to the symmetric tensors homogeneous of degree n.

(b) For $i > S$, $\mathcal{R}^i(Z) = 0$. For $0 \leq i \leq S$ and for any finite-dimensional $(\mathfrak{k},K)$ module V,

$$\dim \operatorname{Hom}_K(V,\mathcal{R}^i(Z)) \leq \sum_{n=0}^{\infty} \dim \operatorname{Hom}_{L\cap K}(H_i(\mathfrak{u}\cap\mathfrak{k},V), S^n(\mathfrak{u}\cap\mathfrak{p})\otimes_{\mathbb{C}} Z^{\#}) . \tag{8.18b}$$

We shall give the proof only for $\mathcal{L}_i(Z)$, the proof for $\mathcal{R}^i(Z)$ being similar. The argument will involve some detailed computations with universal enveloping algebras. Specifically we shall filter $P^{\mathfrak{g},L\cap K}_{\overline{\mathfrak{q}},L\cap K}(Z^{\#}_{\overline{\mathfrak{q}}})$ by certain $(\mathfrak{k},L\cap K)$ submodules and make computations with the successive quotients. For this purpose, let $n \geq 0$ and define

$$U_n(\mathfrak{g}) = \text{span in } U(\mathfrak{g}) \text{ of all } uX_1\cdots X_r \text{ with } r\leq n,\ X_i\in\mathfrak{p},\ u\in U(\mathfrak{k})$$

$$U_n(\mathfrak{u}) = \text{span in } U(\mathfrak{u}) \text{ of all } uX_1\cdots X_r \text{ with } r\leq n,\ X_i\in\mathfrak{u}\cap\mathfrak{p},\ u\in U(\mathfrak{u}\cap\mathfrak{k}) .$$

Also let $U_{-1}(\mathfrak{g}) = U_{-1}(\mathfrak{u}) = 0$.

<u>Lemma 8.8.</u> For $n \geq 0$,

$$U_n(\mathfrak{g}) \subseteq U_n(\mathfrak{u})\otimes_{\mathbb{C}} U(\overline{\mathfrak{q}}) \tag{8.19}$$

if the right side is identified with a subset of $U(\mathfrak{g})$ via the multiplication map.

We defer the proof of Lemma 8.8 to the end of the section. The next lemma enables us to work with $U(\mathfrak{g})$ rather than $R(\mathfrak{g}, L\cap K)$ in the proof of Theorem 8.7. We defer its proof, too.

Lemma 8.9. Suppose W is a $(\bar{\mathfrak{q}}, L\cap K)$ module with action π. Define an action of $L\cap K$ on $U(\mathfrak{g})\otimes_{U(\bar{\mathfrak{q}})} W$ by $\mathrm{Ad}\otimes\pi$. Then

$$R(\mathfrak{g}, L\cap K)\otimes_{R(\bar{\mathfrak{q}}, L\cap K)} W \cong U(\mathfrak{g})\otimes_{U(\bar{\mathfrak{q}})} W \qquad (8.20)$$

as $(\mathfrak{g}, L\cap K)$ modules under the mapping $(u\otimes c)\otimes w \to u\otimes\pi(c)w$ for $u\in U(\mathfrak{g})$, $c\in C_{L\cap K}$, and $w\in W$.

Proof of Theorem 8.7a. By (8.11a),

$$\mathcal{F}^{\mathfrak{k},K}_{\mathfrak{g},K}\mathcal{L}_i(Z) = \mathcal{F}^{\mathfrak{k},K}_{\mathfrak{g},K}\Pi_i(P^{\mathfrak{g},L\cap K}_{\bar{\mathfrak{q}},L\cap K}(Z^{\#}_{\bar{\mathfrak{q}}})) \cong \Pi^K_i\mathcal{F}^{\mathfrak{k},L\cap K}_{\mathfrak{g},L\cap K}(P^{\mathfrak{g},L\cap K}_{\bar{\mathfrak{q}},L\cap K}(Z^{\#}_{\bar{\mathfrak{q}}})) .$$

Thus the same argument that proves (8.15a) and (8.15b) also proves

$$\mathrm{Hom}_K(\mathcal{L}_i(Z),V) \cong \mathrm{Ext}^i_{\mathfrak{k},L\cap K}(P^{\mathfrak{g},L\cap K}_{\bar{\mathfrak{q}},L\cap K}(Z^{\#}_{\bar{\mathfrak{q}}}),V) . \qquad (8.21)$$

Recall from Proposition 8.2 that $\mathfrak{u}$ is the sum of the eigenspaces for positive eigenvalues for ad of an element X_0 of $\mathfrak{h}_{\mathbb{R}}$ and that X_0 is forced to be in the center of $\mathfrak{l}$. Since $L\cap K$ is connected, $\mathrm{Ad}(L\cap K)$ fixes X_0. Therefore X_0 acts as a scalar in every $L\cap K$ type, and $Z^{\#}$ is fully reducible under X_0. Since X_0 commutes with $\mathfrak{l}$, $Z^{\#}$ splits into the finite or countable direct sum of its eigenspaces under X_0, each of which is an $(\mathfrak{l}, L\cap K)$ module. Since V is finite-dimensional, the Ext's on the right of

(8.21) split as a corresponding sum, and the Hom's on the right of (8.18a) split as a corresponding sum. Thus we may assume that X_0 acts as a scalar in $Z^{\#}$.

Put

$$M = P_{\mathfrak{q}, L\cap K}^{\mathfrak{g}, L\cap K}(Z_{\mathfrak{q}}^{\#}) \cong U(\mathfrak{g}) \otimes_{U(\overline{\mathfrak{q}})} Z_{\mathfrak{q}}^{\#} ,$$

the isomorphism following from Lemma 8.9. For $n \geq -1$, let

$$M_n = \text{image in } M \text{ of } U_n(\mathfrak{g}) \otimes_{\mathbb{C}} Z_{\mathfrak{q}}^{\#}$$

as a $(\mathfrak{k}, L\cap K)$ module. Let us see that

$$M_n \cong U_n(\mathfrak{u}) \otimes_{\mathbb{C}} Z_{\mathfrak{q}}^{\#} \tag{8.22}$$

as $(\mathfrak{l}\cap\mathfrak{k}, L\cap K)$ modules; the isomorphism imbeds the right side into $U(\mathfrak{g}) \otimes_{\mathbb{C}} Z_{\mathfrak{q}}^{\#}$ and then passes to the quotient. In fact, Lemma 8.8 gives

$$\begin{aligned} \text{image}(U_n(\mathfrak{u}) \otimes_{\mathbb{C}} Z_{\mathfrak{q}}^{\#}) &\subseteq \text{image}(U_n(\mathfrak{g}) \otimes_{\mathbb{C}} Z_{\mathfrak{q}}^{\#}) \\ &\subseteq \text{image}((U_n(\mathfrak{u}) \otimes_{\mathbb{C}} U(\overline{\mathfrak{q}})) \otimes_{\mathbb{C}} Z_{\mathfrak{q}}^{\#}) . \end{aligned}$$

On the other hand, the end members here are equal, so that

$$M_n \cong \text{image}(U_n(\mathfrak{u}) \otimes_{\mathbb{C}} Z_{\mathfrak{q}}^{\#}) .$$

Since $U_n(\mathfrak{u}) \otimes_{\mathbb{C}} Z_{\mathfrak{q}}^{\#}$ maps into M in one-one fashion, (8.22) follows.

From (8.22) we obtain

$$M/M_n \cong (U(\mathfrak{u})/U_n(\mathfrak{u})) \otimes_{\mathbb{C}} Z_{\mathfrak{q}}^{\#} ,$$

and $\mathfrak{l}\cap\mathfrak{k}$ acts by the tensor product action, with ad in the first factor. Here each member of $U(\mathfrak{u})/U_n(\mathfrak{u})$ has at least $n+1$ factors from $\mathfrak{u}$ in every term of its expansion and thus has only large eigenvalues under ad X_0. Since X_0 acts as a scalar in $Z_{\mathfrak{q}}^{\#}$, the eigenvalues of M/M_n are $\geq c(n)$, with $\lim_{n\to\infty} c(n) = +\infty$. Now

$$\mathrm{Ext}^i_{\mathfrak{k}, L\cap K}(M/M_n, V) \cong H^i(\mathfrak{k}, L\cap K; \mathrm{Hom}_{\mathbb{C}}(M/M_n, V)_{L\cap K})$$

may be computed from the complex

$$\mathrm{Hom}_{L\cap K}(\Lambda^i(\mathfrak{k}/(\mathfrak{l}\cap\mathfrak{k})), \mathrm{Hom}_{\mathbb{C}}(M/M_n, V)_{L\cap K}),$$

and the cochains must respect eigenvalues under X_0. Since M/M_n has only large eigenvalues, we see that

$$\mathrm{Ext}^i_{\mathfrak{k}, L\cap K}(M/M_N, V) = 0 \tag{8.23}$$

for some N and all i. Plugging (8.23) into the long exact sequence of Theorem 5.28d for $\mathrm{Ext}^i_{\mathfrak{k}, L\cap K}(\cdot, V)$, namely

$$\cdots \longrightarrow \mathrm{Ext}^{i-1}_{\mathfrak{k}, L\cap K}(M_N, V) \longrightarrow \mathrm{Ext}^i_{\mathfrak{k}, L\cap K}(M/M_N, V)$$
$$\longrightarrow \mathrm{Ext}^i_{\mathfrak{k}, L\cap K}(M, V) \longrightarrow \mathrm{Ext}^i_{\mathfrak{k}, L\cap K}(M_N, V) \longrightarrow \cdots,$$

we conclude

$$\mathrm{Ext}^i_{\mathfrak{k}, L\cap K}(M, V) \cong \mathrm{Ext}^i_{\mathfrak{k}, L\cap K}(M_N, V) \tag{8.24}$$

for this N and all i.

From the long exact sequence for $\mathrm{Ext}(\cdot, V)$, it follows trivially that

$$\dim \mathrm{Ext}^i(A,V) \leq \dim \mathrm{Ext}^i(A/B,V) + \dim \mathrm{Ext}^i(B,V).$$

If C is a submodule of B, we can apply this inequality similarly to $\mathrm{Ext}^i(B,V)$ and proceed inductively. The result is a subadditivity formula for filtrations. If A has a finite filtration, then $\dim \mathrm{Ext}^i(A,V)$ is $\leq$ the sum of $\dim \mathrm{Ext}^i$ for the various subquotients. Applying this fact to the filtration

$$M_N \supseteq M_{N-1} \supseteq \cdots \supseteq M_0 \supseteq M_{-1} = 0$$

and using (3.14), we obtain

$$\begin{aligned} \dim \mathrm{Ext}^i_{\mathfrak{k}, L\cap K}(M,V) &\leq \sum_{n=0}^{N} \dim \mathrm{Ext}^i_{\mathfrak{k}, L\cap K}(M_n/M_{n-1},V) \\ &= \sum_{n=0}^{\infty} \dim \mathrm{Ext}^i_{\mathfrak{k}, L\cap K}(M_n/M_{n-1},V) . \end{aligned} \tag{8.25}$$

Next let us establish the $(\mathfrak{k}, L\cap K)$ isomorphism

$$M_n/M_{n-1} \cong U(\mathfrak{k}) \otimes_{U(\bar{\mathfrak{q}}\cap\mathfrak{k})} (S^n(\mathfrak{u}\cap\mathfrak{p}) \otimes_{\mathbb{C}} Z^{\#}_{\mathfrak{q}}) \tag{8.26}$$

for an action by $\bar{\mathfrak{q}} \cap \mathfrak{k}$ on $S^n(\mathfrak{u}\cap\mathfrak{p})$ described as follows: We have an $L\cap K$ isomorphism

$$\mathfrak{u}\cap\mathfrak{p} \cong \mathfrak{p}/(\bar{\mathfrak{q}}\cap\mathfrak{p}) ,$$

and $\bar{\mathfrak{q}}\cap\mathfrak{k}$ acts in the expected way on the right side. Therefore $\bar{\mathfrak{q}}\cap\mathfrak{k}$ acts on $\mathfrak{u}\cap\mathfrak{p}$ and also on $S^n(\mathfrak{u}\cap\mathfrak{p})$. We denote the action of X in $\bar{\mathfrak{q}}\cap\mathfrak{k}$ on p in $S^n(\mathfrak{u}\cap\mathfrak{p})$ by $X\cdot p$. If $\mathfrak{s}$ denotes symmetrization, this action has the property that

$$\mathfrak{s}(X\cdot p) \equiv (\mathrm{ad}\, X)\mathfrak{s}(p) \mod U_{n-1}(\mathfrak{g}) . \tag{8.27}$$

To establish (8.26) we map

$$U(\mathfrak{k}) \otimes_{\mathbb{C}} (S^n(\mathfrak{u} \cap \mathfrak{p}) \otimes_{\mathbb{C}} Z_{\mathfrak{q}}^{\#}) \to M_n/M_{n-1} \tag{8.28}$$

by

$$u \otimes (p \otimes z) \to u\mathfrak{s}(p) \otimes z \, .$$

If X is in $\bar{\mathfrak{q}} \cap \mathfrak{k}$, then

$$\begin{aligned} &uX \otimes (p \otimes z) - u \otimes X(p \otimes z) \\ &\qquad = uX \otimes (p \otimes z) - u \otimes (X \cdot p \otimes z) - u \otimes (p \otimes Xz) \\ &\qquad \to uX\mathfrak{s}(p) \otimes z - u\mathfrak{s}(X \cdot p) \otimes z - u\mathfrak{s}(p) \otimes Xz \\ &\qquad \subseteq uX\mathfrak{s}(p) \otimes z - u(\operatorname{ad} X)\mathfrak{s}(p) \otimes z - u\mathfrak{s}(p) \otimes Xz + uU_{n-1}(\mathfrak{g}) \otimes z \\ &\qquad \subseteq u\mathfrak{s}(p)X \otimes z - u\mathfrak{s}(p) \otimes Xz + M_{n-1} \\ &\qquad = M_{n-1} \, . \end{aligned}$$

Therefore (8.28) descends to a map from right to left in (8.26), and this map is clearly $\mathfrak{k}$ linear. The map is compatible with the vector space isomorphisms

$$\begin{aligned} U(\mathfrak{k}) \otimes_{U(\bar{\mathfrak{q}} \cap \mathfrak{k})} (S^n(\mathfrak{u} \cap \mathfrak{p}) \otimes_{\mathbb{C}} Z_{\mathfrak{q}}^{\#}) &\cong U(\mathfrak{u} \cap \mathfrak{k}) \otimes_{\mathbb{C}} S^n(\mathfrak{u} \cap \mathfrak{p}) \otimes_{\mathbb{C}} Z_{\mathfrak{q}}^{\#} \\ &\cong U_n(\mathfrak{u})/U_{n-1}(\mathfrak{u}) \otimes Z_{\mathfrak{q}}^{\#} \cong M_n/M_{n-1} \, , \end{aligned}$$

and thus the map in (8.26) is one-one onto.

Combining (8.21), (8.25), and (8.26) and using Lemma 8.9 for $\mathfrak{k}$ instead of $\mathfrak{g}$, we obtain

$$\dim \operatorname{Hom}_K(\mathcal{L}_i(Z), V) \le \sum_{n=0}^{\infty} \dim \operatorname{Ext}^i_{\mathfrak{k}, L \cap K}(P^{\mathfrak{k}, L\cap K}_{\bar{\mathfrak{q}} \cap \mathfrak{k}, L \cap K}(S^n(\mathfrak{u} \cap \mathfrak{p}) \otimes_{\mathbb{C}} Z_{\mathfrak{q}}^{\#}), V)$$

for the nontrivial action of $\bar{\mathfrak{u}} \cap \mathfrak{k}$ on $S^n(\mathfrak{u} \cap \mathfrak{p})$. By Shapiro's Lemma (7.51), the right side here is

$$= \sum_{n=0}^{\infty} \dim \operatorname{Ext}^{i}_{\bar{\mathfrak{q}}\cap\mathfrak{k},\, L\cap K}(S^n(\mathfrak{u}\cap\mathfrak{p})\otimes_{\mathbb{C}} Z^{\#}_{\mathfrak{q}}, V) . \tag{8.29}$$

The $(\bar{\mathfrak{q}}\cap\mathfrak{k}, L\cap K)$ module $S^n(\mathfrak{u}\cap\mathfrak{p})$ is filtered according to the eigenvalues of $\operatorname{ad} X_0$, since $\bar{\mathfrak{u}}\cap\mathfrak{k}$ decreases eigenvalues. We apply our subadditivity formula for Ext to (8.29): The respective quotients have 0 action by $\bar{\mathfrak{u}}\cap\mathfrak{k}$, and we can reassemble the direct sum to see that (8.29) is

$$\leq \sum_{n=0}^{\infty} \dim \operatorname{Ext}^{i}_{\bar{\mathfrak{q}}\cap\mathfrak{k},\, L\cap K}(S^n(\mathfrak{u}\cap\mathfrak{p})\otimes_{\mathbb{C}} Z^{\#}_{\mathfrak{q}}, V) \tag{8.30}$$

with trivial $\bar{\mathfrak{u}}\cap\mathfrak{k}$ action on $S^n(\mathfrak{u}\cap\mathfrak{p})$.

We have now reached a step in the argument analogous to (8.15c). To complete the proof, we argue as for (8.15d) that

$$\operatorname{Ext}^{i}_{\bar{\mathfrak{q}}\cap\mathfrak{k},\, L\cap K}(W,V) \cong \operatorname{Hom}_{\mathfrak{l}\cap\mathfrak{k},\, L\cap K}(W, H^i(\bar{\mathfrak{u}}\cap\mathfrak{k}, V))$$

for any $(\bar{\mathfrak{q}}\cap\mathfrak{k}, L\cap K)$ module W for which $\bar{\mathfrak{u}}\cap\mathfrak{k}$ acts trivially. Applying this isomorphism with $W = S^n(\mathfrak{u}\cap\mathfrak{p})\otimes_{\mathbb{C}} Z^{\#}_{\mathfrak{q}}$, we obtain (8.18a). Since $H^i(\bar{\mathfrak{u}}\cap\mathfrak{k}, V) = 0$ for $i > S$, we see from (8.18a) that $\mathcal{L}_i(Z) = 0$ for $i > S$, by the same argument as at the end of Proposition 8.6. This completes the proof of the theorem.

<u>Proof of Lemma 8.8.</u> First let us prove that

$$U(\bar{\mathfrak{q}})\mathfrak{g} \subseteq U(\bar{\mathfrak{q}}) + \mathfrak{u}U(\bar{\mathfrak{q}}) . \tag{8.31}$$

To do so, we prove that $U^m(\bar{\mathfrak{q}})\mathfrak{g} \subseteq U(\bar{\mathfrak{q}}) + \mathfrak{u}U(\bar{\mathfrak{q}})$ by induction on m. For $m=0$, we have $\mathfrak{g} = \bar{\mathfrak{q}} \oplus \mathfrak{u}$, and the assertion follows. Assume the assertion for $m-1$ and let $\bar{q}_1 \cdots \bar{q}_m X$ be given with each $\bar{q}_j$ in $\bar{\mathfrak{q}}$ and with X in $\mathfrak{g}$. Then we have

$$\bar{q}_1\cdots\bar{q}_m X = \bar{q}_1\cdots\bar{q}_{m-1}X\bar{q}_m + \bar{q}_1\cdots\bar{q}_{m-1}X'$$

with $X' = [\bar{q}_m, X]$. We apply the inductive hypothesis to $\bar{q}_1\cdots\bar{q}_{m-1}X$ and to $\bar{q}_1\cdots\bar{q}_{m-1}X'$, and the assertion follows for stage m. This proves (8.31).

Now we prove the assertion of the lemma by induction on n. For $n=0$ the assertion is that $U(\mathfrak{k}) \subseteq U(\mathfrak{u}\cap\mathfrak{k})\otimes_{\mathbb{C}} U(\bar{\mathfrak{q}})$, and this follows from the Poincaré-Birkhoff-Witt Theorem since $\mathfrak{k} = (\mathfrak{u}\cap\mathfrak{k})\oplus(\bar{\mathfrak{q}}\cap\mathfrak{k})$. Assume the assertion for all stages $< n$, and let $uX_1\cdots X_n$ be given with u in $U(\mathfrak{k})$ and all X_j in $\mathfrak{p}$. By inductive hypothesis $uX_1\cdots X_{n-1}$ is a sum of terms $u'X_1'\cdots X_r'\bar{q}'$ with $u'\in U(\mathfrak{u}\cap\mathfrak{k})$, $X_j'\in\mathfrak{u}\cap\mathfrak{p}$, $\bar{q}'\in U(\bar{\mathfrak{q}})$, and $r \leq n-1$. Hence $uX_1\cdots X_n$ is a sum of terms

$$u'X_1'\cdots X_r'\bar{q}'X_n. \tag{8.32}$$

Applying (8.31) to $\bar{q}'X_n$, we can write (8.32) as a sum of terms

$$u'X_1'\cdots X_r'\bar{q}_1' + u'X_1'\cdots X_r'u_0\bar{q}_2' \tag{8.33}$$

with u_0 in $\mathfrak{u}$ and $\bar{q}_1'$ and $\bar{q}_2'$ in $U(\bar{\mathfrak{q}})$. Let us write $u_0 = u_0' + u_0''$ with $u_0'\in\mathfrak{u}\cap\mathfrak{k}$ and $u_0''\in\mathfrak{u}\cap\mathfrak{p}$. Then the terms

$$u'(X_1'\cdots X_r')\bar{q}_1' \quad\text{and}\quad u'(X_1'\cdots X_r'u_0'')\bar{q}_2'$$

in (8.33) are of the required form, and we have only to treat

$$u'X_1'\cdots X_r'u_0'\bar{q}_2'.$$

However, u_0' is in $\mathfrak{u}\cap\mathfrak{k}$ and normalizes $\mathfrak{u}\cap\mathfrak{p}$. Hence we can move it to the left one step at a time in the above expression and exhibit the expression as a sum

$$u'u_0'X_1''\cdots X_r''\bar{q}_2' + \sum_{j=0}^{r-1} u'X_1'\cdots X_j'X_{j+1}''\cdots X_r''\bar{q}_2'$$

with $X_j'' = [X_j', u_0']$, and this is of the required form. This completes the induction.

Proof of Lemma 8.9. Once the map is well defined, it is clear that it is a $\mathfrak{g}$ map, and it is an $L \cap K$ map since Lemma 7.25 implies that $L \cap K$ acts on the left of $R(\mathfrak{g}, L \cap K)$ by $\mathrm{Ad} \otimes \ell$. To check that the map implementing (8.19) is one-one onto, it is enough to check that the vector space isomorphism $\mathfrak{g} = \mathfrak{u} \oplus \bar{\mathfrak{q}}$ leads to an isomorphism

$$U(\mathfrak{u}) \otimes_{\mathbb{C}} R(\bar{\mathfrak{q}}, L \cap K) \cong R(\mathfrak{g}, L \cap K) \tag{8.34}$$

as right $R(\bar{\mathfrak{q}}, L \cap K)$ modules under the mapping

$$u \otimes (\bar{q} \otimes c) \to u\bar{q} \otimes c ;$$

here we regard the algebra $U(\mathfrak{u})$ as a trivial right $R(\bar{\mathfrak{q}}, L \cap K)$ module. In fact, the map implementing (8.34) respects the $R(\bar{\mathfrak{q}}, L \cap K)$ structures. And on the level of vector space isomorphisms the map is the composition of

$$\begin{aligned} U(\mathfrak{u}) \otimes_{\mathbb{C}} R(\bar{\mathfrak{q}}, L \cap K) &= U(\mathfrak{u}) \otimes_{\mathbb{C}} (U(\bar{\mathfrak{q}}) \otimes_{\mathfrak{l} \cap \mathfrak{k}} C_{L \cap K}) \\ &\cong (U(\mathfrak{u}) \otimes_{\mathbb{C}} U(\bar{\mathfrak{q}})) \otimes_{\mathfrak{l} \cap \mathfrak{k}} C_{L \cap K} \\ &\cong U(\mathfrak{g}) \otimes_{\mathfrak{l} \cap \mathfrak{k}} C_{L \cap K} \\ &= R(\mathfrak{g}, L \cap K) . \end{aligned}$$

Hence the map is an isomorphism.

Thus we are to check that the map implementing (8.34) is

well defined. We check first that it is well defined from $R(\mathfrak{g}, L\cap K)\otimes_{\mathbb{C}} W$ to $U(\mathfrak{g})\otimes_{U(\overline{\mathfrak{q}})} W$. In fact, let $u\in U(\mathfrak{g})$, $X\in \mathfrak{l}\cap \mathfrak{k}$, $c\in C_{L\cap K}$, and $w\in W$. Then

$$(uX\otimes c)\otimes w \to uX\otimes \pi(c)w \equiv u\otimes \pi(X)\pi(c)w$$

and

$$(u\otimes \ell(X)c)\otimes w \to u\otimes \pi(\ell(X)c)w\,.$$

We readily check that $\pi(X)\pi(c)w = \pi(\ell(X)c)w$, and hence our map is well defined on $R(\mathfrak{g}, L\cap K)\otimes_{\mathbb{C}} W$.

To see that the map descends to $R(\mathfrak{g}, L\cap K)\otimes_{R(\overline{\mathfrak{q}}, L\cap K)} W$, let X be in $\overline{\mathfrak{q}}$. By Proposition 7.6, we have $(u\otimes c)X = u(\mathrm{Ad}(\cdot)X)\otimes c$. In terms of an inner product on $\overline{\mathfrak{q}}$ and an orthonormal basis X_i, this equality means

$$\begin{aligned}
&(u\otimes c)X\otimes w - (u\otimes c)\otimes \pi(X)w \\
&\quad = \sum_i (uX_i\otimes (\mathrm{Ad}(\cdot)X, X_i)c(\cdot))\otimes w - (u\otimes c)\otimes \pi(X)w \\
&\quad \to \sum_i uX_i\otimes \pi\{(\mathrm{Ad}(\cdot)X, X_i)c(\cdot)\}w - u\otimes \pi(c)\pi(X)w \\
&\quad = \sum_i u\otimes \pi(X_i)\pi\{(\mathrm{Ad}(\cdot)X, X_i)c(\cdot)\}w - u\otimes \pi(c)\pi(X)w \\
&\quad = u\otimes\Big\{\sum_i \int_K (\mathrm{Ad}(k)X, X_i)c(k)\pi(X_i)\pi(k)w\,dk - \int_K c(k)\pi(k)\pi(X)w\,dk\Big\} \\
&\quad = u\otimes\Big\{\int_K c(k)\pi(\mathrm{Ad}(k)X)\pi(k)w\,dk - \int_K c(k)\pi(k)\pi(X)w\,dk\Big\} \\
&\quad = 0\,.
\end{aligned}$$

4. First reduction below the middle dimension

Under additional conditions on the $(\mathfrak{l}, L\cap K)$ module Z, $\mathcal{L}_i(Z)$ and $\mathcal{R}^i(Z)$ are 0 for $i < S$. To get at such results, we try to get started with (8.12) and (8.13) in the context of $(\mathfrak{g}, K)$ modules and mappings, without settling for some analog on the $\mathfrak{k}$ level. It will be a little easier to work with $\mathcal{R}^i(Z)$ and (8.13) than with $\mathcal{L}_i(Z)$ and (8.12) because of the presence of the pseudo-forgetful functor $\mathcal{F}^\vee$ in (8.12). Thus we shall examine $\mathcal{R}^i(Z)$ and (8.13), later handling $\mathcal{L}_i(Z)$ by an easy duality argument. The first step is to reduce matters to facts about $\mathfrak{u}$ homology.

Theorem 8.10. If $i = n$ is the smallest integer such that $\mathcal{R}^i(Z) \neq 0$ and if

$$\mathrm{Hom}_{L\cap K}(H_i(\mathfrak{u}, \mathcal{R}^n(Z)), Z^\# \otimes_{\mathbb{C}} \Lambda^{n-i}(\mathfrak{l}\cap\mathfrak{k})^*) = 0 \quad \text{for } i < n,$$

then

$$\mathrm{Hom}_{L\cap K}(H_n(\mathfrak{u}, \mathcal{R}^n(Z)), Z^\#) \neq 0 .$$

Remarks. The integer n is merely the name of the smallest i in the hypothesis and is not to be confused with the parameter defining $\mathfrak{g}$. Anyway the theorem says that if $H_i(\mathfrak{u}, \mathcal{R}^n(Z))$ has no $L\cap K$ types in common with $Z^\# \otimes_{\mathbb{C}} \Lambda^{n-i}(\mathfrak{l}\cap\mathfrak{k})^*$ for $i < n$, then $H_n(\mathfrak{u}, \mathcal{R}^n(Z))$ does have an $L\cap K$ type in common with $Z^\#$. The proof requires some preparation.

Proposition 8.11. Let $\mathcal{C}$, $\mathcal{C}'$, and $\mathcal{C}''$ be good categories, and suppose $\mathcal{C}$ has enough projectives and enough injectives. Let $G : \mathbf{C} \to \mathbf{C}'$ be a covariant left exact functor that carries injectives to injectives, and let $F : \mathbf{C}' \to \mathbf{C}''$ be a covariant left exact functor, so that $F\,G : \mathbf{C} \to \mathbf{C}''$ is covariant and left exact. Suppose W is a module in $\mathbf{C}$ such that $G(W)$ is injective and

$$G^1(W) = \ldots = G^{n-1}(W) = 0 .$$

Then

$$(F\circ G)^n(W) \cong F(G^n(W)) . \qquad (8.35)$$

Proof. We proceed by induction on n, starting with the case $n = 1$. We are to prove that $G(W)$ injective implies

$$(F\circ G)^1(W) \cong F(G^1(W)) . \qquad (8.36)$$

Since $\mathbf{C}$ has enough injectives, W maps one-one into some injective I, so that we have an exact sequence

$$0 \longrightarrow W \longrightarrow I \longrightarrow J \longrightarrow 0 \qquad (8.37)$$

for some J. By Theorem 5.28c, we obtain a long exact sequence

$$0 \longrightarrow G(W) \longrightarrow G(I) \longrightarrow G(J) \longrightarrow G^1(W) \longrightarrow G^1(I) = 0 , \qquad (8.38)$$

the equality following since $0 \to I \to I \to 0$ is an injective resolution of I. In this sequence, $G(W)$ and $G(I)$ are both injective, by assumption, and Lemma 5.22 shows that

$$G(I) \cong G(W) \oplus K \qquad (8.39)$$

for the module $K \cong G(I)/\text{image } G(W)$. Applying F to (8.39) gives

$$F(G(I)) \cong F(G(W)) \oplus F(K),$$

which we need only in the form

$$F(K) \cong F(G(I))/F(G(W)). \tag{8.40}$$

From (8.38) the sequence

$$0 \longrightarrow K \longrightarrow G(J) \longrightarrow G^1(W) \longrightarrow 0 \tag{8.41}$$

is exact, and (8.39) shows K is injective. Lemma 5.22 shows that (8.41) splits, and hence

$$G(J) \cong K \oplus G^1(W). \tag{8.42}$$

Applying F to (8.42) gives

$$F(G(J)) \cong F(K) \oplus F(G^1(W)).$$

From (8.40) we obtain

$$F(G(J)) \cong F(G(I))/F(G(W)) \oplus F(G^1(W)). \tag{8.43}$$

Meanwhile we can use Theorem 5.28c with (8.37) to obtain a long exact sequence

$$0 \longrightarrow F(G(W)) \longrightarrow F(G(I)) \longrightarrow F(G(J))$$
$$\longrightarrow (F\circ G)^1(W) \longrightarrow (F\circ G)^1(I) = 0.$$

This sequence identifies the quotient

$$F(G(J))/(F(G(I))/F(G(W)))$$

as isomorphic to $(F\circ G)^1(W)$. Combining this isomorphism with (8.43), we obtain (8.36).

For the inductive step, we use dimension shifting. Thus let us assume (8.35) for stage $n-1$. We may assume $n > 1$. From (8.37) and Theorem 5.28c, we obtain a long exact sequence

$$0 = G^{n-1}(I) \longrightarrow G^{n-1}(J) \longrightarrow G^n(W) \longrightarrow G^n(I) = 0 .$$

Thus

$$G^n(W) \cong G^{n-1}(J)$$

and

$$F(G^n(W)) \cong F(G^{n-1}(J)) . \tag{8.44}$$

Similarly from (8.37) and Theorem 5.28c, we obtain a long exact sequence

$$0 = (F\circ G)^{n-1}(I) \longrightarrow (F\circ G)^{n-1}(J) \longrightarrow (F\circ G)^n(W) \longrightarrow (F\circ G)^n(I) = 0 .$$

Thus

$$(F\circ G)^n(W) \cong (F\circ G)^{n-1}(J) . \tag{8.45}$$

To relate (8.44) and (8.45), we shall use our inductive hypothesis. Since $n > 1$, $G^1(W) = 0$ by assumption. Combining this fact with (8.42) and (8.39), we see that

$$G(I) \cong G(W) \oplus G(J) .$$

Thus $G(J)$ is isomorphic with a summand of the injective $G(I)$ and hence is injective. Next, if $k \geq 1$, Theorem 5.28c gives us a long exact sequence

$$0 = G^k(I) \longrightarrow G^k(J) \longrightarrow G^{k+1}(W) \longrightarrow G^{k+1}(I) = 0 .$$

Thus

$$G^k(J) \cong G^{k+1}(W) ,$$

and this is 0 by our assumption if also $k+1 \leq n-1$, i.e., if $1 \leq k \leq n-2$. Hence we can apply our inductive hypothesis and obtain

$$(F \circ G)^{n-1}(J) \cong F(G^{n-1}(J)) .$$

Combining this isomorphism with (8.44) and (8.45), we obtain (8.35). This completes the induction.

Proposition 8.12. The functor Γ from $\mathcal{C}(\mathfrak{g}, L \cap K)$ to $\mathcal{C}(\mathfrak{g}, K)$ carries injectives to injectives.

Proof. By Corollary 7.21b it is enough to consider standard injectives $W = I_{\mathfrak{l} \cap \mathfrak{k}, L\cap K}^{\mathfrak{g}, L\cap K}(M)$. By Proposition 7.16

$$\Gamma(W) = I_{\mathfrak{g}, L\cap K}^{\mathfrak{g}, K}(I_{\mathfrak{l} \cap \mathfrak{k}, L\cap K}^{\mathfrak{g}, L\cap K}(M)) \cong I_{\mathfrak{l} \cap \mathfrak{k}, L\cap K}^{\mathfrak{g}, K}(M) ,$$

and the right side is injective by Lemma 7.8 and Corollary 7.19.

Proof of Theorem 8.10. We are guided by the steps of (8.13). For (a), Frobenius reciprocity (7.46) gives us an isomorphism

$$\operatorname{Hom}_{\mathfrak{g}, K}(V, \Gamma(W)) \cong \operatorname{Hom}_{\mathfrak{g}, L\cap K}(V, W)$$

natural in W, and we write this as a natural isomorphism of functors

$$\mathrm{Hom}_{\mathfrak{g},K}(V,\cdot)\circ\Gamma \cong \mathrm{Hom}_{\mathfrak{g},L\cap K}(V,\cdot)\ . \tag{8.46}$$

If the integer n (the least one for which $\mathcal{R}^n(Z) \neq 0$) is 0, we apply this identity with $V = \mathcal{R}^0(Z)$ and obtain

$$\mathrm{Hom}_{\mathfrak{g},K}(\mathcal{R}^0(Z),\mathcal{R}^0(Z)) \cong \mathrm{Ext}^0_{\mathfrak{g},L\cap K}(\mathcal{R}^0(Z), I^{\mathfrak{g},L\cap K}_{\mathfrak{q},L\cap K}(Z^{\#}_{\mathfrak{q}})). \tag{8.47a}$$

If the integer n is > 0, we apply Proposition 8.11 with $G = \Gamma$, $F = \mathrm{Hom}_{\mathfrak{g},K}(V,\cdot)$, and $W = I^{\mathfrak{g},L\cap K}_{\mathfrak{q},L\cap K}(Z^{\#}_{\mathfrak{q}})$. Formula (8.46) identifies $F\circ G$, Proposition 8.12 shows G carries injectives to injectives, and $G(W)$ is injective since it is 0. The result with $V = \mathcal{R}^n(Z)$ is

$$\mathrm{Hom}_{\mathfrak{g},K}(\mathcal{R}^n(Z),\mathcal{R}^n(Z)) \cong \mathrm{Ext}^n_{\mathfrak{g},L\cap K}(\mathcal{R}^n(Z), I^{\mathfrak{g},L\cap K}_{\mathfrak{q},L\cap K}(Z^{\#}_{\mathfrak{q}})). \tag{8.47b}$$

Combining (8.47) with Shapiro's Lemma (Theorem 7.24), we obtain

$$\mathrm{Hom}_{\mathfrak{g},K}(\mathcal{R}^n(Z),\mathcal{R}^n(Z)) \cong \mathrm{Ext}^n_{\mathfrak{q},L\cap K}(\mathcal{R}^n(Z), Z^{\#}_{\mathfrak{q}})\ . \tag{8.48}$$

Let $\mathfrak{q}_1 = (\mathfrak{l}\cap\mathfrak{k})\oplus\mathfrak{u}$. Later in this section we shall prove that

$$\mathrm{Ext}^i_{\mathfrak{q}_1,L\cap K}(\mathcal{R}^n(Z), Z^{\#}_{\mathfrak{q}}\otimes_{\mathbb{C}}\textstyle\bigwedge^{n-i}(\mathfrak{l}\cap\mathfrak{k})^*_{\mathfrak{q}_1}) = 0 \quad \text{for} \quad i < n \tag{8.49}$$

implies that

$$\mathrm{Ext}^n_{\mathfrak{q},L\cap K}(\mathcal{R}^n(Z), Z^{\#}_{\mathfrak{q}}) \text{ maps one-one into } \mathrm{Ext}^n_{\mathfrak{q}_1,L\cap K}(\mathcal{R}^n(Z), Z^{\#}_{\mathfrak{q}})\ . \tag{8.50}$$

We readily check that

$$\mathrm{Hom}_{\mathfrak{q}_1, L\cap K}(\cdot, Z_{\mathfrak{q}}^{\#} \otimes_{\mathbb{C}} \Lambda^{n-i}(\mathfrak{l} \cap \mathfrak{k})^*_{\mathfrak{q}_1})$$

$$\cong \mathrm{Hom}_{\mathfrak{l}\cap\mathfrak{k}, L\cap K}(\cdot, Z^{\#} \otimes_{\mathbb{C}} \Lambda^{n-i}(\mathfrak{l} \cap \mathfrak{k})^*) \circ ((\cdot) \otimes_{\mathfrak{u}} \mathbb{C})$$

on $\mathcal{C}(\mathfrak{q}_1, L\cap K)$. The derived functors of $(\cdot)\otimes_{\mathfrak{u}}\mathbb{C}$ are $H_i(\mathfrak{u},\cdot)$, by the proof of Proposition 7.33a, and $\mathrm{Hom}_{\mathfrak{l}\cap\mathfrak{k}, L\cap K}(\cdot, Z^{\#} \otimes_{\mathbb{C}} \Lambda^{n-i}(\mathfrak{l} \cap \mathfrak{k})^*)$ is exact by Lemma 7.8 and Proposition 5.10b. Thus Proposition 5.16b gives

$$\mathrm{Ext}^j_{\mathfrak{q}_1, L\cap K}(\cdot, Z_{\mathfrak{q}}^{\#} \otimes_{\mathbb{C}} \Lambda^{n-i}(\mathfrak{l} \cap \mathfrak{k})^*_{\mathfrak{q}_1})$$

$$\cong \mathrm{Hom}_{L\cap K}(\cdot, Z^{\#} \otimes_{\mathbb{C}} \Lambda^{n-i}(\mathfrak{l} \cap \mathfrak{k})^*) \circ H_j(\mathfrak{u},\cdot). \qquad (8.51)$$

Since our hypothesis is that

$$\mathrm{Hom}_{L\cap K}(H_i(\mathfrak{u},\mathcal{R}^n(Z)), Z^{\#} \otimes_{\mathbb{C}} \Lambda^{n-i}(\mathfrak{l} \cap \mathfrak{k})^*) = 0 \quad \text{for } i < n,$$

the case $j = i$ of (8.51) shows that (8.49) is satisfied. Thus (8.50) holds, and we interpret this by (8.51) as saying that

$$\mathrm{Ext}^n_{\mathfrak{q}, L\cap K}(\mathcal{R}^n(Z), Z_{\mathfrak{q}}^{\#}) \text{ maps one-one into } \mathrm{Hom}_{L\cap K}(H_n(\mathfrak{u},\mathcal{R}^n(Z)), Z^{\#}). \qquad (8.52)$$

Since the identity mapping is an element of the left member of (8.48), $\mathrm{Ext}^n_{\mathfrak{q}, L\cap K}(\mathcal{R}^n(Z), Z_{\mathfrak{q}}^{\#})$ is nonzero. Thus (8.52) says that $\mathrm{Hom}_{L\cap K}(H_n(\mathfrak{u},\mathcal{R}^n(Z)), Z^{\#})$ is nonzero.

This proves Theorem 8.10 except for the fact that (8.49) implies (8.50). This implication is handled by the lemma below.

Lemma 8.13. Let $\mathfrak{q}_1 = (\mathfrak{l} \cap \mathfrak{k}) \oplus \mathfrak{u}$, and let V and W be in $\mathcal{C}(\mathfrak{q}_1, L \cap K)$. If

$$\operatorname{Ext}^i_{\mathfrak{q}_1, L\cap K}(V, W \otimes_{\mathbb{C}} \Lambda^{n-i}(\mathfrak{l} \cap \mathfrak{k})^*_{\mathfrak{q}_1}) = 0 \quad \text{for } i < n, \tag{8.53}$$

then

$$\operatorname{Ext}^n_{\mathfrak{q}, L\cap K}(V, W) \text{ maps one-one into } \operatorname{Ext}^n_{\mathfrak{q}_1, L\cap K}(V, W).$$

Remark. For $n = 0$, the result says that $\operatorname{Hom}_{\mathfrak{q}, L\cap K}(V, W)$ maps one-one into $\operatorname{Hom}_{\mathfrak{q}_1, L\cap K}(V, W)$. The mapping is restriction of scalars, and the result is clear in this case. The lemma generalizes the result to the higher derived functors.

Proof. The space $\mathfrak{s} = (\mathfrak{l} \cap \mathfrak{p}) \oplus \mathfrak{u}$ is an $(L \cap K)$-invariant complement to $\mathfrak{l} \cap \mathfrak{k}$ in $\mathfrak{q}$, and $\mathfrak{u}$ is an $(L \cap K)$-invariant complement to $\mathfrak{l} \cap \mathfrak{k}$ in $\mathfrak{q}_1$. Thus (7.63) and Proposition 7.23 allow us to compute $\operatorname{Ext}^n_{\mathfrak{q}, L\cap K}(V, W)$ and $\operatorname{Ext}^n_{\mathfrak{q}_1, L\cap K}(V, W)$ as the nth cohomology of the respective complexes

$$\operatorname{Hom}_{L\cap K}(\Lambda^n \mathfrak{s}, \operatorname{Hom}_{\mathbb{C}}(V, W)_{L\cap K}) \tag{8.54a}$$

and

$$\operatorname{Hom}_{L\cap K}(\Lambda^n \mathfrak{u}, \operatorname{Hom}_{\mathbb{C}}(V, W)_{L\cap K}). \tag{8.54b}$$

Since $\mathfrak{u} \subseteq \mathfrak{s}$, there is an obvious restriction map $\omega \to \omega_{\mathfrak{u}}$ from (8.54a) to (8.54b), and inspection of (7.63b) shows that this restriction map commutes with the coboundary operator. Thus we are to show that this restriction map, on the cohomology level in degree n, is one-one.

We shall write X_α for a typical basis vector in $\mathfrak{u}$ and

Y_β for a typical basis vector in $\mathfrak{l} \cap \mathfrak{p}$. It will be important to notice that $[\mathfrak{u}, \mathfrak{l} \cap \mathfrak{p}] \subseteq \mathfrak{u}$. We abbreviate

$$X_{\alpha_1} \wedge \ldots \wedge X_{\alpha_i} \wedge Y_{\beta_1} \wedge \ldots \wedge Y_{\beta_j}$$

as

$$X_A \wedge Y_B \quad \text{with} \quad |A| = i \quad \text{and} \quad |B| = j .$$

We shall write the respective coboundary operators for (8.54a) and (8.54b) as d and $d_{\mathfrak{u}}$.

We proceed by induction on n. For $n=0$ the spaces (8.54a) and (8.54b) are identical. The condition $d\omega = 0$ for ω in (8.54a) is that $X(\omega) = 0$ and $Y(\omega) = 0$ for all X and Y, while the condition $d_{\mathfrak{u}}\omega = 0$ for ω in (8.54b) is that $X(\omega) = 0$ for all X. Thus cocycles map in one-one fashion to cocycles. Since there are no coboundaries, the map is one-one on cohomology.

Assume inductively that the result holds through stage $n-1$. Let ω be in (8.54a) with $d\omega = 0$, and suppose that $\omega_{\mathfrak{u}}$ is a coboundary. We are to show that $\omega = d\eta$ for some η in

$$\mathrm{Hom}_{L\cap K}(\wedge^{n-1}\mathfrak{s}, \mathrm{Hom}_{\mathbb{C}}(V,W)_{L\cap K}) . \tag{8.55}$$

To do so, we show by induction on k that ω is of the form

$$\omega = \omega^k + d\eta^k \tag{8.56}$$

where ω^k is in (8.54a), η^k is in (8.55a), and

$$\omega^k(X_A \wedge Y_B) = 0 \quad \text{whenever} \quad |A| = n-i , \quad |B| = i , \quad i \leq k . \tag{8.57}$$

The case $k=n$ will have $\omega = d\eta^n$, and thus we can take $\eta = \eta^n$.

First let $k=0$. Since $\omega_{\mathfrak{u}}$ is a coboundary, we can find $\eta_{\mathfrak{u}}$ in

$$\mathrm{Hom}_{L\cap K}(\Lambda^{n-1}\mathfrak{u}, \mathrm{Hom}_{\mathbb{C}}(V,W)_{L\cap K}) \tag{8.58}$$

such that $\omega_{\mathfrak{u}} = d_{\mathfrak{u}}\eta_{\mathfrak{u}}$. Extend $\eta_{\mathfrak{u}}$ to a member η^0 of (8.55) by making it 0 on all basis vectors $X_A \wedge Y_B$ with $|B| \geq 1$. This extension η^0 is still an $L\cap K$ map, since $\mathfrak{l}\cap\mathfrak{p}$ is $(L\cap K)$-invariant. Then $\omega_{\mathfrak{u}} = d_{\mathfrak{u}}\eta_{\mathfrak{u}}$ implies that ω and $d\eta^0$ agree on basis elements X_A with $|A| = n$. Hence $\omega^0 = \omega - d\eta^0$ satisfies (8.57) for $i=0$, and (8.56) holds for $k=0$.

Now suppose inductively that $k > 0$ and that (8.56) holds at stage $k-1$:

$$\omega = \omega^{k-1} + d\eta^{k-1} \tag{8.59}$$

and

$$\omega^{k-1}(X_A \wedge Y_B) = 0 \quad \text{whenever } |A| = n-i,\ |B| = i,\ i \leq k-1. \tag{8.60}$$

Let us rewrite (8.60) as follows. We have

$$\Lambda^n \mathfrak{s} \cong \sum_{i=0}^{n} \oplus\ (\Lambda^{n-i}\mathfrak{u} \otimes_{\mathbb{C}} \Lambda^i(\mathfrak{l}\cap\mathfrak{t}))\,.$$

Since each summand is $(L\cap K)$-invariant, we have a corresponding decomposition of (8.54a) and can write

$$\omega^{k-1} = \sum_{i=0}^{n} \omega_i^{k-1} \tag{8.61a}$$

with

$$\omega_i^{k-1} \in \operatorname{Hom}_{L\cap K}(\Lambda^{n-i}\mathfrak{u} \otimes_{\mathbb{C}} \Lambda^i(\mathfrak{l}\cap\mathfrak{k}), \operatorname{Hom}_{\mathbb{C}}(V,W)_{L\cap K}). \tag{8.61b}$$

Condition (8.60) says that

$$\omega_0^{k-1} = \dots = \omega_{k-1}^{k-1} = 0. \tag{8.62}$$

Applying d to (8.59), we see that $d\omega^{k-1} = 0$. Inspection of the formula (7.63b) for d shows that the decomposition (8.61) for $d\omega_i^{k-1}$ is

$$d\omega_i^{k-1} = (d\omega_i^{k-1})_i + (d\omega_i^{k-1})_{i+1}.$$

Taking (8.62) into account, we see that

$$(d\omega_k^{k-1})_k = 0. \tag{8.63}$$

Under the isomorphism

$$\begin{aligned}&\operatorname{Hom}_{L\cap K}(\Lambda^{n-k}\mathfrak{u} \otimes_{\mathbb{C}} \Lambda^k(\mathfrak{l}\cap\mathfrak{k}), \operatorname{Hom}_{\mathbb{C}}(V,W)_{L\cap K})\\ &\qquad\cong \operatorname{Hom}_{L\cap K}(\Lambda^{n-k}\mathfrak{u}, \operatorname{Hom}_{\mathbb{C}}(\Lambda^k(\mathfrak{l}\cap\mathfrak{k}), \operatorname{Hom}_{\mathbb{C}}(V,W)_{L\cap K}))\end{aligned} \tag{8.64}$$

of (7.31), ω_k^{k-1} maps to $\tilde{\omega}_k^{k-1}$ given by

$$\tilde{\omega}_k^{k-1}(X_A)(Y_B) = \omega_k^{k-1}(X_A \wedge Y_B) \quad \text{for } |A| = n-k, \ |B| = k.$$

We regard $\Lambda^k(\mathfrak{l}\cap\mathfrak{k})$ as a $(\mathfrak{q}_1, L\cap K)$ module by having $\mathfrak{u}$ act as 0, and then the right side of (8.64), by Proposition 7.23, is the $(n-k)^{\text{th}}$ member of the standard complex for

$$\begin{aligned}&H^{n-k}(\mathfrak{q}_1, L\cap K; \operatorname{Hom}_{\mathbb{C}}(\Lambda^k(\mathfrak{l}\cap\mathfrak{k}), \operatorname{Hom}_{\mathbb{C}}(V,W)_{L\cap K}))\\ &\qquad\cong H^{n-k}(\mathfrak{q}_1, L\cap K; \operatorname{Hom}_{\mathbb{C}}(V, W\otimes_{\mathbb{C}} \Lambda^k(\mathfrak{l}\cap\mathfrak{k})^*)_{L\cap K})\end{aligned}$$

$$\cong \operatorname{Ext}^{n-k}_{\mathfrak{q}_1, L\cap K}(V, W \otimes_{\mathbb{C}} \Lambda^k(\mathfrak{l} \cap \mathfrak{k})^*) . \tag{8.65}$$

Let us compute $d_{\mathfrak{u}}\tilde{\omega}^{k-1}_k$. We have

$$d_{\mathfrak{u}}\tilde{\omega}^{k-1}_k(X_{\alpha_1} \wedge \cdots \wedge X_{n-k+1})(Y_B)$$

$$= \Sigma\ (-1)^{j+1}\, X_{\alpha_j}(\tilde{\omega}^{k-1}_k(X_{\alpha_1} \wedge \cdots \wedge \hat{X}_{\alpha_j} \wedge \cdots \wedge X_{n-k+1}))(Y_B)$$

$$+ \sum_{r<s} (-1)^{r+s}\, \tilde{\omega}^{k-1}_k([X_{\alpha_r}, X_{\alpha_s}] \wedge X_{\alpha_1} \wedge \cdots \wedge \hat{X}_{\alpha_r} \wedge \cdots \wedge \hat{X}_{\alpha_s} \wedge \cdots \wedge X_{n-k+1})(Y_B)$$

$$= \Sigma\ (-1)^{j+1}\, X_{\alpha_j}(\tilde{\omega}^{k-1}_k(X_{\alpha_1} \wedge \cdots \wedge \hat{X}_{\alpha_j} \wedge \cdots \wedge X_{n-k+1})(Y_B))$$

$$+ \sum_{r<s} (-1)^{r+s}\, \tilde{\omega}^{k-1}_k([X_{\alpha_r}, X_{\alpha_s}] \wedge X_{\alpha_1} \wedge \cdots \wedge \hat{X}_{\alpha_r} \wedge \cdots \wedge \hat{X}_{\alpha_s} \wedge \cdots \wedge X_{n-k+1})(Y_B)$$

since $X_j \in \mathfrak{u}$ implies $X_j(Y_B) = 0$

$$= \Sigma\ (-1)^{j+1}\, X_{\alpha_j}(\omega^{k-1}_k(X_{\alpha_1} \wedge \cdots \wedge \hat{X}_{\alpha_j} \wedge \cdots \wedge X_{n-k+1} \wedge Y_B)$$

$$+ \sum_{r<s} (-1)^{r+s}\, \omega^{k-1}_k([X_{\alpha_r}, X_{\alpha_s}] \wedge X_{\alpha_1} \wedge \cdots \wedge \hat{X}_{\alpha_r} \wedge \cdots \wedge \hat{X}_{\alpha_s} \wedge \cdots \wedge X_{n-k+1} \wedge Y_B)$$

$$= (d\omega^{k-1}_k)_k(X_{\alpha_1} \wedge \cdots \wedge X_{n-k+1} \wedge Y_B) , \tag{8.66}$$

and this is 0 by (8.63). Since $k > 0$, the right member of (8.65) is 0 by assumption. Thus there exists $\tilde{\eta}$ in

$$\operatorname{Hom}_{L\cap K}(\Lambda^{n-k-1}\mathfrak{u}, \operatorname{Hom}_{\mathbb{C}}(\Lambda^k(\mathfrak{l} \cap \mathfrak{k}), \operatorname{Hom}_{\mathbb{C}}(V, W)_{L\cap K}))$$

with $d_{\mathfrak{u}}\tilde{\eta} = \tilde{\omega}_k^{k-1}$. (If $n-k-1 = -1$, we understand $\tilde{\eta} = 0$.) Let $\bar{\eta}_k$ be the corresponding member of

$$\mathrm{Hom}_{L\cap K}(\Lambda^{n-k-1}\mathfrak{u} \otimes_{\mathbb{C}} \Lambda^k(\mathfrak{l} \cap \mathfrak{k}), \mathrm{Hom}_{\mathbb{C}}(V,W)_{L\cap K}) \tag{8.67}$$

and define $\bar{\eta}$ in (8.55) to be $\bar{\eta}_k$ on (8.67) and to be 0 on

$$\mathrm{Hom}_{L\cap K}(\Lambda^{n-i-1}\mathfrak{u} \otimes_{\mathbb{C}} \Lambda^i(\mathfrak{l} \cap \mathfrak{k}), \mathrm{Hom}_{\mathbb{C}}(V,W)_{L\cap K}) \quad \text{for } i \neq k. \tag{8.68}$$

The same computation as for (8.66) shows that

$$(d\bar{\eta})_k(X_A \wedge Y_B) = (d_{\mathfrak{u}}\tilde{\eta})(X_A)(Y_B) = \tilde{\omega}_k^{k-1}(X_A)(Y_B) = \omega_k^{k-1}(X_A \wedge Y_B)$$

when $|A| = n-k$ and $|B| = k$. Since $\bar{\eta}$ is 0 on (8.68), $(d\bar{\eta})_i = 0 = \omega_i^{k-1}$ for $i < k$. Thus

$$\omega^{k-1} = \omega_k^{k-1} + \sum_{i>k} \omega_i^{k-1} = d\bar{\eta} + \sum_{i>k} \omega_i^{k-1}.$$

Adding $d\eta^{k-1}$ to both sides and using (8.59), we obtain

$$\omega = \Big(\sum_{i>k} \omega_i^{k-1}\Big) + d(\eta^{k-1} + \bar{\eta}).$$

This is of the form (8.56), and the proof is complete.

5. Second reduction below the middle dimension

The next step is to relate $H_i(\mathfrak{u}, \mathcal{R}^n(Z))$ to $H_{i'}(\mathfrak{u} \cap \mathfrak{k}, \mathcal{R}^n(Z))$ with $i' \leq i$. The latter homology spaces depend only on K behavior, and we can handle them. In particular, in the special case that $\Delta(\mathfrak{l} \cap \mathfrak{k})$ is empty (so that $L \cap K$ reduces to T, $\mathfrak{u} \cap \mathfrak{k}$ reduces to $\mathfrak{n} \cap \mathfrak{k}$, and $\mathfrak{q} \cap \mathfrak{k} = \mathfrak{l} \cap \mathfrak{k} \oplus \mathfrak{u} \cap \mathfrak{k}$ reduces to $\mathfrak{b} \cap \mathfrak{k} = \mathfrak{h} \oplus \mathfrak{n} \cap \mathfrak{k}$), $H_{i'}(\mathfrak{u} \cap \mathfrak{k}, \cdot)$

is given explicitly by the version of Kostant's Theorem in Theorem 6.12. In combination with Theorem 8.10, the reduction to $\mathfrak{u} \cap \mathfrak{k}$ homology will say that $\mathcal{R}^n(Z) \neq 0$ for some $n < S$ has implications for the $L \cap K$ types that occur in $H_n(\mathfrak{u}, \mathcal{R}^n(Z))$.

If V is in $\mathcal{C}(\mathfrak{q}, L \cap K)$, then the homology $H_j(\mathfrak{u}, V)$ is in $\mathcal{C}(\mathfrak{l}, L \cap K)$. The irreducible representations of $L \cap K$ are parametrized by the integral forms λ on $\mathfrak{h}$ that are dominant for $\Delta^+(\mathfrak{l} \cap \mathfrak{k}) = \Delta(\mathfrak{l} \cap \mathfrak{k}) \cap \Delta^+$; let us denote by τ_λ the representation corresponding to λ. The space $H_j(\mathfrak{u}, V)$ is the direct sum of its $L \cap K$ isotypic spaces for the various τ_λ's, and the multiplicity of τ_λ is given by the dimension of $\mathrm{Hom}_{L \cap K}(\tau_\lambda, H_j(\mathfrak{u}, V))$, according to Corollary 3.8.

Similar remarks apply to $H_j(\mathfrak{u} \cap \mathfrak{k}, V)$, which is in $\mathcal{C}(\mathfrak{l} \cap \mathfrak{k}, L \cap K)$. It too is the direct sum of its $L \cap K$ isotypic subspaces for the various τ_λ's, and the multiplicity of τ_λ is given by the dimension of $\mathrm{Hom}_{L \cap K}(\tau_\lambda, H_j(\mathfrak{u} \cap \mathfrak{k}, V))$.

According to (7.62) and to the style of proof in Proposition 7.33, we can compute the homology $H_j(\mathfrak{u}, V)$ from the standard complex whose j^{th} term is $\Lambda^j \mathfrak{u} \otimes_{\mathbb{C}} V$, and we can compute $H_j(\mathfrak{u} \cap \mathfrak{k}, V)$ from the standard complex whose j^{th} term is $\Lambda^j(\mathfrak{u} \cap \mathfrak{k}) \otimes_{\mathbb{C}} V$. Meanwhile we have an inclusion

$$\Lambda^j(\mathfrak{u} \cap \mathfrak{k}) \otimes_{\mathbb{C}} V \subseteq \Lambda^j \mathfrak{u} \otimes_{\mathbb{C}} V,$$

and this inclusion commutes with the ∂ operators, as we see by inspection of (7.62b) (since ∂ deletes factors and since $[\mathfrak{u} \cap \mathfrak{k}, \mathfrak{u} \cap \mathfrak{k}] \subseteq \mathfrak{u} \cap \mathfrak{k}$). Moreover this inclusion commutes with the action of $L \cap K$. Hence we obtain a canonical map

$$H_j(\mathfrak{u} \cap \mathfrak{k}, V) \to H_j(\mathfrak{u}, V)$$

that respects the action of $L \cap K$. Consequently there is a canonical map

$$\mathrm{Hom}_{L \cap K}(\tau_\lambda, H_j(\mathfrak{u} \cap \mathfrak{k}, V)) \to \mathrm{Hom}_{L \cap K}(\tau_\lambda, H_j(\mathfrak{u}, V))$$

associated to each integral form λ on $\mathfrak{h}$ that is $\Delta^+(\mathfrak{l} \cap \mathfrak{k})$ dominant.

Theorem 8.14. Let V be in $\mathcal{C}(\mathfrak{q}, L \cap K)$, let $k \geq 0$ be an integer, and let λ be an integral form on $\mathfrak{h}$ dominant for $\Delta^+(\mathfrak{l} \cap \mathfrak{k})$. Suppose for all $j < k$ that

$$\mathrm{Hom}_{L \cap K}(\tau_\lambda \otimes_{\mathbb{C}} \Lambda^{k-j}(\mathfrak{u} \cap \mathfrak{p})^*, H_j(\mathfrak{u} \cap \mathfrak{k}, V)) = 0 .$$

Then the canonical map

$$\mathrm{Hom}_{L \cap K}(\tau_\lambda, H_k(\mathfrak{u} \cap \mathfrak{k}, V)) \to \mathrm{Hom}_{L \cap K}(\tau_\lambda, H_k(\mathfrak{u}, V))$$

is onto.

Proof. We proceed by induction on k, the case $k = 0$ being trivial. Let $k > 0$ and assume that the result holds through stage $k-1$. We are to show the following: If ω in the τ_λ subspace of $\Lambda^k \mathfrak{u} \otimes_{\mathbb{C}} V$ is given with $\partial\omega = 0$, then there exist ω_K in the τ_λ subspace of $\Lambda^k(\mathfrak{u} \cap \mathfrak{k}) \otimes_{\mathbb{C}} V$ and ω' in the τ_λ subspace of $\Lambda^{k+1}\mathfrak{u} \otimes_{\mathbb{C}} V$ such that $\omega = \omega_K + \partial\omega'$.

We shall decompose ω a little and then carry out an inner induction according to properties of the decomposition.

We have

$$\Lambda^k \mathfrak{u} \otimes_{\mathbb{C}} V \cong \sum_{j=0}^{k} \oplus \{(\Lambda^{k-j}(\mathfrak{u} \cap \mathfrak{k}) \otimes_{\mathbb{C}} \Lambda^j(\mathfrak{u} \cap \mathfrak{p})) \otimes_{\mathbb{C}} V\} \qquad (8.69)$$

with each summand $(L \cap K)$-invariant. Let $j = n$ be the largest index for which ω has nonzero component in the jth term of the right side of (8.69). We shall induct upward on n. If $n = 0$, then ω is already in $\Lambda^k(\mathfrak{u} \cap \mathfrak{k}) \otimes_{\mathbb{C}} V$. So we can just take $\omega_K = \omega$ and $\omega' = 0$.

Let $n > 0$ and assume that we have handled all stages through $n-1$. Let ω be given with decomposition according to (8.69) as

$$\omega = (\omega_0 + \dots + \omega_{n-1}) + \omega_n = \omega' + \omega_n .$$

The space $\Lambda^n(\mathfrak{u} \cap \mathfrak{p})$ is $(L \cap K)$-invariant, and we let $\Lambda^n(\mathfrak{u} \cap \mathfrak{p})^{(\mu)}$ be the isotypic subspace of type τ_μ. Then

$$\Lambda^n(\mathfrak{u} \cap \mathfrak{p}) = \sum_\mu \oplus (\Lambda^n(\mathfrak{u} \cap \mathfrak{p})^{(\mu)}),$$

and we have a corresponding decomposition

$$(\Lambda^{k-n}(\mathfrak{u} \cap \mathfrak{k}) \otimes_{\mathbb{C}} \Lambda^n(\mathfrak{u} \cap \mathfrak{p})) \otimes_{\mathbb{C}} V = \sum_\mu (\Lambda^{k-n}(\mathfrak{u} \cap \mathfrak{k}) \otimes_{\mathbb{C}} \Lambda^n(\mathfrak{u} \cap \mathfrak{p})^{(\mu)}) \otimes_{\mathbb{C}} V . \qquad (8.70)$$

Let ω_n decompose according to (8.70) as $\omega_n = \Sigma_\mu \, \omega_n^{(\mu)}$.

With k and n fixed, we carry out a further induction as follows: We fix a lexicographic ordering on the real linear span of the weights, suuposing that it is compatible with Δ^+ and that $(\text{span}\, \Delta_L)^\perp$ comes before $\text{span}\, \Delta_L$ in the ordering. Our further induction is then downward on the smallest μ such that $\omega_n^{(\mu)} \neq 0$. Let this smallest μ be

μ_0. Since $\Lambda(\mathfrak{u} \cap \mathfrak{p})$ has only finitely many weights, we may assume inductively that we have already handled all cases where the smallest μ is $> \mu_0$. Lumping terms in (8.70), write $\omega_n = \omega_n^{(\mu_0)} + \omega''$.

Let us abbreviate a typical monomial

$$X_{\alpha_1} \wedge \ldots \wedge X_{\alpha_{k-n}} \quad \text{in } \Lambda^{k-n}(\mathfrak{u} \cap \mathfrak{k})$$

as X_A and a typical monomial

$$Y_{\beta_1} \wedge \ldots \wedge Y_{\beta_n} \quad \text{in } \Lambda^n(\mathfrak{u} \cap \mathfrak{p})$$

as Y_B. Then we can write

$$\partial = \partial_X + \partial_Y + \partial_{XX} + \partial_{XY} + \partial_{YY},$$

where ∂_X refers to terms where one X_α is deleted, ∂_{XY} refers to terms where one X_α and one Y_β are deleted (and $[X_\alpha, Y_\beta]$ is inserted), and so on.

The operator ∂ carries (8.69) into

$$\Lambda^{k-1}\mathfrak{u} \otimes_{\mathbb{C}} V \cong \sum_{j=0}^{k-1} \oplus \{(\Lambda^{k-j-1}(\mathfrak{u} \cap \mathfrak{k}) \otimes_{\mathbb{C}} \Lambda^j(\mathfrak{u} \cap \mathfrak{p})) \otimes_{\mathbb{C}} V\}. \qquad (8.71)$$

The operators ∂_Y and ∂_{YY} carry the jth term of (8.69) to the $(j-1)^{st}$ term of (8.71), while ∂_X, ∂_{XX}, and ∂_{XY} carry j^{th} to j^{th}. In particular, the operators ∂_X, ∂_{XX}, and ∂_{XY} carry (8.70) into

$$(\Lambda^{k-n-1}(\mathfrak{u} \cap \mathfrak{k}) \otimes_{\mathbb{C}} \Lambda^n(\mathfrak{u} \cap \mathfrak{p})) \otimes_{\mathbb{C}} V$$
$$= \sum_{\mu} (\Lambda^{k-n-1}(\mathfrak{u} \cap \mathfrak{k}) \otimes_{\mathbb{C}} \Lambda^n(\mathfrak{u} \cap \mathfrak{p})^{(\mu)}) \otimes_{\mathbb{C}} V. \qquad (8.72)$$

The operators ∂_X and ∂_{XX} carry the μth term of (8.70) to the μth term of (8.72), since the Y part is unchanged. Let us see that ∂_{XY} raises the index μ.

In fact, let

$$X_{\alpha_1} \wedge \cdots \wedge X_{\alpha_{k-n}} \wedge Y_{\beta_1} \wedge \cdots \wedge Y_{\beta_n} \otimes v \qquad (8.73)$$

be a monomial in the expression of an element in the μth term of (8.70). We may assume that $\beta_1 + \cdots + \beta_n$ is a weight of τ_μ, so that

$$\mu = \beta_1 + \cdots + \beta_n + \Sigma\, a_i\gamma_i\,, \quad \gamma_i \in \Delta_L\,.$$

When we apply ∂_{XY} to (8.73), we get a sum of terms

$$(-1)^{i+j}[X_{\alpha_i}, Y_{\beta_j}] \wedge X_{\alpha_1} \wedge \cdots \wedge \hat{X}_{\alpha_i} \wedge \cdots \wedge X_{\alpha_{k-n}}$$
$$\wedge\, Y_{\beta_1} \wedge \cdots \wedge \hat{Y}_{\beta_j} \wedge \cdots \wedge Y_{\beta_n} \otimes v \qquad (8.74)$$

lying in (8.72). If (8.74) has nonzero projection to the μ'^{th} term of (8.72), then

$$\mu' = (\alpha_i + \beta_j) + \beta_1 + \cdots + \hat{\beta}_j + \cdots + \beta_n + \Sigma\, b_i\gamma_i\,, \quad \gamma_i \in \Delta_L\,.$$

Hence

$$\mu' = \mu + \alpha_i + \Sigma\, c_i\gamma_i\,, \quad \gamma_i \in \Delta_L\,.$$

Since α_i is positive and is not in Δ_L, α_i is greater than $-\Sigma\, c_i\gamma_i$ in our ordering. Thus μ' is greater than μ, and the effect of ∂_{XY} has been to raise the index μ.

Bearing these matters in mind, we apply ∂ to ω:

$$\begin{aligned} 0 = \partial\omega &= \partial(\omega_n^{(\mu_0)} + \omega' + \omega'') \\ &= \partial_X\omega_n^{(\mu_0)} + \partial_Y\omega_n^{(\mu_0)} + \partial_{XX}\omega_n^{(\mu_0)} + \partial_{XY}\omega_n^{(\mu_0)} + \partial_{YY}\omega_n^{(\mu_0)} + \partial\omega' + \partial\omega'' . \end{aligned} \tag{8.75}$$

Of the terms on the right side of (8.75), the second, fifth, and sixth have 0 component in (8.72), and the fourth and seventh have 0 component in

$$(\Lambda^{k-n-1}(\mathfrak{u}\cap\mathfrak{k}) \otimes_{\mathbb{C}} \Lambda^n(\mathfrak{u}\cap\mathfrak{p})^{(\mu_0)}) \otimes_{\mathbb{C}} V . \tag{8.76}$$

Meanwhile the first and third terms are completely in (8.76). Therefore

$$\partial_X\omega_n^{(\mu_0)} + \partial_{XX}\omega_n^{(\mu_0)} = 0 . \tag{8.77}$$

In the space

$$(\Lambda^{k-n}(\mathfrak{u}\cap\mathfrak{k}) \otimes_{\mathbb{C}} \Lambda^n(\mathfrak{u}\cap\mathfrak{p})^{(\mu_0)}) \otimes_{\mathbb{C}} V \tag{8.78}$$

let us regard $\Lambda^n(\mathfrak{u}\cap\mathfrak{p})^{(\mu_0)}$ as a $(\mathfrak{q}\cap\mathfrak{k}, L\cap K)$ module by having $\mathfrak{u}\cap\mathfrak{k}$ act trivially. Then $\Lambda^n(\mathfrak{u}\cap\mathfrak{p})^{(\mu_0)} \otimes_{\mathbb{C}} V$ is in $\mathcal{C}(\mathfrak{q}\cap\mathfrak{k}, L\cap K)$. If we regroup (8.78) as

$$\Lambda^{k-n}(\mathfrak{u}\cap\mathfrak{k}) \otimes_{\mathbb{C}} (\Lambda^n(\mathfrak{u}\cap\mathfrak{p})^{(\mu_0)} \otimes_{\mathbb{C}} V) , \tag{8.79}$$

then (8.79) is the $(k-n)^{th}$ member of the complex from which we compute

$$H_*(\mathfrak{u}\cap\mathfrak{k}, \Lambda^n(\mathfrak{u}\cap\mathfrak{p})^{(\mu_0)} \otimes_{\mathbb{C}} V) ,$$

and $\partial_X + \partial_{XX}$ is just the usual boundary operator ∂_K of this complex (see (7.62)), since $\mathfrak{u}\cap\mathfrak{k}$ acts trivially on $\Lambda^n(\mathfrak{u}\cap\mathfrak{p})^{(\mu_0)}$. Hence we can regard $\omega_n^{(\mu_0)}$ as a member $\widetilde{\omega}_n^{(\mu_0)}$

of (8.79), and (8.77) gives

$$\partial_K \tilde{\omega}_n^{(\mu_0)} = 0. \tag{8.80}$$

Since $n > 0$, the assumption in the theorem is that

$$\mathrm{Hom}_{L\cap K}(\tau_\lambda \otimes_{\mathbb{C}} \Lambda^n(\mathfrak{u}\cap\mathfrak{p})^*, H_{k-n}(\mathfrak{u}\cap\mathfrak{k}, V)) = 0.$$

Then

$$\mathrm{Hom}_{L\cap K}(\tau_\lambda, \mathrm{Hom}_{\mathbb{C}}(\Lambda^n(\mathfrak{u}\cap\mathfrak{p})^*, H_{k-n}(\mathfrak{u}\cap\mathfrak{k}, V))) = 0,$$

and the finite-dimensionality of $\Lambda^n(\mathfrak{u}\cap\mathfrak{p})^*$ allows us to write

$$\mathrm{Hom}_{L\cap K}(\tau_\lambda, \Lambda^n(\mathfrak{u}\cap\mathfrak{p}) \otimes_{\mathbb{C}} H_{k-n}(\mathfrak{u}\cap\mathfrak{k}, V)) = 0.$$

A summand of the left side is

$$\mathrm{Hom}_{L\cap K}(\tau_\lambda, \Lambda^n(\mathfrak{u}\cap\mathfrak{p})^{(\mu_0)} \otimes_{\mathbb{C}} H_{k-n}(\mathfrak{u}\cap\mathfrak{k}, V)).$$

Since $\mathfrak{u}\cap\mathfrak{k}$ acts trivially on $\Lambda^n(\mathfrak{u}\cap\mathfrak{p})^{(\mu_0)}$ by definition, we can put $\Lambda^n(\mathfrak{u}\cap\mathfrak{p})^{(\mu_0)}$ inside the homology and get

$$\mathrm{Hom}_{L\cap K}(\tau_\lambda, H_{k-n}(\mathfrak{u}\cap\mathfrak{k}, \Lambda^n(\mathfrak{u}\cap\mathfrak{p})^{(\mu_0)} \otimes_{\mathbb{C}} V)) = 0.$$

Taking (8.80) into account, we see that

$$\tilde{\omega}_n^{(\mu_0)} = \partial_K \tilde{\eta} \tag{8.81}$$

for some $\tilde{\eta}$ in the τ_λ subspace of

$$\Lambda^{k-n+1}(\mathfrak{u}\cap\mathfrak{k}) \otimes_{\mathbb{C}} (\Lambda^n(\mathfrak{u}\cap\mathfrak{p})^{(\mu_0)} \otimes_{\mathbb{C}} V).$$

If η is the corresponding member of the τ_λ subspace of

$$(\Lambda^{k-n+1}(\mathfrak{u}\cap\mathfrak{k}) \otimes_{\mathbb{C}} \Lambda^n(\mathfrak{u}\cap\mathfrak{p})^{(\mu_0)}) \otimes_{\mathbb{C}} V \subseteq \Lambda^{k+1}\mathfrak{u} \otimes_{\mathbb{C}} V,$$

then (8.81) says

$$\omega_n^{(\mu_0)} = \partial_X\eta + \partial_{XX}\eta = \partial\eta - \partial_Y\eta - \partial_{XY}\eta - \partial_{YY}\eta . \qquad (8.82)$$

As earlier, the terms $\partial_Y\eta$ and $\partial_{XY}\eta$ and $\partial_{YY}\eta$ all have 0 component in (8.78). Thus (8.82) gives

$$\omega_n^{(\mu_0)} = \partial\eta - \omega''' ,$$

and ω''' has 0 component in (8.78). Adding $\omega' + \omega''$, we obtain

$$\omega = \partial\eta + (\omega' + \omega'' - \omega''') ,$$

and $\omega' + \omega'' - \omega'''$ has 0 component in (8.78). By inductive hypothesis

$$\omega' + \omega'' - \omega''' = \omega_K + \partial\eta'$$

with ω_K in $\Lambda^k(\mathfrak{u} \cap \mathfrak{k}) \otimes_{\mathbb{C}} V$, η' in $\Lambda^{k+1}\mathfrak{u} \otimes_{\mathbb{C}} V$, and ω_K and η' both in the τ_λ subspace. Thus

$$\omega = \omega_K + \partial(\eta + \eta') ,$$

and the induction is complete.

6. Vanishing below the middle dimension

We can combine the reduction steps of §§4-5 with Kostant's Theorem and Poincaré duality to obtain a vanishing theorem for $\mathcal{R}^i(Z)$ when i is less than the middle dimension $S = \dim(\mathfrak{u} \cap \mathfrak{k})$. In any expression $\Delta(\cdots)$ for a set of roots, let $\delta(\cdots)$ denote half the sum of the roots in the set.

Theorem 8.15. There exists a constant $C = C(\mathfrak{q})$ such that $\mathcal{R}^i(Z) = 0$ for $i < S$ whenever all $L \cap K$ types τ_μ of Z satisfy

$$\langle \mu, \alpha \rangle \geq C \quad \text{for all} \quad \alpha \in \Delta(\mathfrak{u} \cap \mathfrak{k}) .$$

Remarks. The proof needs Kostant's Theorem (Theorem 6.12) in a version applicable to the parabolic $\mathfrak{q} \cap \mathfrak{k}$ of $\mathfrak{k}$. The version of Kostant's Theorem in Theorem 6.12 is not general enough, applying only to $\mathfrak{b} \cap \mathfrak{k}$, and consequently we shall prove Theorem 8.15 only when $L \cap K = T$. However, we proceed as far as possible with an argument that is applicable in the general case.

Proof when $L \cap K = T$. Let $i = n$ be the least integer such that $\mathcal{R}^i(Z) \neq 0$. By Theorem 8.10 and Corollary 3.8, there exists an integer $k \leq n$ with

$$\mathrm{Hom}_{L \cap K}(Z^{\#} \otimes_{\mathbb{C}} \Lambda^{n-k}(\mathfrak{l} \cap \mathfrak{k})^*, H_k(\mathfrak{u}, \mathcal{R}^n(Z))) \neq 0 .$$

Thus we can find an $L \cap K$ type $\tau_{\mu'}$ in $Z^{\#} \otimes_{\mathbb{C}} \Lambda^{n-k}(\mathfrak{l} \cap \mathfrak{k})^*$ such that

$$\mathrm{Hom}_{L \cap K}(\tau_{\mu'}, H_k(\mathfrak{u}, \mathcal{R}^n(Z))) \neq 0 .$$

By Theorem 8.14, there exists $j \leq k$ with

$$\mathrm{Hom}_{L \cap K}(\tau_{\mu'} \otimes_{\mathbb{C}} \Lambda^{k-j}(\mathfrak{u} \cap \mathfrak{p})^*, H_j(\mathfrak{u} \cap \mathfrak{k}, \mathcal{R}^n(Z))) \neq 0 .$$

Thus we can find an $L \cap K$ type $\tau_{\mu''}$ in $\tau_{\mu'} \otimes_{\mathbb{C}} \Lambda^{k-j}(\mathfrak{u} \cap \mathfrak{p})^*$ such that

$$\mathrm{Hom}_{L \cap K}(\tau_{\mu''}, H_j(\mathfrak{u} \cap \mathfrak{k}, \mathcal{R}^n(Z))) \neq 0 .$$

By Poincaré duality ((6.30) and Proposition 7.34), we obtain

$$\mathrm{Hom}_{L\cap K}(\tau_{\mu''}, H^{S-j}(\mathfrak{u}\cap\mathfrak{k}, \mathcal{R}^n(Z))\otimes_{\mathbb{C}}\Lambda^S(\mathfrak{u}\cap\mathfrak{k})^*) \neq 0 .$$

The cohomology in this expression is the sum of contributions from the K-irreducible summands of $\mathcal{R}^n(Z)$. Thus we can apply Kostant's Theorem and conclude that

$$\mu'' + 2\delta(\mathfrak{u}\cap\mathfrak{k}) = w(\eta+\delta_K) - \delta_K$$

for some Δ_K^+ dominant form η and for some w of length $S-j$ in the Weyl group of Δ_K. Meanwhile μ'' and μ' are related by

$$\mu'' = \mu' - \sum_{i=1}^{k-j}\beta_i ,$$

where the β_i are distinct roots contributing to $\mathfrak{u}\cap\mathfrak{p}$. And μ' is related to some $L\cap K$ type τ_λ occurring in Z by

$$\mu' = \lambda + 2\delta(\mathfrak{u}) - \sum\gamma_i ,$$

where the γ_i are distinct roots in $\Delta_{L\cap K}$, at most $n-k$ of them occurring. Putting these facts together, we obtain

$$w(\eta+\delta_K) = \lambda + [2\delta(\mathfrak{u}) - \sum\gamma_i - \sum\beta_i + 2\delta(\mathfrak{u}\cap\mathfrak{k}) + \delta_K]. \tag{8.83}$$

If $n < S$, then $S-j$ is > 0 and w cannot be 1. Hence there exists $\alpha\in\Delta(\mathfrak{u}\cap\mathfrak{k})$ with

$$\langle w(\eta+\delta_K), \alpha\rangle < 0 . \tag{8.84}$$

If we take C to be the maximum value of

$$|\langle 2\delta(\mathfrak{u}) - \sum\gamma_i - \sum\beta_i + 2\delta(\mathfrak{u}\cap\mathfrak{k}) + \delta_K, \alpha'\rangle|$$

over all systems $\Sigma \gamma_i$, all systems $\Sigma \beta_i$, and all α' in $\Delta(\mathfrak{u} \cap \mathfrak{k})$, then $\langle \lambda, \alpha \rangle \geq C$ implies that the right side of (8.83) has inner product ≥ 0 with α, in contradiction to (8.84). We conclude $n \geq S$, and this is the conclusion of the theorem.

Let λ be an integral form on $\mathfrak{h}$ that is orthogonal to all members of Δ_L. Then λ is the (highest) weight of a one-dimensional representation $\mathbb{C}_\lambda$ of L (by the Theorem of the Highest Weight for $\mathfrak{l}$). We define

$$A_{\mathfrak{q}}(\lambda) = \mathcal{L}_S(\mathbb{C}_\lambda) \quad \text{and} \quad A^{\mathfrak{q}}(\lambda) = \mathcal{R}^S(\mathbb{C}_\lambda). \tag{8.85}$$

Theorem 8.15 says that $\mathcal{R}^i(\mathbb{C}_\lambda) = 0$ for $i \neq S$ if $\langle \lambda, \alpha \rangle \geq C$ for all $\alpha \in \Delta(\mathfrak{u} \cap \mathfrak{k})$. This conclusion will allow us in §8 to compute the multiplicities of all K types in $A^{\mathfrak{q}}(\lambda)$. (Notice that the two conditions $\langle \lambda, \beta \rangle = 0$ for $\beta \in \Delta_L$ and $\langle \lambda, \alpha \rangle \geq C$ for $\alpha \in \Delta(\mathfrak{u} \cap \mathfrak{k})$ are compatible since we can define λ to have inner product 0 with the simple roots that are in Δ_L and to have large inner product with the simple roots that are outside Δ_L.)

We mentioned in §4 that a vanishing theorem for $\mathcal{L}_i(Z)$ comes from an easy duality argument. In fact, we start with any $(\mathfrak{q}, L \cap K)$ module W and observe that

$$\begin{aligned} I^{\mathfrak{g},K}_{\mathfrak{q},L\cap K}(W^c) &= \mathrm{Hom}_{R(\mathfrak{q},L\cap K)}(R(\mathfrak{g},K), \mathrm{Hom}_{\mathbb{C}}(W,\mathbb{C})_{L\cap K})_K \\ &\cong \mathrm{Hom}_{\mathbb{C}}(R(\mathfrak{g},K) \otimes_{R(\mathfrak{q},L\cap K)} W, \mathbb{C})_K \\ &= P^{\mathfrak{g},K}_{\mathfrak{q},L\cap K}(W)^c \end{aligned}$$

naturally in W. By Proposition 7.16 we can write this isomorphism as a natural isomorphism of functors

$$I^{\mathfrak{g},K}_{\mathfrak{g},L\cap K}\circ(I^{\mathfrak{g},L\cap K}_{\mathfrak{q},L\cap K}\circ\mathrm{dual}) \cong \mathrm{dual}\circ P^{\mathfrak{g},K}_{\mathfrak{g},L\cap K}\circ P^{\mathfrak{q},L\cap K}_{\mathfrak{q},L\cap K}.$$

Using Propositions 5.16 and 5.17 (with Proposition 7.9, Proposition 7.15, and Corollary 7.19 as tools for checking the hypotheses), we obtain

$$\Gamma^{i}\circ(I^{\mathfrak{g},L\cap K}_{\mathfrak{q},L\cap K}\circ\mathrm{dual}) \cong \mathrm{dual}\circ\Pi_{i}\circ P^{\mathfrak{g},L\cap K}_{\mathfrak{q},L\cap K}.$$

Therefore

$$\Gamma^{i}(I^{\mathfrak{g},L\cap K}_{\mathfrak{q},L\cap K}(V^{c}_{\mathfrak{q}})) \cong (\Pi_{i}(P^{\mathfrak{g},L\cap K}_{\mathfrak{q},L\cap K}(V_{\mathfrak{q}})))^{c}.$$

for any V in $\mathcal{C}(\mathfrak{l},L\cap K)$. Taking $V = Z^{\#}$ and replacing $\mathfrak{q}$ by $\bar{\mathfrak{q}}$ gives

$$\mathcal{L}_{i}(Z)^{c} \cong \Gamma^{i}(I^{\mathfrak{g},L\cap K}_{\bar{\mathfrak{q}},L\cap K}((Z^{\#})^{c}_{\bar{\mathfrak{q}}})). \tag{8.86}$$

The right side is $\mathcal{R}^{i}_{\bar{\mathfrak{q}}}(Z^{c})$, namely $\mathcal{R}^{i}(Z^{c})$ computed relative to $\bar{\mathfrak{q}}$ rather than $\mathfrak{q}$. If it is 0, then (8.86) says $\mathcal{L}_{i}(Z)$ is 0.

So, for example, if $Z = \mathbb{C}_{\lambda}$ as above, then $Z^{c} = \mathbb{C}_{-\lambda}$. Theorem 8.15 says that $\mathcal{R}^{i}_{\bar{\mathfrak{q}}}(\mathbb{C}_{-\lambda}) = 0$ for $i < S$ if $\langle -\lambda,\alpha\rangle \geq C$ for all $\alpha \in \Delta(\bar{\mathfrak{u}}\cap\mathfrak{k})$, i.e., if $\langle\lambda,\alpha\rangle \geq C$ for all $\alpha \in \Delta(\mathfrak{u}\cap\mathfrak{k})$. We conclude that $\mathcal{L}_{i}(\mathbb{C}_{\lambda}) = 0$ for $i \neq S$ if $\langle\lambda,\alpha\rangle \geq C$ for all $\alpha\in\Delta(\mathfrak{u}\cap\mathfrak{k})$. This conclusion will allow us in §8 to compute the multiplicities of all K types in $A_{\mathfrak{q}}(\lambda)$.

7. Effect on infinitesimal character

Under the assumption that Z has an infinitesimal character (i.e., the center $Z(\mathfrak{l})$ of $U(\mathfrak{l})$ acts by scalars), $\mathcal{L}_i(Z)$ and $\mathcal{R}^i(Z)$ turn out to have infinitesimal characters (i.e., $Z(\mathfrak{g})$ acts by scalars), and the latter infinitesimal characters are given by simple formulas.

Let $\mathfrak{n} = \Sigma_{\alpha\in\Delta^+}\mathfrak{g}_\alpha$ and $\bar{\mathfrak{n}} = \Sigma_{\alpha\in\Delta^+}\mathfrak{g}_{-\alpha}$. In §1 we recalled the definition of the Harish-Chandra homomorphism. We write

$$U(\mathfrak{g}) = U(\mathfrak{h}) \oplus (U(\mathfrak{g})\mathfrak{n} + \bar{\mathfrak{n}}U(\mathfrak{g})) ,$$

and we let $\gamma' = \gamma'_{\mathfrak{n}}$ be the projection of $Z(\mathfrak{g})$ to the first coordinate. Then we let $\sigma = \sigma_{\mathfrak{n}}$ be the algebra isomorphism of $U(\mathfrak{h})$ that sends 1 into 1 and extends the map $\sigma_{\mathfrak{n}}(H) = H - \delta(H)1$ of $\mathfrak{h}$ into $U(\mathfrak{h})$ to all of $U(\mathfrak{h})$. The <u>Harish-Chandra homomorphism</u> is $\gamma = \sigma_{\mathfrak{n}}\circ\gamma'_{\mathfrak{n}}$, and the Weyl group invariance in Theorem 6.17 implies that γ is independent of the choice of Δ^+. We shall often find it convenient to write γ in the form

$$\gamma = \sigma_{\bar{\mathfrak{n}}}\circ\gamma'_{\bar{\mathfrak{n}}} . \tag{8.87}$$

Similarly there is a Harish-Chandra homomorphism for $\mathfrak{l}$, given as $\gamma_{\mathfrak{l}} : Z(\mathfrak{l}) \to U(\mathfrak{h})$ by

$$\gamma_{\mathfrak{l}} = \sigma_{\mathfrak{l}\cap\bar{\mathfrak{n}}}\circ\gamma'_{\mathfrak{l}\cap\bar{\mathfrak{n}}} , \tag{8.88}$$

where $\sigma_{\mathfrak{l}\cap\bar{\mathfrak{n}}}$ extends

$$\sigma_{\mathfrak{l}\cap\bar{\mathfrak{n}}}(H) = H + \delta_L(H)1 .$$

We define a relative Harish-Chandra homomorphism $\mu_{\mathfrak{g}/\mathfrak{l}} : Z(\mathfrak{g}) \to Z(\mathfrak{l})$ by

$$\mu_{\mathfrak{g}/\mathfrak{l}} = \gamma_{\mathfrak{l}}^{-1} \circ \gamma .$$

This is well defined since image γ is equal to the Weyl group invariants for $\mathfrak{g}$, which are contained in the Weyl group invariants for $\mathfrak{l}$. Meanwhile we can write

$$U(\mathfrak{g}) = U(\mathfrak{l}) \oplus (\mathfrak{u}U(\mathfrak{g}) + U(\mathfrak{g})\bar{\mathfrak{u}}) \tag{8.89}$$

and let $\mu'_{\mathfrak{g}/\mathfrak{l}}$ be the projection to the $U(\mathfrak{l})$ coordinate. Let us define also a mapping $\sigma_{\bar{\mathfrak{u}}}' : Z(\mathfrak{l}) \to Z(\mathfrak{l})$. We write $\mathfrak{l} = Z_{\mathfrak{l}} \oplus [\mathfrak{l},\mathfrak{l}]$, where $Z_{\mathfrak{l}}$ is the center of $\mathfrak{l}$ and is contained in $\mathfrak{h}$ since $\mathfrak{h}$ is maximal abelian in $\mathfrak{g}$. Then $U(\mathfrak{l}) = U(Z_{\mathfrak{l}}) \otimes_{\mathbb{C}} U([\mathfrak{l},\mathfrak{l}])$ and $U(Z_{\mathfrak{l}})$ is central, so that

$$Z(\mathfrak{l}) = U(Z_{\mathfrak{l}}) \otimes_{\mathbb{C}} Z([\mathfrak{l},\mathfrak{l}]) . \tag{8.90}$$

We define $\sigma_{\bar{\mathfrak{u}}} = \sigma_{\bar{\mathfrak{u}},1} \otimes 1$, where

$$\sigma_{\bar{\mathfrak{u}},1}(H) = H + \delta(\mathfrak{u})(H)1$$

is extended from a linear map $Z_{\mathfrak{l}} \to U(Z_{\mathfrak{l}})$ to an algebra isomorphism of $U(Z_{\mathfrak{l}})$ sending 1 into 1.

Lemma 8.16. The Harish-Chandra mappings satisfy

(a) $\mu'_{\mathfrak{g}/\mathfrak{l}}(Z(\mathfrak{g})) \subseteq Z(\mathfrak{l})$

(b) $\gamma'_{\bar{\mathfrak{n}}} = \gamma'_{\mathfrak{l}\cap\bar{\mathfrak{n}}} \circ \mu'_{\mathfrak{g}/\mathfrak{l}}$ on $Z(\mathfrak{g})$

(c) $\gamma'_{\mathfrak{l}\cap\bar{\mathfrak{n}}} \circ \sigma_{\bar{\mathfrak{u}}} = \sigma_{\bar{\mathfrak{u}}} \circ \gamma'_{\mathfrak{l}\cap\bar{\mathfrak{n}}}$ on $Z(\mathfrak{l})$

(d) $\mu_{\mathfrak{g}/\mathfrak{l}} = \sigma_{\bar{\mathfrak{u}}} \circ \mu'_{\mathfrak{g}/\mathfrak{l}}$ on $Z(\mathfrak{g})$.

Proof. (a) Temporarily take $\mu'_{\mathfrak{g}/\mathfrak{l}}$ to be defined on all of $U(\mathfrak{g})$. If ℓ is in $U(\mathfrak{l})$ and x is in $U(\mathfrak{g})$, we readily check that (8.89) and the inclusions

$$[\mathfrak{l},\mathfrak{u}] \subseteq \mathfrak{u} \quad \text{and} \quad [\mathfrak{l},\bar{\mathfrak{u}}] \subseteq \bar{\mathfrak{u}}$$

imply

$$\mu'_{\mathfrak{g}/\mathfrak{l}}(\ell x) = \ell\mu'_{\mathfrak{g}/\mathfrak{l}}(x) \quad \text{and} \quad \mu'_{\mathfrak{g}/\mathfrak{l}}(x\ell) = \mu'_{\mathfrak{g}/\mathfrak{l}}(x)\ell .$$

If x is actually in $Z(\mathfrak{g})$, then we obtain

$$\ell\mu'_{\mathfrak{g}/\mathfrak{l}}(x) = \mu'_{\mathfrak{g}/\mathfrak{l}}(\ell x) = \mu'_{\mathfrak{g}/\mathfrak{l}}(x\ell) = \mu'_{\mathfrak{g}/\mathfrak{l}}(x)\ell ,$$

and (a) follows.

(b) If we follow the decomposition (8.89) with the decomposition according to

$$U(\mathfrak{l}) = U(\mathfrak{h}) \oplus ((\mathfrak{l} \cap \mathfrak{n})U(\mathfrak{l}) + U(\mathfrak{l})(\mathfrak{l} \cap \bar{\mathfrak{n}})) , \tag{8.91}$$

we obtain a decomposition consistent with

$$U(\mathfrak{g}) = U(\mathfrak{h}) \oplus (\mathfrak{n}U(\mathfrak{g}) + U(\mathfrak{g})\bar{\mathfrak{n}}) , \tag{8.92}$$

and this is (b).

(c) If z is in $Z([\mathfrak{l},\mathfrak{l}])$, then $\gamma'(z)$ is in $U(\mathfrak{h} \cap [\mathfrak{l},\mathfrak{l}])$ and $\sigma_{\bar{\mathfrak{u}}}$ is the identity on this. Also $\sigma_{\bar{\mathfrak{u}}}(z) = 1$, and the result is that (c) holds on $Z([\mathfrak{l},\mathfrak{l}])$. On $U(Z_\mathfrak{l})$, $\gamma'_{\mathfrak{l}\cap\bar{\mathfrak{n}}}$ is the identity, and thus (c) holds on $U(Z_\mathfrak{l})$ as well. Hence (c) holds on $Z(\mathfrak{l})$ by (8.90).

(d) We have

$$\begin{aligned} \gamma &= \sigma_{\bar{\mathfrak{n}}}\circ\gamma'_{\bar{\mathfrak{n}}} && \text{by (8.87)} \\ &= \sigma_{\mathfrak{l}\cap\bar{\mathfrak{n}}}\circ\sigma_{\bar{\mathfrak{u}}}\circ\gamma'_{\mathfrak{l}\cap\bar{\mathfrak{n}}}\circ\mu'_{\mathfrak{g}/\mathfrak{l}} && \text{by (b)} \end{aligned}$$

$$= \sigma_{\mathfrak{l}\cap\overline{\mathfrak{n}}} \circ \gamma'_{\mathfrak{l}\cap\overline{\mathfrak{n}}} \circ \sigma_{\overline{\mathfrak{u}}} \circ \mu'_{\mathfrak{g}/\mathfrak{l}} \qquad \text{by (a) and (c)}$$

$$= \gamma_{\mathfrak{l}} \circ (\sigma_{\overline{\mathfrak{u}}} \circ \mu'_{\mathfrak{g}/\mathfrak{l}}) \qquad \text{by (8.88).}$$

Applying $\gamma_{\mathfrak{l}}^{-1}$ on the left, we obtain the result (d).

Let λ be in $\mathfrak{h}^*$. We say that the $(\mathfrak{l}, L\cap K)$ module Z has <u>infinitesimal character</u> λ if $Z(\mathfrak{l})$ acts by scalars on Z according to the formula

$$zv = \lambda(\gamma_{\mathfrak{l}}(z))v$$

for $z \in Z(\mathfrak{l})$ and $v \in Z$. We have already defined infinitesimal characters for $(\mathfrak{g}, K)$ modules in analogous fashion.

<u>Theorem 8.17</u>. If the $(\mathfrak{l}, L\cap K)$ module Z has infinitesimal character λ, then the $(\mathfrak{g}, K)$ modules $\mathcal{L}_i(Z)$ and $\mathcal{R}^i(Z)$ have infinitesimal character $\lambda + \delta(\mathfrak{u})$.

Proof. We give the argument for $\mathcal{L}_i(Z)$, and the argument for $\mathcal{R}^i(Z)$ is similar. First we check readily that

$$Z^{\#} = Z \otimes_{\mathbb{C}} \wedge^{R+S}\mathfrak{u}$$

has infinitesimal character $\lambda + 2\delta(\mathfrak{u})$. For purposes of understanding the $U(\mathfrak{g})$ action, Lemma 8.9 says we can make the identification

$$P^{\mathfrak{g}, L\cap K}_{\mathfrak{q}, L\cap K}(Z^{\#}_{\mathfrak{q}}) = R(\mathfrak{g}, L\cap K) \otimes_{R(\overline{\mathfrak{q}}, L\cap K)} Z^{\#}_{\mathfrak{q}} \cong U(\mathfrak{g}) \otimes_{U(\overline{\mathfrak{q}})} Z^{\#}_{\mathfrak{q}}.$$

If z is in $Z(\mathfrak{g})$, write

$$z = \mu'_{\mathfrak{g}/\mathfrak{l}}(z) + \sum x_i \overline{u}_i, \qquad x_i \in U(\mathfrak{g}), \quad \overline{u}_i \in \overline{\mathfrak{u}},$$

according to the decomposition (8.89). (Here the $uU(\mathfrak{g})$ terms are not needed, by the same argument as for Lemma 6.16b.) Then

$$\begin{aligned}
z(x \otimes v) &= zx \otimes v = xz \otimes v \\
&= x\mu'_{\mathfrak{g}/I}(z) \otimes v + \Sigma\, xx_i \bar{u}_i \otimes v \\
&= x \otimes \mu'_{\mathfrak{g}/I}(z)v + \Sigma\, xx_i \otimes \bar{u}_i v \\
&= x \otimes \mu'_{\mathfrak{g}/I}(z)v \qquad \text{since } \bar{u} \text{ annihilates } Z^{\#}_{\mathfrak{q}} \\
&= (\lambda + 2\delta(u))(\gamma_I(\mu'_{\mathfrak{g}/I}(z)))(x \otimes v) \\
&= (\lambda + 2\delta(u))(\sigma_{I \cap \bar{n}} \circ \gamma'_{I \cap \bar{n}} \circ \mu'_{\mathfrak{g}/I}(z))(x \otimes v) \qquad \text{by (8.88)} \\
&= ((\lambda + 2\delta(u))(\sigma^{-1}_{\bar{u}} \circ \sigma_{\bar{n}} \circ \gamma'_{\bar{n}}(z))(x \otimes v) \qquad \text{by Lemma 8.16b} \\
&= ((\lambda + 2\delta(u))(\sigma^{-1}_{\bar{u}} \circ \gamma(z))(x \otimes v) \qquad \text{by (8.87)} \\
&= (\lambda + \delta(u))(\gamma(z))(x \otimes v) .
\end{aligned}$$

So z acts on the $(\mathfrak{g}, L \cap K)$ module $P^{\mathfrak{g}, L\cap K}_{\mathfrak{q}, L\cap K}(Z^{\#}_{\mathfrak{q}})$ with infinitesimal character $\lambda + \delta(u)$.

Now suppose z is in $Z(\mathfrak{g})$ and P is a $(\mathfrak{g}, L \cap K)$ module. Since z is central, the action of z on P is a $(\mathfrak{g}, L \cap K)$ map of P into itself. We claim that the action of z on $\Pi(P)$ is just $\Pi(z)$. In fact,

$$\Pi(P) = R(\mathfrak{g}, K) \otimes_{R(\mathfrak{g}, L\cap K)} P$$

with $\Pi(z)$ given as the map $1 \otimes z$. Meanwhile z acts on $\Pi(P)$ by $z \otimes 1$, and it is easy to check that these actions are equal by using (7.21a) and the fact that $Ad(K)$ fixes z.

Let $\bar{P} = P^{\mathfrak{g}, L\cap K}_{\mathfrak{q}, L\cap K}(Z^{\#}_{\mathfrak{q}})$, and let z act by a scalar $c(z)$. Let

$$0 \longleftarrow \overline{P} \longleftarrow P_0 \longleftarrow P_1 \longleftarrow \cdots$$

be a projective resolution of $\overline{P}$. The diagram

$$\begin{array}{ccccccccc} 0 & \longleftarrow & \overline{P} & \longleftarrow & P_0 & \longleftarrow & P_1 & \longleftarrow & \cdots \\ & & \downarrow z & & \downarrow z & & \downarrow z & & \\ 0 & \longleftarrow & \overline{P} & \longleftarrow & P_0 & \longleftarrow & P_1 & \longleftarrow & \cdots \end{array} \tag{8.93}$$

is commutative since the horizontal maps are $U(\mathfrak{g})$ maps, and it is a diagram of $(\mathfrak{g}, L\cap K)$ maps since z is central. Applying Π and using the result of the previous paragraph, we obtain a commutative diagram

$$\begin{array}{ccccccc} 0 & \longleftarrow & \Pi(P_0) & \longleftarrow & \Pi(P_1) & \longleftarrow & \cdots \\ & & \downarrow z & & \downarrow z & & \\ 0 & \longleftarrow & \Pi(P_0) & \longleftarrow & \Pi(P_1) & \longleftarrow & \cdots \end{array} \tag{8.94}$$

The diagram (8.93) exhibits $\{z\}$ as a chain map over z, and it is clear that scalar multiplication by $c(z)$ is another chain map over z, since z acts by $c(z)$ on $\overline{P}$. By Proposition 5.4, (8.93) is homotopic to

$$\begin{array}{ccccccccc} 0 & \longleftarrow & \overline{P} & \longleftarrow & P_0 & \longleftarrow & P_1 & \longleftarrow & \cdots \\ & & \downarrow z & & \downarrow c(z) & & \downarrow c(z) & & \\ 0 & \longleftarrow & \overline{P} & \longleftarrow & P_0 & \longleftarrow & P_1 & \longleftarrow & \cdots \end{array}$$

Applying Π, we see that (8.94) is homotopic to

$$\begin{array}{ccccccc} 0 & \longleftarrow & \Pi(P_0) & \longleftarrow & \Pi(P_1) & \longleftarrow & \cdots \\ & & \downarrow c(z) & & \downarrow c(z) & & \\ 0 & \longleftarrow & \Pi(P_0) & \longleftarrow & \Pi(P_1) & \longleftarrow & \cdots \end{array}$$

Passing to homology and using Proposition 5.1, we see that z acts by $c(z)$ on $\Pi_i(\overline{P})$. Combining this fact with the fact that $P^{\mathfrak{g},L\cap K}_{\mathfrak{q},L\cap K}(Z^{\#}_{\mathfrak{q}})$ has infinitesimal character $\lambda+\delta(\mathfrak{u})$, we obtain the conclusion that $\mathcal{L}_i(Z)$ has infinitesimal character $\lambda+\delta(\mathfrak{u})$.

8. Effect on multiplicities of K types

In this section we shall assume that Z is an $(\mathfrak{l},L\cap K)$ module that has an infinitesimal character and whose $L\cap K$ types all have finite multiplicity. When we have a vanishing theorem for $\mathcal{L}_i(Z)$ or $\mathcal{R}^i(Z)$ below the middle dimension, we shall see that it is possible to make an exact computation of the multiplicities of the K types in $\mathcal{L}_S(Z)$ or $\mathcal{R}^S(Z)$. Recall from Corollary 3.8 that the multiplicity of a K type τ_Λ (acting on V) in a $(\mathfrak{g},K)$ module U is given by

$$\dim \operatorname{Hom}_K(V,U) = \dim \operatorname{Hom}_K(U,V).$$

The technique for computing the multiplicity of τ_Λ in $\mathcal{L}_S(Z)$ will be to consider the alternating sum

$$\sum_{i\geq 0} (-1)^i \dim \operatorname{Hom}_K(\mathcal{L}_i(Z),V). \tag{8.95}$$

With this alternating sum, we will find that the inequalities in the proof of Theorem 8.7 become equalities, and an exact formula results. Since (by Theorem 8.7) only the Sth term in (8.95) is nonzero when there is vanishing below the middle dimension, we obtain an exact formula for $\dim \mathrm{Hom}_K(\mathcal{L}_S(Z),V)$.

The idea of seeking results for an alternating sum of dimensions (or other quantities for which an analog of (8.98) below holds), rather than the individual dimensions is called the Euler-Poincaré principle. A precise formulation of this principle that will be adequate for our purposes is in the following proposition and corollary.

Proposition 8.18. Let $F : \mathbf{C} \to \mathbf{C}'$ be a functor between good categories, suppose $\mathbf{C}$ has enough projectives and enough injectives, and let the derived functors of F be denoted F_i or F^i, say F^i for definiteness. Suppose

$$0 \longrightarrow A \longrightarrow B \longrightarrow C \longrightarrow 0 \qquad (8.96)$$

is an exact sequence in $\mathbf{C}$ such that $F^i(A)$, $F^i(B)$, and $F^i(C)$ are finite-dimensional for all i. If $F^i = 0$ for i sufficiently large, then

$$\sum_{i \geq 0} (-1)^i \{\dim F^i(B) - \dim F^i(A) - \dim F^i(C)\} = 0. \qquad (8.97)$$

Proof. If U, V, and W are finite-dimensional vector spaces such that

$$0 \longrightarrow U \xrightarrow{\varphi} V \longrightarrow W \longrightarrow 0$$

is exact, then $W \cong V/\varphi(U)$ and consequently

$$- \dim U + \dim V - \dim W = 0 \tag{8.98}$$

We claim that if

$$0 \longrightarrow V_1 \xrightarrow{\varphi_1} V_2 \xrightarrow{\varphi_2} \cdots \xrightarrow{\varphi_{n-1}} V_n \longrightarrow 0 \tag{8.99}$$

is an exact sequence of finite-dimensional vector spaces, then

$$\sum_{i=1}^{n} (-1)^i \dim V_i = 0. \tag{8.100}$$

In fact, the exactness of (8.99) implies that the sequences

$$0 \longrightarrow V_1 \xrightarrow{\varphi_1} V_2 \xrightarrow{\varphi_2} \text{image } \varphi_2 \longrightarrow 0$$

$$0 \longrightarrow \text{image } \varphi_2 \xrightarrow{\text{inc}} V_3 \xrightarrow{\varphi_3} \text{image } \varphi_3 \longrightarrow 0$$

$$0 \longrightarrow \text{image } \varphi_3 \xrightarrow{\text{inc}} V_4 \xrightarrow{\varphi_4} \text{image } \varphi_4 \longrightarrow 0$$

$$\vdots$$

$$0 \longrightarrow \text{image } \varphi_{n-2} \xrightarrow{\text{inc}} V_{n-1} \xrightarrow{\varphi_{n-1}} V_n \longrightarrow 0$$

are exact. Thus (8.98) gives

$$- \dim V_1 + \dim V_2 - \dim \text{image } \varphi_2 = 0$$

$$\dim \text{image } \varphi_2 - \dim V_3 + \dim \text{image } \varphi_3 = 0$$

$$- \dim \text{image } \varphi_3 + \dim V_4 - \dim \text{image } \varphi_4 = 0$$

$$\vdots$$

$$- (-1)^{n-1} \dim \text{image } \varphi_2 + (-1)^{n-1} \dim V_{n-1} - (-1)^{n-1} \dim V_n = 0.$$

Adding these relations yields (8.100).

To prove the proposition, we take the long exact sequence

(Theorem 5.28) obtained for the derived functors of F applied to (8.96). Applying (8.100) to this sequence, we obtain the required formula (8.97).

Corollary 8.19. Suppose $F : \mathbf{C} \to \mathbf{C}'$ is a functor between good categories such that $\mathbf{C}$ has enough projectives and enough injectives, and suppose that the derived functors F_i or F^i of F, say F^i, satisfy $F^i = 0$ for i sufficiently large. Let

$$M_N \supseteq M_{N-1} \supseteq \cdots \supseteq M_0 \supseteq M_{-1} = 0$$

be a sequence of modules in $\mathbf{C}$, and suppose that $\dim F^i(M_n/M_{n-1}) < \infty$ for $0 \leq n \leq N$ and for all i. Then

$$\sum_{i \geq 0} (-1)^i \dim F^i(M_N) = \sum_{i \geq 0} \sum_{n=0}^{N} (-1)^i \dim F^i(M_n/M_{n-1}) ,$$

all the terms and the sums being finite.

Proof. We proceed by induction on N, the case $N = 0$ being obvious. Suppose $\dim F^i(M_{N-1}) < \infty$ for all i and

$$\sum_{i \geq 0} (-1)^i \dim F^i(M_{N-1}) = \sum_{i \geq 0} \sum_{n=0}^{N-1} (-1)^i \dim F^i(M_n/M_{n-1}) . \tag{8.101}$$

Now $F^i(M_N)$ lies in between $F^i(M_{N-1})$ and $F^i(M_N/M_{N-1})$ in an exact sequence, and hence

$$\dim F^i(M_N) \leq \dim F^i(M_{N-1}) + \dim F^i(M_N/M_{N-1}) < \infty .$$

Thus $F^i(M_N)$ is finite-dimensional, and

$$0 \longrightarrow M_{N-1} \longrightarrow M_N \longrightarrow M_N/M_{N-1} \longrightarrow 0$$

is an exact sequence to which Proposition 8.18 applies. The proposition gives

$$\sum_{i \geq 0} (-1)^i (\dim F^i(M_N) - \dim F^i(M_{N-1})) = \sum_{i \geq 0} (-1)^i \dim F^i(M_N/M_{N-1}) .$$

Adding this equality to (8.101) gives the result of the corollary.

<u>Theorem 8.20</u>. Suppose that Z is an $(\mathfrak{l}, L \cap K)$ module having an infinitesimal character and having all $L \cap K$ types of finite multiplicity. Then $\mathcal{L}_i(Z)$ and $\mathcal{R}^i(Z)$ have all K types of finite multiplicity. In fact, if V is a finite-dimensional $(\mathfrak{k}, K)$ module, then

$$\sum_{i=0}^{S} (-1)^i \dim \operatorname{Hom}_K(\mathcal{L}_i(Z), V)$$
$$= \sum_{i=0}^{S} \sum_{n=0}^{\infty} (-1)^i \dim \operatorname{Hom}_{L\cap K}(S^n(\mathfrak{u} \cap \mathfrak{p}) \otimes_{\mathbb{C}} Z^{\#}, H^i(\bar{\mathfrak{u}} \cap \mathfrak{k}, V))$$

and

$$\sum_{i=0}^{S} (-1)^i \dim \operatorname{Hom}_K(V, \mathcal{R}^i(Z))$$
$$= \sum_{i=0}^{S} \sum_{n=0}^{\infty} (-1)^i \dim \operatorname{Hom}_{L\cap K}(H_i(\mathfrak{u} \cap \mathfrak{k}, V), S^n(\mathfrak{u} \cap \mathfrak{p}) \otimes_{\mathbb{C}} Z^{\#})$$

with all terms 0 for n sufficiently large.

Proof. We give the proof only for $\mathcal{L}_i(Z)$, the proof for $\mathcal{R}^i(Z)$ being similar. The idea will be to go over the argument for Theorem 8.7, obtaining equalities in place of inequalities when we take alternating sums. From (8.21) we have

$$\mathrm{Hom}_K(\mathcal{L}_i(Z), V) \cong \mathrm{Ext}^i_{\mathfrak{l}, L\cap K}(P^{\mathfrak{g}, L\cap K}_{\mathfrak{q}, L\cap K}(Z^{\#}_{\mathfrak{q}}), V) . \tag{8.102}$$

Recall once more from Proposition 8.2 that $\mathfrak{u}$ is the sum of the eigenspaces for positive eigenvalues for ad of an element X_0 of $\mathfrak{h}_{\mathbb{R}}$ and that X_0 is in the center of $\mathfrak{l}$. Since Z has an infinitesimal character, X_0 acts as a scalar in $Z^{\#}_{\mathfrak{q}}$.

As in Theorem 8.7, put

$$M = P^{\mathfrak{g}, L\cap K}_{\mathfrak{q}, L\cap K}(Z^{\#}_{\mathfrak{q}}) \cong U(\mathfrak{g}) \otimes_{U(\bar{\mathfrak{q}})} Z^{\#}_{\mathfrak{q}} .$$

For $n \geq -1$, let

$$\begin{aligned} M_n &= \text{image in } M \text{ of } \; U_n(\mathfrak{g}) \otimes_{\mathbb{C}} Z^{\#}_{\mathfrak{q}} \\ &\cong U_n(\mathfrak{u}) \otimes_{\mathbb{C}} Z^{\#}_{\mathfrak{q}} \end{aligned}$$

as a $(\mathfrak{l}, L\cap K)$ module. By (8.24) there exists N_0 such that

$$\mathrm{Ext}^i_{\mathfrak{l}, L\cap K}(M, V) \cong \mathrm{Ext}^i_{\mathfrak{l}, L\cap K}(M_N, V) \tag{8.103}$$

for $N \geq N_0$ and for all i.

Temporarily let us ignore the finiteness assumption in Corollary 8.19 and use the corollary anyway. With $N \geq N_0$ we apply the corollary to the sequence

$$M_N \supseteq M_{N-1} \supseteq \cdots \supseteq M_0 \supseteq M_{-1} = 0$$

in $\mathcal{C}(\mathfrak{l}, L\cap K)$ and to the functor $F = \mathrm{Hom}_{\mathfrak{l}, L\cap K}(\cdot, V)$. By (8.103) the result is

$$\sum_{i \geq 0} (-1)^i \dim \operatorname{Ext}^i_{\mathfrak{k}, L\cap K}(M, V)$$

$$= \sum_{i \geq 0} \sum_{n=0}^{N} (-1)^i \dim \operatorname{Ext}^i_{\mathfrak{k}, L\cap K}(M_n/M_{n-1}, V), \tag{8.104}$$

and (8.26) shows this is

$$= \sum_{i \geq 0} \sum_{n=0}^{N} (-1)^i \dim \operatorname{Ext}^i_{\mathfrak{k}, L\cap K}(U(\mathfrak{k}) \otimes_{U(\bar{\mathfrak{q}}\cap\mathfrak{k})} (S^n(\mathfrak{u}\cap\mathfrak{p}) \otimes_{\mathbb{C}} Z^{\#}_{\mathfrak{q}}), V)$$

for a certain nontrivial action of $\bar{\mathfrak{u}} \cap \mathfrak{k}$ on $S^n(\mathfrak{u}\cap\mathfrak{p})$. By Shapiro's Lemma (7.51), the right side is

$$= \sum_{i \geq 0} \sum_{n=0}^{N} (-1)^i \dim \operatorname{Ext}^i_{\bar{\mathfrak{q}}\cap\mathfrak{k}, L\cap K}(S^n(\mathfrak{u}\cap\mathfrak{p}) \otimes_{\mathbb{C}} Z^{\#}_{\mathfrak{q}}, V). \tag{8.105}$$

The $(\bar{\mathfrak{q}}\cap\mathfrak{k}, L\cap K)$ module $S^n(\mathfrak{u}\cap\mathfrak{p})$ is filtered according to the eigenvalues of X_0, since $\bar{\mathfrak{u}}\cap\mathfrak{k}$ decreases eigenvalues, and the respective quotients have 0 action by $\bar{\mathfrak{u}}\cap\mathfrak{k}$. Applying Corollary 8.19 to this filtration (while ignoring the finiteness assumption temporarily) and to the functor $F = \operatorname{Hom}_{\bar{\mathfrak{q}}\cap\mathfrak{k}, L\cap K}(\cdot \otimes_{\mathbb{C}} Z^{\#}_{\mathfrak{q}}, V)$ and reassembling the quotients as a direct sum afterward, we see that (8.105) is

$$= \sum_{i \geq 0} \sum_{n=0}^{N} (-1)^i \dim \operatorname{Ext}^i_{\bar{\mathfrak{q}}\cap\mathfrak{k}, L\cap K}(S^n(\mathfrak{u}\cap\mathfrak{p}) \otimes_{\mathbb{C}} Z^{\#}_{\mathfrak{q}}, V) \tag{8.106}$$

for the trivial action of $\bar{\mathfrak{u}}\cap\mathfrak{k}$ on $S^n(\mathfrak{u}\cap\mathfrak{p})$. By the same argument as for (8.15d), this expression is

$$= \sum_{i \geq 0} \sum_{n=0}^{N} (-1)^i \dim \operatorname{Hom}_{L\cap K}(S^n(\mathfrak{u}\cap\mathfrak{p}) \otimes_{\mathbb{C}} Z^{\#}, H^i(\bar{\mathfrak{u}}\cap\mathfrak{k}, V)). \tag{8.107}$$

Since the left side of (8.104) is independent of N, we can replace the N by infinity here. Taking (8.102) into account, we obtain the desired formula from the equality of (8.104) and (8.107), since Theorem 8.7 shows that the terms for $i > S$ are all 0.

We still have to check the finiteness conditions. In (8.105) let the filtration of $S^n(\mathfrak{u} \cap \mathfrak{p})$ be

$$S^n(\mathfrak{u} \cap \mathfrak{p}) = S_{N'} \supseteq \cdots \supseteq S_0 \supseteq S_{-1} = 0 .$$

Then $\bar{\mathfrak{u}} \cap \mathfrak{k}$ acts trivially on S_k/S_{k-1} and also on $S_k/S_{k-1} \otimes_{\mathbb{C}} Z_{\mathfrak{q}}^{\#}$. Hence

$$\begin{aligned} \mathrm{Ext}^i_{\mathfrak{q}\cap\mathfrak{k}, L\cap K}(S_k/S_{k-1} \otimes_{\mathbb{C}} Z_{\mathfrak{q}}^{\#}, V) &\cong \mathrm{Hom}_{L\cap K}(S_k/S_{k-1} \otimes_{\mathbb{C}} Z^{\#}, H^i(\bar{\mathfrak{u}} \cap \mathfrak{k}, V)) \\ &\cong \mathrm{Hom}_{L\cap K}(Z^{\#}, \mathrm{Hom}_{\mathbb{C}}(S_k/S_{k-1}), H^i(\bar{\mathfrak{u}} \cap \mathfrak{k}, V)) . \end{aligned} \tag{8.108}$$

The $\mathrm{Hom}_{\mathbb{C}}$ factor of this $\mathrm{Hom}_{L\cap K}$ is finite-dimensional, and each $L\cap K$ type in $Z^{\#}$ has finite multiplicity. Hence (8.108) is finite-dimensional. Thus Corollary 8.19 applies for passing from (8.106) to (8.105). Part of the conclusion of the corollary is that each term of (8.105) is finite. These terms are the same as the ones on the right side of (8.104), and thus the finiteness assumption is satisfied for the application of Corollary 8.19 to obtain (8.104). Part of the conclusion of the corollary is that the terms on the left side of (8.104) are finite, and thus $\dim \mathrm{Hom}_K(\mathcal{L}_i(Z), V)$ is finite for each i. This completes the proof.

Corollary 8.21. Suppose that Z is an $(\mathfrak{l}, L\cap K)$ module having an infinitesimal character and having all $L\cap K$ types of finite multiplicity. Let $\tau_{\Lambda'}$ be an irreducible representation of K acting in a space V.

(a) If $\mathcal{L}_i(Z) = 0$ for $0 \leq i < S$, then the multiplicity of $\tau_{\Lambda'}$ in $\mathcal{L}_S(Z)$ is

$$\sum_{j=0}^{S} \sum_{n=0}^{\infty} (-1)^j \dim \operatorname{Hom}_{L\cap K}(S^n(\mathfrak{u} \cap \mathfrak{p}) \otimes_{\mathbb{C}} Z^{\#}, H^{S-j}(\bar{\mathfrak{u}} \cap \mathfrak{k}, V)) .$$

(b) If $\mathcal{R}^i(Z) = 0$ for $0 \leq i < S$, then the multiplicity of $\tau_{\Lambda'}$ in $\mathcal{R}^S(Z)$ is

$$\sum_{j=0}^{S} \sum_{n=0}^{\infty} (-1)^j \dim \operatorname{Hom}_{L\cap K}(H_{S-j}(\mathfrak{u} \cap \mathfrak{k}, V), S^n(\mathfrak{u} \cap \mathfrak{p}) \otimes_{\mathbb{C}} Z^{\#}) .$$

Proof. All terms on the left in Theorem 8.20 are 0 except when $i = S$. On the right side we set $j = S - i$.

Corollary 8.22. Let λ be an integral form on $\mathfrak{h}$ that is orthogonal to all members of Δ_L, so that λ is the (highest) weight of a one-dimensional representation of L. Let $C = C(\mathfrak{q})$ be the constant in Theorem 8.15, and let $\tau_{\Lambda'}$ be an irreducible representation of K acting in a space V. If $\langle \lambda, \alpha \rangle \geq C$ for all $\alpha \in \Delta(\mathfrak{u} \cap \mathfrak{k})$, then the multiplicities of $\tau_{\Lambda'}$ in $A_{\mathfrak{q}}(\lambda)$ and $A^{\mathfrak{q}}(\lambda)$ are given by

$$\sum_{j=0}^{S} \sum_{n=0}^{\infty} (-1)^j \dim \operatorname{Hom}_{L\cap K}(S^n(\mathfrak{u} \cap \mathfrak{p}) \otimes_{\mathbb{C}} \mathbb{C}_{\lambda+2\delta(\mathfrak{u})}, H^{S-j}(\bar{\mathfrak{u}} \cap \mathfrak{k}, V))$$

and

$$\sum_{j=0}^{S} \sum_{n=0}^{\infty} (-1)^j \dim \operatorname{Hom}_{L\cap K}(H_{S-j}(\mathfrak{u} \cap \mathfrak{k}, V), S^n(\mathfrak{u} \cap \mathfrak{p}) \otimes_{\mathbb{C}} \mathbb{C}_{\lambda+2\delta(\mathfrak{u})})$$

respectively. Moreover $A_{\mathfrak{q}}(\lambda)$ and $A^{\mathfrak{q}}(\lambda)$ have infinitesimal character $\lambda + \delta$.

Proof. We begin with infinitesimal characters. In $\mathbb{C}_\lambda$, a member z of $Z(\mathfrak{l})$ acts by the scalar

$$\lambda(\gamma'_{\mathfrak{l}\cap\mathfrak{n}}(z)) = \lambda(\sigma^{-1}_{\mathfrak{l}\cap\mathfrak{n}}\circ\gamma_{\mathfrak{l}}(z)) = (\lambda+\delta_L)(\gamma_{\mathfrak{l}}(z)).$$

Thus $Z = \mathbb{C}_\lambda$ has infinitesimal character $\lambda+\delta_L$. By Theorem 8.17, $A_{\mathfrak{q}}(\lambda) = \mathcal{L}_S(\mathbb{C}_\lambda)$ has infinitesimal character $(\lambda+\delta_L+\delta(\mathfrak{u})) = \lambda+\delta$. And so does $A^{\mathfrak{q}}(\lambda)$.

Now $\mathbb{C}_\lambda$ has an infinitesimal character, and its sole $L\cap K$ type has multiplicity one. Theorem 8.15 and the remarks after it show that $\mathcal{L}_i(\mathbb{C}_\lambda) = 0$ and $\mathcal{R}^i(C_\lambda) = 0$ for $0 \leq i < S$ if $\langle\lambda,\alpha\rangle \geq C$ when $\alpha\in\Delta(\mathfrak{u}\cap\mathfrak{k})$. So Corollary 8.21 applies and gives the desired result.

In the special case that $L\cap K = T$, we can use Kostant's Theorem and Poincaré duality to make the formulas more explicit in Corollary 8.22. For $A^{\mathfrak{q}}(\lambda)$, we have

$$\begin{aligned}
&\operatorname{Hom}_{L\cap K}(H_{S-j}(\mathfrak{u}\cap\mathfrak{k},V), S^n(\mathfrak{u}\cap\mathfrak{p})\otimes_{\mathbb{C}}\mathbb{C}_{\lambda+2\delta(\mathfrak{u})}) \\
&\qquad\cong \operatorname{Hom}_{L\cap K}(H^j(\mathfrak{u}\cap\mathfrak{k},V)\otimes_{\mathbb{C}}\textstyle\bigwedge^S(\mathfrak{u}\cap\mathfrak{k}), S^n(\mathfrak{u}\cap\mathfrak{p})\otimes_{\mathbb{C}}\mathbb{C}_{\lambda+2\delta(\mathfrak{u})}) \\
&\qquad\cong \operatorname{Hom}_{L\cap K}(H^j(\mathfrak{u}\cap\mathfrak{k},V)\otimes_{\mathbb{C}}\mathbb{C}_{-\lambda-2\delta(\mathfrak{u}\cap\mathfrak{p})}, S^n(\mathfrak{u}\cap\mathfrak{p})) \qquad (8.109)
\end{aligned}$$

with $L\cap K = T$ and $\mathfrak{u}\cap\mathfrak{k} = \mathfrak{n}\cap\mathfrak{k}$. The highest weight of the K representation on V being Λ', the version of Kostant's Theorem in Theorem 6.12 says that $H^j(\mathfrak{u}\cap\mathfrak{k},V)$ is the sum of the T types $w(\Lambda'+\delta_K)-\delta_K$ for $w\in W_K$ of length j, each T type occurring with multiplicity one. So the first argument in the Hom on the right side of (8.109) is the sum of the T types

$$w(\Lambda' + \delta_K) - \delta_K - \lambda - 2\delta_n$$

for $\ell(w) = j$, each with multiplicity one.

For ν integral on $\mathfrak{h}$, let $\mathcal{P}(\nu)$ be the number of ways that ν can be written as $\Sigma_{\alpha \in \Delta_n^+} n_\alpha \alpha$ with all n_α integers ≥ 0. The ways with $\Sigma n_\alpha = n$ correspond to occurrences of the T type ν in $S^n(\mathfrak{u} \cap \mathfrak{p})$. Consequently, under the hypotheses on λ in Corollary 8.22, the multiplicity of the K type $\tau_{\Lambda'}$ in $A^{\mathfrak{q}}(\lambda)$ is

$$\sum_{w \in W_K} (-1)^{\ell(w)} \mathcal{P}(w(\Lambda' + \delta_K) - \delta_K - \lambda - 2\delta_n) . \qquad (8.110)$$

This formula is called the <u>Blattner multiplicity formula.</u>

NOTES

Chapter I

Examples of closed linear groups beyond those in §3 may be found in pp. 4-6 of Knapp [1986]. For more detailed information about such examples, see Chapter I of Chevalley [1946] and Chapter IX of Helgason [1962].

The books by Chevalley [1946] and F. Warner [1983] contain a more systematic treatment of manifolds, as well as an introduction to Lie groups. See Appendix A of Knapp [1986] for an outline of the results with no proofs. The present notes give a general definition of "Lie group," but the emphasis is on closed linear groups. One can exhibit Lie groups that cannot be realized as closed linear groups, but they play no role in these notes.

The general theory of Lie groups contains a correspondence between a class of subgroups (of a Lie group) and Lie subalgebras, as a consequence of theorems about differential equations. The treatment here associates a Lie subalgebra to any closed subgroup, but more subgroups than the closed ones must be allowed to obtain groups corresponding to each Lie subalgebra.

A more thorough treatment of Lie groups provides more information about homomorphisms. We have seen how to pass from homomorphisms of Lie groups to homomorphisms of Lie algebras. One can also pass backwards, at least locally, and one can associate a global group homomorphism to an arbitrary Lie algebra homomorphism if the domain group is connected and simply connected.

The vectors in the subspace V_0 defined in §8 are called C^∞ vectors. For further discussion of them, see §§3-4 of

Chapter III of Knapp [1986]. Theorem 8.9 of that book gives a circumstance where $\mathfrak{g}$-invariant subspaces of infinite-dimensional representations control the G-invariant subspaces.

Chapter II

This chapter consists of two interwoven themes—elementary Lie algebra constructions and graduate level multilinear algebra. These respective themes appear separately in Chapters I and V of Jacobson [1962] and in Chapter XVI of Lang [1984], respectively.

A tensor product $A \otimes B$ can be defined for any two abelian groups A and B, not just for vector spaces. Then the tensor product $A \otimes_R B$ in §8 makes sense for any ring R if A is a right R module and B is a left R module.

In §4 the symmetric and exterior algebras are quotients of the tensor algebra. Alternatively they can be identified with subspaces; the operation of multiplication must then be followed by a symmetric or antisymmetric sum operation. Spivak [1965] carries through such a development of the exterior algebra in his pp. 75-84. See Greub, Halperin, and Vanstone [1972] for a detailed treatment.

More general algebraic treatments of the Poincaré-Birkhoff-Witt Theorem of §6 appear in Chapter V of Jacobson [1962] and Chapter 2 of Dixmier [1974]. Helgason [1962] and Knapp [1986] treat this theorem analytically.

Symmetrization is used systematically as a tool in Varadarajan [1974]. Further treatments of symmetrization appear in Helgason [1962], Dixmier [1974], and Knapp [1986]. The last of these books gives on page 721 some history of the Poincaré-Birkhoff-Witt Theorem and the symmetrization mapping.

Chapter III

The first six sections are abridged versions of portions of Chapters I, II, and IV of Knapp [1986]. That book should be consulted for historical references and further reading, as well as for the theory of roots that allows one to generalize the material of §§3-6 to all compact connected Lie groups.

Section 7 is the first section of motivation for the algebraic constructions of representations that come later. It uses a small amount of complex-variable theory. In particular, complex manifolds (defined using charts and holomorphic functions in the setting of §I.4) play a role. The book Wallach [1973] develops more systematically and more generally the theory that is in this section, and the Borel-Weil Theorem appears as Corollary 6.3.9. A quick proof of the Borel-Weil Theorem using a little structure theory of noncompact semisimple groups appears in Chapter V of Knapp [1986]; this book gives historical references for the theorem.

The motivation of §7 continues in §VI.1 and §VII.1. See the Notes for Chapter VII for further references.

Chapter IV

The definition of the exterior derivative operator in §1 is the simple constructive one in Chapter V of Chevalley [1946] and Chapter 4 of Spivak [1965]. Chevalley gives a derivation of (4.5) with more geometric content. The more advanced treatment in Lang [1962] gives an invariant definition of d directly on a manifold and then derives, on p. 84, our formula (4.13) as the explicit formula implementing the definition. For a treatment of the theorem of de Rham, see Singer and Thorpe [1976].

In the theory of groups, Schreier was the first one to try to determine all groups with a given normal subgroup and a given quotient. Chapter 15 of Hall [1959] gives an account of this work and how it leads to cohomology of groups. For a more modern treatment, see Chapter IV of Brown [1982]. It is natural to expect a similar theory for Lie algebras, and indeed some important structure theorems result from the theory for Lie algebras. These theorems clarify the role of semisimple Lie algebras among all finite-dimensional Lie algebras. (A finite-dimensional Lie algebra is <u>semisimple</u> if it has no nonzero solvable ideals, a solvable ideal being one with a chain of subideals decreasing to 0 and having successive quotients abelian.) The basic results are two lemmas of Whitehead, which in cohomological terms say that $H^1(\mathfrak{g},V) = H^2(\mathfrak{g},V) = 0$ if $\mathfrak{g}$ is semisimple over a field of characteristic 0 and V is finite-dimensional. From $H^1(\mathfrak{g},V) = 0$, one finds that finite-dimensional representations of semisimple Lie algebras are direct sums of irreducible representations. From $H^2(\mathfrak{g},V) = 0$, one obtains Levi's theorem that any finite-dimensional Lie algebra is the semidirect product of a semisimple Lie algebra and a solvable Lie algebra. A second application of $H^1(\mathfrak{g},V) = 0$ gives Harish-Chandra's uniqueness theorem that any two Levi decompositions are conjugate. An account of these matters appears in Jacobson [1962], pp. 79-80 and 83-96.

Lie algebra cohomology as defined in §3 was introduced by Chevalley and Eilenberg [1948], p. 115. They proved our Proposition 4.1. Except for §4, much of the remainder of the chapter is due to Koszul [1950].

The first hint that any kind of homology or cohomology can be computed flexibly as in §4 is in the work of W. Hurewicz and of H. Hopf. Hurewicz considered the following situation: M is a finite connected simplicial complex, U is its universal cover, and G is the fundamental group of M. Suppose that U is contractible. The group G acts freely on the group $C_*(U)$ of simplicial chains of U. The boundary operator then gives us an exact sequence

$$0 \longleftarrow \mathbb{Z} \longleftarrow C_0(U) \longleftarrow C_1(U) \longleftarrow C_2(U) \longleftarrow \cdots$$

of modules with each $C_j(U)$ a free G module. Applying $\otimes_G$, we obtain the complex

$$0 \longleftarrow C_0(M) \longleftarrow C_1(M) \longleftarrow C_2(M) \longleftarrow \cdots .$$

Now the homology $H_0(M)$ is just $\mathbb{Z}$, and $H_1(M)$ is just G modulo its commutator; thus $H_0(M)$ and $H_1(M)$ depend only on G. What Hurewicz showed is that all higher $H_i(M)$ depend only on G; he did not address existence of such spaces for given G. Hopf clarified the situation and drew attention to it by making an explicit calculation: Dropping all assumptions on U other than its simple connectivity, he gave a formula for $H_2(M)/\{\text{"spherical homology classes"}\}$ in terms of G. Later he obtained a result for higher-degree homology. (See Steenrod [1968], p. 1282, for reviews of three papers by Hopf, and see Brown [1982] for the effect on cohomology of groups.) During the 1940's, these developments led to the realization that homology and cohomology can be calculated from suitable abstract resolutions, a fact that now lies at the heart of the subject of homological algebra. The present notes take up homological algebra in Chapter V.

Chapter V

The origins of homological algebra are discussed in the Notes for Chapter IV. The classic book on the subject is Cartan and Eilenberg [1956], which deals systematically with the category of all left R modules, R being a ring with identity. We have followed the philosophy of those authors, working only with categories of modules, though our theory in Chapter VII has required us to allow certain subcategories of modules as the basic things to study. A reasonable level of generality that incorporates additional examples that arise in other branches of mathematics is that of all "abelian categories." Abelian categories are what are studied in Hilton and Stammbach [1971].

The results of this chapter are all standard ones, explicit or implicit in the two above-mentioned books. We have limited the definition of derived functors to the situations in the table on page 233, insisting on one-sided exactness for our functor and matching projective/injective for our resolution to properties of the functor. Applications of homological algebra normally follow this convention, but the two books seek greater generality and impose no such limitation.

Chapter VI

The results of §§1-3 are standard facts about the homological algebra of $U(\mathfrak{g})$ modules, and most of them are explicit or implicit in Cartan and Eilenberg [1956], particularly Chapter XIII. Poincaré duality is due to Koszul [1950].

Kostant's Theorem originally appeared in Kostant [1961], and our proof is taken from Vogan [1981]. The version of the theorem in Kostant's paper is considerably more general than our Theorem 6.12. It allows any compact Lie group, connected or not, in place of $U(n)$, and it allows $\mathfrak{n}$ to be the nilpotent part of any parabolic subalgebra containing the complexified Lie algebra of a maximal torus.

The Harish-Chandra isomorphism was first proved in Harish-Chandra [1951]. See Varadarajan [1974], pp. 333-341, for a clean proof valid for an semisimple Lie algebra $\mathfrak{g}$. Verma modules were introduced by Harish-Chandra in his 1951 paper and were so named because of the later work Verma [1968]. Theorem 6.19 is due to Casselman and Osborne [1975].

Chapter VII

Hochschild [1956] introduced relative homological algebra with respect to a pair $(\mathfrak{g},\mathfrak{k})$ satisfying some properties. Meanwhile the theory of group representations began to give realizations of irreducible unitary representations of semisimple Lie groups in various cohomology spaces. The first example was the generalization in Bott [1957] of the (analytic form of the) Borel-Weil Theorem. This is a theorem about finite-dimensional representations of compact Lie groups.

<u>Discrete series</u> are representations that occur as irreducible direct summands of $L^2(G)$. After Harish-Chandra [1966] parametrized the discrete series of noncompact semisimple groups, people began to look for global realizations of these representations. Griffiths proved a vanishing theorem about cohomology spaces for $L^2(\Gamma\backslash G/K)$, where G/K is a

Hermitian symmetric space and Γ is a cocompact discrete subgroup of G; this theorem was known widely but was first published only in the paper of Griffiths and Schmid [1969]. With the Griffiths result as a starting point, Langlands [1966] was led to conjecture a specific cohomological setting for realizing discrete series. In a series of papers ending with Schmid [1976], Schmid proved the Langlands conjecture. For an exposition, see Williams [1982].

There were a few other examples of representations realized in cohomology spaces, particularly in Dolbeault cohomology spaces. But the difficulty in getting a general theory beyond the case of the discrete series was always in proving that image $\bar{\partial}$ is closed, so that $\ker \bar{\partial}/\text{image } \bar{\partial}$ is a reasonable space.

Aware of this problem and reasoning as with the examples of §VI.1 and §VII.1, Zuckerman [1978] was led to formulate everything algebraically and so to bypass the analytic problem whether image $\bar{\partial}$ is closed. He defined a functor Γ much as in the examples of §1 and developed a theory of its derived functors. The relevant category was all $(\mathfrak{g}, K)$ modules, and the right setting was the relative homological algebra of Hochschild [1956], or something very close to it.

The notion of $(\mathfrak{g}, K)$ module had evolved from work by Harish-Chandra [1953] that connected infinite-dimensional representations of semisimple groups with representations of their Lie algebras. The algebra $R(\mathfrak{g}, K)$ as developed in §3 came later. Much of the material in §§4-8 is substantially in Hochschild's and Zuckerman's work and appears in written form in Chapter I of Borel and Wallach [1980], in Enright and Wallach [1980], and in §§6.1-6.2 of Vogan [1981]. However, Γ is given a definition in several stages, starting from the geometric definition of §1, and the P and I functors are defined only when the compact group does not change.

In the course of the development Zuckerman proposed a certain duality theorem for the derived functors of Γ, though a full proof was given only later—and by other people. But together with P. Trauber, Zuckerman did give a number of ideas toward a proof. One of these was to use the convolution algebra of all K-finite distributions on G that are supported on K. This algebra is what $R(\mathfrak{g},K)$ amounts to when $(\mathfrak{g},K)$ comes from a group G. Knapp and Vogan [1986] pursued this construction, and some of the details are in §3. Use of $R(\mathfrak{g},K)$ allows one to give a unified definition of Γ, P, and I, as well as to define the functor Π dual to Zuckerman's Γ.

Poincaré duality (Theorem 7.31) is proved in Borel and Wallach [1980] under an "admissibility" assumption. The proof here shows this assumption to be unnecessary.

Chapter VIII

This chapter applies more widely than to $U(m,n)$. The necessary structure theory to generalize §1 may be found in Vogan [1981] or Knapp [1986]. The group G should be allowed to have considerable disconnectedness, and the structure theory appropriate for such generality is summarized in Vogan [1987].

Cohomological induction was introduced by Zuckerman [1978]. Actually Zuckerman used only the functor Γ, not Π, and thus his results were confined to $\mathcal{R}^i(Z)$. The results of §2 for $\mathcal{R}^i$ are all due to Zuckerman and are written down in §6.3 of Vogan [1981].

Vogan [1981] gives a proof of Theorem 8.7b by pursuing an algebraic analog of the "order-of-vanishing filtration" in the analytic setting. Our proof of Theorem 8.7a is a dualized version of the Vogan argument and is taken from Knapp and Vogan [1986].

Zuckerman's approach to vanishing below the middle dimension was to combine an irreducibility result, his duality theorem, and vanishing above the middle dimension. Vogan [1981] approaches vanishing below the middle dimension through a "series of tricks which are quite difficult to motivate." We have chosen a more direct approach.

Concealed in our approach are several spectral sequences. A *spectral sequence* is a device that gives information about the homology or cohomology of a complex when the modules of the complex have compatible filtrations by submodules. Their theory is explained briefly in §VII.2 of Brown [1982] and in more detail in Chapter VIII of Hilton and Stammbach [1971]. Actually the general theory of spectral sequences appears to accomplish nothing in the present book any more quickly than is accomplished by manual computation; the theory merely gives a common framework for describing several situations. The simplest example of a spectral sequence arises when we have an exact sequence of complexes as in (5.29); in this case, use of the spectral sequence yields the same information as the long exact sequence of Theorem 5.20. Another class of spectral sequences arises with the derived functors of the composition of two functors (the Grothendieck spectral sequence); the setting is as in Proposition 5.17, except that the inside functor F is not assumed exact. In this case, the spectral sequence gives partial information about the derived functors of the composition. When F is exact, the spectral sequence "degenerates," and the result is the explicit formula in Proposition 5.17.

Proposition 8.11 is another situation of derived functors of a composition. The spectral sequence degenerates in this case, and the proposition gives an explicit formula.

Lemma 8.13 and Theorem 8.14 involve comparisons of homology or cohomology of a Lie algebra with that of a subalgebra. They are variants of Theorem 3.12 of Vogan [1979]. Hochschild and Serre [1953] introduced a spectral sequence for this situation. Setting up matters in the Hochschild-Serre format requires roughly the same amount of work as the proofs we give.

In §6, if $\mathfrak{l} = \mathfrak{h}$ and if λ is Δ^+ dominant, then $A^{\mathfrak{q}}(\lambda)$ is a discrete series representation, and all discrete series representations arise this way. This conclusion follows under the additional assumption $|\langle\lambda,\alpha\rangle|$ large from Corollary 8.22 and from a theorem of Schmid (see Propositions 4.1 and 7.13 of

Vogan [1979]). For other λ that are Δ^+ dominant, the result follows by a translation principle (see Vogan [1981]). The multiplicity formula for the K types of a discrete series representation is the original Blattner formula and was first established by Hecht and Schmid [1975]. See pp. 735-736 of Knapp [1986] for further discussion and history of the formula. The algebraic formulation in §8 is due to Zuckerman [1978].

REFERENCES

Borel, A., and N. Wallach, Continuous Cohomology, Discrete Subgroups, and Representations of Reductive Groups, Princeton University Press, Princeton, 1980.

Bott, R., Homogeneous vector bundles, Ann. of Math. 66 (1957), 203-248.

Brown, K. S., Cohomology of Groups, Springer-Verlag, New York-Heidelberg-Berlin, 1982.

Cartan, H., and S. Eilenberg, Homological Algebra, Princeton University Press, Princeton, 1956.

Casselman, W., and M. S. Osborne, The $\mathfrak{n}$-cohomology of representations with an infinitesimal character, Compositio Math. 31 (1975), 219-227.

Chevalley, C., Theory of Lie Groups I, Princeton University Press, Princeton, 1946.

Chevalley, C., and S. Eilenberg, Cohomology theory of Lie groups and Lie algebras, Trans. Amer. Math. Soc. 63 (1948), 85-124.

Dixmier, J., Algèbres Enveloppantes, Gauthier-Villars, Paris, 1974.

Enright, T. J., and N. R. Wallach, Notes on homological algebra and representations of Lie algebras, Duke Math J. 47 (1980), 1-15.

Greub, W., S. Halperin, and R. Vanstone, Connections, Curvature, and Cohomology, Vol. I, Academic Press, New York, 1972.

Griffiths, P., and W. Schmid, Locally homogeneous complex manifolds, Acta Math. 123 (1969), 253-302.

Hall, M., The Theory of Groups, Macmillan, New York, 1959.

Harish-Chandra, On some applications of the universal enveloping algebra of a semisimple Lie algebra, Trans. Amer. Math. Soc. 70 (1951), 28-96.

Harish-Chandra, Representations of a semisimple Lie group on a Banach space I, Trans. Amer. Math. Soc. 75 (1953), 185-243.

Harish-Chandra, Discrete series for semisimple Lie groups II. Explicit determination of the characters, Acta Math. 116 (1966), 1-111.

Hecht, H., and W. Schmid, A proof of Blattner's conjecture, Invent. Math. 31 (1975), 129-154.

Helgason, S., Differential Geometry and Symmetric Spaces, Academic Press, New York, 1962.

Hilton, P. J., and U. Stammbach, A Course in Homological Algebra, Springer-Verlag, New York-Heidelberg-Berlin, 1971.

Hochschild, G., Relative homological algebra, Trans. Amer. Math. Soc. 82 (1956), 246-269.

Hochschild, G., and J.-P. Serre, Cohomology of Lie algebras, Ann. of Math. 57 (1953), 591-603.

Jacobson, N., Lie Algebras, Interscience, New York, 1962.

Knapp, A. W., Representation Theory of Semisimple Groups: An Overview Based on Examples, Princeton University Press, Princeton, 1986.

Knapp, A. W., and D. A. Vogan, Duality theorems in relative Lie algebra cohomology, duplicated notes, State University of New York at Stony Brook and Massachusetts Institute of Technology, 1986.

Kostant, B., Lie algebra cohomology and the generalized Borel-Weil Theorem, Ann. of Math. 74 (1961), 329-387.

Koszul, J.-L., Homologie et cohomologie des algèbres de Lie, Bull. Soc. Math. France 78 (1950), 65-127.

Lang, S., Introduction to Differentiable Manifolds, Interscience, New York, 1962.

Lang, S., Algebra, 2nd edition, Addison-Wesley, Reading, Mass., 1984.

Langlands, R. P., Dimension of spaces of automorphic forms, Algebraic Groups and Discontinuous Subgroups, Proc. Symp. in Pure Math. 9 (1966), 253-257, American Mathematical Society, Providence, R. I.

Schmid, W., L^2-cohomology and the discrete series, Ann. of Math. 103 (1976), 375-394.

Singer, I. M., and J. A. Thorpe, Lecture Notes on Elementary Topology and Geometry, Springer-Verlag, New York-Heidelberg-Berlin, 1976.

Spivak, M., Calculus on Manifolds, Benjamin/Cummings, Menlo Park, Calif., 1965.

Steenrod, N. E., Reviews of Papers in Algebraic and Differential Topology, Topological Groups, and Homological Algebra, Part II, American Mathematical Society, Providence, R. I., 1968.

Varadarajan, V. S., Lie Groups, Lie Algebras, and Their Representations, Prentice-Hall, Englewood Cliffs, N. J., 1974; 2nd edition, Springer-Verlag, New York, 1984.

Verma, D.-N., Structure of certain induced representations of complex semisimple Lie algebras, Bull. Amer. Math. Soc. 74 (1968), 160-166 and 628.

Vogan, D. A., The algebraic structure of the representation of semisimple Lie groups I, Ann. of Math. 109 (1979), 1-60.

Vogan, D. A., Representations of Real Reductive Lie Groups, Birkäuser, Boston 1981.

Vogan, D. A., Unitary Representations of Reductive Lie Groups, Princeton University Press, Princeton, 1987.

Wallach, N. R., Harmonic Analysis on Homogeneous Spaces, Marcel Dekker, New York, 1973.

Warner, F. W., Foundations of Differentiable Manifolds and Lie Groups, Springer-Verlag, New York-Berlin-Heidelberg-Tokyo, 1983.

Williams, F. L., Frobenius reciprocity and Lie group representations on $\bar\partial$ cohomology spaces, L'Enseignement Math. 28 (1982), 3-30.

Zuckerman, G. J., Construction of representations via derived functors, lectures, Institute for Advanced Study, 1978.

INDEX OF NOTATION

In this list, Latin, German, and script symbols appear together and are followed by Greek symbols and non-letters.

INDEX

Library of Congress Cataloging-in-Publication Data

Knapp, Anthony W.
Lie groups, lie algebras, and cohomology.

(Mathematical notes ; 34)
Bibliography: p.
Includes index.
1. Lie groups. 2. Lie algebras. 3. Homology theory.
I. Title. II. Series: Mathematical notes
(Princeton University Press) ; 34.
QA387.K57 1988 512′.55 88-2514
ISBN 0-691-08498-X (pbk.)